SI Base Units

Base Quantity	Name of Unit	Symbol
Length	Meter	m
Mass	Kilogram	kg
Time	Second	s
Electrical current	Ampere	A
Temperature	Kelvin	K
Amount of substance	Mole	mol
Luminous intensity	Candela	cd

Derived Units in the SI System

Physical Quantity	Name	Symbol	Units
Energy	Joule	J	$kg\ m^2\ s^{-2}$
Force	Newton	N	$kg\ m\ s^{-2}$
Power	Watt	W	$kg\ m^2\ s^{-3}$
Electric charge	Coulomb	C	$A\ s$
Electrical resistance	Ohm	Ω	$kg\ m^2\ s^{-3}\ A^{-2}$
Electrical potential difference	Volt	V	$kg\ m^2\ s^{-3}\ A^{-1}$
Electrical capacity	Farad	F	$A^2\ s^4\ kg^{-1}\ m^{-2}$
Frequency	Hertz	Hz	s^{-1}

Some Commonly Used Non-SI Units

Unit	Quantity	Symbol	Conversion Factor
Angstrom	Length	Å	$1\ Å = 10^{-10}\ m = 100\ pm$
Calorie	Energy	cal	$1\ cal = 4.184\ J$
Debye	Dipole moment	D	$1\ D = 3.3356 \times 10^{-30}\ C\ m$
Gauss	Magnetic field	G	$1\ G = 10^{-4}\ T$
Liter	Volume	L	$1\ L = 10^{-3}\ m^3$

Prefixes Used with SI Units

Prefix	Symbol	Meaning
Tera-	T	1,000,000,000,000, or 10^{12}
Giga-	G	1,000,000,000, or 10^9
Mega-	M	1,000,000, or 10^6
Kilo-	k	1,000, or 10^3
Deci-	d	1/10, or 10^{-1}
Centi-	c	1/100, or 10^{-2}
Milli-	m	1/1,000, or 10^{-3}
Micro-	μ	1/1,000,000, or 10^{-6}
Nano-	n	1/1,000,000,000, or 10^{-9}
Pico-	p	1/1,000,000,000,000, or 10^{-12}

Problems and Solutions

to accompany

RAYMOND CHANG

PHYSICAL CHEMISTRY
for the Chemical and Biological Sciences

Problems and Solutions

to accompany

RAYMOND CHANG

PHYSICAL CHEMISTRY
for the Chemical and Biological Sciences

Helen O. Leung
MOUNT HOLYOKE COLLEGE

and

Mark D. Marshall
AMHERST COLLEGE

University Science Books
Sausalito, California

University Science Books
55D Gate Five Road
Sausalito, CA 94965

Fax: (415) 332-5393
www.uscibooks.com

Production manager: *Susanna Tadlock*
Designer: *Robert Ishi*
Compositor: *Eigentype*
Printer & binder: *Maple-Vail Book Manufacturing Group*

This book is printed on acid-free paper.

ISBN 1-891389-11-4
Library of Congress Control Number: 00-132154

Printed in the United States of America
10 9 8 7 6 5 4 3 2

Contents

Preface

Some advice we commonly give to students is that the best, if not the only, way to learn chemistry is to work through problems. Among instructors at all levels it is also widely known that one does not truly learn a subject until you have to teach it. To these we now know to add, "You don't really know how to teach physical chemistry until you work through one-thousand problems and write the solutions for publication." Of course it helps that those problems come from an author who possesses the wide-ranging command and deep understanding of physical chemistry that Raymond Chang does. Reading his textbook and working all the problems in it have revealed to us connections previously undiscovered and have improved our own mastery of the material. Professor Chang reviewed all the solutions as we wrote them, and our electronic discussions with him throughout the process were extremely valuable. We were privileged to work with him, once again for one of us and for the other, for the first time.

It is important to the authors of this solutions manual and to Professor Chang that it is written by faculty members who actively teach physical chemistry. This allows the book to adopt the same conversational style that we use in our own classrooms, with answers explained and assumptions discussed. Only for the most straightforward, often definitional problems, are the answers merely quoted. Our goal is that the manual combine the vitality and excitement of modern physical chemistry with strong, effective pedagogy.

We have made every effort to ensure consistency in the values of physical constants used, both internally and with those given in the textbook. Likewise, choices for symbols and units were made with care and hopefully consistency. We do have a word to say about significant figures, since every measurement has an associated uncertainty that must be considered when comparing results from different experiments or those of experiment with predictions from theory. The appropriate number of significant figures is determined by the data given in each problem. So, for example, if a mass is given as 1.78 g, then the answer to the problem should contain three significant figures. Quantities given as "1 mole," or "5 L" are taken to be exact. Intermediate calculations are done retaining one extra digit, and the final answer is rounded to the correct number of significant figures in the final step. If a result from a previous part of a problem is used in a subsequent part, the value with the extra digit is used. In such cases, the answer to the earlier part would look similar to "45.678 kJ mol^{-1} = 45.68 kJ mol^{-1}," where the first value given is used later in the problem and the "correct" answer is the second. Of course, every rule has its exceptions, and we did choose to be somewhat more flexible in problems involving exponentiation and logarithms. (Do note that only those digits to the right of the decimal point in a common logarithm are significant. Those to the left merely set the value of the exponent in scientific notation.)

We thank Jane Ellis of University Science Books for keeping us on task and for helping to assemble the production team that transformed our LaTeX manuscript into a well-designed publication. Many prefaces close with the authors expressing gratitude for understanding spouses who provided moral support during the preparation of the manuscript, even though they had no direct connection with the project. In this case, however, each author is grateful for a partner who is not only a wonderful spouse, but a most convivial coauthor as well.

Helen O. Leung, SOUTH HADLEY, MA

Mark D. Marshall, AMHERST, MA

Problems and Solutions

to accompany

RAYMOND CHANG

PHYSICAL CHEMISTRY
for the Chemical and Biological Sciences

CHAPTER 2

The Gas Laws

PROBLEMS AND SOLUTIONS

2.1 Classify each of the following properties as intensive or extensive: force, pressure (P), volume (V), temperature (T), mass, density, molar mass, molar volume ($\overline{V}$).

Intensive properties: P, T, density, molar mass, $\overline{V}$

Extensive properties: force, V, mass

2.2 Some gases, such as NO_2 and NF_2, do not obey Boyle's law at any pressure. Explain.

Boyle's law applies only at constant n and constant T. Both NO_2 and NF_2 undergo association reactions:

$$2NO_2 \rightleftharpoons N_2O_4$$

$$2NF_2 \rightleftharpoons N_2F_4$$

so n is not constant for these gases.

2.3 An ideal gas originally at 0.85 atm and 66°C was allowed to expand until its final volume, pressure, and temperature were 94 mL, 0.60 atm, and 45°C, respectively. What was its initial volume?

Because n is kept constant, the ideal gas equation $PV = nRT$ can be rewritten as

$$\frac{PV}{T} = \text{constant}$$

or

$$\frac{P_1 V_1}{T_1} = \frac{P_2 V_2}{T_2}$$

Therefore,

$$V_1 = \frac{P_2 V_2}{T_2} \frac{T_1}{P_1} = \frac{(0.60 \text{ atm})\,(94 \text{ mL})}{(273 + 45)\,\text{K}} \frac{(273 + 66)\,\text{K}}{0.85 \text{ atm}} = 71 \text{ mL}$$

2.4 Some ballpoint pens have a small hole in the main body of the pen. What is the purpose of this hole?

As the pen is used, and the ink leaves the pen, the gas inside expands to fill the volume left by the ink. As the volume increases, the pressure inside the pen decreases. The hole is needed to equalize the pressure to allow for easy flow of ink.

2.5 Starting with the ideal-gas equation, show how you can calculate the molar mass of a gas from a knowledge of its density.

Let m, $\mathcal{M}$, and ρ be the mass, the molar mass, and the density of the gas, respectively.

$$PV = nRT = \frac{m}{\mathcal{M}} RT$$

$$\mathcal{M} = \frac{m}{V} \frac{RT}{P} = \frac{\rho RT}{P}$$

2.6 At STP (standard temperature and pressure), 0.280 L of a gas weighs 0.400 g. Calculate the molar mass of the gas.

Following the procedure in Problem 2.5,

$$\mathcal{M} = \frac{mRT}{VP} = \frac{(0.400 \text{ g})\left(0.08206 \text{ L atm K}^{-1} \text{ mol}^{-1}\right)(273 \text{ K})}{(0.280 \text{ L})\,(1.00 \text{ atm})} = 32.0 \text{ g mol}^{-1}$$

2.7 Ozone molecules in the stratosphere absorb much of the harmful radiation from the sun. Typically, the temperature and partial pressure of ozone in the stratosphere are 250 K and 1.0×10^{-3} atm, respectively. How many ozone molecules are present in 1.0 L of air under these conditions? Assume ideal-gas behavior.

First calculate the number of moles of ozone molecules, from which the number of ozone molecules can be obtained.

$$n = \frac{PV}{RT} = \frac{\left(1.0 \times 10^{-3} \text{ atm}\right)(1.0 \text{ L})}{\left(0.08206 \text{ L atm K}^{-1} \text{ mol}^{-1}\right)(250 \text{ K})} = 4.87 \times 10^{-5} \text{ mol}$$

$$\text{Number of ozone molecules} = \left(4.87 \times 10^{-5} \text{ mol}\right)\left(\frac{6.022 \times 10^{23} \text{ molecules}}{1 \text{ mol}}\right)$$

$$= 3.0 \times 10^{19} \text{ molecules}$$

2.8 Calculate the density of HBr in $g\,L^{-1}$ at 733 mmHg and 46°C. Assume ideal-gas behavior.

Use the expression found in Problem 2.5 for the density of an ideal gas in terms of its pressure, molar mass, and temperature.

$$\rho = \frac{P\mathcal{M}}{RT} = \frac{(733 \text{ mmHg})\left(\frac{1 \text{ atm}}{760 \text{ mmHg}}\right)(80.91 \text{ g mol}^{-1})}{(0.08206 \text{ L atm K}^{-1} \text{ mol}^{-1})(273 + 46)\,\text{K}} = 2.98 \text{ g L}^{-1}$$

2.9 Dissolving 3.00 g of an impure sample of $CaCO_3$ in an excess of HCl acid produced 0.656 L of CO_2 (measured at 20°C and 792 mmHg). Calculate the percent by mass of $CaCO_3$ in the sample.

The relevant reaction is

$$CaCO_3(s) + 2HCl(aq) \longrightarrow CaCl_2(aq) + H_2O(l) + CO_2(g)$$

Thus, the number of moles of carbonate in the sample is equal to the number of moles of CO_2 evolved, which can be determined from the ideal gas law:

$$n_{CO_2} = \frac{PV}{RT} = \frac{(792 \text{ mm Hg})\left(\frac{1 \text{ atm}}{760 \text{ mm Hg}}\right)(0.656 \text{ L})}{(0.08206 \text{ L atm K}^{-1} \text{ mol}^{-1})(20 + 273)\,\text{K}}$$

$$= 2.843 \times 10^{-2} \text{ mol} = n_{CaCO_3}$$

$$m_{CaCO_3} = (2.843 \times 10^{-2} \text{ mol})(100.1 \text{ g mol}^{-1}) = 2.846 \text{ g}$$

$$\%CaCO_3 = \frac{m_{CaCO_3}}{\text{mass of sample}} \times 100\% = \frac{2.846 \text{ g}}{3.00 \text{ g}} \times 100\% = 94.9\%$$

2.10 The saturated vapor pressure of mercury is 0.0020 mmHg at 300 K and the density of air at 300 K is 1.18 g L^{-1}. **(a)** Calculate the concentration of mercury vapor in air in $mol\,L^{-1}$. **(b)** What is the number of parts per million (ppm) by mass of mercury in air?

(a) An expression for the concentration of mercury vapor can be obtained by rearranging the ideal gas equation, $PV = nRT$.

$$\text{Concentration} = \frac{n}{V} = \frac{P}{RT}$$

$$= \frac{(0.0020 \text{ mmHg})\left(\frac{1 \text{ atm}}{760 \text{ mmHg}}\right)}{(0.08206 \text{ L atm K}^{-1} \text{ mol}^{-1})(300 \text{ K})}$$

$$= 1.07 \times 10^{-7} \text{ mol L}^{-1}$$

$$= 1.1 \times 10^{-7} \text{ mol L}^{-1}$$

(b) In 1 L,

$$n_{Hg} = 1.07 \times 10^{-7} \text{ mol}$$

$$m_{Hg} = \left(1.07 \times 10^{-7} \text{ mol}\right)\left(200.6 \text{ g mol}^{-1}\right) = 2.15 \times 10^{-5} \text{ g}$$

$$m_{air} = 1.18 \text{ g}$$

$$m_{total} = m_{air} + m_{Hg} \approx m_{air} = 1.18 \text{ g}$$

$$\text{ppm of Hg in air} = \frac{m_{Hg}}{m_{total}} \times 10^6 = \frac{2.15 \times 10^{-5} \text{ g}}{1.18 \text{ g}} \times 10^6 = 18$$

2.11   A very flexible balloon with a volume of 1.2 L at 1.0 atm and 300 K is allowed to rise to the stratosphere, where the temperature and pressure are 250 K and 3.0×10^{-3} atm, respectively. What is the final volume of the balloon? Assume ideal-gas behavior.

Use the PVT relationship in Problem 2.3:

$$\frac{P_1 V_1}{T_1} = \frac{P_2 V_2}{T_2}$$

$$V_2 = \frac{P_1 V_1}{T_1} \frac{T_2}{P_2} = \frac{(1.0 \text{ atm})(1.2 \text{ L})}{300 \text{ K}} \frac{250 \text{ K}}{3.0 \times 10^{-3} \text{ atm}} = 3.3 \times 10^2 \text{ L}$$

A very large final volume, indeed.

Actually, due to the tension in the balloon's surface, the pressure inside any balloon is somewhat greater than that of outside.

2.12   Sodium bicarbonate ($NaHCO_3$) is called baking soda because when heated, it releases carbon dioxide gas, which causes cookies, doughnuts, and bread to rise during baking. (a) Calculate the volume (in liters) of CO_2 produced by heating 5.0 g of $NaHCO_3$ at 180°C and 1.3 atm. (b) Ammonium bicarbonate (NH_4HCO_3) has also been used as a leavening agent. Suggest one advantage and one disadvantage of using NH_4HCO_3 instead of $NaHCO_3$ for baking.

(a) The following reaction takes place when sodium bicarbonate is heated:

$$2NaHCO_3(s) \longrightarrow Na_2CO_3(s) + H_2O(g) + CO_2(g)$$

$$n_{NaHCO_3} = \frac{5.0 \text{ g}}{84.0 \text{ g mol}^{-1}} = 5.95 \times 10^{-2} \text{ mol}$$

$$n_{CO_2} = n_{NaHCO_3} \left(\frac{1 \text{ mol } CO_2}{2 \text{ mol } NaHCO_3}\right)$$

$$= \left(5.95 \times 10^{-2} \text{ mol } NaHCO_3\right)\left(\frac{1 \text{ mol } CO_2}{2 \text{ mol } NaHCO_3}\right) = 2.98 \times 10^{-2} \text{ mol}$$

$$V_{CO_2} = \frac{n_{CO_2}RT}{P} = \frac{(2.98 \times 10^{-2}\ \text{mol})\,(0.08206\ \text{L atm K}^{-1}\ \text{mol}^{-1})\,(273 + 180)\ \text{K}}{1.3\ \text{atm}} = 0.85\ \text{L}$$

(b) Ammonium bicarbonate decomposes upon heating according to the following equation:

$$NH_4HCO_3(s) \longrightarrow NH_3(g) + H_2O(g) + CO_2(g)$$

The advantage in using the ammonium salt is that more gas is produced per gram of reactant. (The molar mass of NH_4HCO_3 is 79.1 g mol^{-1}, smaller than that of $NaHCO_3$.) The disadvantage is that one of the gases is ammonia. The strong odor of ammonia would not make the ammonium salt a good choice for baking.

2.13   A common, non-SI unit for pressure is pounds per square inch (psi). Show that 1 atm = 14.7 psi. An automobile tire is inflated to 28.0 psi gauge pressure when cold, at 18°C. **(a)** What will the pressure be if the tire is heated to 32°C by driving the car? **(b)** What percent of the air in the tire would have to be let out to reduce the pressure to the original 28.0 psi? Assume that the volume of the tire remains constant with temperature. (A tire gauge measures not the pressure of the air inside but its excess over the external pressure, which is 14.7 psi.)

Use the appropriate conversion factors to write

$$1\ \text{atm}\left(\frac{1.01325 \times 10^5\ \text{N m}^{-2}}{1\ \text{atm}}\right)\left(\frac{1\ \text{lb force}}{4.448\ \text{N}}\right)\left(\frac{1\ \text{m}}{39.37\ \text{in}}\right)^2 = 14.7\ \text{lb force in}^{-2}$$

or, using the common abbreviation for "lb force m^{-2}", 1 atm = 14.7 psi.

(a) Charles' Law allows the use of any consistent set of units for the pressure, but remember that the tire gauge measures the pressure above external pressure and add 14.7 psi.

At constant n and V,

$$\frac{P_1}{T_1} = \frac{P_2}{T_2}$$

$$P_2 = \frac{P_1}{T_1}T_2 = \frac{(28.0 + 14.7)\ \text{psi}}{(273 + 18)\ \text{K}}(273 + 32)\ \text{K} = 44.8\ \text{psi}$$

The gauge pressure would read (44.8 − 14.7) psi = 30.1 psi.

(b) From part **(a)**, the pressure was shown to increase by 2.1 psi due to the increase in temperature. So the pressure must be reduced by this amount, or (remember to use the total pressure, not just the gauge pressure)

$$\frac{2.1\ \text{psi}}{44.8\ \text{psi}} \times 100\% = 4.7\%$$

At constant T and V, the pressure is proportional to the amount (number of moles) of gas present. Thus to reduce the pressure by 4.7%, the amount of gas must be reduced by the same amount. 4.7% of the gas must be let out.

2.14 **(a)** What volume of air at 1.0 atm and 22°C is needed to fill a 0.98-L bicycle tire to a pressure of 5.0 atm at the same temperature? (Note that 5.0 atm is the gauge pressure, which is the difference between the pressure in the tire and atmospheric pressure. Initially, the gauge pressure in the tire was 0 atm.) **(b)** What is the total pressure in the tire when the gauge reads 5.0 atm? **(c)** The tire is pumped with a hand pump full of air at 1.0 atm; compressing the gas in the cylinder adds all the air in the pump to the air in the tire. If the volume of the pump is 33% of the tire's volume, what is the gauge pressure in the tire after 3 full strokes of the pump?

(a) Enough gas (air) must be added to increase the pressure in the tire by 5.0 atm, since it starts at a gauge pressure of 0.0 atm. This is the same amount of gas whose volume is desired at ambient conditions. At constant n and T

$$P_1 V_1 = P_2 V_2$$

Letting the pressure and volume in the tire be P_1 and V_1, respectively, and denoting the same quantities at ambient conditions as P_2 and V_2,

$$V_2 = \frac{(5.0 \text{ atm})(0.98 \text{ L})}{1.0 \text{ atm}} = 4.9 \text{ L}$$

(b) When the gauge pressure reads 5.0 atm, the total pressure in the tire is 6.0 atm.

(c) Since the tire and the pump are at the same temperature, the amount of gas each contains at 1.0 atm is proportional to their volumes. Thus three strokes of the pump will add to the tire 99% of the amount of gas that the tire originally contains at 1.0 atm. At constant V and T,

$$\frac{P_1}{n_1} = \frac{P_2}{n_2}$$

Here, $n_2 = 1.99 n_1$,

$$P_2 = \frac{(1.0 \text{ atm})}{n_1}(1.99 n_1) = 1.99 \text{ atm}$$

When the total pressure in the tire is 1.99 atm, the gauge pressure reads 0.99 atm.

2.15 A student breaks a thermometer and spills most of the mercury (Hg) onto the floor of a laboratory that measures 15.2 m long, 6.6 m wide, and 2.4 m high. **(a)** Calculate the mass of mercury vapor (in grams) in the room at 20°C. **(b)** Does the concentration of mercury vapor exceed the air quality regulation of 0.050 mg Hg m^{-3} of air? **(c)** One way to treat small quantities of spilled mercury is to spray powdered sulfur over the metal. Suggest a physical and a chemical reason for this treatment. The vapor pressure of mercury at 20°C is 1.7×10^{-6} atm.

(a) The volume occupied by Hg, V, is the same as the volume of the room. Therefore,

$$V = (15.2 \text{ m}) (6.6 \text{ m}) (2.4 \text{ m}) = 2.41 \times 10^2 \text{ m}^3 = 2.41 \times 10^5 \text{ L}$$

$$n_{\text{Hg}} = \frac{PV}{RT} = \frac{\left(1.7 \times 10^{-6} \text{ atm}\right) \left(2.41 \times 10^5 \text{ L}\right)}{(0.08206 \text{ L atm K}^{-1} \text{ mol}^{-1}) (273 + 20) \text{ K}} = 1.70 \times 10^{-2} \text{ mol}$$

$$m_{\text{Hg}} = \left(1.70 \times 10^{-2} \text{ mol}\right) \left(200.6 \text{ g mol}^{-1}\right) = 3.4 \text{ g}$$

(b)

$$\text{Concentration of Hg} = \frac{\text{mass of Hg}}{V_{\text{room}}} = \left(\frac{3.4 \text{ g}}{2.4 \times 10^2 \text{ m}^3}\right)\left(\frac{1000 \text{ mg}}{1 \text{ g}}\right) = 14 \text{ mg m}^{-3}$$

This concentration far exceeds the air quality regulation of 0.050 mg Hg m^{-3} of air.

(c) Physical: the sulfur powder covers the Hg surface, thus retarding the rate of evaporation. Chemical: sulfur reacts slowly with Hg to form HgS. HgS has no measurable vapor pressure.

2.16 Nitrogen forms several gaseous oxides. One of them has a density of 1.27 g L^{-1} measured at 764 mmHg and 150°C. Write the formula of the compound.

The density of a gas is related to its molar mass as shown in Problem 2.5:

$$\mathcal{M} = \frac{\rho RT}{P} = \frac{\left(1.27 \text{ g L}^{-1}\right)\left(0.08206 \text{ L atm K}^{-1} \text{ mol}^{-1}\right)(273 + 150) \text{ K}}{(764 \text{ mmHg})\left(\frac{1 \text{ atm}}{760 \text{ mmHg}}\right)} = 43.9 \text{ g mol}^{-1}$$

Some nitrogen oxides and their molar masses are NO: 30.0 g mol^{-1}; N_2O: 44.0 g mol^{-1}; NO_2: 46.0 g mol^{-1}.

The nitrogen oxide is N_2O.

2.17 Nitrogen dioxide (NO_2) cannot be obtained in a pure form in the gas phase because it exists as a mixture of NO_2 and N_2O_4. At 25°C and 0.98 atm, the density of this gas mixture is 2.7 g L^{-1}. What is the partial pressure of each gas?

Calculate the average molar mass, $\mathcal{M}_{\text{mix}}$, of the mixture using the relation between $\mathcal{M}$, ρ, T, and P derived in Problem 2.5. This average molar mass is related to the mole fraction of NO_2, x_{NO_2}, and the mole fraction of N_2O_4, $x_{N_2O_4}$. The mole fractions can then be used to calculate the partial pressures of the gases.

$$\mathcal{M}_{\text{mix}} = \frac{\rho_{\text{mix}} RT}{P_{\text{mix}}} = \frac{\left(2.7 \text{ g L}^{-1}\right)\left(0.08206 \text{ L atm K}^{-1} \text{ mol}^{-1}\right)(273 + 25) \text{ K}}{0.98 \text{ atm}} = 67.4 \text{ g mol}^{-1}$$

$$x_{NO_2}\mathcal{M}_{NO_2} + x_{N_2O_4}\mathcal{M}_{N_2O_4} = \mathcal{M}_{\text{mix}} = 67.4 \text{ g mol}^{-1}$$

The sum of all mole fractions is unity, that is, $x_{NO_2} + x_{N_2O_4} = 1$, or $x_{N_2O_4} = 1 - x_{NO_2}$. Using this relation, the above equation becomes

$$x_{NO_2}\mathcal{M}_{NO_2} + \left(1 - x_{NO_2}\right)\mathcal{M}_{N_2O_4} = 67.4 \text{ g mol}^{-1}$$

$$x_{NO_2}\left(46.01 \text{ g mol}^{-1}\right) + \left(1 - x_{NO_2}\right)\left(92.02 \text{ g mol}^{-1}\right) = 67.4 \text{ g mol}^{-1}$$

$$46.01x_{NO_2} + 92.02 - 92.02x_{NO_2} = 67.4$$

$$46.01x_{NO_2} = 24.6$$

$$x_{NO_2} = 0.535$$

Therefore, $x_{N_2O_4} = 1 - x_{NO_2} = 1 - 0.535 = 0.465$, and the partial pressures are

$$P_{NO_2} = x_{NO_2} P_{mix} = (0.535)(0.98 \text{ atm}) = 0.52 \text{ atm}$$

$$P_{N_2O_4} = x_{N_2O_4} P_{mix} = (0.465)(0.98 \text{ atm}) = 0.46 \text{ atm}$$

2.18 An ultra-high-vacuum pump can reduce the pressure of air from 1.0 atm to 1.0×10^{-12} mmHg. Calculate the number of air molecules in a liter at this pressure and 298 K. Compare your results with the number of molecules in 1.0 L at 1.0 atm and 298 K. Assume ideal-gas behavior.

When $P = 1.0 \times 10^{-12}$ mmHg:

$$n = \frac{PV}{RT} = \frac{(1.0 \times 10^{-12} \text{ mmHg})\left(\frac{1 \text{ atm}}{760 \text{ mmHg}}\right)(1.0 \text{ L})}{(0.08206 \text{ L atm K}^{-1} \text{ mol}^{-1})(298 \text{ K})} = 5.38 \times 10^{-17} \text{ mol}$$

Number of molecules $= (5.38 \times 10^{-17} \text{ mol})\left(\dfrac{6.022 \times 10^{23} \text{ molecules}}{1 \text{ mol}}\right) = 3.2 \times 10^7$ molecules

When $P = 1.0$ atm:

$$n = \frac{PV}{RT} = \frac{(1.0 \text{ atm})(1.0 \text{ L})}{(0.08206 \text{ L atm K}^{-1} \text{ mol}^{-1})(298 \text{ K})} = 0.0409 \text{ mol}$$

Number of molecules $= (0.0409 \text{ mol})\left(\dfrac{6.022 \times 10^{23} \text{ molecules}}{1 \text{ mol}}\right) = 2.5 \times 10^{22}$ molecules

The number of molecules present when $P = 1.0$ atm is 7.8×10^{14} times greater than when $P = 1.0 \times 10^{-12}$ mmHg.

2.19 An air bubble with a radius of 1.5 cm at the bottom of a lake where the temperature is 8.4°C and the pressure is 2.8 atm rises to the surface, where the temperature is 25.0°C and the pressure is 1.0 atm. Calculate the radius of the bubble when it reaches the surface. Assume ideal-gas behavior. [*Hint*: The volume of a sphere is given by $(4/3)\pi r^3$, where r is the radius.]

The volume of the bubble at the bottom of the lake:

$$V_1 = \frac{4}{3}\pi r^3 = \frac{4}{3}\pi (1.5 \text{ cm})^3 \left(\frac{1 \text{ L}}{1000 \text{ cm}^3}\right) = 1.41 \times 10^{-2} \text{ L}$$

The volume of the bubble at the surface:

$$\frac{P_1 V_1}{T_1} = \frac{P_2 V_2}{T_2} \quad (\text{ Problem 2.3 })$$

$$V_2 = \frac{P_1 V_1}{T_1}\frac{T_2}{P_2} = \frac{(2.8 \text{ atm})(1.41 \times 10^{-2} \text{ L})}{(273.2 + 8.4) \text{ K}}\frac{(273.2 + 25.0) \text{ K}}{1.0 \text{ atm}} = 4.18 \times 10^{-2} \text{ L}$$

The radius of the bubble when it reaches the surface can now be calculated:

$$V_2 = \left(4.18 \times 10^{-2}\,\text{L}\right)\left(\frac{1000\,\text{cm}^3}{1\,\text{L}}\right) = \frac{4}{3}\pi r^3$$

$$r^3 = 9.98\,\text{cm}^3$$

$$r = 2.2\,\text{cm}$$

2.20   The density of dry air at 1.00 atm and 34.4°C is 1.15 g L^{-1}. Calculate the composition of air (percent by mass) assuming that it contains only nitrogen and oxygen and behaves like an ideal gas. (*Hint*: First calculate the "molar mass" of air, then the mole fractions, and then the mass fractions of O_2 and N_2.)

This problem is similar to Problem 2.17. From the density of air, calculate its molar mass, $\mathcal{M}_{\text{air}}$ (see Problem 2.5), which in turn yields the mole fraction of oxygen, x_{O_2}, and the mole fraction of nitrogen, x_{N_2}. Once the mole fractions are obtained, the composition of air can be calculated.

$$\mathcal{M}_{\text{air}} = \frac{\rho_{\text{air}}RT}{P_{\text{air}}} = \frac{\left(1.15\,\text{g L}^{-1}\right)\left(0.08206\,\text{L atm K}^{-1}\,\text{mol}^{-1}\right)\left(273.2 + 34.4\right)\text{K}}{1.00\,\text{atm}} = 29.03\,\text{g mol}^{-1}$$

$$x_{O_2}\mathcal{M}_{O_2} + x_{N_2}\mathcal{M}_{N_2} = \mathcal{M}_{\text{air}} = 29.03\,\text{g mol}^{-1}$$

The sum of all mole fractions is unity, that is, $x_{O_2} + x_{N_2} = 1$, or $x_{N_2} = 1 - x_{O_2}$. Use this relation in the above equation,

$$x_{O_2}\mathcal{M}_{O_2} + \left(1 - x_{O_2}\right)\mathcal{M}_{N_2} = 29.03\,\text{g mol}^{-1}$$

$$x_{O_2}\left(32.00\,\text{g mol}^{-1}\right) + \left(1 - x_{O_2}\right)\left(28.02\,\text{g mol}^{-1}\right) = 29.03\,\text{g mol}^{-1}$$

$$32.00x_{O_2} + 28.02 - 28.02x_{O_2} = 29.03$$

$$3.98x_{O_2} = 1.01$$

$$x_{O_2} = 0.254$$

Therefore, $x_{N_2} = 1 - x_{O_2} = 1 - 0.254 = 0.746$.

In 1 mol of air, there are 0.25 mol of O_2 and 0.75 mol of N_2. The corresponding masses are therefore:

$$\text{mass of } O_2 = \left(0.254\,\text{mol}\right)\left(32.00\,\text{g mol}^{-1}\right) = 8.13\,\text{g}$$

$$\text{mass of } N_2 = \left(0.746\,\text{mol}\right)\left(28.02\,\text{g mol}^{-1}\right) = 20.9\,\text{g}$$

Therefore,

$$\% \ O_2 \text{ by mass} = \frac{8.13\,\text{g}}{8.13\,\text{g} + 20.9\,\text{g}} \times 100\% = 28\%$$

$$\% \ N_2 \text{ by mass} = 1 - \% \ O_2 \text{ by mass} = 1 - 28\% = 72\%$$

2.21   A gas that evolved during the fermentation of glucose has a volume of 0.78 L when measured at 20.1°C and 1.0 atm. What was the volume of this gas at the fermentation temperature of 36.5°C? Assume ideal-gas behavior.

At constant n and P,

$$\frac{V_1}{T_1} = \frac{V_2}{T_2}$$

$$V_2 = \frac{V_1}{T_1} T_2 = \frac{0.78 \text{ L}}{(273.2 + 20.1) \text{ K}} (273.2 + 36.5) \text{ K} = 0.82 \text{ L}$$

2.22   Two bulbs of volumes V_A and V_B are connected by a stopcock. The number of moles of gases in the bulbs are n_A and n_B, and initially the gases are at the same pressure, P, and temperature, T. Show that the final pressure of the system, after the stopcock has been opened, is equal to P. Assume ideal-gas behavior.

Gas A and gas B both obey the ideal gas equation, that is, before the stopcock is open,

$$PV_A = n_A RT$$

$$PV_B = n_B RT$$

When the stopcock is open,

$$V_{\text{total}} = V_A + V_B$$

$$n_{\text{total}} = n_A + n_B$$

The total pressure is

$$P_{\text{total}} = \frac{n_{\text{total}} RT}{V_{\text{total}}} = \frac{(n_A + n_B) RT}{V_A + V_B} = \frac{n_A RT + n_B RT}{V_A + V_B}$$

From above, $n_A RT = PV_A$ and $n_B RT = PV_B$. Upon substitution into the expression for P_{total},

$$P_{\text{total}} = \frac{PV_A + PV_B}{V_A + V_B} = \frac{P(V_A + V_B)}{V_A + V_B} = P$$

2.23   The composition of dry air at sea level is 78.03% N_2, 20.99% O_2, and 0.033% CO_2 by volume. **(a)** Calculate the average molar mass of this air sample. **(b)** Calculate the partial pressures of N_2, O_2, and CO_2 in atm. (At constant temperature and pressure, the volume of a gas is directly proportional to the number of moles of the gas.)

(a) At constant P and T, $n \propto V$. Thus, % by volume = mol %.

$$\begin{aligned}\text{Average molar mass} &= \left(\text{mol \% N}_2\right)\mathcal{M}_{N_2} + \left(\text{mol \% O}_2\right)\mathcal{M}_{O_2} + \left(\text{mol \% CO}_2\right)\mathcal{M}_{CO_2}\\ &= (78.03\%)\left(28.02 \text{ g mol}^{-1}\right) + (20.99\%)\left(32.00 \text{ g mol}^{-1}\right)\\ &\quad + (0.033\%)\left(44.01 \text{ g mol}^{-1}\right)\\ &= 29 \text{ g mol}^{-1}\end{aligned}$$

(b) The atmospheric pressure at sea level is 1.0 atm.

$$P_{N_2} = x_{N_2}\,(1.0 \text{ atm}) = (0.7803)\,(1.0 \text{ atm}) = 0.78 \text{ atm}$$

$$P_{O_2} = x_{O_2}\,(1.0 \text{ atm}) = (0.2099)\,(1.0 \text{ atm}) = 0.21 \text{ atm}$$

$$P_{CO_2} = x_{CO_2}\,(1.0 \text{ atm}) = \left(3.3 \times 10^{-4}\right)(1.0 \text{ atm}) = 3.3 \times 10^{-4} \text{ atm}$$

2.24   A mixture containing nitrogen and hydrogen weighs 3.50 g and occupies a volume of 7.46 L at 300 K and 1.00 atm. Calculate the mass percent of these two gases. Assume ideal-gas behavior.

First calculate the total number of moles of the mixture, n_{mix}, which, together with the mass of the mixture, m_{mix}, is used to determine the number of moles of N_2 and the number of moles of H_2, and, consequently, the mass percent of these gases.

$$n_{mix} = \frac{PV}{RT} = \frac{(1.00 \text{ atm})\,(7.46 \text{ L})}{\left(0.08206 \text{ L atm K}^{-1}\text{ mol}^{-1}\right)(300 \text{ K})} = 0.3030 \text{ mol}$$

The mass of the mixture is

$$m_{mix} = n_{N_2}\mathcal{M}_{N_2} + n_{H_2}\mathcal{M}_{H_2} = 3.50 \text{ g}$$

Because $n_{N_2} + n_{H_2} = n_{mix} = 0.303$ mol,

$$n_{H_2} = 0.3030 \text{ mol} - n_{N_2}$$

Therefore,

$$m_{mix} = n_{N_2}\mathcal{M}_{N_2} + \left(0.3030 \text{ mol} - n_{N_2}\right)\mathcal{M}_{H_2} = 3.50 \text{ g}$$

$$n_{N_2}\left(28.02 \text{ g mol}^{-1}\right) + \left(0.3030 \text{ mol} - n_{N_2}\right)\left(2.016 \text{ g mol}^{-1}\right) = 3.50 \text{ g}$$

$$28.02 n_{N_2}\text{ mol}^{-1} + 0.6108 - 2.016 n_{N_2}\text{ mol}^{-1} = 3.50$$

$$26.00 n_{N_2}\text{ mol}^{-1} = 2.889$$

$$n_{N_2} = 0.1111 \text{ mol}$$

$$\text{mass of N}_2 = n_{N_2}\left(28.02 \text{ g mol}^{-1}\right) = (0.1111 \text{ mol})\left(28.02 \text{ g mol}^{-1}\right) = 3.113 \text{ g}$$

$$\text{mass of H}_2 = 3.50 \text{ g} - 3.113 \text{ g} = 0.387 \text{ g}$$

$$\text{mass \% of N}_2 = \frac{3.113 \text{ g}}{3.50 \text{ g}} \times 100\% = 88.9\%$$

$$\text{mass \% of H}_2 = \frac{0.387 \text{ g}}{3.50 \text{ g}} \times 100\% = 11.1\%$$

2.25 The relative humidity in a closed room with a volume of 645.2 m^3 is 87.6% at 300 K, and the vapor pressure of water at 300 K is 0.0313 atm. Calculate the mass of water in the air. [*Hint*: The relative humidity is defined as $(P/P_s) \times 100\%$, where P and P_s are the partial pressure and saturated partial pressure of water, respectively.]

The partial pressure of water can be obtained from the relative humidity:

$$\text{Relative humidity} = \frac{P}{P_s} \times 100\% = 87.6\%$$

$$P = P_s \left(\frac{87.6\%}{100\%}\right) = (0.0313 \text{ atm})\left(\frac{87.6\%}{100\%}\right) = 0.02742 \text{ atm}$$

The number of moles of water can now be calculated:

$$n = \frac{PV}{RT} = \frac{(0.02742 \text{ atm})\left(645.2 \text{ m}^3\right)\left(\frac{1000 \text{ L}}{1 \text{ m}^3}\right)}{\left(0.08206 \text{ L atm K}^{-1} \text{ mol}^{-1}\right)(300 \text{ K})} = 718.6 \text{ mol}$$

$$\text{mass of water vapor} = (718.6 \text{ mol})\left(18.02 \text{ g mol}^{-1}\right) = 1.29 \times 10^4 \text{ g}$$

2.26 Death by suffocation in a sealed container is normally caused not by oxygen deficiency but by CO_2 poisoning, which occurs at about 7% CO_2 by volume. For what length of time would it be safe to be in a sealed room $10 \times 10 \times 20$ ft? [Source: "Eco-Chem," J. A. Campbell, *J. Chem. Educ.* **49**, 538 (1972).]

The source of the excess CO_2 is that which is exhaled and which had as its source the O_2 that was inhaled and metabolized. Thus, to calculate how much CO_2 is added to the room, calculate how much O_2 is depleted. (This assumes a 1:1 molar ratio between CO_2 formed and O_2 used with little hydrogen oxidized to H_2O by inhaled oxygen. The ratio is actually about 1.2 O_2:1 CO_2.)

The air becomes lethal (due to CO_2) after 7% of the O_2 (to 1 sig. fig.) is removed, or

$$(10 \text{ ft})(10 \text{ ft})(20 \text{ ft})\left(\frac{27 \text{ L}}{1 \text{ ft}^3}\right)(0.07) = 3800 \text{ L of O}_2.$$

A person breathes about 0.5 L of air 12 times per minute, and the air is about 20% O_2. Thus,

$$(12 \text{ min}^{-1})(0.5 \text{ L})(0.20) = 1.2 \text{ L O}_2 \text{ min}^{-1}.$$

About 30% of this inhaled O_2 is absorbed in the lungs, so that a person typically uses

$$(0.30)(1.2 \text{ L O}_2 \text{ min}^{-1}) = 0.36 \text{ L O}_2 \text{ min}^{-1}.$$

For a calm, quiet person about half this amount, or 0.2 L min^{-1} would be enough, where we have rounded to 1 sig. fig. Thus one person could last

$$\left(\frac{3800 \text{ L of O}_2}{0.2 \text{ L min}^{-1}}\right)\left(\frac{1 \text{ day}}{1440 \text{ min}}\right) = 13 \text{ days}$$

2.27 A flask contains a mixture of two ideal gases, A and B. Show graphically how the total pressure of the system depends on the amount of A present; that is, plot the total pressure versus the mole fraction of A. Do the same for B on the same graph. The total number of moles of A and B is constant.

$P_A + P_B = x_A P + x_B P = (x_A + x_B) P = P$ where the last step follows because $x_A + x_B = 1$ in a two-component mixture. Thus the total pressure remains constant. The partial pressures show a linear dependence on mole fraction, $P_A = x_A P$ and $P_B = x_B P = (1 - x_A) P$. See graph below.

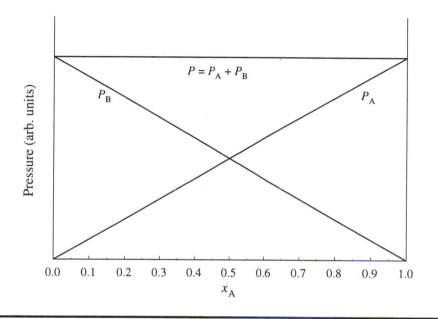

2.28 A mixture of helium and neon gases is collected over water at 28.0°C and 745 mmHg. If the partial pressure of helium is 368 mmHg, what is the partial pressure of neon? (*Note:* Vapor pressure of water at 28°C is 28.3 mmHg.)

P_{Ne} can be determined by rearranging the equation $P_{total} = P_{He} + P_{Ne} + P_{H_2O}$:

$$P_{Ne} = P_{total} - P_{He} - P_{H_2O} = 745 \text{ mmHg} - 368 \text{ mmHg} - 28.3 \text{ mmHg} = 349 \text{ mmHg}$$

2.29 If the barometric pressure falls in one part of the world, it must rise somewhere else. Explain why.

The barometer measures local air pressure, and thinking of the atmosphere as a continuous system, when the pressure decreases in one area it must be compensated by an increase of pressure in another area.

2.30 A piece of sodium metal reacts completely with water as follows:

$$2Na(s) + 2H_2O(l) \longrightarrow 2NaOH(aq) + H_2(g)$$

The hydrogen gas generated is collected over water at 25.0°C. The volume of the gas is 246 mL measured at 1.00 atm. Calculate the number of grams of sodium used in the reaction. (*Note:* Vapor pressure of water at 25°C = 0.0313 atm.)

First calculate the number of moles of H_2 from the ideal gas law, from which the number of moles of Na, and therefore, the mass of Na used in the reaction can be determined.

$$P_{H_2} + P_{H_2O} = P_{total}$$

$$P_{H_2} = P_{total} - P_{H_2O} = 1.00 \text{ atm} - 0.0313 \text{ atm} = 0.969 \text{ atm}$$

$$n_{H_2} = \frac{P_{H_2} V}{RT} = \frac{(0.969 \text{ atm}) (0.246 \text{ L})}{(0.08206 \text{ L atm K}^{-1} \text{ mol}^{-1}) (273.2 + 25.0) \text{ K}} = 9.74 \times 10^{-3} \text{ mol}$$

According to the chemical equation,

$$n_{Na} = n_{H_2} \left(\frac{2 \text{ mol Na}}{1 \text{ mol H}_2} \right) = (9.74 \times 10^{-3} \text{ mol H}_2) \left(\frac{2 \text{ mol Na}}{1 \text{ mol H}_2} \right) = 1.95 \times 10^{-2} \text{ mol}$$

The mass of Na used is

$$m_{Na} = (1.95 \times 10^{-2} \text{ mol}) (22.99 \text{ g mol}^{-1}) = 0.45 \text{ g}$$

2.31   A sample of zinc metal reacts completely with an excess of hydrochloric acid:

$$Zn(s) + 2HCl(aq) \longrightarrow ZnCl_2(aq) + H_2(g)$$

The hydrogen gas produced is collected over water at 25.0°C. The volume of the gas is 7.80 L, and the pressure is 0.980 atm. Calculate the amount of zinc metal in grams consumed in the reaction. (*Note:* Vapor pressure of water at 25°C is 23.8 mmHg.)

This problem is similar to Problem 2.30. First calculate the number of moles of H_2 from the ideal gas law, from which the number of moles of Zn, and therefore, the mass of Zn used in the reaction can be determined.

$$P_{H_2} + P_{H_2O} = P_{total}$$

$$P_{H_2} = P_{total} - P_{H_2O} = 0.980 \text{ atm} - 23.8 \text{ mmHg} \left(\frac{1 \text{ atm}}{760 \text{ mmHg}} \right) = 0.9487 \text{ atm}$$

$$n_{H_2} = \frac{P_{H_2} V}{RT} = \frac{(0.9487 \text{ atm}) (7.80 \text{ L})}{(0.08206 \text{ L atm K}^{-1} \text{ mol}^{-1}) (273.2 + 25.0) \text{ K}} = 0.3024 \text{ mol}$$

According to the chemical equation,

$$n_{Zn} = n_{H_2} \left(\frac{1 \text{ mol Zn}}{1 \text{ mol H}_2} \right) = (0.3024 \text{ mol H}_2) \left(\frac{1 \text{ mol Zn}}{1 \text{ mol H}_2} \right) = 0.3024 \text{ mol}$$

The mass of Zn used is

$$m_{Zn} = (0.3024 \text{ mol}) \left(65.39 \text{ g mol}^{-1}\right) = 19.8 \text{ g}$$

2.32 Helium is mixed with oxygen gas for deep sea divers. Calculate the percent by volume of oxygen gas in the mixture if the diver has to submerge to a depth where the total pressure is 4.2 atm. The partial pressure of oxygen is maintained at 0.20 atm at this depth.

At constant P and T, $n \propto V$. Thus, % by volume = mol %, and mole fraction is directly related to mol %.

$$P_{O_2} = x_{O_2} P_{total}$$

$$x_{O_2} = \frac{P_{O_2}}{P_{total}} = \frac{0.20 \text{ atm}}{4.2 \text{ atm}} = 0.048$$

The mole fraction of O_2 is 0.048, or the percent by volume of $O_2 = 4.8\%$.

2.33 A sample of ammonia (NH_3) gas is completely decomposed to nitrogen and hydrogen gases over heated iron wool. If the total pressure is 866 mmHg, calculate the partial pressures of N_2 and H_2.

The relevant chemical equation is

$$2NH_3(g) \longrightarrow N_2(g) + 3H_2(g)$$

The stoichiometric relationship gives $n_{H_2} = 3n_{N_2}$. At constant T and V, $P \propto n$. That is, $P_{H_2} = 3P_{N_2}$. In addition, it is given that

$$P_{N_2} + P_{H_2} = 866 \text{ mmHg}$$

Therefore,

$$P_{N_2} + 3P_{N_2} = 866 \text{ mmHg}$$

$$4P_{N_2} = 866 \text{ mmHg}$$

$$P_{N_2} = 217 \text{ mmHg}$$

$$P_{H_2} = 3P_{N_2} = 650 \text{ mmHg}$$

2.34 The partial pressure of carbon dioxide in air varies with the seasons. Would you expect the partial pressure in the Northern Hemisphere to be higher in the summer or winter? Explain.

Plant photosynthesis is a major contributor to the seasonal variation of the amount of carbon dioxide in the atmosphere. Thus, in the Northern Hemisphere the partial pressure of CO_2 is higher in the winter when less CO_2 is being utilized in photosynthesis.

2.35 A healthy adult exhales about 5.0×10^2 mL of a gaseous mixture with every breath. Calculate the number of molecules present in this volume at 37°C and 1.1 atm. List the major components of this gaseous mixture.

First calculate the number of moles of molecules, from which the number of molecules can be obtained.

$$n = \frac{PV}{RT} = \frac{(1.1 \text{ atm}) \left(5.0 \times 10^2 \text{ mL}\right) \left(\frac{1 \text{ L}}{1000 \text{ mL}}\right)}{(0.08206 \text{ L atm K}^{-1} \text{ mol}^{-1})(273 + 37) \text{ K}} = 0.0216 \text{ mol}$$

$$\text{Number of molecules} = (0.0216 \text{ mol}) \left(\frac{6.022 \times 10^{23} \text{ molecules}}{1 \text{ mol}}\right) = 1.3 \times 10^{22} \text{ molecules}$$

The major components of exhaled air are CO_2, O_2, N_2, and H_2O.

2.36 Describe how you would measure, by either chemical or physical means (other than mass spectrometry), the partial pressures of a mixture of gases: **(a)** CO_2 and H_2, **(b)** He and N_2.

(a) A measurement of the total pressure of the mixture can be made at known temperature and volume. A chemical separation may then be used to measure the amount of a single component. A good choice is the reaction between CO_2 and sodium hydroxide

$$CO_2(g) + 2NaOH(aq) \longrightarrow Na_2CO_3(aq) + H_2O(l)$$

This leaves only the H_2 gas and water vapor. The partial pressure of H_2 can now be determined under the same conditions of temperature and volume after correcting for the known vapor pressure of water. Finally, the partial pressure of CO_2 is calculated from

$$P_{CO_2} = P_{total} - P_{H_2}$$

(b) In this case there is no convenient chemical means of separation, but there is a significant difference in boiling points that can be utilized. As in part **(a)** the total pressure of the mixture is first measured. The temperature is then lowered until the nitrogen liquefies (b.p. N_2: 77 K). At this point the He is still gaseous (b.p. He: 4 K), and its pressure can be measured. Charles' Law is then used to calculate the pressure of He at the temperature of the original total pressure measurement (assuming a constant volume container). The partial pressure of N_2 is then the difference between the total pressure and helium pressure at this temperature.

2.37 The gas laws are vitally important to scuba divers. The pressure exerted by 33 ft of seawater is equivalent to 1 atm pressure. **(a)** A diver ascends quickly to the surface of the water from a depth of 36 ft without exhaling gas from his lungs. By what factor would the volume of his lungs increase by the time he reaches the surface? Assume that the temperature is constant. **(b)** The partial pressure of oxygen in air is about 0.20 atm. (Air is 20% oxygen by volume.) In deep-sea diving, the composition of air the diver breathes must be changed to maintain this partial pressure. What must the oxygen content (in percent by volume) be when the total pressure exerted on the diver is 4.0 atm?

(a) The pressure experienced by the diver at a depth of 36 ft beneath the surface is

$$P_1 = \text{atmospheric pressure} + \text{pressure exerted by 36 ft of seawater}$$
$$= 1.0 \text{ atm} + (36 \text{ ft})\left(\frac{1 \text{ atm}}{33 \text{ ft}}\right) = 2.09 \text{ atm}$$

At constant n and T,

$$P_1 V_1 = P_2 V_2$$

The ratio between the volume of the diver's lungs at 36 ft under the surface and that at the surface is

$$\frac{V_2}{V_1} = \frac{P_1}{P_2} = \frac{2.09 \text{ atm}}{1.0 \text{ atm}} = 2.1$$

The diver's lungs would increase in volume by a factor of 2.1 by the time he reaches the surface.

(b) At constant T and P, $V \propto n$. Therefore, % of O_2 by volume = mol % of O_2.

$$P_{O_2} = x_{O_2} P_{\text{total}}$$

$$x_{O_2} = \frac{P_{O_2}}{P_{\text{total}}} = \frac{0.20 \text{ atm}}{4.0 \text{ atm}} = 0.050$$

% O_2 by volume = 5.0%.

2.38 A 1.00-L bulb and a 1.50-L bulb, connected by a stopcock, are filled, respectively, with argon at 0.75 atm and helium at 1.20 atm at the same temperature. Calculate the total pressure and the partial pressures of each gas after the stopcock has been opened and the mole fraction of each gas. Assume ideal-gas behavior.

At constant n and T, $P_1 V_1 = P_2 V_2$, where 1 and 2 denote the state before and after the stopcock is opened, respectively.

For Ar,

$$P_2 = \frac{P_1 V_1}{V_2} = \frac{(0.75 \text{ atm})(1.00 \text{ L})}{2.50 \text{ L}} = 0.30 \text{ atm} = P_{\text{Ar}}$$

For He,

$$P_2 = \frac{P_1 V_1}{V_2} = \frac{(1.20 \text{ atm})(1.50 \text{ L})}{2.50 \text{ L}} = 0.720 \text{ atm} = P_{\text{He}}$$

The total pressure is

$$P = P_{\text{Ar}} + P_{\text{He}} = 0.30 \text{ atm} + 0.720 \text{ atm} = 1.02 \text{ atm}$$

The mole fractions are

$$x_{Ar} = \frac{P_{Ar}}{P} = \frac{0.30 \text{ atm}}{1.02 \text{ atm}} = 0.29$$

$$x_{He} = 1 - x_{Ar} = 1 - 0.29 = 0.71$$

2.39   A mixture of helium and neon weighing 5.50 g occupies a volume of 6.80 L at 300 K and 1.00 atm. Calculate the composition of the mixture in mass percent.

This problem is similar to Problem 2.24. First calculate the total number of moles of the mixture, n_{mix}, which, together with the mass of the mixture, m_{mix}, can be used to determine the number of moles of He and the number of moles of Ne, and consequently, the mass % of these gases.

$$n_{mix} = \frac{PV}{RT} = \frac{(1.00 \text{ atm}) (6.80 \text{ L})}{(0.08206 \text{ L atm K}^{-1} \text{ mol}^{-1}) (300 \text{ K})} = 0.2762 \text{ mol}$$

The mass of the mixture is

$$m_{mix} = n_{He}\mathcal{M}_{He} + n_{Ne}\mathcal{M}_{Ne} = 5.50 \text{ g}$$

Because $n_{He} + n_{Ne} = n_{mix} = 0.2762$ mol,

$$n_{Ne} = 0.2762 \text{ mol} - n_{He}$$

Therefore,

$$m_{mix} = n_{He}\mathcal{M}_{He} + (0.2762 \text{ mol} - n_{He})\mathcal{M}_{Ne} = 5.50 \text{ g}$$

$$n_{He}(4.003 \text{ g mol}^{-1}) + (0.2762 \text{ mol} - n_{He})(20.18 \text{ g mol}^{-1}) = 5.50 \text{ g}$$

$$4.003 n_{He} \text{ mol}^{-1} + 5.574 - 20.18 n_{He} \text{ mol}^{-1} = 5.50$$

$$16.18 n_{He} \text{ mol}^{-1} = 0.074$$

$$n_{He} = 4.6 \times 10^{-3} \text{ mol}$$

mass of He $= n_{He}(4.003 \text{ g mol}^{-1}) = (4.6 \times 10^{-3} \text{ mol})(4.003 \text{ g mol}^{-1}) = 1.8 \times 10^{-2} \text{ g}$

mass of Ne $= 5.50 \text{ g} - 2 \times 10^{-2} \text{ g} = 5.482 \text{ g}$

mass % of He $= \dfrac{1.8 \times 10^{-2} \text{ g}}{5.50 \text{ g}} \times 100\% = 0.3\%$

mass % of Ne $= \dfrac{5.482 \text{ g}}{5.50 \text{ g}} \times 100\% = 99.7\%$

2.40   Suggest two demonstrations to show that gases do not behave ideally.

One demonstration is quite common. Namely, the condensation of a gas at low temperatures and/or high pressures to form a liquid, such as was used in Problem 2.36(b). The condensation demonstrates the existence of attractive forces between molecules. A second demonstration would be to plot the compressibility factor, $Z = P\overline{V}/RT$ vs. P. Deviations from unity show that the gas does not behave ideally.

2.41 Which of the following combinations of conditions most influences a gas to behave ideally: **(a)** low pressure and low temperature, **(b)** low pressure and high temperature, **(c)** high pressure and high temperature, and **(d)** high pressure and low temperature.

Choice **(b)**, low pressure and high temperature most favors a gas to behave ideally.

2.42 The van der Waals constants a and b for benzene are 18.00 atm L^2 mol^{-2} and 0.115 L mol^{-1}, respectively. Calculate the critical constants for benzene.

Calculate $\overline{V}_c$ from Eq. 2.21, T_c from Eq. 2.22, and P_c from Eq. 2.23.

$$\overline{V}_c = 3b = 3\left(0.115 \text{ L mol}^{-1}\right) = 0.345 \text{ L mol}^{-1}$$

$$T_c = \frac{8a}{27Rb} = \frac{8\left(18.00 \text{ atm L}^2 \text{ mol}^{-2}\right)}{27\left(0.08206 \text{ L atm K}^{-1} \text{ mol}^{-1}\right)\left(0.115 \text{ L mol}^{-1}\right)} = 565 \text{ K}$$

$$P_c = \frac{a}{27b^2} = \frac{18.00 \text{ atm L}^2 \text{ mol}^{-2}}{27\left(0.115 \text{ L mol}^{-1}\right)^2} = 50.4 \text{ atm}$$

2.43 Using the data shown in Table 2.1, calculate the pressure exerted by 2.500 moles of carbon dioxide confined in a volume of 1.000 L at 450 K. Compare the pressure with that calculated assuming ideal behavior.

For CO_2, $a = 3.60$ atm L^2 mol^{-2} and $b = 0.0427$ L mol^{-1}. Rearrange the van der Waals equation, $\left(P + \frac{an^2}{V^2}\right)(V - nb) = nRT$ to give

$$P = \frac{nRT}{V - nb} - \frac{an^2}{V^2}$$

$$= \frac{(2.500 \text{ mol})(0.08206 \text{ L atm K}^{-1} \text{ mol}^{-1})(450 \text{ K})}{1.000 \text{ L} - (2.500 \text{ mol})(0.0427 \text{ L mol}^{-1})} - \frac{(3.60 \text{ atm L}^2 \text{ mol}^{-2})(2.500 \text{ mol})^2}{(1.000 \text{ L})^2}$$

$$= 80.9 \text{ atm}$$

If CO_2 behaved ideally,

$$P = \frac{nRT}{V} = \frac{(2.500 \text{ mol})\left(0.08206 \text{ L atm K}^{-1} \text{ mol}^{-1}\right)(450 \text{ K})}{1.000 \text{ L}} = 92.3 \text{ atm}$$

The pressure calculated using the van der Waals equation of state is smaller than that calculated using the ideal gas equation. Thus, under these conditions there are net attractive forces between molecules.

2.44   Without referring to a table, select from the following list the gas that has the largest value of b in the van der Waals equation: CH_4, O_2, H_2O, CCl_4, Ne.

The van der Waals constant b is related to the size of the molecule, so look for the largest molecule, which is CCl_4 in this case.

2.45   Referring to Figure 2.4, we see that for He the plot has a positive slope even at low pressures. Explain this behavior.

Positive deviations of the compressibility factor from unity arise from intermolecular repulsion, represented by the excluded volume term, b, in the van der Waals equation. Negative deviations are due to intermolecular attractive forces, which are reflected in the van der Waals constant a. Referring to the Table 2.1, we see that He has a very small value for a. Thus, He atoms have very weak attraction for each other, and the intermolecular forces are on average repulsive even at low pressures. The table suggests that similar behavior can be observed for H_2 and Ne as well, and indeed it is. These plots of compressibility factor versus pressure are temperature dependent. At temperatures below 100 K, both H_2 and Ne will have negative slopes at low pressures, dipping below $Z = 1$. He continues to have a positive slope until the temperature falls below 26 K. (Note: It can be shown, as is done in Problem 2.48, that as $P \rightarrow 0$, the slope of Z versus P approaches the second virial coefficient, B'.)

2.46   At 300 K, the virial coefficients (B) of N_2 and CH_4 are -4.2 cm^3 mol^{-1} and -15 cm^3 mol^{-1}, respectively. Which gas behaves more ideally at this temperature?

According to the equation

$$Z = 1 + \frac{B}{\overline{V}} + \cdots$$

the closer to zero the value of B, the closer is Z to unity, that is, the more ideal is the gas. According to the data given, N_2 behaves more ideally than CH_4.

2.47   Calculate the molar volume of carbon dioxide at 400 K and 30 atm, given that the second virial coefficient (B) of CO_2 is -0.0605 L mol^{-1}. Compare your result with that obtained using the ideal-gas equation.

Ignoring higher order terms,

$$Z = \frac{P\overline{V}}{RT} \approx 1 + \frac{B}{\overline{V}} = 1 + B\frac{P}{RT}$$

Thus,

$$Z = 1 + (-0.0605 \text{ L mol}^{-1})\frac{30 \text{ atm}}{(0.08206 \text{ L atm K}^{-1} \text{ mol}^{-1})(400 \text{ K})}$$

$$= 0.9447$$

Then, since $Z = \frac{P\overline{V}}{RT}$,

$$\overline{V} = \frac{ZRT}{P}$$

$$= \frac{0.9447(0.08206\ \text{L atm K}^{-1}\ \text{mol}^{-1})(400\ \text{K})}{30\ \text{atm}}$$

$$= 1.03\ \text{L mol}^{-1}$$

This value for $\overline{V}$ can be compared with the result from the ideal gas law which shows that the real gas is experiencing the effects of attractive intermolecular forces,

$$\overline{V} = \frac{RT}{P}$$

$$= \frac{(0.08206\ \text{L atm K}^{-1}\ \text{mol}^{-1})(400\ \text{K})}{30\ \text{atm}}$$

$$= 1.09\ \text{L mol}^{-1}$$

2.48 Consider the virial equation $Z = 1 + B'P + C'P^2$, which describes the behavior of a gas at a certain temperature. From the following plot of Z versus P, deduce the signs of B' and C' (< 0, $= 0$, > 0).

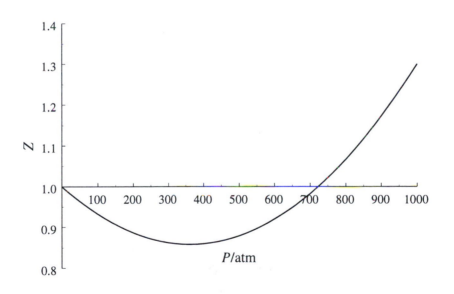

We see that the slope of the Z versus P plot is given by

$$\frac{dZ}{dP} = B' + 2C'P$$

As $P \to 0$, $\frac{dZ}{dP} \to B'$.

Additionally, the curvature of the graph is given by

$$\frac{d^2Z}{dP^2} = 2C'$$

Near $P = 0$ the graph has a negative slope, starting at $Z = 1$ and dipping below 1. It does display positive curvature, with the graph turning around to rise above 1. Thus $B' < 0$ and $C' > 0$.

2.49 The critical temperature and critical pressure of naphthalene are 474.8 K and 40.6 atm, respectively. Calculate the van der Waals constants a and b for naphthalene.

$$a = \frac{27R^2T_c^2}{64P_c} = \frac{27(0.08206 \text{ L atm K}^{-1} \text{ mol}^{-1})^2(474.8 \text{ K})^2}{64(40.6 \text{ atm})} = 15.8 \text{ atm L}^2 \text{ mol}^{-2}$$

$$b = \frac{RT_c}{8P_c} = \frac{(0.08206 \text{ L atm K}^{-1} \text{ mol}^{-1})(474.8 \text{ K})}{8(40.6 \text{ atm})} = 0.120 \text{ L mol}^{-1}$$

2.50 Derive the van der Waals constants a and b in terms of the critical constants by recognizing that at the critical point, $\left(\partial P/\partial \overline{V}\right)_T = 0$ and $\left(\partial^2 P/\partial \overline{V}^2\right)_T = 0$. (This problem requires a knowledge of partial differentiation; see p. 66).

Rearrange the van der Waals equation, $\left(P + \frac{an^2}{V^2}\right)(V - nb) = nRT$ to give

$$P = \frac{nRT}{V - nb} - \frac{an^2}{V^2}$$

Divide the numerator and denominator of the first term by n, and those of the second term by n^2,

$$P = \frac{RT}{\overline{V} - b} - \frac{a}{\overline{V}^2} \tag{2.50.1}$$

Take the first and second derivatives of P with respect to $\overline{V}$,

$$\left(\frac{\partial P}{\partial \overline{V}}\right)_T = -\frac{RT}{\left(\overline{V} - b\right)^2} + \frac{2a}{\overline{V}^3} \tag{2.50.2}$$

$$\left(\frac{\partial^2 P}{\partial \overline{V}^2}\right)_T = \frac{2RT}{\left(\overline{V} - b\right)^3} - \frac{6a}{\overline{V}^4} \tag{2.50.3}$$

At the critical point, $P = P_c$, $\overline{V} = \overline{V}_c$, and $T = T_c$. In addition, $\left(\frac{\partial P}{\partial \overline{V}}\right)_T = 0$ and $\left(\frac{\partial^2 P}{\partial \overline{V}^2}\right)_T = 0$. Therefore, Eqs. 2.50.2 and 2.50.3 become

$$-\frac{RT_c}{\left(\overline{V}_c - b\right)^2} + \frac{2a}{\overline{V}_c^3} = 0 \tag{2.50.4}$$

$$\frac{2RT_c}{\left(\overline{V}_c - b\right)^3} - \frac{6a}{\overline{V}_c^4} = 0 \tag{2.50.5}$$

Rearrange Eqs. 2.50.4 and 2.50.5:

$$\frac{2a}{\overline{V}_c^3} = \frac{RT_c}{(\overline{V}_c - b)^2} \tag{2.50.6}$$

$$\frac{6a}{\overline{V}_c^4} = \frac{2RT_c}{(\overline{V}_c - b)^3} \tag{2.50.7}$$

Now divide Eq. 2.50.6 by Eq. 2.50.7:

$$\frac{\frac{2a}{\overline{V}_c^3}}{\frac{6a}{\overline{V}_c^4}} = \frac{\frac{RT_c}{(\overline{V}_c - b)^2}}{\frac{2RT_c}{(\overline{V}_c - b)^3}}$$

$$\frac{\overline{V}_c}{3} = \frac{\overline{V}_c - b}{2}$$

$$2\overline{V}_c = 3\overline{V}_c - 3b$$

$$\overline{V}_c = 3b \tag{2.50.8}$$

Substitute Eq. 2.50.8 into 2.50.6:

$$\frac{2a}{(3b)^3} = \frac{RT_c}{(3b - b)^2}$$

$$T_c = \frac{2a}{27b^3}\frac{4b^2}{R}$$

$$= \frac{8a}{27Rb} \tag{2.50.9}$$

At the critical point, Eq. 2.50.1 becomes

$$P_c = \frac{RT_c}{\overline{V}_c - b} - \frac{a}{\overline{V}_c^2} \tag{2.50.10}$$

P_c can be solved for by substituting Eqs. 2.50.8 and 2.50.9 into Eq. 2.50.10:

$$P_c = \frac{R\left(\frac{8a}{27Rb}\right)}{3b - b} - \frac{a}{(3b)^2}$$

$$= \frac{4a}{27b^2} - \frac{a}{9b^2}$$

$$= \frac{a}{27b^2}$$

2.51 From the relationships among the van der Waals constants and the critical constants, show that $Z_c = P_c\overline{V}_c/RT_c = 0.375$, where Z_c is the compressibility factor at the critical point.

$$Z_c = \frac{P_c \overline{V}_c}{R T_c} = \frac{\left(\frac{a}{27b^2}\right)(3b)}{R\left(\frac{8a}{27Rb}\right)} = \frac{3}{8} = 0.375$$

This suggests all substances have $Z_c = 0.375$. Is this true?

2.52 A CO_2 fire extinguisher is located on the outside of a building in Massachusetts. During the winter months, one can hear a sloshing sound when the extinguisher is gently shaken. In the summertime, there is often no sound when it is shaken. Explain. Assume that the extinguisher has no leaks and that it has not been used.

The critical temperature for CO_2 is $T_c = 304.2\ \text{K} = 31.0°C = 88°F$. On a hot summer day, such as the one at this writing with $T > 90°F$, CO_2 is a supercritical fluid. During more temperate seasons, the CO_2 in the extinguisher is present as liquid with vapor above it and can slosh around.

2.53 A barometer with a cross-sectional area of $1.00\ \text{cm}^2$ at sea level measures a pressure of 76.0 cm of mercury. The pressure exerted by this column of mercury is equal to the pressure exerted by all the air on $1\ \text{cm}^2$ of Earth's surface. Given that the density of mercury is $13.6\ \text{g cm}^{-3}$ and the average radius of Earth is 6371 km, calculate the total mass of Earth's atmosphere in kilograms. (*Hint*: The surface area of a sphere is $4\pi r^2$, where r is the radius of the sphere.)

The total mass of Earth's atmosphere can be determined from the mass of air on $1\ \text{cm}^2$ of Earth's surface and the surface area of the Earth.

$$\text{Mass of air on 1 cm}^2 \text{ of Earth's surface} = \text{Mass of 76.0 cm mercury/cm}^2$$
$$= (76.0\ \text{cm})\left(13.6\ \text{g cm}^{-3}\right) = 1.034 \times 10^3\ \text{g cm}^{-2}$$

$$\text{Surface area of the Earth} = 4\pi r^2 = 4\pi \left(6371 \times 10^3\ \text{m}\right)^2 \left(\frac{100\ \text{cm}}{1\ \text{m}}\right)^2 = 5.1006 \times 10^{18}\ \text{cm}^2$$

$$\text{Mass of atmosphere} = \left(1.034 \times 10^3\ \text{g cm}^{-2}\right)\left(5.1006 \times 10^{18}\ \text{cm}^2\right)$$
$$= 5.27 \times 10^{21}\ \text{g} = 5.27 \times 10^{18}\ \text{kg}$$

2.54 It has been said that every breath we take, on average, contains molecules once exhaled by Wolfgang Amadeus Mozart (1756 – 1791). The following calculations demonstrate the validity of this statement. **(a)** Calculate the total number of molecules in the atmosphere. (*Hint*: Use the result from Problem 2.53 and $29.0\ \text{g mol}^{-1}$ as the molar mass of air.) **(b)** Assuming the volume of every breath (inhale or exhale) is 500 mL, calculate the number of molecules exhaled in each breath at 37°C, which is the body temperature. **(c)** If Mozart's life span was exactly 35 years, how many molecules did he exhale in that period (given that an average person breathes 12 times per minute)? **(d)** Calculate the fraction of molecules in the atmosphere that were exhaled by Mozart. How many of Mozart's molecules do we inhale with each breath of air? Round your answer to one significant digit. **(e)** List three important assumptions in these calculations.

(a) The number of air molecules in the atmosphere can be calculated from the number of moles of air in the atmosphere.

$$\text{Number of moles of air} = \frac{\text{mass of air}}{\text{molar mass of air}}$$

$$= \frac{5.27 \times 10^{21} \text{ g}}{29.0 \text{ g mol}^{-1}} = 1.817 \times 10^{20} \text{ mol}$$

$$\text{Number of air molecules} = \left(1.817 \times 10^{20} \text{ mol}\right) \left(\frac{6.022 \times 10^{23} \text{ molecules}}{1 \text{ mol}}\right)$$

$$= 1.09 \times 10^{44} \text{ molecules}$$

(b) For each breath,

$$\text{Moles of molecules inhaled or exhaled} = \frac{PV}{RT}$$

$$= \frac{(1.00 \text{ atm}) (500 \text{ mL}) \left(\frac{1 \text{ L}}{1000 \text{ mL}}\right)}{\left(0.08206 \text{ L atm K}^{-1} \text{ mol}^{-1}\right) (273 + 37) \text{ K}}$$

$$= 1.966 \times 10^{-2} \text{ mol}$$

$$\text{Number of molecules inhaled or exhaled} = \left(1.966 \times 10^{-2} \text{ mol}\right) \left(\frac{6.022 \times 10^{23} \text{ molecules}}{1 \text{ mol}}\right)$$

$$= 1.18 \times 10^{22} \text{ molecules}$$

(c) The number of molecules Mozart exhaled in his life span can be determined by his life span and the number of molecules exhaled/breath calculated in part (b).

$$\text{Number of minutes in Mozart's life span} = (35 \text{ yr}) \left(\frac{365 \text{ day}}{1 \text{ yr}}\right) \left(\frac{24 \text{ hr}}{1 \text{ day}}\right) \left(\frac{60 \text{ min}}{1 \text{ hr}}\right)$$

$$= 1.84 \times 10^7 \text{ min}$$

$$\text{Number of breaths in Mozart's life span} = (12 \text{ breaths/min}) \left(1.84 \times 10^7 \text{ min}\right)$$

$$= 2.21 \times 10^8 \text{ breaths}$$

$$\text{Number of molecules exhaled by Mozart}$$
$$= \left(2.21 \times 10^8 \text{ breaths}\right) \left(1.18 \times 10^{22} \text{ molecules/breath}\right)$$
$$= 2.6 \times 10^{30} \text{ molecules}$$

(d) The fraction of molecules in the atmosphere that was exhaled by Mozart can be determined by data in parts (a) and (c):

$$\text{Fraction of molecules in the atmosphere that was exhaled by Mozart}$$

$$= \frac{\text{number of molecules exhaled by Mozart}}{\text{number of air molecules in the atmosphere}} = \frac{2.6 \times 10^{30} \text{ molecules}}{1.09 \times 10^{44} \text{ molecules}} = 2.4 \times 10^{-14}$$

In a single breath, 1.18×10^{22} molecules are inhaled. Therefore,

$$\text{Number of Mozart's molecules we inhale/breath} = \left(1.18 \times 10^{22} \text{ molecules}\right)\left(2.4 \times 10^{-14}\right)$$

$$= 3 \times 10^8 \text{ molecules}$$

(e) Aside from the estimates of typical breathing rates and volumes, which do not count as major assumptions here, there are some serious assumptions that have been made. These include (1) that the molecules in Mozart's breath have been homogeneously distributed throughout the atmosphere, (2) that Mozart did not exhale the same molecules in different breaths, that is, each breath involved a different 1.18×10^{22} molecules, and (3) that all of the exhaled molecules are still in the atmosphere and have not been removed by incorporation into living matter.

2.55   A stockroom supervisor measured the contents of a partially filled 25.0-gallon acetone drum on a day when the temperature was 18.0°C and atmospheric pressure was 750 mmHg, and found that 15.4 gallons of the solvent remained. After tightly sealing the drum, an assistant dropped the drum while carrying it upstairs to the organic laboratory. The drum was dented and its internal volume was decreased to 20.4 gallons. What is the total pressure inside the drum after the accident? The vapor pressure of acetone at 18.0°C is 400 mmHg. (*Hint*: At the time the drum was sealed, the pressure inside the drum, which is equal to the sum of the pressures of air and acetone, was equal to the atmospheric pressure.)

First calculate the pressure of air sealed in the drum before it was dented, $P_{\text{air, i}}$. Then from Boyle's law calculate the pressure of air in the drum after it was dented, $P_{\text{air, f}}$. The partial pressure (or vapor pressure) of acetone remains constant as T is constant.

$$P_{\text{air, i}} + P_{\text{acetone}} = 750 \text{ mmHg}$$
$$P_{\text{air, i}} = 750 \text{ mmHg} - P_{\text{acetone}} = 750 \text{ mmHg} - 400 \text{ mmHg} = 350 \text{ mmHg}$$

The volumes occupied by air before and after the drum was dented are, respectively,

$$V_i = 25.0 \text{ gallons} - 15.4 \text{ gallons} = 9.6 \text{ gallons}$$
$$V_f = 20.4 \text{ gallons} - 15.4 \text{ gallons} = 5.0 \text{ gallons}$$

At constant n_{air} and T,

$$P_{\text{air, i}} V_i = P_{\text{air, f}} V_f$$

$$P_{\text{air, f}} = \frac{P_{\text{air, i}} V_i}{V_f} = \frac{(350 \text{ mmHg}) (9.6 \text{ gallons})}{5.0 \text{ gallons}} = 6.72 \times 10^2 \text{ mmHg}$$

The total pressure inside the drum after the accident is

$$P = P_{\text{air, f}} + P_{\text{acetone}} = 6.7 \times 10^2 \text{ mmHg} + 400 \text{ mmHg} = 1.07 \times 10^3 \text{ mmHg}$$

2.56   A relation known as the barometric formula is useful for estimating the change in atmospheric pressure with altitude. (a) Starting with the knowledge that atmospheric pressure decreases with altitude, we have $dP = -\rho g dh$ where ρ is the density of air, g is the acceleration due to gravity

(9.81 m s^{-2}), and P and h are the pressure and height, respectively. Assuming ideal-gas behavior and constant temperature, show that the pressure P at height h is related to the pressure at sea level P_0 $(h = 0)$ by $P = P_0 e^{-g\mathcal{M}h/RT}$. (*Hint*: For an ideal gas, $\rho = P\mathcal{M}/RT$, where $\mathcal{M}$ is the molar mass.) **(b)** Calculate the atmospheric pressure at a height of 5.0 km, assuming the temperature is constant at 5.0°C, given that the average molar mass of air is 32.0 g mol^{-1}.

(a) To understand the origin of the statement $dP = -\rho g\, dh$, consider the difference in pressure between the two heights h and $h + dh$. This difference is due to the weight (force due to gravity) of the section of air of volume $A\, dh$, where A is the area over which this weight is distributed. The pressure difference will be weight/area. The density, ρ is the connection between volume and mass, and since this section of air has mass $m = \rho A\, dh$, the force is $F = \rho g A\, dh$. Finally, we see that the pressure difference $dP = F/A = \rho g A\, dh/A = \rho g\, dh$.

To continue on with the solution, substitute $\rho = \frac{P\mathcal{M}}{RT}$ into the dP expression:

$$dP = -\frac{P\mathcal{M}}{RT}g\,dh$$

Arrange variables containing P on the left hand side of the equation, and constants and the differential dh on the right hand side of the equation:

$$\frac{dP}{P} = -\frac{\mathcal{M}}{RT}g\,dh$$

This equation is then integrated. The limits for h are 0 and h while the corresponding limits for P are P_0 and P.

$$\int_{P_0}^{P} \frac{dP}{P} = -\int_{0}^{h} \frac{\mathcal{M}}{RT}g\,dh$$

$$\ln\frac{P}{P_0} = -\frac{\mathcal{M}}{RT}g\,h$$

$$\frac{P}{P_0} = e^{-g\mathcal{M}h/RT}$$

$$P = P_0 e^{-g\mathcal{M}h/RT}$$

(b) First calculate $g\mathcal{M}h/RT$ while taking care of the units:

$$\frac{g\mathcal{M}h}{RT} = \frac{(9.81 \text{ m s}^{-2})\,(32.00 \text{ g mol}^{-1})\,(5.0 \times 10^3 \text{ m})}{(0.08206 \text{ L atm K}^{-1} \text{ mol}^{-1})\,(273.2 + 5.0)\,\text{K}}$$

$$= \left(\frac{6.9 \times 10^4 \text{ m}^2 \text{ g s}^{-2}}{\text{L atm}}\right)\left(\frac{1 \text{ kg}}{1000 \text{ g}}\right)\left(\frac{1 \text{ N}}{1 \text{ kg m s}^{-2}}\right)\left(\frac{1 \text{ Pa}}{1 \text{ N m}^{-2}}\right)$$

$$\times \left(\frac{1000 \text{ L}}{1 \text{ m}^3}\right)\left(\frac{1 \text{ atm}}{1.01325 \times 10^5 \text{ Pa}}\right)$$

$$= 0.679$$

$$P = P_0 e^{-g\mathcal{M}h/RT} = (1.0 \text{ atm})\,e^{-0.679} = 0.51 \text{ atm}$$

2.57   In terms of the hard-sphere gas model, molecules are assumed to possess finite volume, but there is no interaction among the molecules. **(a)** Compare the $P - V$ isotherm for an ideal gas and that for a hard-sphere gas. **(b)** Let b be the effective volume of the gas. Write an equation of state for this gas. **(c)** From this equation, derive an expression for $Z = P\overline{V}/RT$ for the hard-sphere gas and make a plot of Z versus P for two values of T (T_1 and T_2, $T_2 > T_1$). Be sure to indicate the value of the intercepts on the Z axis. **(d)** Plot Z versus T for fixed P for an ideal gas and for the hard-sphere gas.

(a) For any given pressure the volume of a hard sphere gas will be greater than that of the ideal gas for which there is no excluded volume. This difference will be greatest at high pressure and become smaller, approaching zero as the pressure approaches zero. Furthermore, the molar volume can never get smaller than b, where b is the effective volume of a mole of the gas. See graph below.

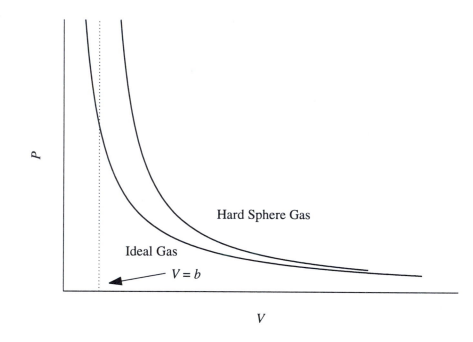

(b) The equation of state would be similar to that of the van der Waals gas, but with a value of zero for a, since there are no intermolecular attractions. Thus,

$$P\left(\overline{V} - b\right) = RT$$

(c) By definition, $Z = \frac{P\overline{V}}{RT}$, and from part **(b)**, $P\left(\overline{V} - b\right) = P\overline{V} - Pb = RT$. Thus,

$$\frac{P\overline{V}}{RT} - \frac{Pb}{RT} = 1$$

or,

$$Z = 1 + \frac{b}{RT}P$$

The graph of Z versus P is seen to be a straight line with a slope of $\frac{b}{RT}$ and as $P \to 0$, $Z \to 1$. This results in the plot below for $T_2 > T_1$.

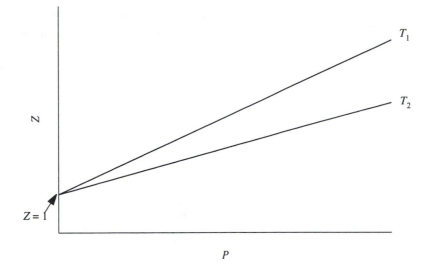

(d) For the ideal gas, $Z = 1$ for all combinations of pressure and temperature, but for the hard sphere gas, part **(c)** shows

$$Z = 1 + \frac{Pb}{R}\frac{1}{T}$$

This is illustrated for fixed P in the plot below.

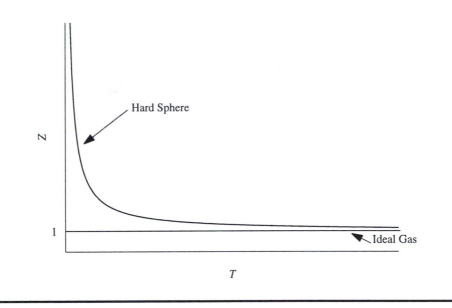

2.58 One way to gain a physical understanding of b in the van der Waals equation is to calculate the "excluded volume." Assume that the distance of closest approach between two similar spherical molecules is the sum of their radii ($2r$). **(a)** Calculate the volume around each molecule into which the center of another molecule cannot penetrate. **(b)** From your result in **(a)**, calculate the excluded volume for one mole of molecules, which is the constant b. How does this compare with the sum of the volumes of 1 mole of the same molecules?

(a) We see from the figure that two hard spheres of radius r cannot approach each other more closely than $2r$ (measured from the centers). Thus there is a sphere of radius $2r$ surrounding each hard sphere from which other hard spheres are excluded. The excluded volume/pair of molecules $= \frac{4}{3}\pi (2r)^3 = \frac{32}{3}\pi r^3 = 8\left(\frac{4}{3}\pi r^3\right)$, or eight times the volume of an individual molecule.

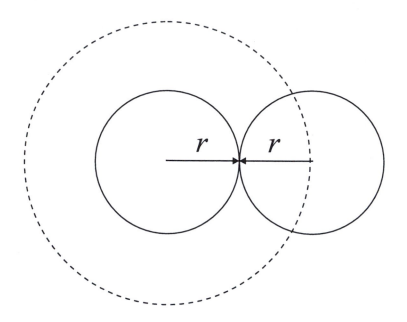

(b) The result in part **(a)** is for a pair of molecules, so the excluded volume/molecule $= \frac{1}{2}\left(\frac{32}{3}\pi r^3\right) = \frac{16}{3}\pi r^3$; Therefore, the excluded volume/mol $= \frac{16}{3}N_A\pi r^3$ where N_A is Avogadro's number. The sum of the volumes of a mole of molecules (treated as hard spheres of radius r) is $\frac{4}{3}N_A\pi r^3$. The excluded volume is four times the volume of the spheres themselves.

2.59   You may have witnessed a demonstration in which a burning candle standing in water is covered by an upturned glass. The candle goes out and the water rises in the glass. The explanation usually given for this phenomenon is that the oxygen in the glass is consumed by combustion, leading to a decrease in volume and hence the rise in the water level. However, the loss of oxygen is only a minor consideration. **(a)** Using $C_{12}H_{26}$ as the formula for paraffin wax, write a balanced equation for the combustion. Based on the nature of the products, show that the predicted rise in water level due to the removal of oxygen is far less than the observed change. **(b)** Devise a chemical process that would allow you to measure the volume of oxygen in the trapped air. (*Hint*: Use steel wool.) **(c)** What is the main reason for the water rising in the glass after the flame is extinguished?

(a) The combustion reaction is

$$C_{12}H_{26}(s) + \frac{37}{2}O_2(g) \longrightarrow 12CO_2(g) + 13H_2O(l)$$

Thus even though O_2 is being used up, some CO_2 gas is being formed to take its place. The mole ratio is $\frac{12}{18.5} \approx \frac{2}{3}$, which is proportional to the volumes involved. Thus, $\frac{2}{3}$ of the volume of O_2 used is replaced by CO_2. Some CO_2 does dissolve in the water, but this effect is small.

(b) A reaction that does not result in a gaseous product must be used. A suitable one is the oxidation of iron to iron oxide, or the process of rusting. If a piece of steel wool is placed in the glass, it will rust over time, using most of the O_2 in the glass. The water level will be observed to rise by about 20% of the air space originally present. (Diffusion of air into the glass is not important.)

(c) While the candle is burning the air trapped inside the glass is warmed by the flame and expands. After the flame goes out the air cools, and its pressure drops. The water level inside the glass rises due to the atmospheric pressure outside.

2.60 Express the van der Waals equation in the form of Equation 2.12. Derive relationships between the van der Waals constants (a and b) and the virial coefficients(B, C, and D), given that

$$\frac{1}{1-x} = 1 + x + x^2 + x^3 + \cdots \quad |x| < 1$$

The van der Waals equation can be rearranged to give (see Problem 2.50)

$$P = \frac{RT}{\overline{V} - b} - \frac{a}{\overline{V}^2}$$

Substitute this expression into $Z = \frac{P\overline{V}}{RT}$:

$$Z = \left(\frac{RT}{\overline{V} - b} - \frac{a}{\overline{V}^2} \right) \frac{\overline{V}}{RT}$$

$$= \frac{\overline{V}}{\overline{V} - b} - \frac{a}{\overline{V} RT}$$

$$= \frac{1}{1 - \frac{b}{\overline{V}}} - \frac{a}{\overline{V} RT}$$

Because $b < \overline{V}$, $\frac{b}{\overline{V}} < 1$, the following expression applies:

$$\frac{1}{1-x} = 1 + x + x^2 + x^3 + \cdots$$

to approximate $\frac{1}{1-\frac{b}{\overline{V}}}$ in the expression for Z:

$$Z = 1 + \frac{b}{\overline{V}} + \frac{b^2}{\overline{V}^2} + \frac{b^3}{\overline{V}^3} + \cdots - \frac{a}{\overline{V} RT}$$

$$= 1 + \left(b - \frac{a}{RT} \right) \frac{1}{\overline{V}} + \frac{b^2}{\overline{V}^2} + \frac{b^3}{\overline{V}^3} + \cdots$$

In terms of the virial coefficients, Z can be written as

$$Z = 1 + \frac{B}{\overline{V}} + \frac{C}{\overline{V}^2} + \frac{D}{\overline{V}^3} + \cdots$$

The following expressions are obtained when the coefficients for $\frac{1}{\overline{V}}$, $\frac{1}{\overline{V}^2}$, and $\frac{1}{\overline{V}^3}$ are compared in these expressions for Z.

$$B = b - \frac{a}{RT}$$
$$C = b^2$$
$$D = b^3$$

2.61   The Boyle temperature is the temperature at which the coefficient B is zero. Therefore, a real gas behaves like an ideal gas at this temperature. **(a)** Give a physical interpretation of this behavior. **(b)** Using your result for B for the van der Waals equation in Problem 2.60, calculate the Boyle temperature for argon, given that $a = 1.345 \text{ atm L}^2 \text{ mol}^{-2}$ and $b = 3.22 \times 10^{-2} \text{ L mol}^{-1}$.

(a) At the Boyle temperature, the attractive forces are equal to the repulsive forces. Since the molecules do not exert any net forces on each other, the gas behaves as if it were an ideal gas and Boyle's law holds.

(b) At the Boyle temperature T_b,

$$B = b - \frac{a}{RT_b} = 0$$

$$T_b = \frac{a}{Rb} = \frac{1.345 \text{ atm L}^2 \text{ mol}^{-2}}{(0.08206 \text{ L atm K}^{-1} \text{ mol}^{-1})(3.22 \times 10^{-2} \text{ L mol}^{-1})} = 509 \text{ K}$$

The experimental Boyle temperature for Ar is 412 K.

2.62   From Equations 2.12 and 2.13 show that $B' = B/RT$ and $C' = (C - B^2)/(RT)^2$. (*Hint:* From Equation 2.12 first obtain expressions for P and P^2. Next, substitute these expressions into Equation 2.13.)

Substitute $Z = \frac{P\overline{V}}{RT}$ into Eq. 2.12:

$$\frac{P\overline{V}}{RT} = 1 + \frac{B}{\overline{V}} + \frac{C}{\overline{V}^2} + \frac{D}{\overline{V}^3} + \cdots$$

Rearranging this equation to yield P and keeping terms up to $\frac{1}{\overline{V}^3}$, the following expression is obtained:

$$P = \left(\frac{RT}{\overline{V}}\right)\left(1 + \frac{B}{\overline{V}} + \frac{C}{\overline{V}^2} + \frac{D}{\overline{V}^3} + \cdots\right) = \frac{RT}{\overline{V}} + \frac{BRT}{\overline{V}^2} + \frac{CRT}{\overline{V}^3} + \cdots$$

Square the expression for P and keep terms up to $\frac{1}{\overline{V}^3}$:

$$P^2 = \frac{R^2T^2}{\overline{V}^2} + \frac{2BR^2T^2}{\overline{V}^3} + \cdots$$

Substitute the expressions for P and P^2 into Eq. 2.13:

$$Z = 1 + B'P + C'P^2$$

$$= 1 + B'\left(\frac{RT}{\overline{V}} + \frac{BRT}{\overline{V}^2} + \frac{CRT}{\overline{V}^3} + \cdots\right) + C'\left(\frac{R^2T^2}{\overline{V}^2} + \frac{2BR^2T^2}{\overline{V}^3} + \cdots\right)$$

$$= 1 + \frac{B'RT}{\overline{V}} + \frac{B'BRT + C'R^2T^2}{\overline{V}^2} + \frac{B'CRT + 2BC'R^2T^2}{\overline{V}^3}$$

The following expression is obtained when the coefficients for $\frac{1}{V}$ in the above expression for Z and Eq. 2.12 are compared:

$$B = B'RT$$

Thus,

$$B' = \frac{B}{RT}$$

The following expression is obtained when the coefficients for $\frac{1}{V^2}$ in the above expression for Z and Eq. 2.12 are compared:

$$C = B'BRT + C'R^2T^2$$

Thus,

$$C' = \frac{C - B'BRT}{R^2T^2}$$

Substitute $B' = \frac{B}{RT}$ into the C' expression:

$$C' = \frac{C - \frac{B}{RT}BRT}{R^2T^2} = \frac{C - B^2}{(RT)^2}$$

2.63 Estimate the distance (in Å) between molecules of water vapor at 100°C and 1.0 atm. Assume ideal-gas behavior. Repeat the calculation for liquid water at 100°C, given that the density of water at 100°C is 0.96 g cm^{-3}. Comment on your results. (The diameter of a H_2O molecule is approximately 3 Å. 1 Å=10^{-8} cm.)

Use the ideal gas law to calculate the number density of molecules for the vapor.

$$\frac{n}{V} = \frac{P}{RT} = \frac{1.0\text{ atm}}{(0.08206\text{ L atm K}^{-1}\text{ mol}^{-1})(100+273)\text{K}} = 0.03267\text{ mol L}^{-1}$$

Now convert moles to molecules and liters to cubic meters

$$(0.03267\text{ mol L}^{-1})(6.022\times10^{23}\text{ mol}^{-1})\left(\frac{1000\text{ L}}{1\text{ m}^3}\right) = 1.967\times10^{25}\text{ m}^{-3}$$

This is the number of ideal gas molecules in a cube 1 meter on each side. Assuming an equal distribution of molecules along the three mutually perpendicular directions defined by the cube, a linear density in one direction may be found

$$(1.967\times10^{25}\text{ m}^{-3})^{1/3} = 2.70\times10^8\text{ m}^{-1}$$

which is the number of molecules on a line 1 meter in length. The distance between each molecule is given by

$$\frac{1}{2.70\times10^8\text{ m}^{-1}} = 3.7\times10^{-9}\text{ m} = 37\text{ Å}$$

The water molecules are separated by over 12 times their diameter.

A similar calculation is done for liquid water. Here the density, or mass per unit volume, is connected to molecules per unit volume by the molar mass as shown below, using N for the number of water molecules and m for their mass.

$$\frac{N}{V} = \frac{N_A n}{V} = N_A \frac{m}{V\mathcal{M}} = N_A \frac{\rho}{\mathcal{M}}$$

For liquid water,

$$N_A \frac{\rho}{\mathcal{M}} = (6.022 \times 10^{23}\ \mathrm{mol}^{-1}) \left(\frac{0.96\ \mathrm{g\,cm}^{-3}}{18.02\ \mathrm{g\,mol}^{-1}} \right) \left(\frac{100\ \mathrm{cm}}{1\ \mathrm{m}} \right)^3 = 3.21 \times 10^{28}\ \mathrm{m}^{-3}$$

This is the number of liquid water molecules in 1 cubic meter. From this point the calculation is just like the previous one, and the space between molecules is found using the same assumptions

$$\frac{1}{(3.21 \times 10^{28}\ \mathrm{m}^{-3})^{1/3}} = 3.1 \times 10^{-10}\ \mathrm{m} = 3.1\ \text{Å}$$

To one significant figure, the molecules are touching each other in the liquid phase.

CHAPTER 3

Kinetic Theory of Gases

PROBLEMS AND SOLUTIONS

3.1 Apply the kinetic theory of gases to explain Boyle's law, Charles' law, and Dalton's law.

Boyle's Law: In the kinetic theory of gases, pressure is proportional to the number of molecular collisions with the container walls per unit time. As the volume is deceased, the density of the gas increases, and the rate with which the molecules collide with the wall likewise increases. One way to think about this is that in a smaller container, each molecule has a shorter distance to travel before running into a wall, and thus collisions with the wall happen at a greater rate.

Charles' Law: In the kinetic theory of gases, pressure is proportional to the momentum transfer taking place at the container walls averaged over all molecule-wall collisions. This, in turn, is proportional to the average kinetic energy of the molecules given by $\frac{3}{2}k_BT$. Thus, pressure is proportional to temperature, which is the alternative statement of Charles' Law.

Dalton's Law: For ideal gases there are no intermolecular (or interatomic) interactions. The molecules of each gas in a mixture will therefore strike the walls of the container independently, and the total pressure will simply be the sum of pressure due to each component (type of molecule) in the mixture.

3.2 Is temperature a microscopic or macroscopic concept? Explain.

Temperature is a macroscopic concept, since one of the postulates of the kinetic theory of gases is that it deals with a very large number of molecules. Temperature is proportional to the average kinetic energy of the molecules in the system, and for this average to be meaningful, it must be taken over a large number of molecules.

3.3 In applying the kinetic theory to gases, we have assumed that the walls of the container are elastic for molecular collisions. Actually, whether these collisions are elastic or inelastic makes no difference as long as the walls are at the same temperature as the gas. Explain.

With the walls at the same temperature as the gas, they each have the same average (kinetic) energy. Thus the net transfer of energy from one to the other is zero.

3.4 If 2.0×10^{23} argon (Ar) atoms strike 4.0 cm^2 of wall per second at a $90°$ angle to the wall when moving with a speed of $45,000 \text{ cm s}^{-1}$, what pressure (in atm) do they exert on the wall?

$F =$ Force exerted by Ar atoms

$= $ (Force exerted by 1 Ar atom) (Number of Ar atoms)

$= $ (Change in momentum for 1 Ar atom/time) (Number of Ar atoms)

$= \left(\dfrac{2mv}{1 \text{ s}}\right)\left(2.0 \times 10^{23}\right)$

$= \dfrac{2\left[(39.95 \text{ amu})\left(1.661 \times 10^{-27} \text{ kg amu}^{-1}\right)\right]\left[(45,000 \text{ cm s}^{-1})\left(\frac{1 \text{ m}}{100 \text{ cm}}\right)\right]}{1 \text{ s}}\left(2.0 \times 10^{23}\right)$

$= 11.9 \text{ N}$

$\dfrac{F}{A} =$ Pressure exerted on the wall by Ar atoms

$= \dfrac{11.9 \text{ N}}{(4.0 \text{ cm}^2)\left(\frac{1 \text{ m}}{100 \text{ cm}}\right)^2}$

$= \left(2.98 \times 10^4 \text{ Pa}\right)\left(\dfrac{1 \text{ atm}}{1.01325 \times 10^5 \text{ Pa}}\right) = 0.29 \text{ atm}$

3.5 A square box contains He at $25°C$. If the atoms are colliding with the walls perpendicularly (at $90°$) at the rate of 4.0×10^{22} times per second, calculate the force and the pressure exerted on the wall given that the area of the wall is 100 cm^2 and the speed of the atoms is 600 m s^{-1}.

$F =$ Force exerted by He atoms

$= $ (Change in momentum for 1 He atom/time) (Number of He atoms)

$= \left(\dfrac{2mv}{1 \text{ s}}\right)\left(4.0 \times 10^{22}\right)$

$= \dfrac{2\left[(4.003 \text{ amu})\left(1.661 \times 10^{-27} \text{ kg amu}^{-1}\right)\right]\left(600 \text{ m s}^{-1}\right)}{1 \text{ s}}\left(4.0 \times 10^{22}\right)$

$= 0.319 \text{ N}$

$\dfrac{F}{A} =$ Pressure exerted on the wall by He atoms

$= \dfrac{0.319 \text{ N}}{(100 \text{ cm}^2)\left(\frac{1 \text{ m}}{100 \text{ cm}}\right)^2}$

$= (31.9 \text{ Pa})\left(\dfrac{1 \text{ atm}}{1.01325 \times 10^5 \text{ Pa}}\right) = 3.1 \times 10^{-4} \text{ atm}$

3.6 Calculate the average translational kinetic energy for a N_2 molecule and for 1 mole of N_2 at 20°C.

For one molecule, $\overline{E}_{trans} = \frac{3}{2}k_B T$.

For one mole of molecules, $\overline{E}_{trans} = \frac{3}{2}RT$.

Therefore, for a N_2 molecule,

$$\overline{E}_{trans} = \frac{3}{2}\left(1.381 \times 10^{-23}\ \text{J K}^{-1}\right)(273 + 20)\ \text{K} = 6.07 \times 10^{-21}\ \text{J}$$

whereas for a mole of N_2 molecules,

$$\overline{E}_{trans} = \frac{3}{2}\left(8.314\ \text{J K}^{-1}\ \text{mol}^{-1}\right)(273 + 20)\ \text{K} = 3.65 \times 10^{3}\ \text{J mol}^{-1}$$

3.7 To what temperature must He atoms be cooled so that they have the same v_{rms} as O_2 at 25°C?

The desired relationship is $v_{rms,\,He} = v_{rms,\,O_2}$, or $\frac{v_{rms,\,He}}{v_{rms,\,O_2}} = 1$. Write the v_{rms}'s in this relationship in terms of $\mathcal{M}$ and T:

$$\frac{v_{rms,\,He}}{v_{rms,\,O_2}} = \frac{\sqrt{\frac{3RT_{He}}{\mathcal{M}_{He}}}}{\sqrt{\frac{3RT_{O_2}}{\mathcal{M}_{O_2}}}} = \sqrt{\frac{\mathcal{M}_{O_2}}{\mathcal{M}_{He}}\frac{T_{He}}{T_{O_2}}} = 1$$

$$\frac{\mathcal{M}_{O_2}}{\mathcal{M}_{He}}\frac{T_{He}}{T_{O_2}} = 1$$

$$T_{He} = \frac{\mathcal{M}_{He}}{\mathcal{M}_{O_2}}T_{O_2} = \left(\frac{4.003\ \text{g mol}^{-1}}{32.00\ \text{g mol}^{-1}}\right)(273 + 25)\ \text{K} = 37.3\ \text{K}$$

3.8 The c_{rms} of CH_4 is 846 m s^{-1}. What is the temperature of the gas?

The temperature of the gas can be calculated by rearranging $c_{rms} = \sqrt{\frac{3RT}{\mathcal{M}}}$

$$\frac{3RT}{\mathcal{M}} = c_{rms}^2$$

$$T = \frac{c_{rms}^2 \mathcal{M}}{3R} = \frac{\left(846\ \text{m s}^{-1}\right)^2 \left(16.04 \times 10^{-3}\ \text{kg mol}^{-1}\right)}{3\left(8.314\ \text{J K}^{-1}\ \text{mol}^{-1}\right)} = 460\ \text{K}$$

3.9 Calculate the value of the c_{rms} of ozone molecules in the stratosphere, where the temperature is 250 K.

$$c_{rms} = \sqrt{\frac{3RT}{\mathcal{M}}} = \sqrt{\frac{3\left(8.314\ \text{J K}^{-1}\ \text{mol}^{-1}\right)(250\ \text{K})}{48.00 \times 10^{-3}\ \text{kg mol}^{-1}}} = 360\ \text{m s}^{-1}$$

3.10 At what temperature will He atoms have the same c_{rms} value as N_2 molecules at 25°C? Solve this problem without calculating the value of c_{rms} for N_2.

This problem is similar to Problem 3.7.

$$T_{He} = \frac{\mathcal{M}_{He}}{\mathcal{M}_{N_2}} T_{N_2} = \left(\frac{4.003\ \text{g mol}^{-1}}{28.02\ \text{g mol}^{-1}}\right)(273 + 25)\ \text{K} = 42.6\ \text{K}$$

3.11 List the conditions used for deriving the Maxwell speed distribution.

The Maxwell speed distribution relies on just three conditions, although each is very important. These are, (1) the gas is an ideal gas (no intermolecular forces and molecules with zero volume), (2) the system is at thermal equilibrium, and (3) there is a very large number of molecules.

3.12 Plot the speed distribution function for **(a)** He, O_2, and UF_6 at the same temperature and **(b)** CO_2 at 300 K and 1000 K.

(a) In the plot, note that the heavier the molecules, the narrower the speed distribution and the smaller the most probable speed; whereas the lighter the molecules, the wider the speed distribution and the greater the most probable speed.

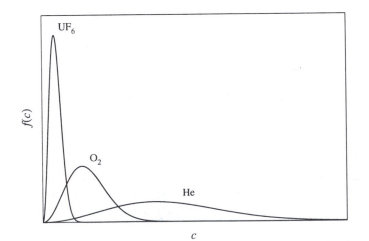

(b) In the plot, note that the lower the temperature, the narrower the speed distribution and the smaller the most probable speed; whereas the higher the temperature, the wider the speed distribution and the greater the most probable speed.

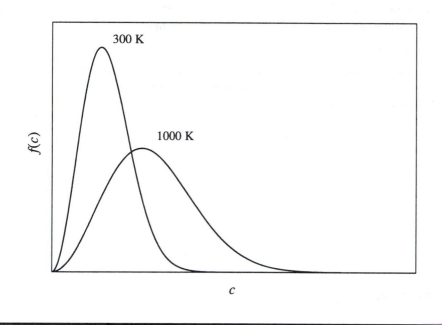

3.13 Account for the maximum in the Maxwell speed distribution curve (Figure 3.4) by plotting the following two curves on the same graph: (1) c^2 versus c and (2) $e^{-mc^2/2k_BT}$ versus c. Use neon (Ne) at 300 K for the plot in (2).

The maximum in the Maxwell speed distribution arises because c^2 increases with c whereas e^{-mc^2/k_BT} decreases with c. At small values of c, the term c^2 dominates, thus the speed distribution curve increases with c. However, at large values of c, the term e^{-mc^2/k_BT} dominates and the distribution curve decreases with increasing value of c.

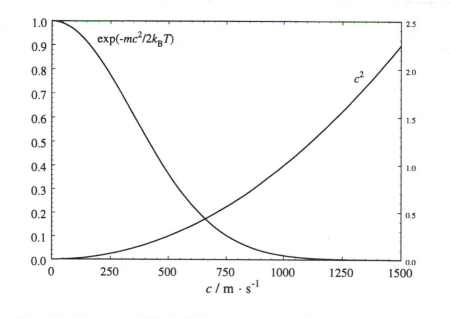

3.14 A N_2 molecule at 20°C is released at sea level to travel upwards. Assuming that the temperature is constant and that the molecule does not collide with other molecules, how far would it travel (in meters) before coming to rest? Do the same calculation for a He atom. [*Hint*: To calculate the altitude, h, the molecule will travel, equate its kinetic energy with the potential energy, mgh, where m is the mass and g the acceleration due to gravity (9.81 m s^{-2}).]

This problem is an application of the law of conservation of energy. In this case, the total energy of the molecule remains the same, but it is changed from one form (kinetic energy) to another (potential energy). Therefore,

$$\text{kinetic energy} = \text{potential energy}$$

$$\frac{3}{2}k_B T = mgh$$

$$h = \frac{3k_B T}{2mg}$$

For a N_2 molecule,

$$h = \frac{3\left(1.381 \times 10^{-23}\ \text{J K}^{-1}\right)(273 + 20)\ \text{K}}{2\,(28.02\ \text{amu})\left(1.661 \times 10^{-27}\ \text{kg amu}^{-1}\right)\left(9.81\ \text{m s}^{-2}\right)} = 1.33 \times 10^4\ \text{m}$$

For a He atom,

$$h = \frac{3\left(1.381 \times 10^{-23}\ \text{J K}^{-1}\right)(273 + 20)\ \text{K}}{2\,(4.003\ \text{amu})\left(1.661 \times 10^{-27}\ \text{kg amu}^{-1}\right)\left(9.81\ \text{m s}^{-2}\right)} = 9.31 \times 10^4\ \text{m}$$

A He atom, being 7 times lighter than a N_2 molecule, would travel 7 times higher than a N_2 molecule.

3.15 The speeds of 12 particles (in cm s^{-1}) are 0.5, 1.5, 1.8, 1.8, 1.8, 1.8, 2.0, 2.5, 2.5, 3.0, 3.5, and 4.0. Find **(a)** the average speed, **(b)** the root-mean-square speed, and **(c)** the most probable speed of these particles. Explain your results.

(a) The average speed for the particles is

$$\bar{c} = \frac{\sum\limits_{i=1}^{12} c_i}{N}$$

$$= \frac{(0.5 + 1.5 + 1.8 + 1.8 + 1.8 + 1.8 + 2.0 + 2.5 + 2.5 + 3.0 + 3.5 + 4.0)\ \text{cm s}^{-1}}{12}$$

$$= 2.2\ \text{cm s}^{-1}$$

(b) The mean-square speed for the particles is

$$\overline{c^2} = \frac{\sum\limits_{i=1}^{12} c_i^2}{N}$$

$$= \frac{\left(0.5^2 + 1.5^2 + 1.8^2 + 1.8^2 + 1.8^2 + 1.8^2 + 2.0^2 + 2.5^2 + 2.5^2 + 3.0^2 + 3.5^2 + 4.0^2\right) \text{cm}^2 \text{ s}^{-2}}{12}$$

$$= 5.77 \text{ cm}^2 \text{ s}^{-2}$$

The root-mean-square speed for the particles is

$$c_{\text{rms}} = \sqrt{\overline{c^2}} = 2.4 \text{ cm s}^{-1}$$

(c) $c_{\text{mp}} = 1.8 \text{ cm s}^{-1}$, as this is the speed that appears most frequently.

As expected, $c_{\text{rms}} > \overline{c}$. However, because 12 particles do not constitute a macroscopic system, c_{mp} can be greater or smaller than c_{rms} or $\overline{c}$.

3.16 At a certain temperature, the speeds of six gaseous molecules in a container are 2.0 m s^{-1}, 2.2 m s^{-1}, 2.6 m s^{-1}, 2.7 m s^{-1}, 3.3 m s^{-1}, and 3.5 m s^{-1}. Calculate the root-mean-square speed and the average speed of the molecules. These two average values are close to each other, but the root-mean-square value is always the larger of the two. Why?

The average speed of the molecules is

$$\overline{c} = \frac{\sum\limits_{i=1}^{6} c_i}{N}$$

$$= \frac{(2.0 + 2.2 + 2.6 + 2.7 + 3.3 + 3.5) \text{ m s}^{-1}}{6}$$

$$= 2.7 \text{ m s}^{-1}$$

The mean-square speed for the molecules is

$$\overline{c^2} = \frac{\sum\limits_{i=1}^{6} c_i^2}{N}$$

$$= \frac{\left(2.0^2 + 2.2^2 + 2.6^2 + 2.7^2 + 3.3^2 + 3.5^2\right) \text{ m}^2 \text{ s}^{-2}}{6}$$

$$= 7.67 \text{ m}^2 \text{ s}^{-2}$$

The root-mean-square speed for the molecules is

$$c_{\text{rms}} = \sqrt{\overline{c^2}} = 2.8 \text{ m s}^{-1}$$

The rms average is always larger than the straight average because squaring favors (more heavily weights) the larger values and biases the result towards the larger values.

3.17   The following diagram shows the Maxwell speed distribution curves for a certain gas at two different temperatures (T_1 and T_2). Calculate the value of T_2.

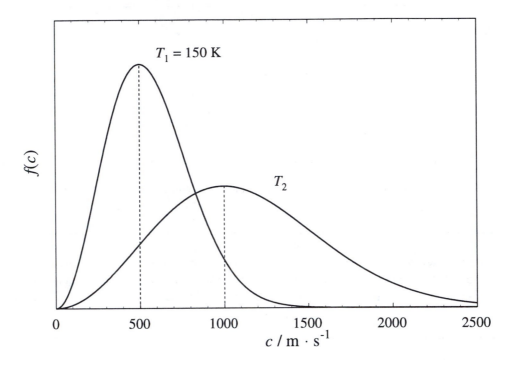

At T_1, the most probable speed, $c_{mp1} = 500 \text{ m s}^{-1}$.

At T_2, the most probable speed, $c_{mp2} = 1000 \text{ m s}^{-1}$.

$$\frac{c_{mp2}}{c_{mp1}} = \frac{\sqrt{\frac{2RT_2}{M}}}{\sqrt{\frac{2RT_1}{M}}} = \sqrt{\frac{T_2}{T_1}}$$

$$\frac{T_2}{T_1} = \left(\frac{c_{mp2}}{c_{mp1}}\right)^2$$

$$T_2 = \left(\frac{c_{mp2}}{c_{mp1}}\right)^2 T_1 = \left(\frac{1000 \text{ m s}^{-1}}{500 \text{ m s}^{-1}}\right)^2 (150 \text{ K}) = 600 \text{ K}$$

3.18   Following the procedure in the chapter to find the value of $\bar{c}$, derive an expression for c_{rms}. (*Hint*: You need to consult the *Handbook of Chemistry and Physics* to evaluate definite integrals.)

By definition,

$$\overline{c^2} = \int_0^\infty c^2 f(c)dc$$

$$= \int_0^\infty c^2 \left[4\pi c^2 \left(\frac{m}{2\pi k_B T} \right)^{\frac{3}{2}} e^{-\frac{mc^2}{2k_B T}} \right] dc$$

$$= 4\pi \left(\frac{m}{2\pi k_B T} \right)^{\frac{3}{2}} \int_0^\infty c^4 e^{-\frac{mc^2}{2k_B T}} dc$$

From a table of integrals,

$$\int_0^\infty x^{2n} e^{-ax^2} dx = \frac{(1)\,(3)\,(5)\cdots(2n-1)}{2^{n+1} a^n} \sqrt{\frac{\pi}{a}}$$

This is applied to the $\overline{c^2}$ integral by identifying

$$x = c$$

$$n = 2$$

$$a = \frac{m}{2k_B T}$$

and resulting in,

$$\overline{c^2} = 4\pi \left(\frac{m}{2\pi k_B T} \right)^{\frac{3}{2}} \frac{(1)\,(3)}{8 \left(\frac{m}{2k_B T} \right)^2} \sqrt{\frac{\pi}{\frac{m}{2k_B T}}}$$

$$= 4\pi \left(\frac{m}{2\pi k_B T} \right)^{\frac{3}{2}} \frac{3\sqrt{\pi}}{8} \left(\frac{2k_B T}{m} \right)^{\frac{5}{2}}$$

$$= \frac{3k_B T}{m} \left(\frac{N_A}{N_A} \right)$$

$$= \frac{3RT}{M}$$

Consequently,

$$c_{\text{rms}} = \sqrt{\overline{c^2}} = \sqrt{\frac{3RT}{M}}$$

3.19 Derive an expression for c_{mp}, following the procedure described in the chapter.

The most probable speed can be calculated by setting $\frac{df(c)}{dc} = 0$ and then solving for c_{mp}.

$$f(c) = 4\pi c^2 \left(\frac{m}{2\pi k_B T} \right)^{\frac{3}{2}} e^{-\frac{mc^2}{2k_B T}}$$

$$= 4\pi \left(\frac{m}{2\pi k_B T} \right)^{\frac{3}{2}} c^2 e^{-\frac{mc^2}{2k_B T}}$$

In differentiating $f(c)$ with respect to c, care must be taken to apply the chain rule to the two terms c^2 and $e^{-\frac{mc^2}{2k_BT}}$.

$$\frac{df(c)}{dc} = 4\pi \left(\frac{m}{2\pi k_B T}\right)^{\frac{3}{2}} \left[2ce^{-\frac{mc^2}{2k_B T}} + c^2 e^{-\frac{mc^2}{2k_B T}} \left(-\frac{2mc}{2k_B T}\right)\right]$$

$$= 4\pi \left(\frac{m}{2\pi k_B T}\right)^{\frac{3}{2}} ce^{-\frac{mc^2}{2k_B T}} \left(2 - \frac{mc^2}{k_B T}\right)$$

When $\frac{df(c)}{dc} = 0$, $c = c_{mp}$. The above equation becomes

$$4\pi \left(\frac{m}{2\pi k_B T}\right)^{\frac{3}{2}} c_{mp} e^{-\frac{mc_{mp}^2}{2k_B T}} \left(2 - \frac{mc_{mp}^2}{k_B T}\right) = 0$$

$$2 - \frac{mc_{mp}^2}{k_B T} = 0$$

Solving for c_{mp}^2,

$$c_{mp}^2 = \frac{2k_B T}{m}\left(\frac{N_A}{N_A}\right)$$

$$= \frac{2RT}{M}$$

or,

$$c_{mp} = \sqrt{\frac{2RT}{M}}$$

3.20   Calculate the values of c_{rms}, c_{mp}, and $\bar{c}$ for argon at 298 K.

$$c_{rms} = \sqrt{\frac{3RT}{M}} = \sqrt{\frac{3\left(8.314\ \text{J K}^{-1}\ \text{mol}^{-1}\right)(298\ \text{K})}{39.95 \times 10^{-3}\ \text{kg mol}^{-1}}} = 431\ \text{m s}^{-1}$$

$$c_{mp} = \sqrt{\frac{2RT}{M}} = \sqrt{\frac{2\left(8.314\ \text{J K}^{-1}\ \text{mol}^{-1}\right)(298\ \text{K})}{39.95 \times 10^{-3}\ \text{kg mol}^{-1}}} = 352\ \text{m s}^{-1}$$

$$\bar{c} = \sqrt{\frac{8RT}{\pi M}} = \sqrt{\frac{8\left(8.314\ \text{J K}^{-1}\ \text{mol}^{-1}\right)(298\ \text{K})}{\pi\left(39.95 \times 10^{-3}\ \text{kg mol}^{-1}\right)}} = 397\ \text{m s}^{-1}$$

3.21   Calculate the value of c_{mp} for C_2H_6 at 25°C. What is the ratio of the number of molecules with a speed of 989 m s^{-1} to the number of molecules with this value of c_{mp}?

$$c_{mp} = \sqrt{\frac{2RT}{\mathcal{M}}} = \sqrt{\frac{2\left(8.314 \text{ J K}^{-1}\text{ mol}^{-1}\right)(273 + 25)\text{ K}}{30.07 \times 10^{-3}\text{ kg mol}^{-1}}} = 406 \text{ m s}^{-1}$$

The number of molecules with speeds between c and $c + dc$ is given by equation 3.10,

$$\frac{dN}{N} = 4\pi c^2 \left(\frac{m}{2\pi k_B T}\right)^{\frac{3}{2}} e^{-\frac{mc^2}{2k_B T}}\, dc$$

Thus, to find the ratio of the number of molecules with speeds between c_2 and $c_2 + dc$, dN_2, to the number with speeds between c_1 and $c_1 + dc$, dN_1, (which is what is meant by the ratio of the number with speed c_2 to that with speed c_1), form the ratio $\frac{dN_2/N}{dN_1/N}$. Notice that the differential element of speed, dc, is the same for both speeds, and that it, with many other common terms, cancels

$$\frac{dN_2/N}{dN_1/N} = \frac{4\pi c_2^2 \left(\frac{m}{2\pi k_B T}\right)^{\frac{3}{2}} e^{-\frac{mc_2^2}{2k_B T}}\, dc}{4\pi c_1^2 \left(\frac{m}{2\pi k_B T}\right)^{\frac{3}{2}} e^{-\frac{mc_1^2}{2k_B T}}\, dc}$$

$$= \frac{c_2^2}{c_1^2} e^{-\frac{m}{2k_B T}\left(c_2^2 - c_1^2\right)}$$

for this problem we have $c_2 = 989 \text{ m s}^{-1}$ and $c_1 = 406 \text{ m s}^{-1}$. Thus,

$$\frac{dN_2}{dN_1} = \left(\frac{989 \text{ m s}^{-1}}{406 \text{ m s}^{-1}}\right)^2 e^{-\left[\frac{30.07\times 10^{-3}\text{ kg mol}^{-1}}{2(8.314 \text{ J K}^{-1}\text{ mol}^{-1})(273+25)\text{K}}\left(989^2 - 406^2\right)\text{ m}^2\text{ s}^{-2}\right]}$$

$$= 0.0427$$

3.22 Derive an expression for the most probable translational energy for an ideal gas. Compare your result with the average translational energy for the same gas.

The problem is similar to Problem 3.19. The most probable translational (or kinetic) energy for an ideal gas can be calculated by setting $\frac{df(E)}{dE} = 0$ and then solving for E_{mp}.

$$f(E) = 2\pi E^{\frac{1}{2}} \left(\frac{1}{\pi k_B T}\right)^{\frac{3}{2}} e^{-\frac{E}{k_B T}}$$

$$= 2\pi \left(\frac{1}{\pi k_B T}\right)^{\frac{3}{2}} E^{\frac{1}{2}} e^{-\frac{E}{k_B T}}$$

In differentiating $f(E)$ with respect to E, care must be taken to apply the chain rule to the two terms $E^{\frac{1}{2}}$ and $e^{-\frac{E}{k_B T}}$.

$$\frac{df(E)}{dE} = 2\pi \left(\frac{1}{\pi k_B T}\right)^{\frac{3}{2}} \left[\frac{1}{2}E^{-\frac{1}{2}}e^{-\frac{E}{k_B T}} + E^{\frac{1}{2}}e^{-\frac{E}{k_B T}}\left(-\frac{1}{k_B T}\right)\right]$$

$$= 2\pi \left(\frac{1}{\pi k_B T}\right)^{\frac{3}{2}} e^{-\frac{E}{k_B T}}\left(\frac{1}{2E^{\frac{1}{2}}} - \frac{E^{\frac{1}{2}}}{k_B T}\right)$$

When $\frac{df(E)}{dE} = 0$, $E = E_{mp}$. The above equation becomes

$$2\pi \left(\frac{1}{\pi k_B T}\right)^{\frac{3}{2}} e^{-\frac{E_{mp}}{k_B T}}\left(\frac{1}{2E_{mp}^{\frac{1}{2}}} - \frac{E_{mp}^{\frac{1}{2}}}{k_B T}\right) = 0$$

$$\frac{1}{2E_{mp}^{\frac{1}{2}}} - \frac{E_{mp}^{\frac{1}{2}}}{k_B T} = 0$$

Multiply the above equation by $2E_{mp}^{\frac{1}{2}}$:

$$1 - \frac{2E_{mp}}{k_B T} = 0$$

$$E_{mp} = \frac{k_B T}{2} = \frac{\overline{E}_{trans}}{3}$$

3.23   Considering the magnitude of molecular speeds, explain why it takes so long (on the order of minutes) to detect the odor of ammonia when someone opens a bottle of concentrated ammonia at the other end of a lab bench.

The ammonia molecules suffer many collisions with air molecules and must diffuse through the room. They do not make a straight line path from one end of the room to the other.

3.24   How does the mean free path of a gas depend on **(a)** the temperature at constant volume, **(b)** the density, **(c)** the pressure at constant temperature, **(d)** the volume at constant temperature, and **(e)** the size of molecules.

The mean free path is given by Equation 3.18 in the text

$$\lambda = \frac{1}{\sqrt{2}\pi d^2 \left(\frac{N}{V}\right)}$$

Although it is possible to answer this question solely by reference to the equation, it is useful to have an understanding of the physical basis for the effects observed. The key physical quantity is the density of the gas.

(a) The mean free path is independent of temperature at constant volume. T does not appear in the equation. (See particularly the discussion following Equation 3.19 in the text.) As the temperature is increased the molecules are moving faster, but the average distance between them is not affected. The mean time between collisions decreases, but the mean distance traveled between collisions remains the same.

(b) As the density increases, the mean free path decreases, since $\frac{N}{V}$ appears in the denominator. In a more dense gas, the molecules are more closely spaced.

(c) As the pressure increases at constant temperature, the mean free path decreases. These conditions lead to a decrease in volume, hence an increase in density. The molecules are being squeezed closer together.

(d) As the volume increases at constant temperature, the mean free path increases. As the molecules move into the expanded volume, they move further apart from each other.

(e) As the size of the molecules increases, the mean free path decreases. The collision diameter, d, appears in the denominator of the equation. Larger molecules do not have to travel as far before they run into each other.

3.25 A bag containing 20 marbles is being shaken vigorously. Calculate the mean free path of the marbles if the volume of the bag is 850 cm^3. The diameter of each marble is 1.0 cm.

$$\lambda = \frac{1}{\sqrt{2}\pi d^2 \left(\frac{N}{V}\right)} = \frac{1}{\sqrt{2}\pi (1.0 \text{ cm})^2 \left(\frac{20}{850 \text{ cm}^3}\right)} = 9.6 \text{ cm}$$

3.26 Calculate the mean free path and the binary number of collisions per liter per second between HI molecules at 300 K and 1.00 atm. The collision diameter of the HI molecules may be taken to be 5.10 Å. Assume ideal-gas behavior.

The ideal gas law is used to calculate $\frac{N}{V}$, which is then used to calculate the mean free path.

$$PV = nRT = \frac{N}{N_A}RT$$

$$\frac{N}{V} = \frac{PN_A}{RT} = \frac{(1.00 \text{ atm}) \left(6.022 \times 10^{23} \text{ mol}^{-1}\right)}{\left(0.08206 \text{ L atm K}^{-1} \text{ mol}^{-1}\right) (300 \text{ K})} = 2.446 \times 10^{22} \text{ L}^{-1} \left(\frac{1000 \text{ L}}{1 \text{ m}^3}\right)$$

$$= 2.446 \times 10^{25} \text{ m}^{-3}$$

$$\lambda = \frac{1}{\sqrt{2}\pi d^2 \left(\frac{N}{V}\right)} = \frac{1}{\sqrt{2}\pi \left(5.10 \times 10^{-10} \text{ m}\right)^2 \left(2.446 \times 10^{25} \text{ m}^{-3}\right)} = 3.53 \times 10^{-8} \text{ m}$$

The binary number of collisions depends on the average molecular speed, which is

$$\bar{c} = \sqrt{\frac{8RT}{\pi \mathcal{M}}} = \sqrt{\frac{8\,(8.314\,\text{J K}^{-1}\,\text{mol}^{-1})\,(300\,\text{K})}{\pi\,(127.9 \times 10^{-3}\,\text{kg mol}^{-1})}} = 222.8\;\text{m s}^{-1}$$

$$Z_{11} = \frac{\sqrt{2}}{2}\pi d^2 \bar{c}\left(\frac{N}{V}\right)^2 = \frac{\sqrt{2}}{2}\pi\,(5.10 \times 10^{-10}\,\text{m})^2\,(222.8\,\text{m s}^{-1})\,(2.446 \times 10^{25}\,\text{m}^{-3})^2$$

$$= (7.702 \times 10^{34}\,\text{m}^{-3}\,\text{s}^{-1})\left(\frac{1\,\text{m}^3}{1000\,\text{L}}\right) = 7.70 \times 10^{31}\;\text{collisions L}^{-1}\,\text{s}^{-1}$$

3.27 Ultra-high vacuum experiments are routinely performed at a total pressure of 1.0×10^{-10} torr. Calculate the mean free path of N_2 molecules at 350 K under these conditions.

This problem is similar to Problem 3.26. The number density of the gas is

$$\frac{N}{V} = \frac{PN_A}{RT} = \frac{(1.0 \times 10^{-10}\,\text{torr})\left(\frac{1\,\text{atm}}{760\,\text{torr}}\right)(6.022 \times 10^{23}\,\text{mol}^{-1})}{(0.08206\,\text{L atm K}^{-1}\,\text{mol}^{-1})\,(350\,\text{K})}$$

$$= 2.76 \times 10^9\;\text{L}^{-1}\left(\frac{1000\,\text{L}}{1\,\text{m}^3}\right) = 2.76 \times 10^{12}\;\text{m}^{-3}$$

The collision diameter of N_2 is given in Table 3.1 as 3.75 Å.

$$\lambda = \frac{1}{\sqrt{2}\pi d^2\left(\frac{N}{V}\right)} = \frac{1}{\sqrt{2}\pi\,(3.75 \times 10^{-10}\,\text{m})^2\,(2.76 \times 10^{12}\,\text{m}^{-3})} = 5.8 \times 10^5\;\text{m} = 580\;\text{km!}$$

The mean free path is so long that there are hardly any collisions between the molecules before they are pumped away.

3.28 Suppose that helium atoms in a sealed container start with the same speed, 2.74×10^4 cm s^{-1}. The atoms are then allowed to collide with one another until the Maxwell distribution is established. What is the temperature of the gas at equilibrium? Assume that there is no heat exchange between the gas and its surroundings.

The total translational energy of the helium atoms can be determined from the initial speed of the atoms. Because energy is conserved, this is also the total translational energy of the atoms after equilibrium is reached. Translational energy is a function of temperature, thus, the latter can be calculated once the former is known.

Suppose there are N helium atoms. Because all the atoms have the same speed, the total translational energy is

$$E_{\text{trans}} = N \left(\frac{1}{2}mv^2 \right) = N \left(\overline{E}_{\text{trans}} \right)$$

$$N \left(\frac{1}{2}mv^2 \right) = N \left(\frac{3}{2}k_{\text{B}}T \right)$$

$$T = \frac{mv^2}{3k_{\text{B}}} = \frac{(4.003 \text{ amu}) \left(1.661 \times 10^{-27} \text{ kg amu}^{-1}\right) \left(2.74 \times 10^4 \text{ cm s}^{-1}\right)^2 \left(\frac{1 \text{ m}}{100 \text{ cm}}\right)^2}{3 \left(1.381 \times 10^{-23} \text{ J K}^{-1}\right)} = 12.0 \text{ K}$$

3.29 Compare the collision number and the mean free path for air molecules at (**a**) sea level ($T = 300$ K and density = 1.2 g L^{-1}) and (**b**) in the stratosphere ($T = 250$ K and density = 5.0×10^{-3} g L^{-1}). The molar mass of air may be taken as 29.0 g and the collision diameter is 3.72 Å.

Finding the collision number,

$$Z_{11} = \frac{\sqrt{2}}{2}\pi d^2 \overline{c} \left(\frac{N}{V} \right)^2$$

and the mean free path,

$$\lambda = \frac{1}{\sqrt{2}\pi d^2 \left(\frac{N}{V} \right)}$$

requires knowledge of the average speed, $\overline{c}$ and density, $\frac{N}{V}$.

Sea level:

$$\overline{c} = \sqrt{\frac{8RT}{\pi \mathcal{M}}}$$

$$= \sqrt{\frac{8(8.314 \text{ J K}^{-1} \text{ mol}^{-1})(300 \text{ K})}{\pi(29.0 \times 10^{-3} \text{ kg mol}^{-1})}}$$

$$= 4.680 \times 10^2 \text{ m s}^{-1}$$

and

$$\frac{N}{V} = \frac{1.2 \text{ g L}^{-1}}{29.0 \text{ g mol}^{-1}} \left(\frac{1000 \text{ L}}{1 \text{ m}^3} \right) \left(\frac{6.022 \times 10^{23} \text{ molecules}}{1 \text{ mol}} \right)$$

$$= 2.49 \times 10^{25} \text{ molecules m}^{-3}$$

Thus,

$$Z_{11} = \frac{\sqrt{2}}{2} \pi d^2 \bar{c} \left(\frac{N}{V} \right)^2$$

$$= \frac{\sqrt{2}}{2} \pi (3.72 \times 10^{-10} \text{ m})^2 (4.680 \times 10^2 \text{ m s}^{-1})(2.49 \times 10^{25} \text{ molecules m}^{-3})^2$$

$$= 8.9 \times 10^{34} \text{ collisions m}^{-3} \text{ s}^{-1}$$

and

$$\lambda = \frac{1}{\sqrt{2} \pi d^2 \left(\frac{N}{V} \right)}$$

$$= \frac{1}{\sqrt{2} \pi (3.72 \times 10^{-10} \text{ m})^2 (2.49 \times 10^{25} \text{ molecules m}^{-3})}$$

$$= 6.5 \times 10^{-8} \text{ m}$$

Stratosphere:

$$\bar{c} = \sqrt{\frac{8RT}{\pi \mathcal{M}}}$$

$$= \sqrt{\frac{8(8.314 \text{ J K}^{-1} \text{ mol}^{-1})(250 \text{ K})}{\pi (29.0 \times 10^{-3} \text{ kg mol}^{-1})}}$$

$$= 4.272 \times 10^2 \text{ m s}^{-1}$$

and

$$\frac{N}{V} = \frac{5.0 \times 10^{-3} \text{ g L}^{-1}}{29.0 \text{ g mol}^{-1}} \left(\frac{1000 \text{ L}}{1 \text{ m}^3} \right) \left(\frac{6.022 \times 10^{23} \text{ molecules}}{1 \text{ mol}} \right)$$

$$= 1.04 \times 10^{23} \text{ molecules m}^{-3}$$

Thus,

$$Z_{11} = \frac{\sqrt{2}}{2} \pi d^2 \bar{c} \left(\frac{N}{V} \right)^2$$

$$= \frac{\sqrt{2}}{2} \pi (3.72 \times 10^{-10} \text{ m})^2 (4.272 \times 10^2 \text{ m s}^{-1})(1.04 \times 10^{23} \text{ molecules m}^{-3})^2$$

$$= 1.4 \times 10^{30} \text{ collisions m}^{-3} \text{ s}^{-1}$$

and

$$\lambda = \frac{1}{\sqrt{2} \pi d^2 \left(\frac{N}{V} \right)}$$

$$= \frac{1}{\sqrt{2} \pi (3.72 \times 10^{-10} \text{ m})^2 (1.04 \times 10^{23} \text{ molecules m}^{-3})}$$

$$= 1.6 \times 10^{-5} \text{ m}$$

3.30 Calculate the values of Z_1 and Z_{11} for mercury (Hg) vapor at 40°C, both at $P = 1.0$ atm and at $P = 0.10$ atm. How do these two quantities depend on pressure?

The number density and average molecular speed need to be determined before calculating Z_1 and Z_{11}. The collision diameter of Hg is 4.26 Å (Table 3.1).

At $P = 1.0$ atm,

$$\frac{N}{V} = \frac{PN_A}{RT} \text{ (See Problem 3.26)}$$

$$= \frac{(1.0 \text{ atm}) (6.022 \times 10^{23} \text{ mol}^{-1})}{(0.08206 \text{ L atm K}^{-1} \text{ mol}^{-1}) (273 + 40) \text{ K}} = 2.34 \times 10^{22} \text{ L}^{-1} \left(\frac{1000 \text{ L}}{1 \text{ m}^3}\right)$$

$$= 2.34 \times 10^{25} \text{ m}^{-3}$$

$$\bar{c} = \sqrt{\frac{8RT}{\pi \mathcal{M}}} = \sqrt{\frac{8 (8.314 \text{ J K}^{-1} \text{ mol}^{-1}) (273 + 40) \text{ K}}{\pi (200.6 \times 10^{-3} \text{ kg})}} = 181.8 \text{ m s}^{-1}$$

$$Z_1 = \sqrt{2}\pi d^2 \bar{c} \frac{N}{V} = \sqrt{2}\pi (4.26 \times 10^{-10} \text{ m})^2 (181.8 \text{ m s}^{-1}) (2.34 \times 10^{25} \text{ m}^{-3})$$

$$= 3.4 \times 10^9 \text{ collisions s}^{-1}$$

$$Z_{11} = \frac{\sqrt{2}}{2}\pi d^2 \bar{c} \left(\frac{N}{V}\right)^2 = \frac{1}{2} Z_1 \left(\frac{N}{V}\right)$$

$$= \frac{1}{2} (3.4 \times 10^9 \text{ collisions s}^{-1}) (2.34 \times 10^{25} \text{ m}^{-3})$$

$$= 4.0 \times 10^{34} \text{ collisions m}^{-3} \text{ s}^{-1}$$

For an ideal gas, $\frac{N}{V} = \frac{PN_A}{RT}$. The results above show $Z_1 \propto P$, whereas $Z_{11} \propto P^2$. A reduction in P to one tenth its original value (from 1.0 atm to 0.10 atm) will likewise reduce Z_1 to one tenth its value at $P = 1.0$ atm, but Z_{11} will decrease to $\frac{1}{10^2} = \frac{1}{100}$ of its value at $P = 1.0$ atm. That is, at $P = 0.10$ atm,

$$Z_1 = 3.4 \times 10^8 \text{ collisions s}^{-1}$$

$$Z_{11} = 4.0 \times 10^{32} \text{ collisions m}^{-3} \text{ s}^{-1}$$

3.31 Account for the difference between the dependence of viscosity on temperature for a liquid and a gas.

The viscosity, η, of a gas *increases* with increasing temperature, but for a liquid the viscosity *decreases* with increasing temperature. In a liquid, the major contributors (by far!) to the viscosity are the intermolecular forces. As the molecules gain kinetic energy through the increase in temperature, they are able to more easily overcome their mutual attractions and the liquid flows more readily. For gases, on the other hand, momentum transfer is the leading contributor to the viscosity. (Note that even ideal gases have non-zero viscosity, so long as finite molecular sizes are included.) As the molecules move faster, they exchange momentum more rapidly, leading to an increase in viscosity.

3.32   Calculate the values of the average speed and collision diameter for ethylene at 288 K. The viscosity of ethylene is 99.8×10^{-7} N s m^{-2} at the same temperature.

$$\bar{c} = \sqrt{\frac{8RT}{\pi \mathcal{M}}} = \sqrt{\frac{8\left(8.314 \text{ J K}^{-1} \text{ mol}^{-1}\right)(288 \text{ K})}{\pi\left(28.05 \times 10^{-3} \text{ kg mol}^{-1}\right)}} = 466.2 \text{ m s}^{-1}$$

$$\eta = \frac{m\bar{c}}{3\sqrt{2}\pi d^2}$$

$$d = \left(\frac{m\bar{c}}{3\sqrt{2}\pi \eta}\right)^{\frac{1}{2}} = \left[\frac{(28.05 \text{ amu})\left(1.661 \times 10^{-27} \text{ kg amu}^{-1}\right)\left(466.2 \text{ m s}^{-1}\right)}{3\sqrt{2}\pi\left(99.8 \times 10^{-7} \text{ N s m}^{-2}\right)}\right]^{\frac{1}{2}}$$

$$= 4.04 \times 10^{-10} \text{ m} = 4.04 \text{ Å}$$

3.33   The viscosity of sulfur dioxide at 21.0°C and 1.0 atm pressure is 1.25×10^{-5} N s m^{-2}. Calculate the collision diameter of the SO$_2$ molecule and the mean free path at the given temperature and pressure.

The collision diameter is related to the viscosity and average speed of the molecules. The first is given, and the second can be calculated readily.

$$\bar{c} = \sqrt{\frac{8RT}{\pi \mathcal{M}}} = \sqrt{\frac{8\left(8.314 \text{ J K}^{-1} \text{ mol}^{-1}\right)(273.2 + 21.0) \text{ K}}{\pi\left(64.07 \times 10^{-3} \text{ kg mol}^{-1}\right)}} = 311.79 \text{ m s}^{-1}$$

$$\eta = \frac{m\bar{c}}{3\sqrt{2}\pi d^2}$$

$$d = \left(\frac{m\bar{c}}{3\sqrt{2}\pi \eta}\right)^{\frac{1}{2}} = \left[\frac{(64.07 \text{ amu})\left(1.661 \times 10^{-27} \text{ kg amu}^{-1}\right)\left(311.79 \text{ m s}^{-1}\right)}{3\sqrt{2}\pi\left(1.25 \times 10^{-5} \text{ N s m}^{-2}\right)}\right]^{\frac{1}{2}}$$

$$= 4.46 \times 10^{-10} \text{ m} = 4.46 \text{ Å}$$

The mean free path is related to the average speed and the number density, which can be calculated following the procedure in Problem 3.26.

$$\frac{N}{V} = \frac{PN_A}{RT}$$

$$= \frac{(1.0 \text{ atm})\left(6.022 \times 10^{23} \text{ mol}^{-1}\right)}{\left(0.08206 \text{ L atm K}^{-1} \text{ mol}^{-1}\right)(273.2 + 21.0) \text{ K}} = 2.49 \times 10^{22} \text{ L}^{-1} \left(\frac{1000 \text{ L}}{1 \text{ m}^3}\right)$$

$$= 2.49 \times 10^{25} \text{ m}^{-3}$$

$$\eta = \frac{1}{3}\left(\frac{N}{V}\right)m\lambda\bar{c}$$

$$\lambda = \frac{3\eta}{\left(\frac{N}{V}\right)m\bar{c}} = \frac{3\left(1.25 \times 10^{-5}\,\mathrm{N\,s\,m^{-2}}\right)}{\left(2.49 \times 10^{25}\,\mathrm{m^{-3}}\right)\left(64.07\,\mathrm{amu}\right)\left(1.661 \times 10^{-27}\,\mathrm{kg\,amu^{-1}}\right)\left(311.79\,\mathrm{m\,s^{-1}}\right)}$$

$$= 4.5 \times 10^{-8}\,\mathrm{m} = 4.5 \times 10^{2}\,\text{Å}$$

3.34 Derive Equation 3.23 from Equation 3.14.

The rate of diffusion (effusion) is proportional to average molecular speed,

$$r \propto \bar{c} = \sqrt{\frac{8RT}{\pi M}}$$

Thus, for two gases of different molar masses, but at the same temperature,

$$\frac{r_1}{r_2} = \frac{\sqrt{\frac{8RT}{\pi M_1}}}{\sqrt{\frac{8RT}{\pi M_2}}} = \sqrt{\frac{M_2}{M_1}}$$

3.35 An inflammable gas is generated in marsh lands and sewage by a certain anaerobic bacterium. A pure sample of this gas was found to effuse through an orifice in 12.6 min. Under identical conditions of temperature and pressure, it takes oxygen 17.8 min to effuse through the same orifice. Calculate the molar mass of the gas, and suggest what this gas might be.

$$\text{Rate of effusion } (r) \propto \frac{1}{\text{Time required for effusion } (t)}$$

Therefore,

$$\frac{t_{O_2}}{t_{gas}} = \frac{r_{gas}}{r_{O_2}} = \sqrt{\frac{M_{O_2}}{M_{gas}}}$$

$$M_{gas} = M_{O_2}\left(\frac{t_{gas}}{t_{O_2}}\right)^2 = \left(32.00\,\mathrm{g\,mol^{-1}}\right)\left(\frac{12.6\,\mathrm{min}}{17.8\,\mathrm{min}}\right)^2 = 16.0\,\mathrm{g\,mol^{-1}}$$

The gas is likely CH_4.

3.36 Nickel forms a gaseous compound of the formula $Ni(CO)_x$. What is the value of x given the fact that under the same conditions of temperature and pressure, methane (CH_4) effuses 3.3 times faster than the compound?

$$\frac{r_{CH_4}}{r_{Ni(CO)_x}} = 3.3 = \sqrt{\frac{M_{Ni(CO)_x}}{M_{CH_4}}}$$

$$M_{Ni(CO)_x} = (3.3)^2 \left(M_{CH_4}\right) = (3.3)^2 \left(16.04 \text{ g mol}^{-1}\right) = 1.75 \times 10^2 \text{ g mol}^{-1}$$

Calculate the expected molar mass of $Ni(CO)_x$ from its chemical formula, and equate the expression with the quantity obtained above.

$$M_{Ni(CO)_x} = 58.69 \text{ g mol}^{-1} + x\left(12.01 \text{ g mol}^{-1} + 16.00 \text{ g mol}^{-1}\right) = 1.75 \times 10^2 \text{ g mol}^{-1}$$

$$28.01x = 1.16 \times 10^2$$

$$x = 4.1 \approx 4$$

3.37 In 2.00 min, 29.7 mL of He effuse through a small hole. Under the same conditions of temperature and pressure, 10.0 mL of a mixture of CO and CO_2 effuse through the hole in the same amount of time. Calculate the percent composition by volume of the mixture.

$$\frac{r_{He}}{r_{mix}} = \sqrt{\frac{M_{mix}}{M_{He}}}$$

$$M_{mix} = \left(\frac{r_{He}}{r_{mix}}\right)^2 M_{He} = \left(\frac{\frac{29.7 \text{ mL}}{2.00 \text{ min}}}{\frac{10.0 \text{ mL}}{2.00 \text{ min}}}\right)^2 \left(4.003 \text{ g mol}^{-1}\right) = 35.31 \text{ g mol}^{-1}$$

Let x_{CO} be the mole fraction of CO, and x_{CO_2} be the mole fraction of CO_2. These mole fractions are related by

$$x_{CO} + x_{CO_2} = 1$$

The effective molar mass of the mixture can be obtained from the mole fractions and molar masses of its components:

$$x_{CO}M_{CO} + x_{CO_2}M_{CO_2} = M_{mix}$$

$$x_{CO}M_{CO} + \left(1 - x_{CO}\right)M_{CO_2} = M_{mix}$$

$$x_{CO}\left(28.01 \text{ g mol}^{-1}\right) + \left(1 - x_{CO}\right)\left(44.01 \text{ g mol}^{-1}\right) = 35.31 \text{ g mol}^{-1}$$

$$28.01x_{CO} + 44.01 - 44.01x_{CO} = 35.31$$

$$16.00x_{CO} = 8.70$$

$$x_{CO} = 0.54$$

At constant P and T, $n \propto V$. Therefore, volume fraction = mole fraction. As a result,

$$\% \text{ of CO by volume} = 54\%$$

$$\% \text{ of CO}_2 \text{ by volume} = 1 - \% \text{ of CO by volume} = 46\%$$

3.38 Uranium-235 can be separated from uranium-238 by the effusion process involving UF_6. Assuming a 50:50 mixture at the start, what is the percentage of enrichment after a single stage of separation?

This problem is the same as the example given in the text.

$$\text{Separation factor} = s = \sqrt{\frac{^{235}UF_6}{^{238}UF_6}} = \sqrt{\frac{M_{^{238}UF_6}}{M_{^{235}UF_6}}} = \sqrt{\frac{238 + (6)(19)}{235 + (6)(19)}} = 1.0043$$

According to this result, $^{235}UF_6$ effuses 1.0043 times faster than $^{235}UF_6$. Therefore, $^{235}UF_6$ is enriched by factor of 0.43% after a single stage of separation.

3.39 An equimolar mixture of H_2 and D_2 effuses through an orifice at a certain temperature. Calculate the composition (in mole fractions) of the gas that passes through the orifice. The molar mass of deuterium is 2.014 g mol^{-1}.

The natural abundance of deuterium (D) is so small that to four significant figures the molar mass of $^1H_2$ is the same as the average molar mass of naturally occuring H_2, but for D_2 the molar mass is 4.028 g mol^{-1}.

$$\frac{r_{H_2}}{r_{D_2}} = \sqrt{\frac{M_{D_2}}{M_{H_2}}} = \sqrt{\frac{4.028 \text{ g mol}^{-1}}{2.016 \text{ g mol}^{-1}}} = 1.414$$

The amount of H_2 that passes through the orifice is 1.414 times more than that of D_2. Therefore,

$$\text{mole fraction of } H_2 \text{ that passes the orifice} = \frac{1.414}{1.414 + 1} = 0.5857$$

$$\text{mole fraction of } D_2 \text{ that passes the orifice} = 1 - \text{mole fraction of } H_2 \text{ that passes the orifice}$$

$$= 0.4143$$

3.40 The rate (r_{eff}) at which molecules confined to a volume V effuse through an orifice of area A is given by $(1/4) n N_A \bar{c} A / V$, where n is the number of moles of the gas. An automobile tire of volume 30.0 L and pressure 1,500 torr is punctured as it runs over a sharp nail. **(a)** Calculate the effusion rate if the diameter of the hole is 1.0 mm. **(b)** How long would it take to lose half of the air in the tire through effusion? Assume a constant effusion rate and constant volume. The molar mass of air is 29.0 g and the temperature is 32.0°C.

(a) The effusion rate can be calculated once n, $\bar{c}$, and A are determined. The ideal gas law is used to calculate n.

$$n = \frac{PV}{RT} = \frac{(1500 \text{ torr}) \left(\frac{1 \text{ atm}}{760 \text{ torr}}\right)(30.0 \text{ L})}{\left(0.08206 \text{ L atm K}^{-1} \text{ mol}^{-1}\right)(273.2 + 32.0) \text{ K}} = 2.364 \text{ mol}$$

$$\bar{c} = \sqrt{\frac{8RT}{\pi \mathcal{M}}} = \sqrt{\frac{8 \left(8.314 \text{ J K}^{-1} \text{ mol}^{-1}\right)(273.2 + 32.0) \text{ K}}{\pi \left(29.0 \times 10^{-3} \text{ kg mol}^{-1}\right)}} = 472.0 \text{ m s}^{-1}$$

$$A = \pi \left(\frac{d}{2}\right)^2 = \pi \left(\frac{1.0 \text{ mm}}{2}\right)^2 \left(\frac{1 \text{ m}}{1000 \text{ mm}}\right)^2 = 7.854 \times 10^{-7} \text{m}^2$$

The effusion rate is

$$r = \frac{nN_A\bar{c}A}{4V} = \frac{(2.364 \text{ mol}) \left(6.022 \times 10^{23} \text{ molecules mol}^{-1}\right)(472.0 \text{ m s}^{-1})\left(7.854 \times 10^{-7}\text{m}^2\right)}{4 (30.0 \text{ L}) \left(\frac{1 \text{ m}^3}{1000 \text{ L}}\right)}$$

$$= 4.39 \times 10^{21} \text{ molecules s}^{-1}$$

(b) The time it would take to lose half of the molecules is

$$t = \frac{\text{\# of molecules effused}}{\text{effusion rate}}$$

$$= \frac{\left(\frac{2.36 \text{ mol}}{2}\right)\left(6.022 \times 10^{23} \text{ molecules mol}^{-1}\right)}{4.39 \times 10^{21} \text{ molecules s}^{-1}}$$

$$= 162 \text{ s} = 2.70 \text{ min}$$

3.41   Calculate the various degrees of freedom for the following molecules: **(a)** Xe, **(b)** HCl, **(c)** CS_2, **(d)** C_2H_2, **(e)** C_6H_6, and **(f)** a hemoglobin molecule containing 9272 atoms.

The following table shows the numbers of degrees of freedom for different types of molecules:

	Translation	Rotation	Vibration	Total
Atoms	3	0	0	3
Linear Molecules	3	2	$3N - 5$	$3N$
Nonlinear Molecules	3	3	$3N - 6$	$3N$

For the molecules of interest, the various degrees of freedom are listed below. Note that HCl, CS_2, and C_2H_2 are linear molecules.

	Translation	Rotation	Vibration	Total
(a) Xe	3	0	0	3
(b) HCl	3	2	1	6
(c) CS_2	3	2	4	9
(d) C_2H_2	3	2	7	12
(e) C_6H_6	3	3	30	36
(f) hemoglobin	3	3	27810	27816

3.42 Explain the equipartition of energy theorem. Why does it fail for diatomic and polyatomic molecules?

The equipartition of energy theorem states that the energy of a molecule is equally divided among all of its degrees of freedom. Based on classical mechanics, the theorem fails for diatomic and polyatomic molecules because it does not take the quantization of energy into account. Specifically, the theorem fails for the electronic and vibrational degrees of freedom of molecules where the spacings between quantized energy levels are much greater than $k_B T$. The theorem may be successfully applied to the translational and rotational degrees of freedom, although for low temperatures, where $k_B T$ becomes smaller than the spacing between quantized rotational energy levels, it will fail for this degree of freedom as well.

3.43 A quantity of 500 joules of energy is delivered to one mole of each of the following gases at 298 K and the same fixed volume: Ar, CH_4, H_2. Which gas will have the highest rise in temperature?

Being a monoatomic gas, Ar has the fewest degrees of freedom and, consequently, the lowest heat capacity. This gas will have the highest rise in temperature.

3.44 Calculate the mean kinetic energy $(\overline{E}_{trans})$ in joules of the following molecules at 350 K: **(a)** He, **(b)** CO_2, and **(c)** UF_6. Explain your results.

Regardless of the identity of the gas, the mean kinetic energy of the molecules is

$$\overline{E}_{trans} = \frac{3}{2}k_B T = \frac{3}{2}(1.381 \times 10^{-23}\ \mathrm{J\,K^{-1}})(350\ \mathrm{K}) = 7.25 \times 10^{-21}\ \mathrm{J}$$

3.45 A sample of neon gas is heated from 300 K to 390 K. Calculate the percent increase in its kinetic energy.

$$\frac{\left(\overline{E}_{trans}\right)_{390}}{\left(\overline{E}_{trans}\right)_{300}} = \frac{\frac{3}{2}k_B\,(390\ \mathrm{K})}{\frac{3}{2}k_B\,(300\ \mathrm{K})} = 1.30$$

Therefore,

$$\%\ \text{increase in kinetic energy} = 30\%$$

3.46 Calculate the value of $\overline{C}_V$ for H_2, CO_2, and SO_2, assuming that only translational and rotational motions contribute to the heat capacities. Compare your results with the values listed in Table 3.3. Explain the differences.

For molecules, only the translational and rotational degrees of freedom fully contribute to the heat capacity. The electronic and vibrational degrees of freedom, due to the large spacing between quantized energy levels, are inaccessible at room temperature (except for partial contributions from low frequency vibrational motions).

H_2 has 3 translational degrees of freedom and 2 rotational degrees of freedom, therefore, the total energy for 1 mole of H_2 is

$$U = \frac{3}{2}RT + \frac{2}{2}RT = \frac{5}{2}RT$$

Because U is molar total energy, the molar heat capacity is

$$\overline{C}_V = \left(\frac{\partial U}{\partial T}\right)_V = \frac{5}{2}R = 20.79 \text{ J K}^{-1}\text{mol}^{-1}$$

The agreement is quite good, indicating no contribution from the single vibrational degree of freedom.

CO_2 is linear, so it has 3 translational degrees of freedom and 2 rotational degrees of freedom, therefore, the total energy for 1 mole of CO_2 is the same as that for 1 mole of H_2. Consequently, $\overline{C}_V$ for CO_2 is also calculated to be 20.79 J K^{-1} mol^{-1}. This is significantly lower than the measured value and is indicative of some vibrational contribution. In fact, CO_2 has two, low-energy bending vibrations into which some thermal energy may be distributed, although not to the extent necessary for the equipartition of energy theorem to be valid.

SO_2 is nonlinear, and has 3 translational degrees of freedom and 3 rotational degrees of freedom, therefore, the total energy for 1 mole of SO_2 is

$$U = \frac{3}{2}RT + \frac{3}{2}RT = 3RT$$

Thus, the calculated molar heat capacity for SO_2 is

$$\overline{C}_V = 3R = 24.94 \text{ J K}^{-1}\text{mol}^{-1}$$

Again the calculated value is significantly lower than that measured for the same reason as for CO_2. In both cases, however, the measured value is lower than that predicted by the equipartition theorem when all degrees of freedom are included.

3.47   One mole of ammonia initially at 5°C is placed in contact with 3 moles of helium initially at 90°C. Given that $\overline{C}_V$ for ammonia is $3R$, if the process is carried out at constant total volume, what is the final temperature of the gases?

To reach equilibrium, ammonia will warm up and helium will cool down until both reach the same final temperature, T_f. According to the law of conservation of energy, the energy given up by helium must be absorbed by ammonia, assuming energy is not transferred to the surroundings. This energy can be calculated using heat capacity. Heat capacity is the energy required to raise the temperature of a given quantity of substance by 1°C or 1 K. Therefore, the energy required to warm up ammonia is

$$n_{NH_3}\overline{C}_{V, NH_3}\left(T_f - 5°C\right)$$

and the energy removed in cooling He is

$$-n_{\text{He}} \overline{C}_{\text{V, He}} \left(T_f - 90°\text{C} \right)$$

(The minus sign indicates that energy is being lost.)

Conservation of energy requires that these energies be equal. Using $\overline{C}_{\text{V, He}} = \frac{3}{2}R$ (He has only translational degrees of freedom, and therefore, $U = \frac{3}{2}RT$.) and the given value for $\overline{C}_{\text{V, NH}_3}$,

$$n_{\text{NH}_3} \overline{C}_{\text{V, NH}_3} \left(T_f - 5°\text{C} \right) = -n_{\text{He}} \overline{C}_{\text{V, He}} \left(T_f - 90°\text{C} \right)$$

$$(1 \text{ mol}) (3R) \left(T_f - 5°\text{C} \right) = (3 \text{ mol}) \left(\frac{3}{2}R \right) \left(90°\text{C} - T_f \right)$$

$$\left(T_f - 5°\text{C} \right) = \frac{3}{2} \left(90°\text{C} - T_f \right)$$

$$T_f - 5°\text{C} = 135°\text{C} - \frac{3}{2}T_f$$

$$\frac{5}{2}T_f = 140°\text{C}$$

$$T_f = 56°\text{C}$$

3.48   The typical energy differences between successive rotational, vibrational, and electronic energy levels are 5.0×10^{-22} J, 0.50×10^{-19} J, and 1.0×10^{-17} J respectively. Calculate the ratios of the numbers of molecules in the two adjacent energy levels (higher to lower) in each case at 298 K.

Let the numbers of molecules in the lower and higher level be N_1 and N_2, respectively. The ratio between the numbers of molecules in these two levels is

$$\frac{N_2}{N_1} = e^{-\frac{\Delta E}{k_{\text{B}}T}} = e^{-\frac{\Delta E}{\left(1.381 \times 10^{-23} \text{ J K}^{-1} \right)(298 \text{ K})}} = e^{-\frac{\Delta E}{4.12 \times 10^{-21} \text{ J}}}$$

Rotational energy levels

$$\frac{N_2}{N_1} = e^{-\frac{5.0 \times 10^{-22} \text{ J}}{4.12 \times 10^{-21} \text{ J}}} = 0.89$$

Vibrational energy levels

$$\frac{N_2}{N_1} = e^{-\frac{0.5 \times 10^{-19} \text{ J}}{4.12 \times 10^{-21} \text{ J}}} = 5.4 \times 10^{-6}$$

Electronic energy levels

$$\frac{N_2}{N_1} = e^{-\frac{1.0 \times 10^{-17} \text{ J}}{4.12 \times 10^{-21} \text{ J}}} = e^{-2427} \approx 0$$

3.49 The first excited electronic energy level of the helium atom is 3.13×10^{-18} J above the ground level. Estimate the temperature at which the electronic motion will begin to make a significant contribution to the heat capacity. That is, at what temperature will the ratio of the population of the first excited state to the ground state be 5.0%?

Find the temperature, T, such that the population ratio, $N_2/N_1 = 0.050$,

$$\frac{N_2}{N_1} = e^{-\frac{\Delta E}{k_B T}} = 0.050$$

$$-\frac{\Delta E}{k_B T} = \ln 0.050$$

$$T = -\frac{\Delta E}{k_B \ln 0.050} = -\frac{3.13 \times 10^{-18} \text{ J}}{\left(1.381 \times 10^{-23} \text{ J K}^{-1}\right)(\ln 0.050)} = 7.6 \times 10^4 \text{ K}$$

3.50 Consider 1 mole each of gaseous He and N_2 at the same temperature and pressure. State which gas (if any) has the greater value: (a) $\bar{c}$, (b) c_{rms}, (c) $\overline{E}_{trans}$, (d) Z_1, (e) Z_{11}, (f) density, (g) mean free path, (h) viscosity.

Both gases are at the same pressure and temperature, hence the same volume and density as well. The two important differences are of size and mass: He is both smaller and less massive than N_2. N_2 has 7 times the mass of He and its diameter is about 1.7 times that of He (See Table 3.1).

(a) Being less massive, He will have the greater $\bar{c}$.

(b) For the same reason as part (a), He will have the greater c_{rms}.

(c) Both gases have the same $\overline{E}_{trans}$, as both are at the same temperature.

(d) $Z_1 \propto \frac{d^2}{\sqrt{M}}$, where the mass dependence arises through the $\bar{c}$ term. Although He is moving $\sqrt{7} = 2.6$ times faster than N_2, the d^2 term contributes a factor of 3 to the N_2, and size outweighs speed to give N_2 the greater Z_1.

(e) $Z_{11} = \frac{1}{2}Z_1\left(\frac{N}{V}\right)$. Since the densities of the two gases are the same, N_2 will have the greater Z_{11} as well.

(f) The densities are the same.

(g) With the greater size, N_2 has the smaller mean free path.

(h) From equation 3.22, $\eta \propto \frac{m\bar{c}}{d^2} \propto \frac{\sqrt{m}}{d^2}$. Thus although He is lighter, its smaller size and consequent larger mean free path leads to He having the greater viscosity.

3.51 The root-mean-square velocity of a certain gaseous oxide is 493 m s^{-1} at 20°C. What is the molecular formula of the compound?

The root-mean-square velocity is used to determine the molar mass of the oxide. Given the known molar mass of oxygen, the molar mass of the other element can be determined, and therefore, its identity.

$$v_{rms} = \sqrt{\frac{3RT}{M}}$$

$$M = \frac{3RT}{v_{rms}^2} = \frac{3\left(8.314 \text{ J K}^{-1}\text{ mol}^{-1}\right)(273 + 20)\text{ K}}{\left(493 \text{ m s}^{-1}\right)^2}$$

$$= 3.01 \times 10^{-2} \text{ kg mol}^{-1} = 30.1 \text{ g mol}^{-1}$$

Since the molar mass of the compound is less than 32 g mol^{-1}, the compound must be a monoxide. Thus, the molar mass of the other element is 30.1 g mol^{-1} − molar mass of O = 14.01 g mol^{-1}, which is the molar mass of N. The oxide is NO.

3.52 At 298 K, the $\overline{C}_V$ of SO_2 is greater than that of CO_2. At very high temperatures (> 1000 K), the $\overline{C}_V$ of CO_2 is greater than that of SO_2. Explain.

At high temperatures, $k_B T$ becomes large enough that the vibrational degrees of freedom of a molecule can fully contribute to the heat capacity. CO_2 is a linear molecule with 3 translational, 2 rotational, and $3(3) - 5 = 4$ vibrational degrees of freedom. Recalling that translational and rotational degrees of freedom each contributes $\frac{R}{2}$ to $\overline{C}_V$ while each vibrational degree of freedom contributes R,

$$\overline{C}_{V,CO_2} = 3\frac{R}{2} + 2\frac{R}{2} + 4R = \frac{13}{2}R$$

SO_2, on the other hand, is not linear. It has 3 translational and 3 rotational degrees of freedom and $3(3) - 6 = 3$ vibrational degrees of freedom. Thus,

$$\overline{C}_{V,SO_2} = 3\frac{R}{2} + 3\frac{R}{2} + 3R = 6R$$

We see that in the linear molecule the substitution of an extra vibrational degree of freedom for the absent rotational degree of freedom leads to a larger $\overline{C}_V$, since vibrational degrees of freedom contribute twice as much to the heat capacity.

At room temperature, however, the vibrational degrees of freedom contribute only slightly, and the extra rotational degree of freedom for the non-linear molecule gives it the larger $\overline{C}_V$.

3.53 Calculate the total translational kinetic energy of the air molecules in a spherical balloon of radius 43.0 cm at 24°C and 1.2 atm. Is this enough energy to heat 200 mL of water from 20°C to 90°C for a cup of tea? (The density of water is 1.0 g cm^{-3} and its specific heat is 4.184 J g^{-1}°C^{-1}.)

The number of moles of air molecules is needed before the total translational kinetic energy can be determined, but to use the ideal gas law to calculate n, the volume occupied by the air molecules must first be calculated.

$$V = \frac{4}{3}\pi r^3 = \frac{4}{3}\pi \, (43.0 \text{ cm})^3 \left(\frac{1 \text{ L}}{1000 \text{ cm}^3}\right) = 333.0 \text{ L}$$

$$n = \frac{PV}{RT} = \frac{(1.2 \text{ atm}) \,(333.0 \text{ L})}{(0.08206 \text{ L atm K}^{-1} \text{ mol}^{-1}) \,(273 + 24) \text{ K}} = 16.4 \text{ mol}$$

$$\overline{E}_{\text{trans}} = n\left(\frac{3}{2}RT\right) = (16.4 \text{ mol})\left(\frac{3}{2}\right)(8.314 \text{ J K}^{-1} \text{ mol}^{-1})\,(273 + 24)\text{ K} = 6.1 \times 10^4 \text{ J}$$

Energy needed to heat 200 g of water from 20°C to 90°C is

$$\left(4.184 \text{ J g}^{-1\circ}\text{C}^{-1}\right)(200 \text{ g})\,(90°\text{C} - 20°\text{C}) = 5.9 \times 10^4 \text{ J}$$

Thus, the total translational kinetic energy is enough to heat 200 g of water from 20°C to 90°C.

3.54 The following apparatus can be used to measure atomic and molecular speed. Suppose that a beam of metal atoms is directed at a rotating cylinder in a vacuum. A small opening in the cylinder allows the atoms to strike a target area. Because the cylinder is rotating, atoms traveling at different speeds will strike the target at different position. In time, a layer of the metal will deposit on the target area, and the variation in its thickness is found to correspond to Maxwell's speed distribution. In one experiment it is found that at 850°C some bismuth (Bi) atoms struck the target at a point 2.80 cm from the spot directly opposite the slit. The diameter of the cylinder is 15.0 cm and it is rotating at 130 revolutions per second. **(a)** Calculate the speed (m s^{-1}) at which the target is moving (*Hint*: The circumference of a circle is given by $2\pi r$, where r is the radius). **(b)** Calculate the time (in seconds) it takes for the target to travel 2.80 cm. **(c)** Determine the speed of the Bi atoms. Compare your result in **(c)** with the c_{rms} value of Bi at 850°C. Comment on the difference.

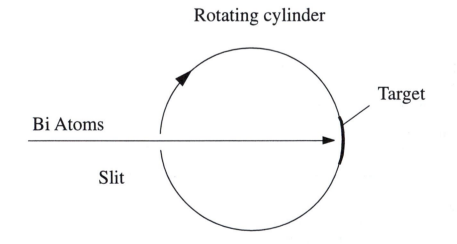

Rotating cylinder

Target

Bi Atoms

Slit

(a) The speed at which a point on the target moves is equal to the product of the circumference of the cylinder and the number of revolutions per second. Calling the speed of the target s,

$$s = \left(\frac{130 \text{ revolutions}}{1 \text{ s}}\right)\left(\frac{2\pi \left(\frac{15.0 \text{ cm}}{2}\right)}{1 \text{ revolution}}\right)\left(\frac{1 \text{ m}}{100 \text{ cm}}\right) = 61.3 \text{ m s}^{-1}$$

(b) The time, t, it takes for a point on the target to move 2.80 cm is given by

$$t = \left(\frac{2.80 \text{ cm}}{61.3 \text{ m s}^{-1}}\right)\left(\frac{1 \text{ m}}{100 \text{ cm}}\right) = 4.57 \times 10^{-4} \text{ s}$$

(c) Since the Bi atoms struck the target at a point 2.80 cm from the spot directly opposite the slit, it took the same amount of time for the atoms to travel the 15.0 cm diameter of the cylinder as it took for the target to move 2.80 cm, as calculated in part **(b)**. Thus the speed, c, of these atoms is

$$c = \left(\frac{15.0 \text{ cm}}{4.57 \times 10^{-4} \text{ s}}\right)\left(\frac{1 \text{ m}}{100 \text{ cm}}\right) = 328 \text{ m s}^{-1}$$

This can be compared with

$$c_{rms} = \sqrt{\frac{3RT}{M}} = \sqrt{\frac{3(8.314 \text{ J K}^{-1} \text{ mol}^{-1})(273+850)\text{K}}{209.0 \times 10^{-3} \text{ kg mol}^{-1}}} = 366 \text{ m s}^{-1}$$

The difference is not unexpected, since c_{rms} is an average speed of the sample of atoms (assuming it is thermally equilibrated) while the speed calculated in part **(c)** is the speed of a particular group of atoms. Nevertheless, the magnitudes of the two speeds are comparable. Furthermore, the measured speed is quite close to

$$c_{mp} = \sqrt{\frac{8RT}{\pi M}} = \sqrt{\frac{8(8.314 \text{ J K}^{-1} \text{ mol}^{-1})(273+850)\text{K}}{\pi(209.0 \times 10^{-3} \text{ kg mol}^{-1})}} = 337 \text{ m s}^{-1}$$

which would be expected to give the most prominent spot in the measurement.

3.55 From your knowledge of heat capacity, explain why hot, humid air is more uncomfortable than hot, dry air and cold, damp air is more uncomfortable than cold, dry air.

H_2O is a non-linear, triatomic molecule with a heat capacity calculated (ignoring vibrational contributions) to be $\overline{C}_V = 3R$, while N_2 and O_2 are diatomic molecules with $\overline{C}_V = \frac{5}{2}R$. The larger heat capacity of water means that more energy is required to effect a given temperature change. Thus, hot, humid air with its greater content of water vapor, is capable of transferring more thermal energy to your body than the diatomic gases, making hot, humid air more uncomfortable than hot, dry air. (Of course, the retardation of the evaporation rate from your skin is another significant reason.) On the other hand, the greater water content in cold damp air is more able to remove thermal energy from your body than cold, dry air can.

3.56 The escape velocity, v, from Earth's gravitational field is given by $(2GM/r)^{1/2}$, where G is the universal gravitational constant (6.67×10^{-11} m^3 kg^{-1} s^{-2}), M is the mass of Earth (6.0×10^{24} kg), and r is the distance from the center of Earth to the object, in meters. Compare the average speeds of He and N_2 molecules in the thermosphere (altitude about 100 km, $T = 250$ K). Which of the two molecules will have a greater tendency to escape? The radius of Earth is 6.4×10^6 m.

Although the molecules are 100 km above the Earth's surface, this quantity is insignificant when compared with the radius of Earth. Therefore, the distance between molecules in the thermosphere and the center of Earth (r) is still 6.4×10^6 m. The escape velocity is

$$v = \sqrt{\frac{2G\mathcal{M}}{r}} = \sqrt{\frac{2\left(6.67 \times 10^{-11}\ \text{m}^3\ \text{kg}^{-1}\ \text{s}^{-2}\right)\left(6.0 \times 10^{24}\ \text{kg}\right)}{6.4 \times 10^6\ \text{m}}} = 1.1 \times 10^4\ \text{m s}^{-1}$$

The average speed of He molecules is

$$\bar{c} = \sqrt{\frac{8RT}{\pi \mathcal{M}}} = \sqrt{\frac{8\left(8.314\ \text{J K}^{-1}\ \text{mol}^{-1}\right)(250\ \text{K})}{\pi \left(4.003 \times 10^{-3}\ \text{kg mol}^{-1}\right)}} = 1.15 \times 10^3\ \text{m s}^{-1}$$

The average speed of N_2 molecules is

$$\bar{c} = \sqrt{\frac{8RT}{\pi \mathcal{M}}} = \sqrt{\frac{8\left(8.314\ \text{J K}^{-1}\ \text{mol}^{-1}\right)(250\ \text{K})}{\pi \left(28.02 \times 10^{-3}\ \text{kg mol}^{-1}\right)}} = 435\ \text{m s}^{-1}$$

Recalling the form of the Maxwell distribution of molecular speeds, there are indeed molecules moving faster than the average, and although the high speed "tail" of the distribution extends out to very large speeds, it does approach the x-axis fairly rapidly. Thus, there will be many more He atoms with sufficient speed to escape the Earth's gravitational field than there are N_2 molecules. Indeed, it is for this reason that there are no appreciable amounts of He or H_2 in the Earth's atmosphere.

3.57 Suppose that you are traveling in a space vehicle on a journey to the moon. The atmosphere in the vehicle consists of 20% oxygen and 80% helium by volume. Before take off someone noticed a small leakage that, if left unchecked, would lead to a continual loss of gas at a rate of 0.050 atm day^{-1} through effusion. If the temperature in the space vehicle is maintained at 22°C and the volume of the vehicle is 1.5×10^4 L, calculate the amounts of helium and oxygen in grams that must be stored on a 10-day journey to allow for the leakage. (*Hint:* First calculate the quantity of gas lost each day, using $PV = nRT$. Note that the rate of effusion is proportional to the pressure of the gas. Assume that effusion does not affect the pressure or the mean free path of the gas in the space vehicle.)

Since the temperature and the volume in the space vehicle are kept constant, the change in the number of moles of gas due to the leak is directly related to the pressure loss,

$$\Delta n_{\text{total}} = \Delta n_{\text{He}} + \Delta n_{\text{O}_2}$$

$$= \frac{V}{RT}\Delta P$$

$$= \frac{1.5 \times 10^4\ \text{L}}{(0.08206\ \text{L atm K}^{-1}\ \text{mol}^{-1})(273 + 22)\text{K}}\,(0.050\ \text{atm})$$

$$= 31.0\ \text{mol}$$

With 4 times as much He as O_2, $P_{He} = 4P_{O_2}$, which must be considered in addition to the different molar masses of the two species in calculating the relative rates of effusion,

$$\frac{r_{He}}{r_{O_2}} = 4\sqrt{\frac{32.00 \text{ g mol}^{-1}}{4.003 \text{ g mol}^{-1}}} = 11.31$$

That is, in any given time period, 11.31 times as much He effuses through the leak as does O_2, or $\Delta n_{He} = 11.31 \Delta n_{O_2}$. From above,

$$\Delta n_{He} + \Delta n_{O_2} = 31.0 \text{ mol}$$

$$11.31 \Delta n_{O_2} + \Delta n_{O_2} = 31.0 \text{ mol}$$

$$12.31 \Delta n_{O_2} = 31.0 \text{ mol}$$

$$\Delta n_{O_2} = 2.52 \text{ mol}$$

and

$$\Delta n_{He} = 28.5 \text{ mol}$$

Thus for each day, $(2.52 \text{ mol})(32.00 \text{ g mol}^{-1}) = 81$ g of O_2 and $(28.5 \text{ mol})(4.003 \text{ g mol}^{-1}) = 1.1 \times 10^2$ g of He are lost. On a 10-day mission, 810 g of O_2 and 1.1×10^3 g of He would have to be stored to allow for this leakage.

3.58   Calculate the ratio of the number of O_3 molecules with a speed of 1300 m s^{-1} at 360 K to the number with that speed at 293 K.

This problem is very similar to Problem 3.21, except that problem wanted a ratio of the numbers of molecules with two different speeds at the same temperature. Here, the ratio of the numbers of molecules with the same speed, but at two different temperatures is desired.

$$\frac{dN_2/N}{dN_1/N} = \frac{4\pi c^2 \left(\frac{m}{2\pi k_B T_2}\right)^{\frac{3}{2}} e^{-\frac{mc^2}{2k_B T_2}} dc}{4\pi c^2 \left(\frac{m}{2\pi k_B T_1}\right)^{\frac{3}{2}} e^{-\frac{mc^2}{2k_B T_1}} dc}$$

$$= \left(\frac{T_1}{T_2}\right)^{3/2} e^{-\frac{mc^2}{2k_B}\left(\frac{1}{T_2} - \frac{1}{T_1}\right)}$$

$$= \left(\frac{293 \text{ K}}{360 \text{ K}}\right)^{\frac{3}{2}} e^{-\frac{(48.00 \text{ amu})(1.661 \times 10^{-27} \text{ kg amu}^{-1})(1300 \text{ m s}^{-1})^2}{2(1.381 \times 10^{-23} \text{ J K}^{-1})}\left(\frac{1}{360 \text{ K}} - \frac{1}{293 \text{ K}}\right)}$$

$$= 16.3$$

3.59   Calculate the collision frequency for 1.0 mole of krypton (Kr) at equilibrium at 300 K and 1.0 atm pressure. Which of the following increases the collision frequency more: **(a)** doubling the temperature at constant pressure or **(b)** doubling the pressure at constant temperature? (*Hint*: Use the collision diameter in Table 3.1.)

The collision frequency can be determined once the average speed and number density are calculated. The collision diameter for Kr is 4.16 Å (Table 3.1).

$$\bar{c} = \sqrt{\frac{8RT}{\pi M}} = \sqrt{\frac{8\left(8.314\ \text{J K}^{-1}\ \text{mol}^{-1}\right)(300\ \text{K})}{\pi\left(83.80 \times 10^{-3}\ \text{kg mol}^{-1}\right)}} = 275.3\ \text{m s}^{-1}$$

$$\frac{N}{V} = \frac{PN_A}{RT}\ \text{(See Problem 3.26)}$$

$$= \frac{(1.0\ \text{atm})\left(6.022 \times 10^{23}\ \text{mol}^{-1}\right)}{\left(0.08206\ \text{L atm K}^{-1}\ \text{mol}^{-1}\right)(300\ \text{K})} = \left(2.45 \times 10^{22}\ \text{L}^{-1}\right)\left(\frac{1000\ \text{L}}{1\ \text{m}^3}\right)$$

$$= 2.45 \times 10^{25}\ \text{m}^{-3}$$

$$Z_1 = \sqrt{2}\pi d^2 \bar{c}\frac{N}{V} = \sqrt{2}\pi\left(4.16 \times 10^{-10}\ \text{m}\right)^2\left(275.3\ \text{m s}^{-1}\right)\left(2.45 \times 10^{25}\ \text{m}^{-3}\right)$$

$$= 5.2 \times 10^9\ \text{collisions s}^{-1}$$

(a) A doubling in temperature will increase the average molecule speed by a factor of $\sqrt{2}$, but will *decrease* the number density by a factor of two, leading to an overall decrease in collision frequency.

(b) Doubling the pressure will double the number density (N/V) of the gas which will in turn double the collision frequency.

3.60 Apply your knowledge of the kinetic theory of gases to the following situations. (a) Two flasks of volumes V_1 and V_2 (where $V_2 > V_1$) contain the same number of helium atoms at the same temperature. (i) Compare the root-mean-square (rms) speeds and average kinetic energies of the helium (He) atoms in the flasks. (ii) Compare the frequency and the force with which the He atoms collide with the walls of their containers. (b) Equal numbers of He atoms are placed in two flasks of the same volume at temperatures T_1 and T_2 (where $T_2 > T_1$). (i) Compare the rms speeds of the atoms in the two flasks. (ii) Compare the frequency and the force with which the He atoms collide with the walls of their containers. (c) Equal numbers of He and neon (Ne) atoms are placed in two flasks of the same volume. The temperature of both gases is 74°C. Comment on the validity of the following statements: (i) The rms speed of He is equal to that of Ne. (ii) The average kinetic energies of the two gases are equal. (iii) The rms speed of each He atom is $1.47 \times 10^3\ \text{m s}^{-1}$.

(a) With the two samples at the same temperature, (i) c_{rms} and average kinetic energies of the atoms in the two flasks are the same. Likewise, (ii) the force with which the He atoms strike the walls of the containers is the same in each flask, but the frequency of collision is greater in the smaller flask, V_1. It will have the higher pressure.

(b) Now at the same volume, but different temperatures, (i) c_{rms} is greater in the flask with the higher temperature, T_2, and (ii) in the flask with the higher temperature, T_2, the atoms collide with the walls both with greater force and with greater frequency than in the lower temperature sample.

(c)(i) The statement is false, since the lighter He atoms have a greater speed.

(c)(ii) True, since the samples are at the same temperature, the average kinetic energies of their atoms are the same.

(c)(iii) True, since

$$c_{rms} = \sqrt{\frac{3RT}{\mathcal{M}}}$$

$$= \sqrt{\frac{3(8.314\,\text{J K}^{-1}\,\text{mol}^{-1})(273 + 74)\text{K}}{4.003 \times 10^{-3}\,\text{kg mol}^{-1}}}$$

$$= 1.47 \times 10^3\,\text{m s}^{-1}$$

CHAPTER 4

The First Law of Thermodynamics

PROBLEMS AND SOLUTIONS

4.1 Explain the term *state function*. Which of the following are state functions? P, V, T, w, q.

A state function is a property that is determined only by the state of a system.

P, V, and T are state functions. q and w are not.

4.2 What is heat? How does heat differ from thermal energy? Under what condition is heat transferred from one system to another?

Heat is the transfer of energy from one object to another as a result of a temperature difference between the two objects. Heat is only transferred between systems when they are at different temperatures.

Thermal energy is that part of the energy of a system that is associated with the random motion (translational, vibrational, and rotational) of atoms and molecules.

4.3 Show that 1 L atm = 101.3 J.

The units conversion can be performed using R expressed in L atm and J, respectively.

$$R = 0.08206 \text{ L atm K}^{-1} \text{ mol}^{-1} = 8.314 \text{ J K}^{-1} \text{ mol}^{-1}$$

$$1 \text{ L atm} = \frac{8.314 \text{ J}}{0.08206} = 101.3 \text{ J}$$

4.4 A 7.24-g sample of ethane occupies 4.65 L at 294 K. **(a)** Calculate the work done when the gas expands isothermally against a constant external pressure of 0.500 atm until its volume is 6.87 L. **(b)** Calculate the work done if the same expansion occurs reversibly.

(a)

$$w = -P_{ex}\Delta V = -(0.500 \text{ atm})(6.87 \text{ L} - 4.65 \text{ L})\left(\frac{101.3 \text{ J}}{1 \text{ L atm}}\right) = -112 \text{ J}$$

(b)

$$w = -nRT \ln\frac{V_2}{V_1} = -\frac{7.24 \text{ g}}{30.07 \text{ g mol}^{-1}}(8.314 \text{ J K}^{-1} \text{ mol}^{-1})(294 \text{ K})\ln\frac{6.87 \text{ L}}{4.65 \text{ L}} = -230 \text{ J}$$

Note that work done in the reversible process in **(b)** is greater in magnitude than that in the irreversible process in **(a)**.

4.5 A 19.2-g quantity of dry ice (solid carbon dioxide) is allowed to sublime (evaporate) in an apparatus like the one shown in Figure 4.1. Calculate the expansion work done against a constant external pressure of 0.995 atm and at a constant temperature of 22°C. Assume that the initial volume of dry ice is negligible and that CO_2 behaves like an ideal gas.

Because the initial volume of dry ice is negligible, $V_1 = 0$. Expansion occurs until the internal pressure is the same as the external pressure, that is, $P_2 = P_{ex}$. The final volume, V_2, occupied by CO_2 can be calculated using the ideal gas law:

$$V_2 = \frac{nRT}{P_2} = \frac{nRT}{P_{ex}}$$

The work done can now be calculated.

$$w = -P_{ex}\Delta V = -P_{ex}(V_2 - V_1) = -P_{ex}V_2$$

$$= -P_{ex}\left(\frac{nRT}{P_{ex}}\right) = -nRT$$

$$= -\left(\frac{19.2 \text{ g}}{44.01 \text{ g mol}^{-1}}\right)(8.314 \text{ J K}^{-1} \text{ mol}^{-1})(295 \text{ K})$$

$$= -1.07 \times 10^3 \text{ J}$$

4.6 Calculate the work done by the reaction

$$Zn(s) + H_2SO_4(aq) \rightarrow ZnSO_4(aq) + H_2(g)$$

when 1 mole of hydrogen gas is collected at 273 K and 1.0 atm. (Neglect volume changes other than the change in gas volume.)

The gas expands until its pressure is the same as the external pressure of 1.0 atm. Furthermore, since volume changes other than gas are neglected, the change in volume of the system is

$$\Delta V = V_{H_2} = \frac{nRT}{P_{H_2}} = \frac{nRT}{P_{ex}}$$

The work done is

$$w = -P_{ex}\Delta V = -P_{ex}V_f = -P_{ex}\left(\frac{nRT}{P_{ex}}\right) = -nRT$$

$$= -(1 \text{ mol})\left(8.314 \text{ J K}^{-1} \text{ mol}^{-1}\right)(273 \text{ K}) = -2.27 \times 10^3 \text{ J}$$

4.7   A truck traveling 60 kilometers per hour is brought to a complete stop at a traffic light. Does this change in velocity violate the law of conservation of energy?

The mechanical (kinetic) energy of the truck is converted to thermal energy as its brakes are warmed through friction. The loss of one form of energy is always balanced by an increase in some other form.

4.8   Some driver's test manuals state that the stopping distance quadruples as the velocity doubles. Justify this statement by using mechanics and thermodynamic arguments.

The kinetic energy of a moving vehicle is given by $E_{kin} = \frac{1}{2}mv^2$. If the velocity doubles, the kinetic energy quadruples. If $v_2 = 2v_1$, then $E_2 = \frac{1}{2}m\left(2v_1\right)^2 = 4E_1$. Assuming kinetic energy is dissipated at a constant rate in the brakes and between the tires and the road as heat through friction, then the stopping distance is proportional to the energy. The doubling in velocity quadruples the kinetic energy and thus the stopping distance.

4.9   Provide a first law analysis for each of the following: **(a)** When a bicycle tire is inflated with a hand pump, the temperature inside rises. You can feel the warming effect at the valve stem. **(b)** Artificial snow is made by quickly releasing a mixture of compressed air and water vapor at about 20 atm from a snow-making machine to the surroundings.

In both cases it is assumed that the process is occurring rapidly enough so that there is no transfer of heat between the system and the surroundings. That is, the process is adiabatic and $q = 0$.

(a) Work is done on the gas (the system) when it is compressed by the bicycle pump. $\Delta U = q + w = w$, since $q = 0$. The internal energy of the gas is increased. For a diatomic, ideal gas, such as N_2 or O_2, which are the major components of air, $\Delta U = \frac{5}{2}nR\Delta T$, where n is the number of moles of gas. Since $\Delta U > 0$, $\Delta T > 0$, also.

(b) This is just the opposite of part **(a)**. The expanding gas does work on the surroundings, so $\Delta U < 0$, and $\Delta T < 0$, or the system cools.

4.10   An ideal gas is compressed isothermally by a force of 85 newtons acting through 0.24 meter. Calculate the values of ΔU and q.

$\Delta U = 0$ for an ideal gas undergoing an isothermal process. Using this information, q can be calculated from w, the work done on the system.

$$w = \text{(force) (distance)} = (85\ \text{N})\ (0.24\ \text{m}) = 20\ \text{J}$$

Because $\Delta U = 0 = q + w$, $q = -w = -20\ \text{J}$.

4.11 Calculate the internal energy of 2 moles of argon gas (assuming ideal behavior) at 298 K. Suggest two ways to increase its internal energy by 10 J.

The only internal energy that argon gas possesses is translational energy. Argon has 3 translational degrees of freedom. Therefore,

$$U = \frac{3}{2}nRT = \frac{3}{2}\ (2\ \text{mol})\ \left(8.314\ \text{J K}^{-1}\ \text{mol}^{-1}\right)\ (298\ \text{K}) = 7.43 \times 10^3\ \text{J}$$

The internal energy of argon gas can be increased by doing work on the gas or heating the gas or a combination of these two processes. The first process can be achieved by compressing the gas adiabatically such that $w = 10$ J and $q = 0$. The second process can be achieved by heating the gas at constant volume such that $q = 10$ J and $w = 0$ J.

4.12 A thermos bottle containing milk is shaken vigorously. Consider the milk as the system. **(a)** Will the temperature rise as a result of the shaking? **(b)** Has heat been added to the system? **(c)** Has work been done on the system? **(d)** Has the system's internal energy changed?

The thermos bottle ensures that $q = 0$, since it prevents the transfer of heat to or from the surroundings.

(a) Energy (mechanical) has been added to the system by shaking. The random motion of the molecules in the milk is increased, and the temperature rises.

(b) The insulation of the thermos bottle prevents the transfer of heat between the system and surroundings.

(c) Work has been done on the system through shaking.

(d) The internal energy of the system has increased.

4.13 A 1.00-mole sample of ammonia at 14.0 atm and 25°C in a cylinder fitted with a movable piston expands against a constant external pressure of 1.00 atm. At equilibrium, the pressure and volume of the gas are 1.00 atm and 23.5 L, respectively. **(a)** Calculate the final temperature of the sample. **(b)** Calculate the values of q, w, and ΔU for the process.

(a) Assume ammonia is an ideal gas. The final temperature is

$$T_2 = \frac{P_2 V_2}{nR} = \frac{(1.00\ \text{atm})\ (23.5\ \text{L})}{(1.00\ \text{mol})\ \left(0.08206\ \text{L atm K}^{-1}\ \text{mol}^{-1}\right)} = 286\ \text{K}$$

(b) The problem provides enough information to calculate w and ΔU directly. Consequently, q can be determined using the first law of thermodynamics.

w is related to the change in volume of the system. The final volume is given, but the initial volume needs to be calculated from the ideal gas law:

$$V_1 = \frac{nRT_1}{P_1} = \frac{(1.00 \text{ mol}) \left(0.08206 \text{ L atm K}^{-1} \text{ mol}^{-1}\right) (298 \text{ K})}{14.0 \text{ atm}} = 1.75 \text{ L}$$

The work done is

$$w = -P_{ex}\Delta V = -(1.00 \text{ atm}) (23.5 \text{ L} - 1.75 \text{ L}) \left(\frac{101.3 \text{ J}}{1 \text{ L atm}}\right) = -2.20 \times 10^3 \text{ J}$$

ΔU is related to C_V, which can be calculated from $\overline{C}_P = 35.66 \text{ J K}^{-1} \text{ mol}^{-1}$ listed in Appendix B. It is assumed that C_V and C_P are independent of temperature.

$$C_V = C_P - nR = (1.00 \text{ mol}) \left(35.66 \text{ J K}^{-1} \text{ mol}^{-1}\right) - (1.00 \text{ mol}) \left(8.314 \text{ J K}^{-1} \text{ mol}^{-1}\right) = 27.35 \text{ J K}^{-1}$$

The change in internal energy is

$$\Delta U = C_V \Delta T = \left(27.35 \text{ J K}^{-1}\right) (286 \text{ K} - 298 \text{ K}) = -3.3 \times 10^2 \text{ J}$$

Using the first law of thermodynamics,

$$q = \Delta U - w = -3.3 \times 10^2 \text{ J} - \left(-2.20 \times 10^3 \text{ J}\right) = 1.87 \times 10^3 \text{ J}$$

4.14 An ideal gas is compressed isothermally from 2.0 atm and 2.0 L to 4.0 atm and 1.0 L. Calculate the values of ΔU and ΔH if the process is carried out **(a)** reversibly and **(b)** irreversibly.

ΔU and ΔH of an ideal gas depend only on T. Therefore, for any isothermal process [either process **(a)** or **(b)**], $\Delta U = 0$ and $\Delta H = 0$.

4.15 Explain the energy changes at the molecular level when liquid acetone is converted to vapor at its boiling point.

The thermal energy supplied as heat is used to overcome the attractive, intermolecular forces between the acetone molecules. The molecules then break free of the condensed, liquid phase and enter the gas phase. Since the added energy is converted to potential and not kinetic energy of the molecules, the temperature remains constant.

4.16 A piece of potassium metal is added to water in a beaker. The reaction that takes place is

$$2K(s) + 2H_2O(l) \rightarrow 2KOH(aq) + H_2(g)$$

Predict the signs of w, q, ΔU, and ΔH.

---

w is negative because the hydrogen gas produced expands and does work on the surroundings.

q is negative because thermal energy is generated in this reaction.

Since $\Delta U = q + w$, it must be negative also.

The process takes place at constant pressure, therefore $q = q_P = \Delta H$. Thus, ΔH is negative.

4.17 At 373.15 K and 1 atm, the molar volume of liquid water and steam are 1.88×10^{-5} m^3 and 3.06×10^{-2} m^3, respectively. Given that the heat of vaporization of water is 40.79 kJ mol^{-1}, calculate the values of ΔH and ΔU for 1 mole in the following process:

$$H_2O(l, \text{373.15 K, 1 atm}) \rightarrow H_2O(g, \text{373.15 K, 1 atm})$$

ΔH for the above process is the heat of vaporization, that is, $\Delta H = 40.79$ kJ for 1 mole of water.

It is necessary to calculate w and q before determining ΔU. Since the process occurs at constant pressure, $q = \Delta H = 40.79$ kJ when 1 mol liquid H$_2$O vaporizes. In the same process,

$$w = -P_{ex}\Delta V$$

$$= -(1.00 \text{ atm}) \left(3.06 \times 10^{-2} \text{ m}^3 - 1.88 \times 10^{-5} \text{ m}^3\right) \left(\frac{1000 \text{ L}}{1 \text{ m}^3}\right) \left(\frac{101.3 \text{ J}}{1 \text{ L atm}}\right)$$

$$= -3.098 \times 10^3 \text{ J}$$

Note that we could have safely ignored the volume of liquid H$_2$O, since it is negligible compared with that of gaseous H$_2$O above.

Using the first law,

$$\Delta U = q + w = 40.79 \text{ kJ} - 3.098 \text{ kJ} = 37.69 \text{ kJ}$$

4.18 Consider a cyclic process involving a gas. If the pressure of the gas varies during the process but returns to the original value at the end, is it correct to write $\Delta H = q_P$?

No, it is not correct. For a cyclic process, $\Delta H = 0$, since H is a state function, but the heat, or q, associated with a process is not a state function and depends on the path. Depending on the path used to achieve the cyclic process it may take on a variety of values.

4.19 Calculate the value of ΔH when the temperature of 1 mole of a monatomic gas is increased from 25°C to 300°C.

ΔH is directly related to change in temperature of a system:

$$\Delta H = C_P \Delta T$$

Assuming that the gas is ideal,

$$C_P = C_V + nR$$

A monatomic gas has only translational degrees of freedom, $C_V = \frac{3}{2}nR$. Therefore, $C_P = \frac{5}{2}nR$ and

$$\Delta H = \frac{5}{2} \left(1 \text{ mol}\right) \left(8.314 \text{ J K}^{-1} \text{ mol}^{-1}\right) (573 \text{ K} - 298 \text{ K}) = 5.72 \times 10^3 \text{ J}$$

4.20   One mole of an ideal gas undergoes an isothermal expansion at 300 K from 1.00 atm to a final pressure while performing 200 J of expansion work. Calculate the final pressure of the gas if the external pressure is 0.20 atm.

$$w = -P_{\text{ex}}\left(V_2 - V_1\right) = -P_{\text{ex}}\left(\frac{nRT}{P_2} - \frac{nRT}{P_1}\right)$$

$$= -P_{\text{ex}}nRT\left(\frac{1}{P_2} - \frac{1}{P_1}\right)$$

$$\frac{1}{P_2} = -\frac{w}{P_{\text{ex}}nRT} + \frac{1}{P_1}$$

$$= -\frac{-200 \text{ J}}{(0.20 \text{ atm})\,(1 \text{ mol})\left(8.314 \text{ J K}^{-1} \text{ mol}^{-1}\right)(300 \text{ K})} + \frac{1}{1.00 \text{ atm}} = 1.401 \text{ atm}^{-1}$$

$$P_2 = 0.71 \text{ atm}$$

4.21   A 6.22-kg piece of copper metal is heated from 20.5°C to 324.3 °C. Given that the specific heat of Cu is 0.385 J g^{-1} °C^{-1}, calculate the heat absorbed (in kJ) by the metal.

$$q = ms\,\Delta T = \left(6.22 \times 10^3 \text{ g}\right)\left(0.385 \text{ J g}^{-1}\text{°C}^{-1}\right)(324.3 \text{ °C} - 20.5 \text{ °C}) = 7.28 \times 10^5 \text{ J}$$

4.22   A 10.0-g sheet of gold with a temperature of 18.0°C is laid flat on a sheet of iron that weighs 20.0 g and has a temperature of 55.6°C. Given that the specific heats of Au and Fe are 0.129 J g^{-1} °C^{-1} and 0.444 J g^{-1} °C^{-1}, respectively, what is the final temperature of the combined metals? Assume that no heat is lost to the surroundings. (*Hint*: The heat gained by the gold must be equal to the heat lost by the iron.)

The final temperature of the sheet of gold is the same as that of the sheet of iron when thermal equilibrium is reached, and is denoted by T_f. Furthermore, the amount of heat gained by gold is the same as that lost by iron, that is,

$$q_{\text{Au}} = -q_{\text{Fe}}$$

The "−" sign is used to indicate that q_{Au} and q_{Fe} are of opposite sign.

The above relation gives

$$m_{Au}s_{Au}\Delta T_{Au} = -m_{Fe}s_{Fe}\Delta T_{Fe}$$

$$(10.0 \text{ g}) \left(0.129 \text{ J g}^{-1}\,^\circ\text{C}^{-1}\right) \left(T_f - 18.0^\circ\text{C}\right) = - (20.0 \text{ g}) \left(0.444 \text{ J g}^{-1}\,^\circ\text{C}^{-1}\right) \left(T_f - 55.6^\circ\text{C}\right)$$

$$\left(1.29 \text{ J}\,^\circ\text{C}^{-1}\right) \left(T_f - 18.0^\circ\text{C}\right) = - \left(8.88 \text{ J}\,^\circ\text{C}^{-1}\right) \left(T_f - 55.6^\circ\text{C}\right)$$

$$T_f - 18.0^\circ\text{C} = -\frac{8.88 \text{ J}\,^\circ\text{C}^{-1}}{1.29 \text{ J}\,^\circ\text{C}^{-1}} \left(T_f - 55.6^\circ\text{C}\right)$$

$$= -6.884 \left(T_f - 55.6^\circ\text{C}\right)$$

$$= -6.884 T_f + 382.8^\circ\text{C}$$

$$7.884 T_f = 400.8^\circ\text{C}$$

$$T_f = 50.8^\circ\text{C}$$

4.23 It takes 330 joules of energy to raise the temperature of 24.6 g of benzene from 21.0°C to 28.7°C at constant pressure. What is the molar heat capacity of benzene at constant pressure?

From the relationship

$$q = C_P \Delta T = n\overline{C}_P \Delta T$$

$\overline{C}_P$ of benzene can be determined:

$$\overline{C}_P = \frac{q}{n\Delta T} = \frac{330 \text{ J}}{\left(\dfrac{24.6 \text{ g}}{78.11 \text{ g mol}^{-1}}\right)(28.7\,^\circ\text{C} - 21.0\,^\circ\text{C})} = 136 \text{ J mol}^{-1}\,^\circ\text{C}^{-1}$$

4.24 The molar heat of vaporization for water is 44.01 kJ mol^{-1} at 298 K and 40.79 kJ mol^{-1} at 373 K. Give a qualitative explanation of the difference in these two values.

At the higher temperature, the water molecules have more kinetic energy and are moving faster. Thus, there are on average fewer hydrogen bonds holding the individual water molecules together. Consequently at the higher temperature, less thermal energy must be supplied to break the remaining intermolecular attractions and allow the molecules to enter the gas phase.

4.25 The constant-pressure molar heat capacity of nitrogen is given by the expression

$$\overline{C}_P = \left(27.0 + 5.90 \times 10^{-3}T - 0.34 \times 10^{-6}T^2\right) \text{ J K}^{-1}\text{ mol}^{-1}$$

Calculate the value of ΔH for heating 1 mole of nitrogen from 25.0°C to 125°C.

Since $\overline{C}_P$ is temperature dependent, it has to be integrated over T to yield ΔH:

$$\Delta H = \int_{298 \text{ K}}^{398 \text{ K}} n \overline{C}_P \, dT$$

$$= \int_{298 \text{ K}}^{398 \text{ K}} (1 \text{ mol}) \left[\left(27.0 + 5.90 \times 10^{-3} T - 0.34 \times 10^{-6} T^2 \right) \text{ J K}^{-1} \text{ mol}^{-1} \right] dT$$

$$= \left[27.0 T + 5.90 \times 10^{-3} \frac{T^2}{2} - 0.34 \times 10^{-6} \frac{T^3}{3} \right]_{298 \text{ K}}^{398 \text{ K}} \text{ J}$$

$$= \left[27.0 \,(398) + 5.90 \times 10^{-3} \frac{398^2}{2} - 0.34 \times 10^{-6} \frac{398^3}{3} \right] \text{ J}$$

$$- \left[27.0 \,(298) + 5.90 \times 10^{-3} \frac{298^2}{2} - 0.34 \times 10^{-6} \frac{298^3}{3} \right] \text{ J}$$

$$= 2.90 \times 10^3 \text{ J}$$

4.26   The heat capacity ratio (γ) for a gas with the molecular formula X_2Y is 1.38. What can you deduce about the structure of the molecule?

The molecule X_2Y is either linear or bent (*e.g.* H_2O or CO_2). Since $\overline{C}_P = \overline{C}_V + R$, the heat capacity ratio may be rewritten as

$$\gamma = \frac{\overline{C}_P}{\overline{C}_V}$$

$$= \frac{\overline{C}_V + R}{\overline{C}_V}$$

This may be solved for $\overline{C}_V$ to give

$$\overline{C}_V = \frac{R}{\gamma - 1}$$

$$= \frac{R}{0.38}$$

$$= 2.6R$$

At typical temperatures the vibrational contribution to the heat capacity is negligible, and for a linear molecule $\overline{C}_V = 2.5R$, while for a bent molecule, $\overline{C}_V = 3.0R$. Better agreement between prediction and measurement is obtained in the case of the linear molecule, and thus the molecule is most likely linear.

4.27   One way to measure the heat capacity ratio (γ) of a gas is to measure the speed of sound in the gas (c), which is given by

$$c = \left(\frac{\gamma RT}{\mathcal{M}}\right)^{1/2}$$

where $\mathcal{M}$ is the molar mass of the gas. Calculate the speed of sound in helium at 25°C.

First calculate γ. Since helium is a monatomic gas with only translational degrees of freedom, $C_V = \frac{3}{2}nR$. Assuming helium is ideal,

$$C_P = C_V + nR = \frac{3}{2}nR + nR = \frac{5}{2}nR$$

Therefore, the heat capacity ratio is

$$\gamma = \frac{C_P}{C_V} = \frac{\frac{5}{2}nR}{\frac{3}{2}nR} = \frac{5}{3}$$

Substitute γ into the expression for c:

$$c = \left(\frac{\gamma RT}{\mathcal{M}}\right)^{1/2} = \left[\frac{\left(\frac{5}{3}\right)\left(8.314\ \mathrm{J\,K^{-1}\,mol^{-1}}\right)(298\ \mathrm{K})}{4.003 \times 10^{-3}\ \mathrm{kg\,mol^{-1}}}\right]^{1/2} = 1.02 \times 10^3\ \mathrm{m\,s^{-1}}$$

4.28   Which of the following gases has the largest $\overline{C}_V$ value at 298 K? He, N_2, CCl_4, HCl.

A gas composed of non-linear, polyatomic molecules is predicted to have a larger $\overline{C}_V$ than monoatomic gases or those made up of linear molecules. In the list above, only CCl_4 is a non-linear molecule, so it should have the largest $\overline{C}_V$.

4.29   **(a)** For most efficient use, refrigerator freezer compartments should be fully packed with food. What is the thermochemical basis for this recommendation? **(b)** Starting at the same temperature, tea and coffee remain hot longer in a thermal flask than soup. Explain.

(a) The heat capacity depends on both the mass and the specific heat, $C = ms$. The more tightly packed the freezer, the greater the mass of food in it. Since the food has a greater heat capacity than the air it displaces, it will take more thermal energy to raise the temperature of the tightly packed freezer. This will require a greater time for the temperature of the freezer contents to rise a given amount.

(b) Water has a larger heat capacity than most solids, such as the meat, vegetables, etc. in soup. Coffee and tea are mostly water, so to the extent that these other ingredients take the place of some water in the soup, coffee or tea would have a larger heat capacity than soup and thus would need to lose more thermal energy for the same temperature change. To the extent that the rate of thermal energy loss from the thermal flask is constant, it will take longer for the coffee or tea to cool down than the soup (unless the soup is a clear broth).

4.30 In the nineteenth century, two scientists named Dulong and Petit noticed that the product of the molar mass of a solid element and its specific heat is approximately $25 \text{ J} \, ^\circ \text{C}^{-1}$. This observation, now called Dulong and Petit's law, was used to estimate the specific heat of metals. Verify the law for aluminum ($0.900 \text{ J g}^{-1} \, ^\circ \text{C}^{-1}$), copper ($0.385 \text{ J g}^{-1} \, ^\circ \text{C}^{-1}$), and iron ($0.444 \text{ J g}^{-1} \, ^\circ \text{C}^{-1}$). The law does not apply to one of the metals. Which one is it? Why?

Al:

$$(26.98 \text{ g}) \left(0.900 \text{ J g}^{-1} \, ^\circ \text{C}^{-1}\right) = 24.3 \text{ J} \, ^\circ \text{C}^{-1}$$

Cu:

$$(63.55 \text{ g}) \left(0.385 \text{ J g}^{-1} \, ^\circ \text{C}^{-1}\right) = 24.5 \text{ J} \, ^\circ \text{C}^{-1}$$

Fe:

$$(55.85 \text{ g}) \left(0.444 \text{ J g}^{-1} \, ^\circ \text{C}^{-1}\right) = 24.8 \text{ J} \, ^\circ \text{C}^{-1}$$

Therefore, Dulong and Petit's law applies to Al, Cu, and Fe. However, since this law applies only to solids, it does not apply to mercury, a liquid. In fact, the product of the molar mass of Hg and its specific heat is

$$(200.6 \text{ g}) \left(0.139 \text{ J g}^{-1} \, ^\circ \text{C}^{-1}\right) = 27.9 \text{ J} \, ^\circ \text{C}^{-1}$$

4.31 The following diagram represents the P–V changes of a gas. Write an expression for the total work done.

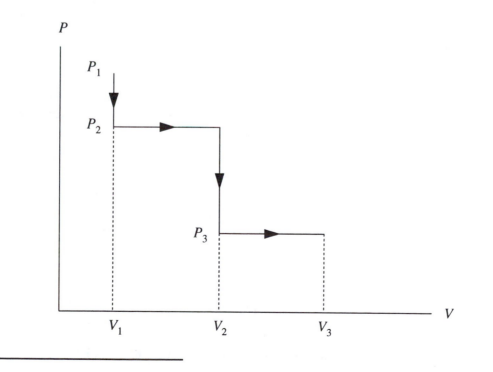

The total work done is the sum of the work done against the constant external pressure P_2 when the system expands from V_1 to V_2, and that against the constant external pressure P_3 when the system expands from V_2 to V_3:

$$w = -P_2 (V_2 - V_1) - P_3 (V_3 - V_2)$$

4.32   The equation of state for a certain gas is given by $P[(V/n) - b] = RT$. Obtain an expression for the maximum work done by the gas in a reversible isothermal expansion from V_1 to V_2.

Write P in terms of n, V, and T:

$$P = \frac{RT}{\frac{V}{n} - b} = \frac{nRT}{V - nb}$$

The maximum work done by the gas undergoing an isothermal expansion is

$$w = -\int_{V_1}^{V_2} P \, dV = -\int_{V_1}^{V_2} \frac{nRT}{V - nb} \, dV$$

$$= -nRT \ln (V - nb)\big|_{V_1}^{V_2}$$

$$= -nRT \ln \frac{V_2 - nb}{V_1 - nb}$$

4.33   Calculate the values of q, w, ΔU, and ΔH for the reversible adiabatic expansion of 1 mole of a monatomic ideal gas from 5.00 m^3 to 25.0 m^3. The temperature of the gas is initially 298 K.

For an adiabatic process, $q = 0$.

ΔU can be calculated once C_V and the final temperature, T_2 are determined. For an ideal monatomic gas, $C_V = \frac{3}{2}nR$, $C_P = C_V + nR = \frac{5}{2}nR$, $\gamma = \frac{C_P}{C_V} = \frac{5}{3}$. T_2 is calculated using the relationship between V and T appropriate for an adiabatic reversible expansion:

$$\left(\frac{V_1}{V_2}\right)^{\gamma - 1} = \frac{T_2}{T_1}$$

$$T_2 = T_1 \left(\frac{V_1}{V_2}\right)^{\gamma - 1} = (298 \text{ K}) \left(\frac{5.00 \text{ m}^3}{25.0 \text{ m}^3}\right)^{(5/3)-1} = 101.9 \text{ K}$$

ΔU is therefore

$$\Delta U = C_V \Delta T = \left(\frac{3}{2}nR\right)(T_2 - T_1)$$

$$= \frac{3}{2}(1 \text{ mol})\left(8.314 \text{ J K}^{-1} \text{ mol}^{-1}\right)(101.9 \text{ K} - 298 \text{ K})$$

$$= -2.45 \times 10^3 \text{ J}$$

According to the first law,

$$w = \Delta U - q = -2.45 \times 10^3 \text{ J} - 0 = -2.45 \times 10^3 \text{ J}$$

ΔH is found from C_P and ΔT:

$$\Delta H = C_P \Delta T = \frac{5}{2} n R \Delta T = \frac{5}{2} (1 \text{ mol}) (8.314 \text{ J K}^{-1} \text{ mol}^{-1}) (101.9 \text{ K} - 298 \text{ K}) = -4.08 \times 10^3 \text{ J}$$

4.34 A quantity of 0.27 mole of neon is confined in a container at 2.50 atm and 298 K and then allowed to expand adiabatically under two different conditions: **(a)** reversibly to 1.00 atm and **(b)** against a constant pressure of 1.00 atm. Calculate the final temperature in each case.

(a) For a reversible adiabatic process, P and V are related as follow:

$$\frac{P_2}{P_1} = \left(\frac{V_1}{V_2} \right)^{\gamma}$$

Assuming neon to be an ideal gas, write V in terms of n, P, and T:

$$\frac{P_2}{P_1} = \left(\frac{\frac{nRT_1}{P_1}}{\frac{nRT_2}{P_2}} \right)^{\gamma} = \left(\frac{T_1}{T_2} \right)^{\gamma} \left(\frac{P_2}{P_1} \right)^{\gamma}$$

Rearrange the equation to give an expression relating T and P:

$$\left(\frac{T_2}{T_1} \right)^{\gamma} = \left(\frac{P_2}{P_1} \right)^{\gamma - 1}$$

$$\frac{T_2}{T_1} = \left(\frac{P_2}{P_1} \right)^{(\gamma - 1)/\gamma}$$

Utilizing the last equation and the fact that γ for an ideal monatomic gas is $\frac{5}{3}$, the final temperature can be calculated:

$$T_2 = T_1 \left(\frac{P_2}{P_1} \right)^{(\gamma - 1)/\gamma} = (298 \text{ K}) \left(\frac{1.00 \text{ atm}}{2.50 \text{ atm}} \right)^{\left(\frac{5}{3} - 1 \right)/\frac{5}{3}} = 207 \text{ K}$$

(b) The process is adiabatic with $q = 0$. Thus, $\Delta U = w$. Since $\Delta U = C_V (T_2 - T_1)$, and $w = -P_{ex} (V_2 - V_1)$,

$$C_V (T_2 - T_1) = -P_{ex} (V_2 - V_1)$$

$$\frac{3}{2} n R (T_2 - T_1) = -P_{ex} \left(\frac{nRT_2}{P_2} - \frac{nRT_1}{P_1} \right)$$

$$\frac{3}{2} (T_2 - T_1) = -P_{ex} \left(\frac{T_2}{P_2} - \frac{T_1}{P_1} \right)$$

The expansion occurs until $P_2 = P_{ex}$. Therefore, the above expression simplifies to

$$\frac{3}{2}(T_2 - T_1) = -T_2 + \frac{P_{ex}T_1}{P_1}$$

$$\frac{5}{2}T_2 = \frac{P_{ex}T_1}{P_1} + \frac{3}{2}T_1 = \left(\frac{P_{ex}}{P_1} + \frac{3}{2}\right)T_1$$

$$T_2 = \frac{2}{5}\left(\frac{P_{ex}}{P_1} + \frac{3}{2}\right)T_1$$

$$= \frac{2}{5}\left(\frac{1.00 \text{ atm}}{2.50 \text{ atm}} + \frac{3}{2}\right)(298 \text{ K}) = 226 \text{ K}$$

4.35   One mole of an ideal monatomic gas initially at 300 K and a pressure of 15.0 atm expands to a final pressure of 1.00 atm. The expansion can occur via any of four different paths: **(a)** isothermal and reversible, **(b)** isothermal and irreversible, **(c)** adiabatic and reversible, and **(d)** adiabatic and irreversible. In irreversible processes, the expansion occurs against an external pressure of 1.00 atm. For each case calculate the values of q, w, ΔU, and ΔH.

(a) When an ideal gas undergoes an isothermal process, $\Delta U = 0$ and $\Delta H = 0$.

$$w = -nRT \ln \frac{P_1}{P_2} = -(1 \text{ mol})(8.314 \text{ J K}^{-1} \text{ mol}^{-1})(300 \text{ K}) \ln \frac{15.0 \text{ atm}}{1.00 \text{ atm}} = -6.75 \times 10^3 \text{ J}$$

$$q = \Delta U - w = -w = 6.75 \times 10^3 \text{ J}$$

(b) When an ideal gas undergoes an isothermal process, $\Delta U = 0$ and $\Delta H = 0$.

$$w = -P_{ex}(V_2 - V_1)$$

V_1 and V_2 can be determined using the ideal gas law:

$$V_1 = \frac{nRT}{P_1} = \frac{(1 \text{ mol})(0.08206 \text{ L atm K}^{-1} \text{ mol}^{-1})(300 \text{ K})}{15.0 \text{ atm}} = 1.641 \text{ L}$$

$$V_2 = \frac{nRT}{P_2} = \frac{(1 \text{ mol})(0.08206 \text{ L atm K}^{-1} \text{ mol}^{-1})(300 \text{ K})}{1.00 \text{ atm}} = 24.62 \text{ L}$$

Therefore,

$$w = -(1.00 \text{ atm})(24.62 \text{ L} - 1.641 \text{ L})\left(\frac{101.3 \text{ J}}{1 \text{ L atm}}\right) = -2.33 \times 10^3 \text{ J}$$

and

$$q = \Delta U - w = -w = 2.33 \times 10^3 \text{ J}$$

(c) $q = 0$ for an adiabatic process.

To determine ΔU and ΔH, T_2 needs to be calculated. Using the same procedure as Problem 4.34(a),

$$T_2 = T_1 \left(\frac{P_2}{P_1}\right)^{(\gamma - 1)/\gamma} = (300 \text{ K}) \left(\frac{1.00 \text{ atm}}{15.0 \text{ atm}}\right)^{\left(\frac{5}{3} - 1\right)/\frac{5}{3}} = 101.6 \text{ K}$$

Now the rest of the quantities can be calculated:

$$\Delta U = C_V \Delta T = \frac{3}{2} (1 \text{ mol}) \left(8.314 \text{ J K}^{-1} \text{ mol}^{-1}\right) (101.6 \text{ K} - 300 \text{ K}) = -2.47 \times 10^3 \text{ J}$$

$$w = \Delta U - q = \Delta U = -2.47 \times 10^3 \text{ J}$$

$$\Delta H = C_P \Delta T = \frac{5}{2} (1 \text{ mol}) \left(8.314 \text{ J K}^{-1} \text{ mol}^{-1}\right) (101.6 \text{ K} - 300 \text{ K}) = -4.12 \times 10^3 \text{ J}$$

(d) $q = 0$ for an adiabatic process.

To determine ΔU and ΔH, T_2 needs to be calculated. Using the same procedure as Problem 4.34(b),

$$T_2 = \frac{2}{5} \left(\frac{P_{ex}}{P_1} + \frac{3}{2}\right) T_1$$

$$= \frac{2}{5} \left(\frac{1.00 \text{ atm}}{15.0 \text{ atm}} + \frac{3}{2}\right) (300 \text{ K}) = 188 \text{ K}$$

Now the rest of the quantities can be calculated:

$$\Delta U = C_V \Delta T = \frac{3}{2} (1 \text{ mol}) \left(8.314 \text{ J K}^{-1} \text{ mol}^{-1}\right) (188 \text{ K} - 300 \text{ K}) = -1.40 \times 10^3 \text{ J}$$

$$w = \Delta U - q = \Delta U = -1.40 \times 10^3 \text{ J}$$

$$\Delta H = C_P \Delta T = \frac{5}{2} (1 \text{ mol}) \left(8.314 \text{ J K}^{-1} \text{ mol}^{-1}\right) (188 \text{ K} - 300 \text{ K}) = -2.33 \times 10^3 \text{ J}$$

4.36   A 0.1375-g sample of magnesium is burned in a constant-volume bomb calorimeter that has a heat capacity of 1769 $\text{J} \,^\circ\text{C}^{-1}$. The calorimeter contains exactly 300 g of water, and the temperature increases by 1.126°C. Calculate the heat given off by the burning magnesium, in kJ g^{-1} and in kJ mol^{-1}. The specific heat of water is 4.184 $\text{J g}^{-1} \,^\circ\text{C}^{-1}$.

Heat given off by the burning magnesium is absorbed by the calorimeter and water. The heat absorbed is calculated using the heat capacity of the calorimeter, specific heat of water, and the temperature rise.

$$\text{Heat gained by calorimeter and water} = \left(1769 \text{ J} \,^\circ\text{C}^{-1}\right) (1.126 \,^\circ\text{C})$$

$$+ (300 \text{ g}) \left(4.184 \text{ J g}^{-1} \,^\circ\text{C}^{-1}\right) (1.126 \,^\circ\text{C})$$

$$= 3.405 \times 10^3 \text{ J}$$

Therefore, heat given off by Mg is

$$\frac{3.405 \times 10^3 \text{ J}}{0.1375 \text{ g}} = 2.48 \times 10^4 \text{ J g}^{-1} = 24.8 \text{ kJ g}^{-1}$$

or

$$(2.48 \times 10^4 \text{ J g}^{-1})(24.31 \text{ g mol}^{-1}) = 6.03 \times 10^5 \text{ J mol}^{-1} = 603 \text{ kJ mol}^{-1}$$

4.37 The enthalpy of combustion of benzoic acid (C_6H_5COOH) is commonly used as the standard for calibrating constant-volume bomb calorimeters; its value has been accurately determined to be $-3226.7 \text{ kJ mol}^{-1}$. **(a)** When 0.9862 g of benzoic acid was oxidized, the temperature rose from 21.84°C to 25.67°C. What is the heat capacity of the calorimeter? **(b)** In a separate experiment, 0.4654 g of α-D-glucose ($C_6H_{12}O_6$) was oxidized in the same calorimeter and the temperature rose from 21.22°C to 22.28°C. Calculate the enthalpy of combustion of glucose, the value of $\Delta_r U$ for the combustion, and the molar enthalpy of formation of glucose.

(a) In a constant-volume calorimeter, the energy of combustion of benzoic acid, $\Delta_r U^\circ$, is the same as heat given off by the reaction, q, which is absorbed by the calorimeter. Therefore, the given value for the enthalpy of combustion must be converted to the energy of combustion. The combustion reaction for benzoic acid is

$$C_6H_5COOH(s) + \frac{15}{2}O_2(g) \rightarrow 7CO_2(g) + 3H_2O(l)$$

and has $\Delta n = -0.5$ mol. From this, $\Delta_r U^\circ$ is found.

$$\Delta_r U^\circ = \Delta_r H^\circ - RT\,\Delta n$$

$$= -3226.7 \text{ kJ mol}^{-1} - (8.314 \text{ JK}^{-1}\text{ mol}^{-1})(298.15 \text{ K})(-0.5 \text{ mol})\left(\frac{1 \text{ kJ}}{1000 \text{ J}}\right)$$

$$= -3225.46 \text{ kJ mol}^{-1}$$

Therefore,

$$q_{\text{calorimeter}} = -q_{\text{benzoic acid}}$$

$$C_{\text{calorimeter}}(25.67\,°C - 21.84\,°C) = -\left(\frac{0.9862 \text{ g}}{122.1 \text{ g mol}^{-1}}\right)(-3225.46 \text{ kJ mol}^{-1})$$

$$C_{\text{calorimeter}} = \frac{26.052 \text{ kJ}}{3.83\,°C} = 6.802 \text{ kJ}\,°C^{-1}$$

(b) In this experiment, the calorimeter absorbs heat transferred from the combustion of glucose.

$$q_{\text{calorimeter}} = -q_{\text{glucose}}$$

$$6.802 \text{ kJ}\,°C^{-1}(22.28\,°C - 21.22\,°C) = -\left(\frac{0.4654 \text{ g}}{180.2 \text{ g mol}^{-1}}\right)(\Delta_r U^\circ)$$

$$\Delta_r U^\circ = -2.79 \times 10^3 \text{ kJ mol}^{-1}$$

To calculate the enthalpy of combustion, the change in number of moles of gases in the reaction has to be determined. The reaction is

$$C_6H_{12}O_6(s) + 6O_2(g) \rightarrow 6CO_2(g) + 6H_2O(l)$$

so $\Delta n = 0$. Therefore,

$$\Delta_r H^\circ = \Delta_r U^\circ + RT\Delta n = \Delta_r U^\circ = -2.79 \times 10^3 \text{ kJ mol}^{-1}$$

The molar enthalpy of formation is related to $\Delta_r H^\circ$ of the combustion reaction.

$$\Delta_r H^\circ = 6\Delta_f \overline{H}^\circ \left[CO_2(g)\right] + 6\Delta_f \overline{H}^\circ \left[H_2O(l)\right] - \Delta_f \overline{H}^\circ \left[C_6H_{12}O_6(s)\right]$$

$$-6\Delta_f \overline{H}^\circ \left[O_2(g)\right]$$

$$\Delta_f \overline{H}^\circ \left[C_6H_{12}O_6(s)\right] = 6\Delta_f \overline{H}^\circ \left[CO_2(g)\right] + 6\Delta_f \overline{H}^\circ \left[H_2O(l)\right] - 6\Delta_f \overline{H}^\circ \left[O_2(g)\right] - \Delta_r H^\circ$$

$$= 6\left(-393.5 \text{ kJ mol}^{-1}\right) + 6\left(-285.8 \text{ kJ mol}^{-1}\right) - 6\left(0 \text{ kJ mol}^{-1}\right)$$

$$- \left(-2.79 \times 10^3 \text{ kJ mol}^{-1}\right)$$

$$= -1.29 \times 10^3 \text{ kJ mol}^{-1}$$

4.38 A quantity of 2.00×10^2 mL of 0.862 M HCl is mixed with 2.00×10^2 mL of 0.431 M $Ba(OH)_2$ in a constant-pressure calorimeter having a heat capacity of 453 J $°C^{-1}$. The initial temperature of the HCl and $Ba(OH)_2$ solutions is the same at 20.48°C. For the process

$$H^+(aq) + OH^-(aq) \rightarrow H_2O(l)$$

the heat of neutralization is -56.2 kJ mol^{-1}. What is the final temperature of the mixed solution?

The final temperature of the solution can be determined once the heat capacity of the calorimeter *plus* solution is known as well as the number of moles of reaction that occurs.

First determine the number of moles of reactants.

$$n_{H^+} = \left(2.00 \times 10^{-1} \text{ L}\right)\left(0.862 \text{ mol L}^{-1}\right) = 0.1724 \text{ mol}$$

$$n_{OH^-} = 2\left(2.00 \times 10^{-1} \text{ L}\right)\left(0.431 \text{ mol L}^{-1}\right) = 0.1724 \text{ mol}$$

There is just enough of each reactant to completely react with the other to form 0.1724 mol of the product, H_2O. The thermal energy released by this reaction, under constant pressure, is

$$q_P = \left(-56.2 \text{ kJ mol}^{-1}\right)(0.1724 \text{ mol}) = -9.689 \text{ kJ}$$

This same thermal energy is used to increase the temperature of the calorimeter and its contents, but since these items are *gaining* thermal energy, the sign is switched.

$$9.689 \text{ kJ} = \left(C_{calorimeter} + m_{solution}s_{solution}\right)\Delta T$$

Two assumptions are now made. Namely, that the densities of the solutions are the same as pure water, and likewise that the specific heat of the solution is the same as that of water. (These assumptions are good to 3-5%.) Under these assumptions, $s_{solution} = 4.184\,\mathrm{J\,g^{-1}\,^{\circ}C^{-1}}$, and

$$m_{solution} = m_{HCl} + m_{Ba(OH)_2}$$

$$= \left(2.00 \times 10^2\ \mathrm{mL}\right)\left(1.00\ \mathrm{g\,mL^{-1}}\right) + \left(2.00 \times 10^2\ \mathrm{mL}\right)\left(1.00\ \mathrm{g\,mL^{-1}}\right)$$

$$= 4.00 \times 10^2\ \mathrm{g}$$

This allows the determination of ΔT,

$$9.689\ \mathrm{kJ} = \left[453\ \mathrm{J\,^{\circ}C^{-1}} + \left(4.00 \times 10^2\ \mathrm{g}\right)\left(4.184\ \mathrm{J\,g^{-1}\,^{\circ}C^{-1}}\right)\right]\left(\tfrac{1\ \mathrm{kJ}}{1000\ \mathrm{J}}\right)\Delta T$$
$$9.689 = \left(2.127\ \mathrm{^{\circ}C^{-1}}\right)\Delta T$$
$$\Delta T_2 = 4.555\mathrm{^{\circ}C}$$

The final temperature of the mixed solution is found from $\Delta T = T_f - T_i$, or $4.555\mathrm{^{\circ}C} = T_f - 20.48\mathrm{^{\circ}C}$ which gives $T_f = 25.04\mathrm{^{\circ}C}$, although the assumptions made limit the accuracy of our answer to $T_f = 25.0\mathrm{^{\circ}C}$.

4.39   When 1 mole of naphthalene ($C_{10}H_8$) is completely burned in a constant-volume bomb calorimeter at 298 K, 5150 kJ of heat is evolved. Calculate the values of $\Delta_r U$ and $\Delta_r H$ for the reaction.

Since volume is constant, $w = 0$, and therefore, $\Delta_r U = q = -5150\ \mathrm{kJ\,mol^{-1}}$.

The change in number of moles of gases needs to be determined before $\Delta_r H$ can be evaluated. From the combustion of 1 mole of naphthalene,

$$C_{10}H_8(s) + 12O_2(g) \rightarrow 10CO_2(g) + 4H_2O(l)$$

$\Delta n = 10 - 12 = -2$. Thus,

$$\Delta_r H = \Delta_r U + RT\,\Delta n = -5150\ \mathrm{kJ\,mol^{-1}} + \left(8.314\ \mathrm{J\,K^{-1}\,mol^{-1}}\right)(298\ \mathrm{K})(-2)\left(\frac{1\ \mathrm{kJ}}{1000\ \mathrm{J}}\right)$$

$$= -5155\ \mathrm{kJ\,mol^{-1}}$$

4.40   Consider the following reaction:

$$2CH_3OH(l) + 3O_2(g) \rightarrow 4H_2O(l) + 2CO_2(g) \qquad \Delta_r H = -1452.8\ \mathrm{kJ\,mol^{-1}}$$

What is the value of $\Delta_r H^{\circ}$ if **(a)** the equation is multiplied throughout by 2, **(b)** the direction of the reaction is reversed so that the products become the reactants and vice versa, **(c)** water vapor instead of liquid water is the product?

(a) $\Delta_r H^{\circ} = 2\left(-1452.8\ \mathrm{kJ\,mol^{-1}}\right) = -2905.6\ \mathrm{kJ\,mol^{-1}}$

(b) $\Delta_r H^{\circ} = -\left(-1452.8\ \mathrm{kJ\,mol^{-1}}\right) = 1452.8\ \mathrm{kJ\,mol^{-1}}$

(c) The reaction

$$2CH_3OH(l) + 3O_2(g) \rightarrow 4H_2O(g) + 2CO_2(g)$$

is a sum of the following equations:

$$2CH_3OH(l) + 3O_2(g) \rightarrow 4H_2O(l) + 2CO_2(g) \qquad \Delta_r H_1^\circ = -1452.8 \text{ kJ mol}^{-1}$$
$$4H_2O(l) \rightarrow 4H_2O(g) \qquad \Delta_r H_2^\circ$$

The standard enthalpy of reaction for vaporization of H_2O is

$$\Delta_r H_2^\circ = 4\Delta_f \overline{H}^\circ \left[H_2O(g) \right] - 4\Delta_f \overline{H}^\circ \left[H_2O(l) \right]$$

$$= 4\left(-241.8 \text{ kJ mol}^{-1} \right) - 4\left(-285.8 \text{ kJ mol}^{-1} \right)$$

$$= 176.0 \text{ kJ mol}^{-1}$$

The standard enthalpy of reaction of $2CH_3OH(l) + 3O_2(g) \rightarrow 4H_2O(g) + 2CO_2(g)$ is

$$\Delta_r H^\circ = \Delta_r H_1^\circ + \Delta_r H_2^\circ = -1452.8 \text{ kJ mol}^{-1} + 176.0 \text{ kJ mol}^{-1} = -1276.8 \text{ kJ mol}^{-1}$$

4.41 Which of the following standard enthalpy of formation values is not zero at 25°C? Na(s), Ne(g), $CH_4(g)$, $S_8(s)$, Hg(l), H(g).

H(g) is not the stable allotropic form of hydrogen at 25°C. Therefore, $\Delta_f H^\circ$ [H(g)] $\neq$ 0. [The stable form is $H_2(g)$.]

4.42 The standard enthalpies of formation of ions in aqueous solution are obtained by arbitrarily assigning a value of zero to H^+ ions; that is, $\Delta_f \overline{H}^\circ$ [$H^+(aq)$] = 0. **(a)** For the following reaction,

$$HCl(g) \rightarrow H^+(aq) + Cl^-(aq) \qquad \Delta_r H^\circ = -74.9 \text{ kJ mol}^{-1}$$

calculate the value of $\Delta_f H^\circ$ for the Cl^- ions. **(b)** The standard enthalpy of neutralization between a HCl solution and a NaOH solution is found to be $-56.2 \text{ kJ mol}^{-1}$. Calculate the standard enthalpy of formation of the hydroxide ion at 25°C.

(a)

$$\Delta_r H^\circ = -74.9 \text{ kJ mol}^{-1} = \Delta_f \overline{H}^\circ \left[H^+(aq) \right] + \Delta_f \overline{H}^\circ \left[Cl^-(aq) \right] - \Delta_f \overline{H}^\circ \left[HCl(g) \right]$$

$$\Delta_f \overline{H}^\circ \left[Cl^-(aq) \right] = -74.9 \text{ kJ mol}^{-1} - \Delta_f \overline{H}^\circ \left[H^+(aq) \right] + \Delta_f \overline{H}^\circ \left[HCl(g) \right]$$

$$= -74.9 \text{ kJ mol}^{-1} - 0 \text{ kJ mol}^{-1} + \left(-92.3 \text{ kJ mol}^{-1} \right)$$

$$= -167.2 \text{ kJ mol}^{-1}$$

(b) The neutralization reaction for 1 mole of H_2O is

$$H^+(aq) + OH^-(aq) \rightarrow H_2O(l) \qquad \Delta_r H^\circ = -56.2 \text{ kJ mol}^{-1}$$

$$\Delta_r H^\circ = -56.2 \text{ kJ mol}^{-1} = \Delta_f \overline{H}^\circ [H_2O(l)] - \Delta_f \overline{H}^\circ [H^+(aq)] - \Delta_f \overline{H}^\circ [OH^-(aq)]$$

$$\Delta_f \overline{H}^\circ [OH^-(aq)] = \Delta_f \overline{H}^\circ [H_2O(l)] - \Delta_f \overline{H}^\circ [H^+(aq)] + 56.2 \text{ kJ mol}^{-1}$$

$$= -285.8 \text{ kJ mol}^{-1} - 0 \text{ kJ mol}^{-1} + 56.2 \text{ kJ mol}^{-1}$$

$$= -229.6 \text{ kJ mol}^{-1}$$

4.43 Determine the amount of heat (in kJ) given off when 1.26×10^4 g of ammonia is produced according to the equation

$$N_2(g) + 3H_2(g) \rightarrow 2NH_3(g) \qquad \Delta_r H^\circ = -92.6 \text{ kJ mol}^{-1}$$

Assume the reaction takes place under standard-state conditions at 25°C.

The equation gives standard enthalpy of reaction (or, since P is constant, the amount of heat given off) when 2 moles of NH_3 are produced. When 1.26×10^4 g of ammonia is produced, the amount of heat given off is

$$\left(\frac{-92.6 \text{ kJ}}{2 \text{ mol NH}_3} \right) \left(\frac{1.26 \times 10^4 \text{ g NH}_3}{17.03 \text{ g mol}^{-1} \text{ NH}_3} \right) = -3.43 \times 10^4 \text{ kJ}$$

4.44 When 2.00 g of hydrazine decomposed under constant-pressure conditions, 7.00 kJ of heat were transferred to the surroundings:

$$3N_2H_4(l) \rightarrow 4NH_3(g) + N_2(g)$$

What is the $\Delta_r H^\circ$ value for the reaction?

The reaction describes the decomposition of 3 moles of hydrazine. Therefore, the amount of heat given must be scaled to this amount of reactant.

$$\Delta_r H^\circ = q_P = \left(\frac{-7.00 \text{ kJ}}{\frac{2.00 \text{ g N}_2\text{H}_4}{32.05 \text{ g mol}^{-1} \text{ N}_2\text{H}_4}} \right) (3.00) = -337 \text{ kJ mol}^{-1}$$

4.45 Consider the reaction

$$N_2(g) + 3H_2(g) \rightarrow 2NH_3(g) \qquad \Delta_r H^\circ = -92.6 \text{ kJ mol}^{-1}$$

If 2.0 moles of N_2 react with 6.0 moles of H_2 to form NH_3, calculate the work done (in joules) against a pressure of 1.0 atm at 25°C. What is the value of $\Delta_r U$ for this reaction? Assume the reaction goes to completion.

Under a constant external pressure,

$$w = -P_{ex}\left(V_2 - V_1\right)$$

where V_1 and V_2 are the volumes occupied by the reactants (8.0 moles) and products (4.0 moles), respectively. Using the ideal gas law,

$$w = -P_{ex}\left(\frac{n_{products}RT}{P} - \frac{n_{reactants}RT}{P}\right) = -P_{ex}\left[\frac{(-4.0)\,RT}{P}\right]$$

where P is the pressure at which the reaction takes place, that is, the same as P_{ex}. Therefore,

$$w = -(-4.0)\,RT = -(-4.0)\left(8.314\,\mathrm{J\,K^{-1}\,mol^{-1}}\right)(298\,\mathrm{K}) = 9.9 \times 10^3\,\mathrm{J\,mol^{-1}}$$

The first law can be used to obtain $\Delta_r U$ once q is calculated. At constant pressure, $q = \Delta_r H$. $\Delta_r H$ given above is for the reaction between 1 mole of N_2 and 3 moles of H_2. The reaction of interest involves twice the amounts of reactants. Therefore,

$$q = 2\left(-92.6\,\mathrm{kJ\,mol^{-1}}\right) = -185.2\,\mathrm{kJ\,mol^{-1}}$$

Thus,

$$\Delta_r U = q + w = -185.2\,\mathrm{kJ\,mol^{-1}} + 9.9\,\mathrm{kJ\,mol^{-1}} = -175.3\,\mathrm{kJ\,mol^{-1}}$$

4.46 The standard enthalpies of combustion of fumaric acid and maleic acid (to form carbon dioxide and water) are $-1336.0\,\mathrm{kJ\,mol^{-1}}$ and $-1359.2\,\mathrm{kJ\,mol^{-1}}$, respectively. Calculate the enthalpy of the following isomerization process:

maleic acid   fumaric acid

The chemical equations and the standard enthalpies of combustion of 1 mole of fumaric acid and 1 mole of maleic acid are given below:

$$\text{fumaric} + 3O_2 \rightarrow 4CO_2 + 2H_2O \qquad \Delta_r H^\circ = -1336.0\,\mathrm{kJ\,mol^{-1}}$$
$$\text{maleic} + 3O_2 \rightarrow 4CO_2 + 2H_2O \qquad \Delta_r H^\circ = -1359.2\,\mathrm{kJ\,mol^{-1}}$$

The isomerization reaction (maleic acid $\rightarrow$ fumaric acid) can be obtained as a combination of these two reactions:

$$4CO_2 + 2H_2O \rightarrow \text{fumaric} + 3O_2 \qquad \Delta_r H^\circ = 1336.0\,\mathrm{kJ\,mol^{-1}}$$
$$\text{maleic} + 3O_2 \rightarrow 4CO_2 + 2H_2O \qquad \Delta_r H^\circ = -1359.2\,\mathrm{kJ\,mol^{-1}}$$

Therefore, the enthalpy of the isomerization process is

$$\Delta_r H = 1336.0 \text{ kJ mol}^{-1} - 1359.2 \text{ kJ mol}^{-1} = -23.2 \text{ kJ mol}^{-1}$$

4.47   From the reaction

$$C_{10}H_8(s) + 12O_2(g) \rightarrow 10CO_2(g) + 4H_2O(l) \qquad \Delta_r H^\circ = -5153.0 \text{ kJ mol}^{-1}$$

and the enthalpies of formation of CO_2 and H_2O (see Appendix B), calculate the enthalpy of formation of naphthalene ($C_{10}H_8$).

$$\Delta_r H^\circ = 10\Delta_f \overline{H}^\circ \left[CO_2(g)\right] + 4\Delta_f \overline{H}^\circ \left[H_2O(l)\right] - \Delta_f \overline{H}^\circ \left[C_{10}H_8(s)\right]$$

$$-12\Delta_f \overline{H}^\circ \left[O_2(g)\right]$$

$$\Delta_f \overline{H}^\circ \left[C_{10}H_8(s)\right] = 10\Delta_f \overline{H}^\circ \left[CO_2(g)\right] + 4\Delta_f \overline{H}^\circ \left[H_2O(l)\right] - 12\Delta_f \overline{H}^\circ \left[O_2(g)\right] - \Delta_r H^\circ$$

$$= 10\left(-393.5 \text{ kJ mol}^{-1}\right) + 4\left(-285.8 \text{ kJ mol}^{-1}\right) - 12\left(0 \text{ kJ mol}^{-1}\right)$$

$$+5153.0 \text{ kJ mol}^{-1}$$

$$= 74.8 \text{ kJ mol}^{-1}$$

4.48   The standard molar enthalpy of formation of molecular oxygen at 298 K is zero. What is its value at 315 K? (*Hint*: Look up the $\overline{C}_P$ value in Appendix B.)

Let $\Delta \overline{H}_1$ and $\Delta \overline{H}_2$ be the enthalpies of formation of molecular oxygen at 298 K and 315 K, respectively.

$$\Delta \overline{H}_2 = \Delta \overline{H}_1 + \overline{C}_P \Delta T = 0 \text{ J mol}^{-1} + \left(29.4 \text{ J K}^{-1} \text{ mol}^{-1}\right)(315 \text{ K} - 298 \text{ K}) = 500 \text{ J mol}^{-1}$$

4.49   Which of the following substances has a nonzero $\Delta_f \overline{H}^\circ$ value at 25°C? Fe(s), $I_2(l)$, $H_2(g)$, Hg(l), $O_2(g)$, C(graphite).

$I_2(l)$ is not the stable allotropic form of iodine at 25°C. Therefore, $\Delta_f \overline{H}^\circ \left[I_2(l)\right] \neq 0$. [The stable form is $I_2(s)$.]

4.50   The hydrogenation for ethylene is

$$C_2H_4(g) + H_2(g) \rightarrow C_2H_6(g)$$

Calculate the change in the enthalpy of hydrogenation from 298 K to 398 K. The values of $\overline{C}_P^o$ for ethylene and ethane are 43.6 J $K^{-1}mol^{-1}$ and 52.7 J $K^{-1}mol^{-1}$, respectively. Assume the heat capacities are temperature independent.

$$\Delta_r H_{398} - \Delta_r H_{298} = \Delta C_P^o \, (398 \text{ K} - 298 \text{ K})$$

$$= \left\{ \overline{C}_P^o \left[C_2H_6(g) \right] - \overline{C}_P^o \left[C_2H_4(g) \right] - \overline{C}_P^o \left[H_2(g) \right] \right\} (10 \text{ K})$$

$$= \left(52.7 \text{ J K}^{-1} \text{ mol}^{-1} - 43.6 \text{ J K}^{-1} \text{ mol}^{-1} - 28.8 \text{ J K}^{-1} \text{ mol}^{-1} \right) (10 \text{ K})$$

$$= -197 \text{ kJ mol}^{-1}$$

4.51 Use the data in Appendix B to calculate the value of $\Delta_r H^o$ for the following reaction at 298 K:

$$N_2O_4(g) \rightarrow 2NO_2(g)$$

What is its value at 350 K? State any assumptions used in your calculation.

At 298 K,

$$\Delta_r H_{298}^o = 2\Delta_f \overline{H}^o \left[NO_2(g) \right] - \Delta_f \overline{H}^o \left[N_2O_4(g) \right]$$

$$= 2 \left(33.9 \text{ kJ mol}^{-1} \right) - 9.7 \text{ kJ mol}^{-1}$$

$$= 58.1 \text{ kJ mol}^{-1}$$

Assuming heat capacities to be temperature independent, the enthalpy of reaction is

$$\Delta_r H_{350}^o = \Delta_r H_{298}^o + \Delta C_P^o \, (350 \text{ K} - 298 \text{ K})$$

$$= \Delta_r H_{298}^o + \left\{ 2\overline{C}_P^o \left[NO_2(g) \right] - \overline{C}_P^o \left[N_2O_4(g) \right] \right\} (52 \text{ K})$$

$$= 58.1 \text{ kJ mol}^{-1} + \left[2 \left(37.9 \text{ J K}^{-1} \text{ mol}^{-1} \right) - 79.1 \text{ J K}^{-1} \text{ mol}^{-1} \right] (52 \text{ K}) \left(\frac{1 \text{ kJ}}{1000 \text{ J}} \right)$$

$$= 57.9 \text{ kJ mol}^{-1}$$

4.52 Calculate the standard enthalpy of formation for diamond, given that

$$\begin{aligned} C(\text{graphite}) + O_2(g) \rightarrow CO_2(g) \qquad & \Delta_r H^o = -393.5 \text{ kJ mol}^{-1} \\ C(\text{diamond}) + O_2(g) \rightarrow CO_2(g) \qquad & \Delta_r H^o = -395.4 \text{ kJ mol}^{-1} \end{aligned}$$

The formation reaction of diamond is

$$C(graphite) \rightarrow C(diamond),$$

which can be thought of as a sum of the reactions:

$$C(graphite) + O_2(g) \rightarrow CO_2(g) \qquad \Delta_r H° = -393.5 \text{ kJ mol}^{-1}$$
$$CO_2(g) \rightarrow C(diamond) + O_2(g) \qquad \Delta_r H° = 395.4 \text{ kJ mol}^{-1}$$

Therefore, the standard enthalpy of formation for diamond is the sum of the standard enthalpies of reaction of the two reactions above:

$$\Delta_f H° \text{ (diamond)} = -393.5 \text{ kJ mol}^{-1} + 395.4 \text{ kJ mol}^{-1} = 1.9 \text{ kJ mol}^{-1}$$

4.53 Photosynthesis produces glucose, $C_6H_{12}O_6$, and oxygen from carbon dioxide and water:

$$6CO_2 + 6H_2O \rightarrow C_6H_{12}O_6 + 6O_2$$

(a) How would you determine the $\Delta_r H°$ value for this reaction experimentally? **(b)** Solar radiation produces approximately 7.0×10^{14} kg glucose a year on Earth. What is the corresponding change in the $\Delta_r H°$ value?

(a) The most straightforward method of determining the $\Delta_r H°$ value for this reaction would be to measure the $\Delta_r H°$ for the reverse reaction, which is the enthalpy of combustion for glucose, in a bomb calorimeter. The desired value is just the negative of the value so determined.

(b) Use standard enthalpies of formation to calculate $\Delta_r H°$,

$$\Delta_r H° = \Delta_f H° \left[C_6H_{12}O_6(s) \right] + 6\Delta_f H° \left[O_2(g) \right] - 6\Delta_f H° \left[CO_2(g) \right] - 6\Delta_f H° \left[H_2O(l) \right]$$
$$= -1274.5 \text{ kJ mol}^{-1} + 6\left(0 \text{ kJ mol}^{-1} \right) - 6\left(-393.5 \text{ kJ mol}^{-1} \right) - 6\left(-285.8 \text{ kJ mol}^{-1} \right)$$
$$= 2801.3 \text{ kJ mol}^{-1}$$

where the value for α-D-glucose is used. This is the enthalpy change for the reaction forming 1 mole of glucose. For each kilogram of glucose,

$$\Delta_r H° = 2801.3 \text{ kJ mol}^{-1} \left(\frac{1 \text{ mol } C_6H_{12}O_6}{180.16 \text{ g } C_6H_{12}O_6} \right) \left(\frac{1000 \text{ g}}{1 \text{ kg}} \right)$$
$$= 1.5549 \times 10^4 \text{ kJ kg}^{-1}$$

With 7.0×10^{14} kg of glucose produced by photosynthesis, the total $\Delta_r H°$ involved is

$$\Delta_r H° = \left(7.0 \times 10^{14} \text{ kg} \right) \left(1.5549 \times 10^4 \text{ kJ kg}^{-1} \right) = 1.1 \times 10^{19} \text{ kJ}$$

4.54 From the following heats of combustion,

$$CH_3OH(l) + \tfrac{3}{2}O_2(g) \rightarrow CO_2(g) + 2H_2O(l) \qquad \Delta_r H° = -726.4 \text{ kJ mol}^{-1}$$
$$C(graphite) + O_2(g) \rightarrow CO_2(g) \qquad \Delta_r H° = -393.5 \text{ kJ mol}^{-1}$$
$$H_2(g) + \tfrac{1}{2}O_2(g) \rightarrow H_2O(l) \qquad \Delta_r H° = -285.8 \text{ kJ mol}^{-1}$$

calculate the enthalpy of formation of methanol (CH_3OH) from its elements:

$$C(graphite) + 2H_2(g) + \tfrac{1}{2}O_2(g) \rightarrow CH_3OH(l)$$

The formation reaction of methanol can be thought of as a sum of the reactions:

$$CO_2(g) + 2H_2O(l) \rightarrow CH_3OH(l) + \tfrac{3}{2}O_2(g) \quad \Delta_r H^\circ = 726.4 \text{ kJ mol}^{-1}$$
$$C(graphite) + O_2(g) \rightarrow CO_2(g) \quad\quad\quad \Delta_r H^\circ = -393.5 \text{ kJ mol}^{-1}$$
$$2H_2(g) + O_2(g) \rightarrow 2H_2O(l) \quad\quad\quad \Delta_r H^\circ = 2\left(-285.8 \text{ kJ mol}^{-1}\right) = -571.6 \text{ kJ mol}^{-1}$$

Therefore, the standard enthalpy of formation of methanol is the sum of the standard enthalpies of reaction of the three reactions above:

$$\Delta_f H^\circ \left[CH_3OH(l)\right] = 726.4 \text{ kJ mol}^{-1} - 393.5 \text{ kJ mol}^{-1} - 571.6 \text{ kJ mol}^{-1} = -238.7 \text{ kJ mol}^{-1}$$

4.55 The standard enthalpy change for the following reaction is 436.4 kJ mol^{-1}:

$$H_2(g) \rightarrow H(g) + H(g)$$

Calculate the standard enthalpy of formation of atomic hydrogen (H).

$$\Delta_r H^\circ = 2\Delta_f \overline{H}^\circ [H(g)] - \Delta_f \overline{H}^\circ \left[H_2(g)\right] = 2\Delta_f \overline{H}^\circ [H(g)]$$

Note that $\Delta_f \overline{H}^\circ \left[H_2(g)\right] = 0$ because $H_2(g)$ is the stable allotrope of hydrogen. Therefore,

$$\Delta_f \overline{H}^\circ [H(g)] = \frac{1}{2}\left(\Delta_r H^\circ\right) = \frac{1}{2}\left(436.4 \text{ kJ mol}^{-1}\right) = 218.2 \text{ kJ mol}^{-1}$$

4.56 Calculate the difference between the values of $\Delta_r H^\circ$ and $\Delta_r U^\circ$ for the oxidation of α-D-glucose at 298 K:

$$C_6H_{12}O_6(s) + 6O_2(g) \rightarrow 6CO_2(g) + 6H_2O(l)$$

$\Delta_r H^\circ$ and $\Delta_r U^\circ$ differ from each other if the number of moles of gases after the reaction is not the same as that before the reaction.

$$\Delta_r H^\circ = \Delta_r U^\circ + RT \Delta n$$

Since $\Delta n = 0$, $\Delta_r H^\circ = \Delta_r U^\circ$, or $\Delta_r H^\circ - \Delta_r U^\circ = 0$.

4.57 Alcohol fermentation is the process in which carbohydrates are broken down into ethanol and carbon dioxide. The reaction is very complex and involves a number of enzyme-catalyzed steps. The overall change is

$$C_6H_{12}O_6(s) \rightarrow 2C_2H_5OH(l) + 2CO_2(g)$$

Calculate the standard enthalpy change for this reaction, assuming that the carbohydrate is α-D-glucose.

$$\Delta_r H^\circ = 2\Delta_f \overline{H}^\circ \left[C_2H_5OH(l) \right] + 2\Delta_f \overline{H}^\circ \left[CO_2(g) \right] - \Delta_f \overline{H}^\circ \left[C_6H_{12}O_6(s) \right]$$

$$= 2\left(-277.0 \text{ kJ mol}^{-1} \right) + 2\left(-393.5 \text{ kJ mol}^{-1} \right) - \left(-1274.5 \text{ kJ mol}^{-1} \right)$$

$$= -66.5 \text{ kJ mol}^{-1}$$

4.58 (a) Explain why the bond enthalpy of a molecule is always defined in terms of a gas-phase reaction. (b) The bond dissociation enthalpy of F_2 is 150.6 kJ mol^{-1}. Calculate the value of $\Delta_f H^\circ$ for F(g).

(a) In the gas phase, molecules are far apart and not affected by intermolecular interactions. The bond dissociation enthalpies so determined thus refer only to the chemical bond between specific atoms and are not influenced by intermolecular interactions.

(b) It is given that $\Delta_r H^\circ = 150.6$ kJ mol^{-1} for the reaction $F_2(g) \rightarrow 2F(g)$. $\Delta_r H^\circ$ is related to the enthalpies of formation of $F_2(g)$ (which is 0) and F(g) in the following manner:

$$\Delta_r H^\circ = 2\Delta_f \overline{H}^\circ \left[F(g) \right] - \Delta_f \overline{H}^\circ \left[F_2(g) \right] = 2\Delta_f \overline{H}^\circ \left[F(g) \right]$$

Therefore,

$$\Delta_f \overline{H}^\circ \left[F(g) \right] = \frac{1}{2}\Delta_r H^\circ = \frac{1}{2}\left(150.6 \text{ kJ mol}^{-1} \right) = 75.3 \text{ kJ mol}^{-1}$$

4.59 From the molar enthalpy of vaporization of water at 373 K and the bond dissociation enthalpies of H_2 and O_2 (see Table 4.4), calculate the average O–H bond enthalpy in water given that

$$H_2(g) + 1/2\, O_2(g) \rightarrow H_2O(l) \qquad \Delta_r H^\circ = -285.8 \text{ kJ mol}^{-1}$$

First calculate the enthalpy of vaporization of H_2O at 298 K ($\Delta_r H^\circ_{298}$). Then assuming heat capacities are temperature independent, calculate the enthalpy of vaporization of H_2O at 373 K ($\Delta_r H^\circ_{373}$).

$$\Delta_r H^\circ_{298} = \Delta_f \overline{H}^\circ \left[H_2O(g) \right] - \Delta_f \overline{H}^\circ \left[H_2O(l) \right]$$

$$= -241.8 \text{ kJ mol}^{-1} - \left(-285.8 \text{ kJ mol}^{-1} \right)$$

$$= 44.0 \text{ kJ mol}^{-1}$$

$$\Delta_r H^o_{373} = \Delta_r H^o_{298} + \Delta C_P \, (373 \text{ K} - 298 \text{ K})$$

$$= \Delta_r H^o_{298} + \left\{ \overline{C}^o_P \left[H_2O(g) \right] - \overline{C}^o_P \left[H_2O(l) \right] \right\} (75 \text{ K})$$

$$= 44.0 \text{ kJ mol}^{-1} + \left(33.6 \text{ J K}^{-1} \text{ mol}^{-1} - 75.3 \text{ J K}^{-1} \text{ mol}^{-1} \right) (75 \text{ K}) \left(\frac{1 \text{ kJ}}{1000 \text{ J}} \right)$$

$$= 40.9 \text{ kJ mol}^{-1}$$

To calculate the average O–H bond enthalpy in water, the enthalpy of reaction for $H_2O(g) \rightarrow 2H(g)$ + $O(g)$ must first be obtained. This reaction can be expressed as a sum of the following reactions:

$$H_2O(g) \rightarrow H_2O(l) \qquad \Delta_r H^o_{373} = -40.9 \text{ kJ mol}^{-1}$$
$$H_2(g) \rightarrow 2H(g) \qquad \Delta_r H^o = 436.4 \text{ kJ mol}^{-1}$$
$$\tfrac{1}{2}O_2(g) \rightarrow O(g) \qquad \Delta_r H^o = \tfrac{498.8}{2} \text{ kJ mol}^{-1} = 249.4 \text{ kJ mol}^{-1}$$
$$H_2O(l) \rightarrow H_2(g) + \tfrac{1}{2} O_2(g) \qquad \Delta_r H^o = 285.8 \text{ kJ mol}^{-1}$$

Thus, the enthalpy of reaction for $2H(g) + O(g) \rightarrow H_2O(g)$ is

$$\Delta_r H^o = -40.9 \text{ kJ mol}^{-1} + 436.4 \text{ kJ mol}^{-1} + 249.4 \text{ kJ mol}^{-1} + 285.8 \text{ kJ mol}^{-1} = 930.7 \text{ kJ mol}^{-1}$$

This enthalpy of reaction represents the bond enthalpy of two moles of O–H bonds. Therefore, the average bond enthalpy is

$$\frac{930.7 \text{ kJ mol}^{-1}}{2} = 465.4 \text{ kJ mol}^{-1}$$

4.60 Use the bond enthalpy values in Table 4.4 to calculate the enthalpy of combustion for ethane,

$$2C_2H_6(g) + 7O_2(g) \rightarrow 4CO_2(g) + 6H_2O(l)$$

Compare your result with that calculated from the enthalpy of formation values of the products and reactants listed in Appendix B.

Calculation of the enthalpy of combustion using bond enthalpies:

Type of bonds broken	Number of bonds broken	Bond enthalpy / kJ·mol^{-1}	Enthalpy change / kJ·mol^{-1}
C—H	12	414	4968
C—C	2	347	694
O=O	7	498.8	3491.6

Type of bonds formed	Number of bonds formed	Bond enthalpy / kJ·mol^{-1}	Enthalpy change / kJ·mol^{-1}
C=O	8	799	6392
O—H	12	460	5520

$$\Delta_r H^o = (4968 + 694 + 3491.6) \text{ kJ mol}^{-1} - (6392 + 5520) \text{ kJ mol}^{-1} = -2758 \text{ kJ mol}^{-1}$$

Calculation of the enthalpy of combustion using enthalpies of formation:

$$\Delta_r H^\circ = 4\Delta_f \overline{H}^\circ \left[CO_2(g) \right] + 6\Delta_f \overline{H}^\circ \left[H_2O(l) \right] - 2\Delta_f \overline{H}^\circ \left[C_2H_6(g) \right] - 7\Delta_f \overline{H}^\circ \left[O_2(g) \right]$$

$$= 4\left(-393.5 \text{ kJ mol}^{-1} \right) + 6\left(-285.8 \text{ kJ mol}^{-1} \right) - 2\left(-84.7 \text{ kJ mol}^{-1} \right) - 7\left(0 \text{ kJ mol}^{-1} \right)$$

$$= -3119.4 \text{ kJ mol}^{-1}$$

The value of $\Delta_r H^\circ$ so calculated is 13% greater than that calculated using bond enthalpies. The value determined using enthalpies of formation is the correct value, since it relies on the first law of thermodynamics. Bond enthalpies are averages determined for similar bonds in many molecules and provide estimates that are typically within 10% of the experimental value for any given, particular reaction.

4.61   A 2.10-mole sample of crystalline acetic acid, initially at 17.0°C, is allowed to melt at 17.0°C and is then heated to 118.1°C (its normal boiling point) at 1.00 atm. The sample is allowed to vaporize at 118.1°C and is then rapidly quenched to 17.0°C, so that it recrystallizes. Calculate the value of $\Delta_r H^\circ$ for the total process as described.

Since the process described is cyclic (that is, the initial and final states are identical), $\Delta_r H^\circ = 0$.

4.62   Predict whether the values of q, w, ΔU, ΔH are positive, zero, or negative for each of the following processes: **(a)** melting of ice at 1 atm and 273 K, **(b)** melting of solid cyclohexane at 1 atm and the normal melting point, **(c)** reversible isothermal expansion of an ideal gas, and **(d)** reversible adiabatic expansion of an ideal gas.

For most substances, the volume increases upon melting, but for ice melting to water under the conditions given, the volume decreases. In all cases, the volume change between the solid and liquid phases is small as is the value of the work associated with the melting. For ideal gases, both internal energy, U, and enthalpy, H, depend only on temperature. Keeping these points in mind results in the following predictions.

	q	w	ΔU	ΔH
(a)	positive	positive	positive	positive
(b)	positive	negative	positive	positive
(c)	positive	negative	zero	zero
(d)	zero	negative	negative	negative

4.63   Einstein's special relativity equation is $E = mc^2$, where E is energy, m is mass, and c is the velocity of light. Does this equation invalidate the conservation of energy, and hence the first law of thermodynamics?

No, special relativity enlarges the definition of energy to include the mass, that is, each mass m has an associated energy $E = mc^2$. Likewise all energy has an associated mass equal to E/c^2. The total energy of the system plus the surroundings is still conserved according to the theory of special relativity, but one needs to include the energy associated with mass as well as the more usual forms of energy. Similarly, the total mass of the system plus surroundings is still conserved, as long as the mass associated with energy is included. Under conditions where it is not necessary to account for effects due to special relativity, energy and mass are separately conserved.

4.64 The convention of arbitrarily assigning a zero enthalpy value to all the (most stable) elements in the standard state and (usually) 298 K is a convenient way of dealing with the enthalpy changes of chemical processes. This convention does not apply to one kind of process, however. What process is it? Why?

In a nuclear process, there are different elements on both sides of the chemical equation, and this convention would not apply.

4.65 Two moles of an ideal gas are compressed isothermally at 298 K from 1.00 atm to 200 atm. Calculate the values of q, w, ΔU, and ΔH for the process if it is carried out **(a)** reversibly and **(b)** by applying an external pressure of 300 atm.

(a) When an ideal gas undergoes an isothermal process, $\Delta U = 0$ and $\Delta H = 0$.

$$w = -nRT \ln \frac{P_1}{P_2} = -(2\text{ mol}) \left(8.314 \text{ J K}^{-1} \text{ mol}^{-1}\right)(298\text{ K}) \ln \frac{1.00\text{ atm}}{200\text{ atm}} = 2.63 \times 10^4 \text{ J}$$

$$q = \Delta U - w = -w = -2.63 \times 10^4 \text{ J}$$

(b) When an ideal gas undergoes an isothermal process, $\Delta U = 0$ and $\Delta H = 0$.

$$w = -P_{ex}\left(V_2 - V_1\right)$$

V_1 and V_2 can be determined using the ideal gas law:

$$V_1 = \frac{nRT}{P_1} = \frac{(2\text{ mol})\left(0.08206 \text{ L atm K}^{-1}\text{ mol}^{-1}\right)(298\text{ K})}{1.00\text{ atm}} = 48.91 \text{ L}$$

$$V_2 = \frac{nRT}{P_2} = \frac{(2\text{ mol})\left(0.08206 \text{ L atm K}^{-1}\text{ mol}^{-1}\right)(298\text{ K})}{200\text{ atm}} = 0.2445 \text{ L}$$

Therefore,

$$w = -(300\text{ atm})(0.2445\text{ L} - 48.91\text{ L})\left(\frac{101.3\text{ J}}{1\text{ L atm}}\right) = 1.48 \times 10^6 \text{ J}$$

and

$$q = \Delta U - w = -w = -1.48 \times 10^6 \text{ J}$$

4.66 The fuel value of hamburger is about 3.6 kcal g^{-1}. If a person eats 1 pound of hamburger for lunch and if none of the energy is stored in his body, estimate the amount of water that would have to be lost in perspiration to keep his body temperature constant. (1 lb = 454 g.)

The fuel value of 1 pound of hamburger is

$$(1 \text{ lb}) \left(\frac{454 \text{ g}}{1 \text{ lb}} \right) (3.6 \text{ kcal g}^{-1}) \left(\frac{4.184 \text{ kJ}}{1 \text{ kcal}} \right) = 6.84 \times 10^3 \text{ kJ}$$

For the vaporization of water at 298 K, $H_2O(l) \longrightarrow H_2O(g)$, $\Delta_{vap}H = 44.01 \text{ kJ mol}^{-1}$. ($\Delta_{vap}H$ is appropriate, since the vaporization is taking place at constant, atmospheric pressure.) Assuming that the entire fuel value of the hamburger is used to vaporize water, it will require that

$$\left(\frac{6.84 \times 10^3 \text{ kJ}}{44.01 \text{ kJ mol}^{-1}} \right) \left(\frac{18.02 \text{ g H}_2\text{O}}{1 \text{ mol H}_2\text{O}} \right) = 2.8 \times 10^3 \text{g H}_2\text{O}$$

be vaporized.

4.67   A quantity of 4.50 g of CaC_2 is reacted with an excess of water at 298 K and atmospheric pressure:

$$CaC_2(s) + 2H_2O(l) \rightarrow Ca(OH)_2(aq) + C_2H_2(g)$$

Calculate the work done in joules by the acetylene gas against the atmospheric pressure.

The work done depends on the external pressure and the change in volume of the system. Because the volumes of compounds in the solid and liquid (or aqueous) phases are negligible compared with those in the gas phase, the volume change of the system is

$$\Delta V = V_2 - V_1 = V_{C_2H_2} = \frac{n_{C_2H_2}RT}{P_2}$$

where $P_2 = P_{ex}$ because the gas expands until it has the same pressure as the external pressure. The amount of acetylene produced is

$$n_{C_2H_2} = \frac{4.50 \text{ g CaC}_2}{64.10 \text{ g mol}^{-1} \text{ CaC}_2} = 0.07020 \text{ mol}$$

The work done by acetylene is

$$w = -P_{ex}\Delta V = -P_{ex} \left(\frac{n_{C_2H_2}RT}{P_2} \right) = -nRT$$

$$= -(0.07020 \text{ mol}) \left(8.314 \text{ J K}^{-1} \text{ mol}^{-1} \right) (298 \text{ K}) = -174 \text{ J}$$

4.68   An oxyacetylene flame is often used in the welding of metals. Estimate the flame temperature produced by the reaction

$$2C_2H_2(g) + 5O_2(g) \rightarrow 4CO_2(g) + 2H_2O(g)$$

Assume that the heat generated from this reaction is all used to heat the products. (*Hint:* First calculate the value of $\Delta_r H°$ for the reaction. Next, look up the heat capacities of the products. Assume that the heat capacities are temperature independent.)

The enthalpy of reaction for the oxidation of acetylene is

$$\Delta_r H^\circ = 4\Delta_f \overline{H}^\circ \left[CO_2(g)\right] + 2\Delta_f \overline{H}^\circ \left[H_2O(g)\right] - 2\Delta_f \overline{H}^\circ \left[C_2H_2(g)\right] - 5\Delta_f \overline{H}^\circ \left[O_2(g)\right]$$

$$= 4\left(-393.5 \text{ kJ mol}^{-1}\right) + 2\left(-241.8 \text{ kJ mol}^{-1}\right) - 2\left(226.6 \text{ kJ mol}^{-1}\right) - 5\left(0 \text{ kJ mol}^{-1}\right)$$

$$= -2510.8 \text{ kJ mol}^{-1}$$

Because the reaction takes place at constant pressure, the enthalpy of reaction is the same as heat released by the reaction, and this heat is absorbed by the products. The initial temperature, T_i, is assumed to be 298 K, and the final temperature of the products, T_f is

$$q = 2510.8 \text{ kJ mol}^{-1} = \left\{4\overline{C}_P^\circ \left[CO_2(g)\right] + 2\overline{C}_P^\circ \left[H_2O(g)\right]\right\} \left(T_f - T_i\right)$$

$$= \left[4\left(37.1 \text{ J mol}^{-1}\text{ K}^{-1}\right) + 2\left(33.6 \text{ J mol}^{-1}\text{ K}^{-1}\right)\right]\left(T_f - 298 \text{ K}\right)\left(\frac{1 \text{ kJ}}{1000 \text{ J}}\right)$$

$$T_f - 298 \text{ K} = 1.165 \times 10^4 \text{ K}$$

$$T_f = 1.19 \times 10^4 \text{ K}$$

The value of $\overline{C}_P^\circ$ for $H_2O(g)$ may be found in a standard reference.

4.69   The $\Delta_f \overline{H}^\circ$ values listed in Appendix B all refer to 1 bar and 298 K. Suppose that a student wants to set up a new table of $\Delta_f \overline{H}^\circ$ values at 1 bar and 273 K. Show how she should proceed on the conversion, using acetone as an example.

The enthalpy of formation of a compound at 1 bar and 273 K, $\Delta_f \overline{H}^\circ$ (273) can be calculated from that at 1 bar and 298 K, $\Delta_f \overline{H}^\circ$ (298) and $\overline{C}_P$ of the compound by assuming that the heat capacity is temperature independent. Using acetone as an example,

$$\Delta_f \overline{H}^\circ (273) = \Delta_f \overline{H}^\circ (298) + \overline{C}_P (273 \text{ K} - 298 \text{ K})$$

$$= -246.8 \text{ kJ mol}^{-1} + \left(126.8 \text{ J K}^{-1}\text{ mol}^{-1}\right)(-25 \text{ K})\left(\frac{1 \text{ kJ}}{1000 \text{ J}}\right)$$

$$= -250.0 \text{ kJ mol}^{-1}$$

4.70   The enthalpies of hydrogenation of ethylene and benzene have been determined at 298 K:

$$C_2H_4(g) + H_2(g) \rightarrow C_2H_6(g) \qquad \Delta_r H^\circ = -132 \text{ kJ mol}^{-1}$$
$$C_6H_6(g) + 3H_2(g) \rightarrow C_6H_{12}(g) \qquad \Delta_r H^\circ = -246 \text{ kJ mol}^{-1}$$

What would be the enthalpy of hydrogenation for benzene if it contained three isolated, unconjugated double bonds? How would you account for the difference between the calculated value based on this assumption and the measured value?

If benzene contained three isolated, unconjugated double bonds, its enthalpy of hydrogenation could be estimated as three times that for the single double bond in ethylene, or $\Delta_r H^\circ_{calc} = 3\left(-132 \text{ kJ mol}^{-1}\right) = -396 \text{ kJ mol}^{-1}$. The difference is $-246 \text{ kJ mol}^{-1} - \left(-396 \text{ kJ mol}^{-1}\right) = 150 \text{ kJ mol}^{-1}$. This implies that benzene is 150 kJ mol^{-1} more stable than the hypothetical molecule with three isolated, unconjugated double bonds. This is attributed to the resonance (electron delocalization) energy of benzene.

4.71   The molar enthalpies of fusion and vaporization of water are 6.01 kJ mol^{-1} and $44.01 \text{ kJ mol}^{-1}$ (at 298 K), respectively. From these values estimate the molar enthalpy of sublimation of ice.

The sublimation process can be considered a sum of the following processes:

$$H_2O(s) \rightarrow H_2O(l) \qquad \Delta_{fus} H^\circ = 6.01 \text{ kJ mol}^{-1}$$
$$H_2O(l) \rightarrow H_2O(g) \qquad \Delta_{vap} H^\circ = 44.01 \text{ kJ mol}^{-1}$$

Thus,

$$\Delta_{sub} H^\circ = 6.01 \text{ kJ mol}^{-1} + 44.01 \text{ kJ mol}^{-1} = 50.02 \text{ kJ mol}^{-1}$$

This value is an approximation, since the value of $\Delta_{vap} H^\circ$ is given at 298 K, whereas ice would not typically be found above 273 K.

4.72   The standard enthalpy of formation at 298 K of HF(aq) is $-320.1 \text{ kJ mol}^{-1}$; OH$^-$($aq$), -229.6 kJ mol^{-1}; F$^-$(aq), $-329.11 \text{ kJ mol}^{-1}$; and H$_2$O($l$), $-285.84 \text{ kJ mol}^{-1}$. **(a)** Calculate the enthalpy of neutralization of HF(aq),

$$HF(aq) + OH^-(aq) \rightarrow F^-(aq) + H_2O(l)$$

(b) Using the value of $-55.83 \text{ kJ mol}^{-1}$ as the enthalpy change from the reaction

$$H^+(aq) + OH^-(aq) \rightarrow H_2O(l)$$

calculate the enthalpy change for the dissociation

$$HF(aq) \rightarrow H^+(aq) + F^-(aq)$$

(a)

$$\Delta_r H^\circ = \Delta_f \overline{H}^\circ \left[F^-(aq)\right] + \Delta_f \overline{H}^\circ \left[H_2O(l)\right] - \Delta_f \overline{H}^\circ \left[HF(aq)\right] - \Delta_f \overline{H}^\circ \left[OH^-(aq)\right]$$

$$= -329.11 \text{ kJ mol}^{-1} + \left(-285.8 \text{ kJ mol}^{-1}\right) - \left(-320.1 \text{ kJ mol}^{-1}\right) - \left(-229.6 \text{ kJ mol}^{-1}\right)$$

$$= -65.2 \text{ kJ mol}^{-1}$$

(b) The dissociation of HF can be considered as a sum of the following equations:

$$HF(aq) + OH^-(aq) \rightarrow F^-(aq) + H_2O(l) \qquad \Delta_r H^\circ = -65.2 \text{ kJ mol}^{-1}$$
$$H_2O(l) \rightarrow H^+(aq) + OH^-(aq) \qquad \Delta_r H^\circ = 55.83 \text{ kJ mol}^{-1}$$

The enthalpy change for the dissociation reaction is therefore

$$\Delta_r H^\circ = -65.2 \text{ kJ mol}^{-1} + 55.83 \text{ kJ mol}^{-1} = -9.4 \text{ kJ mol}^{-1}$$

4.73 It was stated in the chapter that for reactions in condensed phases, the difference between $\Delta_r H$ and $\Delta_r U$ is usually negligibly small. This statement holds for processes carried out under atmospheric conditions. For certain geochemical processes, however, the external pressures may be so great that $\Delta_r H$ and $\Delta_r U$ values can differ by a significant amount. A well-known example is the slow conversion of graphite to diamond under Earth's surface. Calculate the value of the quantity $(\Delta_r H - \Delta_r U)$ for the conversion of 1 mole of graphite to 1 mole of diamond at a pressure of 50,000 atm. The densities of graphite and diamond are 2.25 g cm^{-3} and 3.52 g cm^{-3}, respectively.

At constant pressure,

$$\Delta_r H - \Delta_r U = \Delta\left(P\overline{V}\right) = P\Delta\overline{V}.$$

Thus, the change in volume, $\Delta\overline{V}$, for the conversion process is required to find $\Delta_r H - \Delta_r U$.

The volume of 1 mole, or 12.01 g of graphite is

$$\overline{V}_{\text{graphite}} = \frac{12.01 \text{ g mol}^{-1}}{2.25 \text{ g cm}^{-3}} \left(\frac{1 \text{ L}}{1000 \text{ cm}^3}\right) = 5.338 \times 10^{-3} \text{ L mol}^{-1}$$

The volume of 1 mole, or 12.01 g of diamond is

$$\overline{V}_{\text{diamond}} = \frac{12.01 \text{ g mol}^{-1}}{3.52 \text{ g cm}^{-3}} \left(\frac{1 \text{ L}}{1000 \text{ cm}^3}\right) = 3.412 \times 10^{-3} \text{ L mol}^{-1}$$

Therefore,

$$\Delta_r H - \Delta_r U = (50000 \text{ atm}) \left(3.412 \times 10^{-3} - 5.338 \times 10^{-3}\right) \text{ L mol}^{-1} \left(\frac{101.3 \text{ J}}{1 \text{ L atm}}\right) = -9.78 \times 10^3 \text{ J}$$

4.74 Metabolic activity in the human body releases approximately 1.0×10^4 kJ of heat per day. Assuming the body is 50 kg of water, how fast would the body temperature rise if it were an isolated system? How much water must the body eliminate as perspiration to maintain the normal body temperature (98.6°F)? Comment on your results. The heat of vaporization of water may be taken as 2.41 kJ g^{-1}.

If the body absorbs all the heat released and is an isolated system, then the temperature rise, ΔT is related to q in the following fashion:

$$q_{\text{absorbed}} = C_P\left[H_2O(l)\right]\Delta T$$

$$\Delta T = \frac{q}{C_P\left[H_2O(l)\right]} = \frac{\left(1.0 \times 10^4 \text{ kJ}\right)\left(\frac{1000 \text{ J}}{1 \text{ kJ}}\right)}{\left(\frac{50 \times 10^3 \text{ g}}{18.01 \text{ g mol}^{-1}}\right)\left(75.3 \text{ J K}^{-1} \text{ mol}^{-1}\right)} = 47.8 \text{ K}$$

If the body temperature is to remain constant, then the heat released by metabolic activity must be used for the evaporation of water as perspiration, that is,

$$1.0 \times 10^4 \text{ kJ} = m_{\text{water}} \left(2.41 \text{ kJ g}^{-1}\right)$$

$$m_{\text{water}} = 4.1 \times 10^3 \text{ g}$$

The actual amount of perspiration is less than this because part of the body heat is lost to the surroundings by convection and radiation.

4.75   An ideal gas in a cylinder fitted with a movable piston is adiabatically compressed from V_1 to V_2. As a result, the temperature of the gas rises. Explain what causes the temperature of the gas to rise.

The unidirectionally moving piston transfers linear momentum to the gas molecules as they collide with it. Consequently, the molecules are moving faster after their encounter with the piston. Thus, their kinetic energy is increased, and the temperature of the gas rises.

4.76   Calculate the fraction of the enthalpy of vaporization of water used for the expansion of steam at its normal boiling point.

Treating the steam as an ideal gas, the molar volume of steam at 373 K is calculated as

$$\overline{V} = \frac{RT}{P}$$

$$= \frac{\left(0.08206 \text{ L atm K}^{-1} \text{ mol}^{-1}\right) (373 \text{ K})}{1 \text{ atm}}$$

$$= 30.61 \text{ L mol}^{-1}$$

The work done in the expansion from liquid water to this volume of steam is then found having made the assumption that the volume of the condensed phase is negligible. (One mole of liquid water has a volume of 18 mL, so the approximation is a good one.)

$$\begin{aligned} w &= -P_{\text{ext}} \Delta \overline{V} \\ &= -(1 \text{ atm}) \left(30.61 \text{ L mol}^{-1}\right) \left(\tfrac{101.3 \text{ J}}{1 \text{ L atm}}\right) \left(\tfrac{1 \text{ kJ}}{1000 \text{ J}}\right) \\ &= -3.101 \text{ kJ mol}^{-1} \end{aligned}$$

At 373 K the molar enthalpy of vaporization of water is $\Delta_{\text{vap}} \overline{H}^{\circ} = 40.79 \text{ kJ mol}^{-1}$, so the fraction used for the expansion of steam is

$$\frac{3.101 \text{ kJ mol}^{-1}}{40.79 \text{ kJ mol}^{-1}} = 7.60\%$$

4.77   The combustion of what volume of ethane (C_2H_6), measured at 23.0°C and 752 mmHg, would be required to heat 855 g of water from 25.0°C to 98.0°C?

The combustion reaction is

$$C_2H_6(g) + \frac{7}{2}O_2(g) \longrightarrow 2CO_2(g) + 3H_2O(l)$$

Use ethalpies of formation to determine the enthalpy of combustion.

$$\Delta_r H^\circ = 2\Delta_f H^\circ\left[CO_2(g)\right] + 3\Delta_f H^\circ\left[H_2O(l)\right] - \Delta_f H^\circ\left[C_2H_6(g)\right] - \frac{7}{2}\Delta_f H^\circ\left[O_2(g)\right]$$

$$= 2\left(-393.5\ \text{kJ mol}^{-1}\right) + 3\left(-285.8\ \text{kJ mol}^{-1}\right) - \left(-84.7\ \text{kJ mol}^{-1}\right) - \frac{7}{2}\left(0\ \text{kJ mol}^{-1}\right)$$

$$= -1559.7\ \text{kJ mol}^{-1}$$

Thus each mole of ethane provides 1559.7 kJ of thermal energy upon combustion at constant pressure.

The thermal energy required to heat the water is

$$q = m_{H_2O}s_{H_2O}\Delta T = (855\ \text{g})\left(4.184\ \text{J g}^{-1}\,^\circ\text{C}^{-1}\right)(98.0 - 25.0)\,^\circ\text{C}\left(\frac{1\ \text{kJ}}{1000\ \text{J}}\right) = 261.1\ \text{kJ}$$

Clearly a fraction of a mole of ethane is required. The actual number of moles is

$$n = (261.1\ \text{kJ})\left(\frac{1\ \text{mol ethane}}{1559.7\ \text{kJ}}\right) = 0.1674\ \text{mol ethane}$$

The volume of the ethane at the stated conditions is

$$V = \frac{nRT}{P} = \frac{(0.1674\ \text{mol})\left(0.08206\ \text{L atm K}^{-1}\ \text{mol}^{-1}\right)(23 + 273)\ \text{K}}{(752\ \text{mmHg})\left(\frac{760\ \text{mmHg}}{1\ \text{atm}}\right)} = 4.11\ \text{L}$$

4.78   Calculate the internal energy of a Goodyear blimp filled with helium gas at 1.2×10^5 Pa (compared to the empty blimp). The volume of the inflated blimp is 5.5×10^3 m^3. If all the energy were used to heat 10.0 tons of copper at 21°C, calculate the final temperature of the metal. (*Hint*: 1 ton $= 9.072 \times 10^5$ g.)

The internal energy of a monatomic (ideal) gas is $\frac{3}{2}nRT$. Likewise for an ideal gas, $PV = nRT$. Thus, for an ideal gas, $U = \frac{3}{2}PV$. For the blimp,

$$U = \frac{3}{2}\left(1.2 \times 10^5\ \text{Pa}\right)\left(5.5 \times 10^3\ \text{m}^3\right) = 9.90 \times 10^8\ \text{J}$$

If all this energy is used to heat copper, the temperature change is related to the amount of energy used via $q = m_{Cu}s_{Cu}\Delta T$, giving

$$9.90 \times 10^8\ \text{J} = (10.0\ \text{ton})\left(\frac{9.072 \times 10^5\ \text{g}}{1\ \text{ton}}\right)\left(24.47\ \text{J}\,^\circ\text{C}^{-1}\ \text{mol}^{-1}\right)\left(\frac{1\ \text{mol Cu}}{63.55\ \text{g Cu}}\right)\Delta T$$

$$9.90 \times 10^8 = 3.49 \times 10^6\,^\circ\text{C}^{-1}\Delta T$$

$$283.7^\circ\text{C} = \Delta T = T_f - 21^\circ\text{C}$$

Thus, $T_f = 305^\circ\text{C}$

4.79 Without referring to the chapter, state the conditions for each of the following equations:
(a) $\Delta H = \Delta U + P \Delta V$, (b) $C_P = C_V + nR$, (c) $\gamma = 5/3$, (d) $P_1 V_1^\gamma = P_2 V_2^\gamma$,
(e) $w = n\overline{C}_V (T_2 - T_1)$, (f) $w = -P\Delta V$, (g) $w = -nRT \ln (V_2/V_1)$, (h) $dH = dq$.

(a) Constant pressure,

(b) ideal gas,

(c) monoatomic, ideal gas,

(d) adiabatic, ideal gas, reversible process,

(e) constant volume, $\overline{C}_V$ independent of temperature,

(f) constant external pressure,

(g) ideal gas, isothermal, reversible process,

(h) constant pressure.

4.80 An ideal gas is isothermally compressed from P_1, V_1 to P_2, V_2. Under what conditions would the work done be a minimum? a maximum? Write the expressions for minimum and maximum work done for this process. Explain your reasoning.

The conditions are opposite to those for the isothermal *expansion* of an ideal gas. The minimum work is done in an reversible process because the opposing (internal) pressure is only infinitesimally smaller than the external pressure causing the compression. The work done under these conditions is

$$w_{min} = -nRT \ln \frac{V_2}{V_1}$$

Since $V_2 < V_1$, the work is positive as expected for a compression.

The maximum work would be done in an irreversible process and an external pressure of P_2, giving a value of

$$w_{max} = -P_2 (V_2 - V_1)$$

4.81 Construct a table with the headings q, w, ΔU, and ΔH. For each of the following processes, deduce whether each of the quantities listed is positive (+), negative(−), or zero(0). (a) Freezing of acetone at 1 atm and its normal melting point. (b) Irreversible isothermal expansion of an ideal gas. (c) Adiabatic compression of an ideal gas. (d) Reaction of sodium with water. (e) Boiling of liquid ammonia at its normal boiling point. (f) Irreversible adiabatic expansion of a gas against an external pressure. (g) Reversible isothermal compression of an ideal gas. (h) Heating of a gas at constant volume. (i) Freezing of water at 0°C.

For most substances, the volume increases upon melting. (Ice melting to water is a notable exception.) In all cases, the volume change between the solid and liquid phases is small as is the value of the work associated with the melting. The volume change upon vaporization, however, is very large. For ideal gases, both internal energy U and enthalpy H depend only on temperature. Keeping these points in mind results in the following predictions.

	q	w	ΔU	ΔH
(a)	negative	positive	negative	negative
(b)	positive	negative	zero	zero
(c)	zero	positive	positive	positive
(d)	negative	negative	negative	negative
(e)	positive	negative	positive	positive
(f)	zero	negative	negative	negative
(g)	negative	positive	zero	zero
(h)	positive	zero	positive	positive
(i)	negative	negative	negative	negative

4.82 State whether each of the following statements is true or false: **(a)** $\Delta U \approx \Delta H$ except for gases or high-pressure processes. **(b)** In gas compression, a reversible process does maximum work. **(c)** ΔU is a state function. **(d)** $\Delta U = q + w$ for an open system. **(e)** C_V is temperature independent for gases. **(f)** The internal energy of a real gas depends only on temperature.

(a) True **(b)** False **(c)** False (U is a state function, ΔU is not.) **(d)** False (It is true for a *closed* system.) **(e)** False **(f)** False

4.83 Show that $\left(\partial C_V / \partial V\right)_T = 0$ for an ideal gas.

$$\left(\frac{\partial C_V}{\partial V}\right)_T = \left(\frac{\left(\frac{\partial U}{\partial T}\right)_V}{\partial V}\right)_T = \left[\frac{\partial}{\partial V}\left(\frac{\partial U}{\partial T}\right)_V\right]_T = \left[\frac{\partial}{\partial T}\left(\frac{\partial U}{\partial V}\right)_T\right]_V$$

Since the internal energy of an ideal gas depends only on temperature,

$$\left(\frac{\partial U}{\partial V}\right)_T = 0$$

Thus,

$$\left(\frac{\partial C_V}{\partial V}\right)_T = 0$$

4.84 Calculate the work done during the isothermal, reversible expansion of a van der Waals gas. Account physically for the way in which the coefficients a and b appear in the final expression. *Hint*: You need to apply the Taylor series expansion:

$$\ln(1 - x) = -x - \frac{x^2}{2} \cdots \qquad \text{for } |x| \ll 1$$

to the expression $\ln(V - nb)$. Recall that the a term represents attraction and the b term repulsion.

$$w = -\int_{V_1}^{V_2} P\,dV$$

$$= -\int_{V_1}^{V_2} \left(\frac{nRT}{V-nb} - \frac{an^2}{V^2} \right) dV$$

$$= -nRT \ln \frac{V_2 - nb}{V_1 - nb} - an^2 \left(\frac{1}{V_2} - \frac{1}{V_1} \right)$$

The ln term can be written as

$$\ln \frac{V_2 - nb}{V_1 - nb} = \ln \frac{V_2 \left(1 - \frac{nb}{V_2} \right)}{V_1 \left(1 - \frac{nb}{V_1} \right)}$$

$$= \ln \frac{V_2}{V_1} + \ln \frac{1 - \frac{nb}{V_2}}{1 - \frac{nb}{V_1}}$$

$$= \ln \frac{V_2}{V_1} + \ln \left(1 - \frac{nb}{V_2} \right) - \ln \left(1 - \frac{nb}{V_1} \right)$$

Assume the volume occupied by the gas molecules is much greater than the volume of the molecules, both before and after the expansion, that is, $V_1 \gg nb$ and $V_2 \gg nb$, then $\frac{nb}{V_1} \ll 1$ and $\frac{nb}{V_2} \ll 1$. Under these condition, the Taylor expansion described in the question can be used to simplify the ln terms.

$$\ln \frac{V_2 - nb}{V_1 - nb} = \ln \frac{V_2}{V_1} + \left(-\frac{nb}{V_2} - \frac{1}{2} \frac{n^2 b^2}{V_2^2} + \cdots \right) - \left(-\frac{nb}{V_1} - \frac{1}{2} \frac{n^2 b^2}{V_1^2} + \cdots \right)$$

$$= \ln \frac{V_2}{V_1} - nb \left(\frac{1}{V_2} - \frac{1}{V_1} \right) - \frac{n^2 b^2}{2} \left(\frac{1}{V_2^2} - \frac{1}{V_1^2} \right) + \cdots$$

Substitute the above expression into w.

$$w = -nRT \left[\ln \frac{V_2}{V_1} - nb \left(\frac{1}{V_2} - \frac{1}{V_1} \right) - \frac{n^2 b^2}{2} \left(\frac{1}{V_2^2} - \frac{1}{V_1^2} \right) + \cdots \right] - an^2 \left(\frac{1}{V_2} - \frac{1}{V_1} \right)$$

$$= -nRT \ln \frac{V_2}{V_1} + (bRT - a) n^2 \left(\frac{1}{V_2} - \frac{1}{V_1} \right) + \frac{n^3 b^2 RT}{2} \left(\frac{1}{V_2^2} - \frac{1}{V_1^2} \right) + \cdots$$

The first term in this last equation is just the result for the ideal gas. It is modified by the succeeding terms to account for intermolecular interactions. The second term shows the balance between attractive and repulsive forces. In an expansion, $V_2 > V_1$ so $\left(\frac{1}{V_2} - \frac{1}{V_1} \right) < 0$. If attractive forces dominate, $a > bRT$ and the entire second term is positive, which cancels some of the (negative) work done in the expansion. Because of the attractive forces, some energy must be used to overcome the intermolecular interactions, and not as much work can be done as in the case of the ideal gas.

On the other hand, if $a < bRT$, the entire second term is negative and enhances the (negative) work done in the expansion. In this case the repulsive forces dominate, and the energy released as the molecules move farther apart from each other is available to do more work than the ideal gas could. The higher order terms reinforce this effect for high densities where the repulsive forces are most significant.

4.85   Show that for the adiabatic reversible expansion of an ideal gas

$$T_1^{\overline{C}_V/R} V_1 = T_2^{\overline{C}_V/R} V_2$$

From the text,

$$\left(\frac{V_1}{V_2}\right)^{\gamma-1} = \frac{T_2}{T_1}$$

$$\gamma - 1 = \frac{C_P}{C_V} - 1 = \frac{C_P - C_V}{C_V} = \frac{nR}{C_V} = \frac{R}{\overline{C}_V}$$

Therefore,

$$\left(\frac{V_1}{V_2}\right)^{\gamma-1} = \left(\frac{V_1}{V_2}\right)^{R/\overline{C}_V} = \frac{T_2}{T_1}$$

$$\frac{V_1}{V_2} = \left(\frac{T_2}{T_1}\right)^{\overline{C}_V/R}$$

$$T_1^{\overline{C}_V/R} V_1 = T_2^{\overline{C}_V/R} V_2$$

CHAPTER 5

The Second Law of Thermodynamics

PROBLEMS AND SOLUTIONS

5.1   Determine the probability that all the molecules of a gas will be found in one half of a container when the gas consists of **(a)** 1 molecule, **(b)** 20 molecules, and **(c)** 2 million molecules.

(a) Probability $= \left(\frac{1}{2}\right)^1 = 0.5$. **(b)** Probability $= \left(\frac{1}{2}\right)^{20} = 9.5 \times 10^{-7}$. **(c)** Probability $= \left(\frac{1}{2}\right)^{2 \times 10^6} \approx 0$

5.2   Suppose that your friend told you of the following extraordinary event. A block of metal weighing 500 g was seen rising spontaneously from the table on which it was resting to a height of 1.00 cm above the table. He stated that the metal had absorbed thermal energy from the table which was then used to raise itself against gravitational pull. **(a)** Does this process violate the first law of thermodynamics? **(b)** How about the second law? Assume that the room temperature was 298 K and that the table was large enough so that its temperature was unaffected by this transfer of energy. (*Hint*: First calculate the decrease in entropy as a result of this process and then estimate the probability for the occurrence of such a process. The acceleration due to gravity is 9.81 m s^{-2}.)

(a) The process would not violate the first law of thermodynamics as long as the amount of thermal energy absorbed from the table was equal to the increase in (gravitational) potential energy.

(b) The amount of energy required is found via

$$E = mgh$$
$$= (500 \text{ g}) \left(\frac{1 \text{ kg}}{1000 \text{ g}}\right) (9.81 \text{ m s}^{-2}) (1.00 \text{ cm}) \left(\frac{1 \text{ m}}{100 \text{ cm}}\right)$$
$$= 4.905 \times 10^{-2} \text{ J}$$

The entropy of the table changed because thermal energy was transferred to the block via heat. Were this heat to be transferred reversibly, the entropy change for the table would be

$$\Delta S = \frac{q_{rev}}{T}$$
$$= \frac{-4.905 \times 10^{-2} \text{ J}}{298 \text{ K}}$$
$$= -1.646 \times 10^{-4} \text{ J K}^{-1}$$

since the process occurs at constant temperature. The minus sign is included because thermal energy is leaving the table as heat. The table is assumed large enough so that the change in the state of the table is effectively infinitesimal, and this value is taken to be the entropy change for the process as observed.

Relating the entropy change to the probability that this change happens,

$$\Delta S = k_B \ln \frac{W_2}{W_1}$$

$$-1.646 \times 10^{-4} \, \text{J K}^{-1} = 1.381 \times 10^{-23} \, \text{J K}^{-1} \ln \frac{W_2}{W_1}$$

$$-1.192 \times 10^{19} = \ln \frac{W_2}{W_1}$$

$$\frac{W_2}{W_1} = e^{-1.192 \times 10^{19}} \approx 0$$

The ratio is so small that it is practically zero. That is, the probability of attaining the microstate necessary for this process to occur is effectively zero. The event is practically impossible and would violate the second law of thermodynamics.

5.3 Compare the generation of electricity by a hydroelectric plant to the use of a heat engine. Which method is more efficient? Why?

A hydroelectric plant converts the potential energy due to Earth's gravitational field into electric energy while a heat engine converts thermal energy to electric energy via the exchange of heat. The former is not subject to thermodynamic efficiency considerations because no heat is involved. Thus, the hydroelectric plant is more efficient.

5.4 Convert the P–V diagram for the Carnot cycle to a T–S diagram. What is the area of the enclosed portion?

The temperature and entropy change for each step must be determined before plotting a T–S diagram.

Step 1:

The temperature of the system is kept constant at T_2.

$$\Delta S = S_2 - S_1 = \frac{q_2}{T_2} = R \ln \frac{V_2}{V_1}$$

Step 2:

The temperature of the system drops from T_2 to T_1.

Because $q = 0$, $\Delta S = 0$. Thus, the entropy stays at S_2 throughout this step.

Step 3:

The temperature of the system is kept constant at T_1.

$\Delta S = \frac{q_1}{T_1} = R \ln \frac{V_4}{V_3} = -R \ln \frac{V_2}{V_1}$. The initial entropy is S_2 and $\Delta S = -(S_2 - S_1)$ (see Step 1). Therefore, the final entropy is $S_2 + \Delta S = S_2 - (S_2 - S_1) = S_1$.

Step 4:

The temperature of the system rises from T_1 to T_2.

Because $q = 0$, $\Delta S = 0$. Thus, the entropy stays at S_1 throughout this step.

Using the above information, the T–S diagram is constructed as shown in the figure.

The area of the enclosed portion $= (T_2 - T_1)(S_2 - S_1) = (T_2 - T_1)R \ln \frac{V_2}{V_1}$, which is the same as the magnitude of work done in a Carnot cycle.

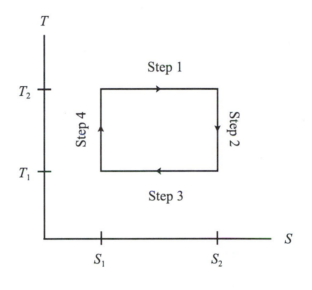

5.5 The internal combustion engine of a 1200-kg car is designed to run on octane (C_8H_{18}), whose enthalpy of combustion is 5510 kJ mol^{-1}. If the car is moving up a slope, calculate the maximum height (in meters) to which the car can be driven on 1.0 gallon of the fuel. Assume that the engine cylinder temperature is 2200°C and the exit temperature is 760°C, and neglect all forms of friction. The mass of 1 gallon of fuel is 3.1 kg. [*Hint*: The work done in moving the car over a vertical distance is mgh, where m is the mass of the car in kg, g the acceleration due to gravity (9.81 m s^{-2}), and h the height in meters.]

To calculate the maximum height the car can be driven, the engine's capacity to do work must first be determined. This work is derived from the heat released by the combustion of octane, but remember the conversion from heat to work is limited by the efficiency of the engine.

$$q_{\text{octane}} = \text{Heat released by combustion of octane}$$

$$= (-5510 \text{ kJ mol}^{-1}) \left(\frac{1 \text{ mol}}{114.22 \times 10^{-3} \text{ kg}} \right) (3.1 \text{ kg}) = -1.50 \times 10^5 \text{ kJ}$$

$$q_{\text{engine}} = \text{heat absorbed by engine} = -q_{\text{octane}} = 1.50 \times 10^5 \text{ kJ}$$

$$\text{Net work done by engine} = \left| w_{engine} \right| = q_{engine} \, (\text{efficiency}) = q_{engine} \left(1 - \frac{T_1}{T_2} \right)$$

$$= \left(1.50 \times 10^5 \text{ kJ} \right) \left(1 - \frac{1033 \text{ K}}{2473 \text{ K}} \right) = 8.73 \times 10^4 \text{ kJ}$$

$$\left| w_{engine} \right| = 8.73 \times 10^7 \text{ J} = mgh = (1200 \text{ kg}) \left(9.81 \text{ m s}^{-2} \right) h$$

$$h = 7.4 \times 10^3 \text{ m}$$

5.6 A heat engine operates between 210°C and 35°C. Calculate the minimum amount of heat that must be withdrawn from the hot source to obtain 2000 J of work.

Let q be the minimum amount of heat that must be withdrawn from the hot source.

$$\text{Efficiency} = \frac{|w|}{q} = 1 - \frac{T_1}{T_2}$$

$$q = \frac{|w|}{1 - \frac{T_1}{T_2}} = \frac{2000 \text{ J}}{1 - \frac{308.2 \text{ K}}{483.2 \text{ K}}} = 5.52 \times 10^3 \text{ J}$$

5.7 Comment on the statement: "Even thinking about entropy increases its value in the universe."

Thinking is an irreversible, biological process. All irreversible processes, and thus thinking, increase the entropy of the universe according to the second law of thermodynamics, $\Delta S_{irrev} > 0$.

5.8 One of the many statements of the second law of thermodynamics is: Heat cannot flow from a colder body to a warmer one without external aid. Assume two systems, 1 and 2, at T_1 and T_2 ($T_2 > T_1$). Show that if a quantity of heat q did flow spontaneously from 1 to 2, the process would result in a decrease in the entropy of the universe. (You may assume that the heat flows very slowly so that the process can be regarded as reversible. Assume also that the loss of heat by system 1 and the gain of heat by system 2 do not affect T_1 and T_2.)

The universe is comprised of the two systems. Since heat flows from system 1 to 2, $q_1 = -q_{rev}$ and $q_2 = q_{rev}$.

$$\Delta S_{univ} = \Delta S_1 + \Delta S_2 = \frac{q_1}{T_1} + \frac{q_2}{T_2} = \frac{-q_{rev}}{T_1} + \frac{q_{rev}}{T_2}$$

$$= q_{rev} \left(\frac{1}{T_2} - \frac{1}{T_1} \right) < 0 \qquad \text{since } T_1 < T_2$$

5.9 A ship in the Indian Ocean takes in the warmer surface water at 32°C to run a heat engine that powers the ship and discharges the used water back to the surface of the sea. Does this scheme violate the second law of thermodynamics? If so, what change would you make to make it work?

Since the heat source and the cold reservoir are the same (the surface of the sea) and consequently have the same temperature, the thermodynamic efficiency of the scheme is zero. The scheme violates the second law of thermodynamics. If the water were to be discharged to a colder layer of the sea, below the surface, then the process would be allowed by thermodynamics.

5.10 Molecules of a gas at any temperature T above absolute zero are in constant motion. Does this "perpetual motion" violate the laws of thermodynamics?

As long as the gas does no work on the surroundings it violates no law of thermodynamics. Were the gas to do work on the surroundings, then the laws of thermodynamics would require that $w = \Delta U - q$ (the first law) and $\Delta S \geq \frac{q}{T}$ for an isothermal process (the second law). This last requirement means that the state of the system must change when work is done, since for an isothermal process doing work on the surroundings, $\Delta U = 0$ (assuming an ideal gas) and $w < 0$ imply $q > 0$. Thus, $\Delta S > 0$ and the system must have changed state, since S is a state function.

5.11 According to the second law of thermodynamics, the entropy of an irreversible process in an isolated system must always increase. On the other hand, it is well known that the entropy of living systems remains small. (For example, the synthesis of highly complex protein molecules from individual amino acids is a process that leads to a decrease in entropy.) Is the second law invalid for living systems? Explain.

Living systems are not isolated systems; they interact with the surroundings. The second law indeed applies to living systems, and requires that decreases in the entropy of a living system be accompanied by increases in the entropy of the surroundings. These increases, in fact, are greater in magnitude than the accompanying decreases in the living system, since $\Delta S_{univ} = \Delta S_{sys} + \Delta S_{surr} > 0$ for irreversible processes.

5.12 On a hot summer day a person tries to cool himself by opening the door of a refrigerator. Is this a wise action, thermodynamically speaking?

No, this is not wise. The refrigerator is not working under ideal conditions, so that it delivers more heat to the room than it extracts. Indeed, even under ideal conditions, the second law of thermodynamics requires an excess of heat delivered to the room over that extracted. The room will get warmer than if the door remained closed.

5.13 The molar heat of vaporization of ethanol is 39.3 kJ mol^{-1}, and the boiling point of ethanol is 78.3°C. Calculate the value of $\Delta_{vap}S$ for the vaporization of 0.50 mole of ethanol.

$$\Delta_{vap} S = \frac{\Delta_{vap} H}{T_b} = \frac{(0.50\ \text{mol})\left(39.3 \times 10^3\ \text{J mol}^{-1}\right)}{351.5\ \text{K}} = 56\ \text{J K}^{-1}$$

5.14   Calculate the values of ΔU, ΔH, and ΔS for the following process:

1 mole of liquid water at 25°C and 1 atm → 1 mol of steam at 100°C and 1 atm

The molar heat of vaporization of water at 373 K is 40.79 kJ mol^{-1}, and the molar heat capacity of water is 75.3 J K^{-1} mol^{-1}. Assume the molar heat capacity to be temperature independent and ideal-gas behavior.

The problem can be solved readily by breaking down the process into two steps, each carried out at 1 atm: (1) $H_2O(l)$ is heated from 25°C to 100°C, then (2) $H_2O(l)$ at 100°C is heated to effect the phase transformation to $H_2O(g)$ at 100°C.

Step 1

$$H_2O(l),\ 25°C \rightarrow H_2O(l),\ 100°C$$

$$\Delta H = C_P \Delta T = (1\ \text{mol})\left(75.3\ \text{J K}^{-1} \text{mol}^{-1}\right)(373.2\ \text{K} - 298.2\ \text{K}) = 5.648 \times 10^3\ \text{J}$$

ΔU is related to ΔH by

$$\Delta U = \Delta H - P \Delta V$$

Since both reactant and product are in the liquid phase, ΔV is negligible. Therefore,

$$\Delta U = \Delta H = 5.648 \times 10^3\ \text{J}$$

$$\Delta S = C_P \ln \frac{T_2}{T_1} = (1\ \text{mol})\left(75.3\ \text{J K}^{-1}\ \text{mol}^{-1}\right) \ln \frac{373.2\ \text{K}}{298.2\ \text{K}} = 16.89\ \text{J K}^{-1}$$

Step 2

$$H_2O(l),\ 100°C \rightarrow H_2O(g),\ 100°C$$

$$\Delta H = n \Delta_{vap} \overline{H} = (1\ \text{mol})\left(40.79\ \text{kJ mol}^{-1}\right) = 40.79\ \text{kJ}$$

To calculate ΔU, the change in volume must first be determined. Since the volume of $H_2O(g)$, V_g is much greater than that of $H_2O(l)$, V_l, the latter is ignored.

$$\Delta U = \Delta H - P\Delta V = \Delta H - P\left(V_g - V_l\right) = \Delta H - PV_g$$

$$= \Delta H - P\frac{nRT}{P} = \Delta H - nRT$$

$$= 40.79 \text{ kJ} - (1 \text{ mol})\left(8.314 \text{ J mol}^{-1}\text{ K}^{-1}\right)(373.2 \text{ K})\left(\frac{1 \text{ kJ}}{1000 \text{ J}}\right)$$

$$= 37.687 \text{ kJ}$$

$$\Delta S = \frac{\Delta H_{vap}}{T_b} = \frac{40.79 \times 10^3 \text{ J}}{373.15 \text{ K}} = 109.31 \text{ J K}^{-1}$$

The values of ΔH, ΔU, and ΔS for the entire process can be obtained by summing the corresponding quantities calculated in the two steps:

$$\Delta H = 5.648 \text{ kJ} + 40.79 \text{ kJ} = 46.44 \text{ kJ}$$

$$\Delta U = 5.648 \text{ kJ} + 37.687 \text{ kJ} = 43.34 \text{ kJ}$$

$$\Delta S = 16.89 \text{ J K}^{-1} + 109.31 \text{ J K}^{-1} = 126.2 \text{ J K}^{-1}$$

5.15 Calculate the value of ΔS in heating 3.5 moles of a monatomic ideal gas from 50°C to 77°C at constant pressure.

At constant pressure,

$$\Delta S = n\overline{C}_P \ln\frac{T_2}{T_1} = \frac{5}{2}nR\ln\frac{T_2}{T_1} = \frac{5}{2}(3.5 \text{ mol})\left(8.314 \text{ J K}^{-1}\text{ mol}^{-1}\right)\ln\frac{350 \text{ K}}{323 \text{ K}} = 5.8 \text{ J K}^{-1}$$

5.16 A quantity of 6.0 moles of an ideal gas is reversibly heated at constant volume from 17°C to 35°C. Calculate the entropy change. What would be the value of ΔS if the heating were carried out irreversibly?

At constant volume, $dq_{rev} = C_V dT$.

$$\Delta S = \int \frac{dq_{rev}}{T} = \int \frac{C_V dT}{T} = C_V \ln\frac{T_2}{T_1} = \frac{3}{2}nR\ln\frac{T_2}{T_1}$$

$$= \frac{3}{2}(6.0 \text{ mol})\left(8.314 \text{ J K}^{-1}\text{ mol}^{-1}\right)\ln\frac{308 \text{ K}}{290 \text{ K}}$$

$$= 4.5 \text{ J K}^{-1}$$

If heating were carried out irreversibly, ΔS is still 4.5 J K^{-1} because S is a state function so that ΔS depends only on the final and initial states. ΔS must be calculated, however, using a reversible pathway.

5.17   One mole of an ideal gas is first, heated at constant pressure from T to $3T$ and second, cooled back to T at constant volume. **(a)** Derive an expression for ΔS for the overall process. **(b)** Show that the overall process is equivalent to an isothermal expansion of the gas at T from V to $3V$, where V is the original volume. **(c)** Show that the value of ΔS for the process in **(a)** is the same as that for **(b)**.

(a) For Step 1,

$$\Delta S_1 = C_P \ln \frac{T_2}{T_1} = \frac{5}{2} n R \ln \frac{3T}{T}$$

$$= \frac{5}{2} R \ln 3$$

For step 2,

$$\Delta S_2 = C_V \ln \frac{T_2}{T_1} = \frac{3}{2} n R \ln \frac{T}{3T} \qquad \text{(See Problem 5.16)}$$

$$= -\frac{3}{2} R \ln 3$$

For the overall process,

$$\Delta S = \Delta S_1 + \Delta S_2 = \frac{5}{2} R \ln 3 - \frac{3}{2} R \ln 3 = R \ln 3$$

(b) Assume initially the system is at P_1, V_1, and T_1. Step 1 carries it to P_2, V_2, and T_2, which in turn is converted to P_3, V_3, and T_3 by Step 2. The relationship between the pressures, volumes, and temperatures are shown in the following.

Since the pressure is constant in Step 1, $P_1 = P_2$. Furthermore, it is given that $T_2 = 3T_1$. V_1 and V_2 are related by

$$\frac{P_1 V_1}{T_1} = \frac{P_2 V_2}{T_2} = \frac{P_1 (V_2)}{3T_1}$$

$$V_2 = 3V_1$$

In Step 2, the volume is kept constant. That is, $V_3 = V_2 = 3V_1$. The temperature decreases to $T_3 = \frac{1}{3} T_2 = T_1$. The pressures are related by

$$\frac{P_2 V_2}{T_2} = \frac{P_3 V_3}{T_3} = \frac{P_3 V_2}{\frac{1}{3} T_2}$$

$$P_3 = \frac{1}{3} P_2 = \frac{1}{3} P_1$$

Therefore, the overall process is $P_1 \ V_1 \ T_1 \rightarrow \frac{1}{3}P_1 \ 3V_1 \ T_1$. This is an isothermal expansion from V to $3V$ where V is the original volume.

(c) ΔS for the overall process in part (b) is

$$\Delta S = nR \ln \frac{3V}{V} = R \ln 3$$

which is the same as ΔS for the process in part (a).

5.18 A quantity of 35.0 g of water at 25.0°C (called A) is mixed with 160.0 g of water at 86.0°C (called B). **(a)** Calculate the final temperature of the system, assuming that the mixing is carried out adiabatically. **(b)** Calculate the entropy change of A, B, and the entire system.

(a) Let the final temperature be T_f and the specific heat of water be s. Since the process is carried out adiabatically, the energy entering A as heat, q_A, is equal in magnitude but opposite in sign to that leaving B as heat, q_B.

$$q_A = -q_B$$

$$(35.0 \text{ g}) \, s \left(T_f - 25.0°C\right) = -(160.0 \text{ g}) \, s \left(T_f - 86.0°C\right)$$

$$T_f - 25.0°C = -4.571\left(T_f - 86.0°C\right)$$

$$= -4.571T_f + 393.1°C$$

$$5.571T_f = 418.1°C$$

$$T_f = 75.0°C$$

(b) Let the entropy change of A be ΔS_A and the entropy change of B be ΔS_B.

$$\Delta S_A = C_p \ln \frac{348.2 \text{ K}}{298.2 \text{ K}} = \left(\frac{35.0 \text{ g}}{18.02 \text{ g mol}^{-1}}\right)(75.3 \text{ J K}^{-1} \text{ mol}^{-1}) \ln \frac{348.2 \text{ K}}{298.2 \text{ K}} = 22.7 \text{ J K}^{-1}$$

$$\Delta S_B = C_p \ln \frac{348.2 \text{ K}}{359.2 \text{ K}} = \left(\frac{160.0 \text{ g}}{18.02 \text{ g mol}^{-1}}\right)(75.3 \text{ J K}^{-1} \text{ mol}^{-1}) \ln \frac{348.2 \text{ K}}{359.2 \text{ K}} = -20.79 \text{ J K}^{-1}$$

$$\Delta S_{\text{total}} = \Delta S_A + \Delta S_B = 22.7 \text{ J K}^{-1} - 20.79 \text{ J K}^{-1} = 1.9 \text{ J K}^{-1}$$

5.19 The heat capacity of chlorine gas is given by

$$\overline{C}_P = (31.0 + 0.008T) \text{ J K}^{-1} \text{ mol}^{-1}$$

Calculate the entropy change when 2 moles of gas are heated from 300 K to 400 K at constant pressure.

$$\Delta S = \int \frac{dq_{\text{rev}}}{T} = \int_{300\,\text{K}}^{400\,\text{K}} \frac{C_p dT}{T}$$

$$= \int_{300\,\text{K}}^{400\,\text{K}} (2\ \text{mol}) \left(\frac{31.0 + 0.008T}{T}\ \text{J K}^{-1}\ \text{mol}^{-1} \right) dT$$

$$= \left[2 \int_{300\,\text{K}}^{400\,\text{K}} \left(\frac{31.0}{T} + 0.008 \right) dT \right]\ \text{J K}^{-1}$$

$$= \left[(2)\, (31.0 \ln T + 0.008T)|_{300}^{400} \right]\ \text{J K}^{-1}$$

$$= 2\, \{[31.0 \ln 400 + 0.008\, (400)] - [31.0 \ln 300 + 0.008\, (300)]\}\ \text{J K}^{-1}$$

$$= 19.4\ \text{J K}^{-1}$$

5.20   A sample of neon (Ne) gas initially at 20°C and 1.0 atm is expanded from 1.2 L to 2.6 L and simultaneously heated to 40°C. Calculate the entropy change for the process.

The number of moles of Ne can be determined using the initial conditions and the ideal gas law.

$$n = \frac{(1.0\ \text{atm})\,(1.2\ \text{L})}{(0.08206\ \text{L atm K}^{-1}\ \text{mol}^{-1})\,(293\ \text{K})} = 4.99 \times 10^{-2}\ \text{mol}$$

The problem can be solved by breaking down the process into 2 steps: (1) isothermal expansion from 1.2 L at 1.0 atm to 2.6 L. The temperature is kept at 20°C; (2) heating at constant volume (2.6 L) from 20°C to 40°C. The entropy changes for these two steps, ΔS_1 and ΔS_2 are

$$\Delta S_1 = nR \ln \frac{V_2}{V_1} = \left(4.99 \times 10^{-2}\ \text{mol} \right) \left(8.314\ \text{J K}^{-1}\ \text{mol}^{-1} \right) \ln \frac{2.6\ \text{L}}{1.2\ \text{L}} = 0.321\ \text{J K}^{-1}$$

$$\Delta S_2 = C_V \ln \frac{T_2}{T_1} = \frac{3}{2} nR \ln \frac{T_2}{T_1} = \frac{3}{2} \left(4.99 \times 10^{-2}\ \text{mol} \right) \left(8.314\ \text{J K}^{-1}\ \text{mol}^{-1} \right) \ln \frac{313\ \text{K}}{293\ \text{K}}$$

$$= 4.11 \times 10^{-2}\ \text{J K}^{-1}$$

The entropy change for the entire process is

$$\Delta S = \Delta S_1 + \Delta S_2 = 0.321\ \text{J K}^{-1} + 4.11 \times 10^{-2}\ \text{J K}^{-1} = 0.36\ \text{J K}^{-1}$$

5.21   One of the early experiments in the development of the atomic bomb was to demonstrate that ^{235}U and not ^{238}U is the fissionable isotope. A mass spectrometer was employed to separate ^{235}UF$_6$ from ^{238}UF$_6$. Calculate the value of ΔS for the separation of 100 mg of the mixture of gas, given that the natural abundances of ^{235}U and ^{238}U are 0.72% and 99.28%, respectively, and that of ^{19}F is 100 %.

The separation process is the reverse of the mixing process. Thus,

$$\Delta S_{\text{separation}} = -\Delta_{\text{mix}}S = R\left(n_A \ln x_A + n_B \ln x_B\right)$$

where A is $^{235}UF_6$ and B is $^{238}UF_6$. The molar masses of ^{235}U and ^{238}U can be obtained from a CRC Handbook, and they are 235.04 and 238.05 g mol^{-1}, respectively. To calculate $\Delta S_{\text{separation}}$, the number of moles and the mole fraction of each compound must first be determined.

In a 100 mg mixture, there are 0.72 mg of $^{235}UF_6$ (A) and 99.28 mg of $^{238}UF_6$ (B).

$$n_A = \frac{0.72 \times 10^{-3} \text{ g}}{349.04 \text{ g mol}^{-1}} = 2.06 \times 10^{-6} \text{ mol}$$

$$n_B = \frac{99.28 \times 10^{-3} \text{ g}}{352.05 \text{ g mol}^{-1}} = 2.8201 \times 10^{-4} \text{ mol}$$

$$x_A = \frac{n_A}{n_A + n_B} = \frac{2.06 \times 10^{-6} \text{ mol}}{2.06 \times 10^{-6} \text{ mol} + 2.8201 \times 10^{-4} \text{ mol}} = 7.25 \times 10^{-3}$$

$$x_B = \frac{n_B}{n_A + n_B} = \frac{2.8201 \times 10^{-4} \text{ mol}}{2.06 \times 10^{-6} \text{ mol} + 2.8201 \times 10^{-4} \text{ mol}} = 0.99275$$

Therefore,

$$\Delta S_{\text{separation}} = \left(8.314 \text{ J K}^{-1} \text{ mol}^{-1}\right)\left[\left(2.06 \times 10^{-6} \text{ mol}\right) \ln 7.25 \times 10^{-3}\right.$$

$$\left. + \left(2.8201 \times 10^{-4} \text{ mol}\right) \ln 0.99275\right]$$

$$= -1.0 \times 10^{-4} \text{ J K}^{-1}$$

5.22   One mole of an ideal gas at 298 K expands isothermally from 1.0 L to 2.0 L **(a)** reversibly and **(b)** against a constant external pressure of 12.2 atm. Calculate the values of ΔS_{sys}, ΔS_{surr}, and ΔS_{univ} in both cases. Are your results consist with the nature of the processes?

(a)

$$\Delta S_{\text{sys}} = nR \ln \frac{V_2}{V_1} = (1 \text{ mol})\left(8.314 \text{ J K}^{-1} \text{ mol}^{-1}\right) \ln \frac{2.0 \text{ L}}{1.0 \text{ L}} = 5.8 \text{ J K}^{-1}$$

$$\Delta S_{\text{surr}} = -5.8 \text{ J K}^{-1}$$

$$\Delta S_{\text{univ}} = 0 \text{ J K}^{-1}$$

(b) ΔS_{sys} is the same above, that is, 5.8 J K^{-1}, since S is a state function, although ΔS has to be calculated using a reversible path.

ΔS_{surr} can be calculated once q_{surr} is determined. The latter quantity is related to q_{sys}, which in turn can be calculated from the work done by the system, w, and the first law of thermodynamics.

$$w = -P_{\text{ex}}\Delta V = -(12.2 \text{ atm})(2.0 \text{ L} - 1.0 \text{ L})\left(\frac{101.3 \text{ J}}{1 \text{ L atm}}\right) = -1.236 \times 10^3 \text{ J}$$

According to the first law, $\Delta U = q + w$. Since $\Delta U = 0$ for an isothermal process, $q_{sys} = q = -w = 1.236 \times 10^3$ J. The entropy change of the surroundings is

$$\Delta S_{surr} = \frac{q_{surr}}{T_{surr}} = \frac{-q_{sys}}{T_{surr}} = \frac{-1.236 \times 10^3 \text{ J}}{298 \text{ K}} = -4.15 \text{ J K}^{-1}$$

Now ΔS_{univ} can be determined:

$$\Delta S_{univ} = 5.8 \text{ J K}^{-1} - 4.15 \text{ J K}^{-1} = 1.7 \text{ J K}^{-1}$$

The results in both parts are consistent with the nature of the processes. Specifically, for a reversible process, $\Delta S_{univ} = 0$, whereas for a spontaneous process, $\Delta S_{univ} > 0$.

5.23   The absolute molar entropies of O_2 and N_2 are 205 J K^{-1} mol^{-1} and 192 J K^{-1} mol^{-1}, respectively at 25°C. What is the entropy of a mixture made up of 2.4 moles of O_2 and 9.2 moles of N_2 at the same temperature and pressure?

The entropy of the mixture, S_f is related to the entropy of mixing, $\Delta_{mix}S$, and the initial entropy of the system, S_i. Before these quantities can be calculated, the mole fractions of O_2 and N_2 need to be determined.

$$x_{O_2} = \frac{2.4 \text{ mol}}{2.4 \text{ mol} + 9.2 \text{ mol}} = 0.207$$

$$x_{N_2} = 1 - 0.207 = 0.793$$

The entropy of mixing is

$$\Delta_{mix}S = -R\left(n_{O_2} \ln x_{O_2} + n_{N_2} \ln x_{N_2}\right)$$

$$= -\left(8.314 \text{ J K}^{-1} \text{ mol}^{-1}\right) [(2.4 \text{ mol}) \ln 0.207 + (9.2 \text{ mol}) \ln 0.793]$$

$$= 49.2 \text{ J K}^{-1}$$

The initial entropy of system is the sum of the entropies of O_2 and N_2:

$$S_i = n_{O_2} \overline{S}_{O_2} + n_{N_2} \overline{S}_{N_2} = (2.4 \text{ mol}) \left(205 \text{ J K}^{-1} \text{ mol}^{-1}\right) + (9.2 \text{ mol}) \left(192 \text{ J K}^{-1} \text{ mol}^{-1}\right)$$

$$= 2.26 \times 10^3 \text{ J K}^{-1}$$

Since $\Delta_{mix}S = S_f - S_i$,

$$S_f = \Delta_{mix}S + S_i = 49.2 \text{ J K}^{-1} + 2.26 \times 10^3 \text{ J K}^{-1} = 2.3 \times 10^3 \text{ J K}^{-1}$$

5.24   A quantity of 0.54 mole of steam initially at 350°C and 2.4 atm undergoes a cyclic process for which $q = -74$ J. Calculate the value of ΔS for the process.

$\Delta S = 0$ for a cyclic process.

5.25 Predict whether the entropy change is positive or negative for each of the following reactions at 298 K:

(a) $4Fe(s) + 3O_2(g) \rightarrow 2Fe_2O_3(s)$ (b) $O(g) + O(g) \rightarrow O_2(g)$
(c) $NH_4Cl(s) \rightarrow NH_3(g) + HCl(g)$ (d) $H_2(g) + Cl_2(g) \rightarrow 2HCl(g)$

(a) Negative, since a gas, which is a highly disordered phase, is being transformed into a more ordered solid.

(b) Negative, two moles of gas are becoming one. This is a more ordered situation.

(c) Positive, an ordered solid is being transformed into two moles of gases.

(d) Near zero, each side of the chemical equation has two moles of gas. It might be expected that the "mixed" molecule HCl is less ordered than the "pure" reactants H_2 and Cl_2 so that the entropy change would more likely be positive than negative, but other factors may also contribute.

5.26 Use the data in Appendix B to calculate the values of $\Delta_r S^\circ$ of the reactions listed in the previous problem.

(a)

$$\Delta_r S^\circ = 2\overline{S}^\circ\left[Fe_2O_3(s)\right] - 4\overline{S}^\circ\left[Fe(s)\right] - 3\overline{S}^\circ\left[O_2(g)\right]$$

$$= 2\left(90.0\,J\,K^{-1}\,mol^{-1}\right) - 4\left(27.2\,J\,K^{-1}\,mol^{-1}\right) - 3\left(205.0\,J\,K^{-1}\,mol^{-1}\right)$$

$$= -543.8\,J\,K^{-1}\,mol^{-1}$$

(b)

$$\Delta_r S^\circ = \overline{S}^\circ\left[O_2(g)\right] - 2\overline{S}^\circ\left[O(g)\right]$$

$$= 205.0\,J\,K^{-1}\,mol^{-1} - 2\left(161.0\,J\,K^{-1}\,mol^{-1}\right)$$

$$= -117.0\,J\,K^{-1}\,mol^{-1}$$

(c)

$$\Delta_r S^\circ = \overline{S}^\circ\left[NH_3(g)\right] + \overline{S}^\circ\left[HCl(g)\right] - \overline{S}^\circ\left[NH_4Cl(s)\right]$$

$$= 192.5\,J\,K^{-1}\,mol^{-1} + 186.5\,J\,K^{-1}\,mol^{-1} - 94.6\,J\,K^{-1}\,mol^{-1}$$

$$= 284.4\,J\,K^{-1}\,mol^{-1}$$

(d)

$$\Delta_r S^\circ = 2\overline{S}^\circ\left[HCl(g)\right] - \overline{S}^\circ\left[H_2(g)\right] - \overline{S}^\circ\left[Cl_2(g)\right]$$

$$= 2\left(186.5\,J\,K^{-1}\,mol^{-1}\right) - \left(130.6\,J\,K^{-1}\,mol^{-1}\right) - \left(223.0\,J\,K^{-1}\,mol^{-1}\right)$$

$$= 19.4\,J\,K^{-1}\,mol^{-1}$$

These results agree with predictions made in Problem 5.25.

5.27 A quantity of 0.35 mole of an ideal gas initially at 15.6°C is expanded from 1.2 L to 7.4 L. Calculate the values of w, q, ΔU, and ΔS if the process is carried out **(a)** isothermally and reversibly, and **(b)** isothermally and irreversibly against an external pressure of 1.0 atm.

(a) $\Delta U = 0$ for an isothermal process. Thus, $q = -w$.

$$w = -nRT \ln \frac{V_2}{V_1}$$

$$= -(0.35 \text{ mol}) \left(8.314 \text{ J K}^{-1} \text{ mol}^{-1}\right) (288.8 \text{ K}) \ln \frac{7.4 \text{ L}}{1.2 \text{ L}}$$

$$= -1.53 \times 10^3 \text{ J}$$

$$q = 1.53 \times 10^3 \text{ J}$$

$$\Delta S = \frac{q_{rev}}{T} = \frac{1.53 \times 10^3 \text{ J}}{288.8 \text{ K}} = 5.3 \text{ J K}^{-1}$$

The final answers for q and w should be rounded to two significant figures, $q = 1.5 \times 10^3$ J and $w = -1.5 \times 10^3$ J.

(b) $\Delta U = 0$ for an isothermal process. Thus, $q = -w$.

$$w = -P_{ex}\Delta V = -(1.0 \text{ atm}) (7.4 \text{ L} - 1.2 \text{ L}) \left(\frac{101.3 \text{ J}}{1 \text{ L atm}}\right) = -6.28 \times 10^2 \text{ J}$$

$$q = 6.28 \times 10^2 \text{ J}$$

$$\Delta S = 5.3 \text{ J K}^{-1} \text{ [as calculated in Part (a).]}$$

Recall that S is a state function, but ΔS has to be calculated using a reversible path. Again, the final answers for q and w should be rounded to two significant figures, $q = 6.3 \times 10^2$ J and $w = -6.3 \times 10^2$ J.

5.28 One mole of an ideal gas is isothermally expanded from 5.0 L to 10 L at 300 K. Compare the entropy changes for the system, surroundings, and the universe if the process is carried out **(a)** reversibly, and **(b)** irreversibly against an external pressure of 2.0 atm.

(a) For the reversible process,

$$\Delta S_{sys} = nR \ln \frac{V_2}{V_1} = (1 \text{ mol}) \left(8.314 \text{ J K}^{-1} \text{ mol}^{-1}\right) \ln \frac{10 \text{ L}}{5.0 \text{ L}} = 5.8 \text{ J K}^{-1}$$

$$\Delta S_{surr} = -5.8 \text{ J K}^{-1}$$

$$\Delta S_{univ} = 0 \text{ J K}^{-1}$$

(b) ΔS_{sys} is the same as above, that is, 5.8 J K^{-1}.

ΔS_{surr} can be calculated once q_{surr} is determined. The latter quantity is related to q_{sys}, which in turn can be calculated from the work done by the system, w, and the first law of thermodynamics.

$$w = -P_{ex}\Delta V = -(2.0 \text{ atm}) (10 \text{ L} - 5.0 \text{ L}) \left(\frac{101.3 \text{ J}}{1 \text{ L atm}}\right) = -1.01 \times 10^3 \text{ J}$$

According to the first law, $\Delta U = q + w$. For an ideal gas, $\Delta U = 0$ for an isothermal process, and $q_{sys} = q = -w = 1.01 \times 10^3$ J. The entropy change of the surroundings is

$$\Delta S_{surr} = \frac{q_{surr}}{T_{surr}} = \frac{-q_{sys}}{T_{surr}} = \frac{-1.01 \times 10^3 \text{ J}}{300 \text{ K}} = -3.4 \text{ J K}^{-1}$$

Therefore,

$$\Delta S_{univ} = 5.8 \text{ J K}^{-1} - 3.4 \text{ J K}^{-1} = 2.4 \text{ J K}^{-1}$$

5.29 The heat capacity of hydrogen may be represented by

$$\overline{C}_P = (1.554 + 0.0022T) \text{ J K}^{-1} \text{ mol}^{-1}$$

Calculate the entropy changes for the system, surroundings, and the universe for the (a) reversible heating, and (b) irreversible heating of 1.0 mole of hydrogen from 300 K to 600 K. [*Hint*: In (b), assume the surroundings to be at 600 K.]

(a) For the reversible process,

$$\Delta S_{sys} = \int \frac{dq_{rev}}{T} = \int_{300 \text{ K}}^{600 \text{ K}} \frac{C_P dT}{T}$$

$$= \int_{300 \text{ K}}^{600 \text{ K}} (1.0 \text{ mol}) \left(\frac{1.554 + 0.0022T}{T} \text{ J K}^{-1} \text{ mol}^{-1} \right) dT$$

$$= \left[\int_{300 \text{ K}}^{600 \text{ K}} \left(\frac{1.554}{T} + 0.0022 \right) dT \right] \text{ J K}^{-1}$$

$$= \left[(1.554 \ln T + 0.0022T)|_{300}^{600} \right] \text{ J K}^{-1}$$

$$= \{[1.554 \ln 600 + 0.0022 (600)] - [1.554 \ln 300 + 0.0022 (300)]\} \text{ J K}^{-1}$$

$$= 1.7 \text{ J K}^{-1}$$

$$\Delta S_{surr} = -1.7 \text{ J K}^{-1}$$

$$\Delta S_{univ} = 0 \text{ J K}^{-1}$$

(b) S is a state function, but ΔS_{sys} has to be calculated using a reversible path. Therefore, it is the same as that calculated above, that is, 1.7 J K^{-1}.

ΔS_{surr} can be calculated once q_{surr} is determined.

$$q_{surr} = -q_{sys} = -\Delta H \qquad (P \text{ is constant})$$

$$= -\int_{300 \text{ K}}^{600 \text{ K}} C_P dT$$

$$= -\int_{300 \text{ K}}^{600 \text{ K}} (1.0 \text{ mol}) \left[(1.554 + 0.0022T) \text{ J K}^{-1} \text{ mol}^{-1} \right] dT$$

$$= -\left(1.554T + 0.0011T^2 \right)_{300}^{600} \text{ J}$$

$$= -\{[1.554 (600) + 0.0011 (600)^2] - [1.554 (300) + 0.0011 (300)^2]\} \text{ J}$$

$$= -7.63 \times 10^2 \, \text{J}$$

$$\Delta S_{\text{surr}} = \frac{q_{\text{surr}}}{T_{\text{surr}}} = \frac{-7.63 \times 10^2 \, \text{J}}{600 \, \text{K}} = -1.3 \, \text{J K}^{-1}$$

Therefore,

$$\Delta S_{\text{univ}} = 1.7 \, \text{J K}^{-1} - 1.3 \, \text{J K}^{-1} = 0.4 \, \text{J K}^{-1}$$

5.30   Consider the reaction

$$N_2(g) + O_2(g) \rightarrow 2NO(g)$$

Calculate the values of $\Delta_r S°$ for the reaction mixture, surroundings, and the universe at 298 K. Why is your result reassuring to Earth's inhabitants?

$$\Delta_r S° = 2\overline{S}° \, [NO(g)] - \overline{S}° \, [N_2(g)] - \overline{S}° \, [O_2(g)]$$

$$= 2 \left(210.6 \, \text{J K}^{-1} \, \text{mol}^{-1} \right) - 191.6 \, \text{J K}^{-1} \, \text{mol}^{-1} - 205.0 \, \text{J K}^{-1} \, \text{mol}^{-1}$$

$$= 24.6 \, \text{J K}^{-1} \, \text{mol}^{-1}$$

$\Delta S°_{\text{surr}}$ is determined from $\Delta_r H°$ and the temperature of the surroundings.

$$\Delta_r H° = 2\Delta_f \overline{H}° \, [NO(g)] - \Delta_f \overline{H}° \, [N_2(g)] - \Delta_f \overline{H}° \, [O_2(g)]$$

$$= 2 \left(90.4 \, \text{kJ mol}^{-1} \right) - 0 \, \text{kJ mol}^{-1} - 0 \, \text{kJ mol}^{-1}$$

$$= 180.8 \, \text{kJ mol}^{-1}$$

$$\Delta H°_{\text{surr}} = -\Delta_r H° = -180.8 \, \text{kJ mol}^{-1}$$

$$\Delta S°_{\text{surr}} = \frac{\Delta H°_{\text{surr}}}{T} = \frac{-180.8 \times 10^3 \, \text{J mol}^{-1}}{298 \, \text{K}} = -607 \, \text{J K}^{-1} \, \text{mol}^{-1}$$

Therefore,

$$\Delta S_{\text{univ}} = 24.7 \, \text{J K}^{-1} \, \text{mol}^{-1} - 607 \, \text{J K}^{-1} \, \text{mol}^{-1} = -582 \, \text{J K}^{-1} \, \text{mol}^{-1}$$

This is not a spontaneous process at 298 K. Therefore, O_2, which is essential to us, does not react with N_2 in the atmosphere at 298 K.

5.31   The $\Delta_f \overline{H}°$ values can be negative, zero, or positive, but the $S°$ values can be only zero or positive. Explain.

Since $S = k_B \ln W$ and W cannot be less than 1, S cannot be negative. On the other hand, the enthalpy change associated with the formation of one mole of a substance from its elements in their standard states may be negative, positive, or zero. That is, the formation reaction can be exothermic, endothermic, or thermoneutral.

5.32 Choose the substance with the greater molar entropy in each of the following pairs: **(a)** $H_2O(l)$, $H_2O(g)$, **(b)** $NaCl(s)$, $CaCl_2(s)$, **(c)** $N_2(0.1 \text{ atm})$, $N_2(1 \text{ atm})$, **(d)** C(diamond), C(graphite), **(e)** $O_2(g)$, $O_3(g)$, **(f)** ethanol (C_2H_5OH), dimethyl ether (C_2H_6O), **(g)** $N_2O_4(g)$, $2NO_2(g)$, **(h)** Fe(s) at 298 K, Fe(s) at 398 K. (Unless otherwise stated, assume temperature is 298 K.)

(a) $H_2O(g)$, a gas has greater entropy than the more ordered liquid.

(b) $CaCl_2(s)$, this is a more complex system than $NaCl(s)$.

(c) N_2 (0.1 atm), at the lower pressure, the gas occupies a larger volume leading to a larger number of microstates for the system.

(d) C(graphite), diamond is a more ordered solid than is graphite.

(e) $O_3(g)$, this is a more complex system than diatomic $O_2(g)$.

(f) Dimethyl ether, ethanol can form hydrogen bonds leading to a more ordered system.

(g) $N_2O_4(g)$, one mole of $N_2O_4(g)$ is a more complex system and has greater entropy than one mole of $NO_2(g)$, although *two* moles of $NO_2(g)$ has greater entropy than one mole of $N_2O_4(g)$.

(h) Fe(s) at 398 K, since it is at a higher temperature.

5.33 A chemist found a discrepancy between the third law entropy and the calculated entropy from statistical thermodynamics for a compound. **(a)** Which value is larger? **(b)** Suggest two reasons that may give rise to this discrepancy.

(a) The value from statistical thermodynamics is larger.

(b) The third law entropy does not account for residual entropy at 0 K, and the discrepancy suggests that the compound does not attain a perfect crystalline form at 0 K. A second reason could be a phase transition that was missed in determining the entropy using the third law.

5.34 Calculate the molar residual entropy of a solid in which the molecules can adopt **(a)** three, **(b)** four, and **(c)** five orientations of equal energy at absolute zero.

(a) $\overline{S}^{\circ} = R \ln 3 = 9.134 \text{ J K}^{-1} \text{ mol}^{-1}$ **(b)** $\overline{S}^{\circ} = R \ln 4 = 11.53 \text{ J K}^{-1} \text{ mol}^{-1}$
(c) $\overline{S}^{\circ} = R \ln 5 = 13.38 \text{ J K}^{-1} \text{ mol}^{-1}$

5.35 Account for the measured residual entropy of $10.1 \text{ J K}^{-1} \text{ mol}^{-1}$ for the CH_3D molecule.

There are four possible choices for the orientation of the deuterated methane molecule in the crystal. That is, compared to the crystalline form of the normal isotopic species of methane, the deuterium substitution could be at any one of the four hydrogens. Assuming each of the four orientations to be of essentially the same energy, they are equally likely. Thus,

$$S_0 = k_B \ln W$$

$$= k_B \ln 4^{N_A}$$

$$= R \ln 4$$

$$= (8.314 \, \text{J} \, \text{K}^{-1} \, \text{mol}^{-1}) \ln 4$$

$$= 11.53 \, \text{J} \, \text{K}^{-1} \, \text{mol}^{-1}$$

This is in excellent agreement with the measured value. The measured value is smaller and suggests that the orientation of the deuterated methane in the crystal is not completely random. That is, there are energy differences between different possible orientations.

5.36 Explain why the value of $\overline{S}^{\,\circ}$(graphite) is greater than that of $\overline{S}^{\,\circ}$(diamond) at 298 K (see Appendix B). Would this inequality hold at 0 K?

Graphite does not have as highly ordered structure as does diamond, and is expected to have the larger entropy. At 0 K, however, both allotropic forms of carbon are expected to have the same entropy, $\overline{S}^{\,\circ} = 0$, presuming there are no crystal defects or impurities.

5.37 Entropy has sometimes been described as "times's arrow" because it is the property that determines the forward direction of time. Explain.

Because the second law of thermodynamics requires that the entropy of the universe is always increasing, then it would be possible to order two points in time by determining the entropy of the universe at those times. The time corresponding to the greater entropy in the universe is the later time.

5.38 State the condition(s) under which the following equations can be applied: **(a)** $\Delta S = \Delta H / T$, **(b)** $S_0 = 0$, **(c)** $dS = C_p dT / T$, **(d)** $dS = dq / T$.

(a) Constant pressure and temperature, reversible process.

(b) Absolute zero, no residual entropy and a pure, crystalline substance.

(c) Constant pressure (Note that when this form is integrated, one must either include the explicit temperature dependence of C_p or assume it to be temperature independent.)

(d) reversible process

5.39 Without referring to any table, predict whether the entropy change is positive, nearly zero, or negative for each of the following reactions:

(a) $N_2(g) + O_2(g) \rightarrow 2NO(g)$ **(b)** $2Mg(s) + O_2(g) \rightarrow 2MgO(s)$
(c) $2H_2O_2(l) \rightarrow 2H_2O(l) + O_2(g)$ **(d)** $H_2(g) + CO_2(g) \rightarrow H_2O(g) + CO(g)$

(a) Nearly zero, there are two moles of gaseous reactants and the same number of gaseous products.

(b) Negative, one mole of gas is incorporated into a more ordered solid.

(c) Positive, a liquid reactant gives rise to a liquid product plus a less ordered gaseous product.

(d) Nearly zero, there are two moles of gaseous reactants and the same number of gaseous products.

5.40 Calculate the entropy change when neon at 25°C and 1.0 atm in a container of volume 0.780 L is allowed to expand to 1.25 L and is simultaneously heated to 85°C. Assume ideal behavior. (*Hint*: Because S is a state function, you can first calculate the value of ΔS for expansion and then calculate the value of ΔS for heating at constant final volume.)

The number of moles of Ne can be determined using the initial conditions and the ideal gas law.

$$n = \frac{(1.0 \text{ atm}) (0.780 \text{ L})}{(0.08206 \text{ L atm K}^{-1} \text{ mol}^{-1}) (298 \text{ K})} = 3.19 \times 10^{-2} \text{ mol}$$

The problem can be solved by breaking down the process into 2 steps: (1) isothermal expansion from 0.780 L at 1.0 atm to 1.25 L. The temperature is kept at 20°C; (2) heating at constant volume (1.25 L) from 25°C to 85°C. The entropy changes for these two steps, ΔS_1 and ΔS_2 are

$$\Delta S_1 = nR \ln \frac{V_2}{V_1} = (3.19 \times 10^{-2} \text{ mol}) (8.314 \text{ J K}^{-1} \text{ mol}^{-1}) \ln \frac{1.25 \text{ L}}{0.780 \text{ L}} = 0.125 \text{ J K}^{-1}$$

$$\Delta S_2 = C_V \ln \frac{T_2}{T_1} = \frac{3}{2} nR \ln \frac{T_2}{T_1} = \frac{3}{2} (3.19 \times 10^{-2} \text{ mol}) (8.314 \text{ J K}^{-1} \text{ mol}^{-1}) \ln \frac{358 \text{ K}}{298 \text{ K}}$$

$$= 7.30 \times 10^{-2} \text{ J K}^{-1}$$

The entropy change for the entire process is

$$\Delta S = \Delta S_1 + \Delta S_2 = 0.125 \text{ J K}^{-1} + 7.30 \times 10^{-2} \text{ J K}^{-1} = 0.20 \text{ J K}^{-1}$$

5.41 Photosynthesis makes use of photons of visible light to bring about chemical changes. Explain why heat energy in the form of infrared photons is ineffective for photosynthesis.

Infrared photons do not have sufficient energy for a single photon to effect the necessary electronic changes necessary for photosynthesis to take place. Rather, their energy corresponds to excitations of molecular vibrations, and the absorption of infrared photons leads to random thermal motion of molecules. Collecting this energy, several photons worth, to effect the electronic change would correspond to the conversion of heat to work and would be limited in efficiency by the second law of thermodynamics. Those plants which could directly capture the concentrated energy in a single photon to effect the first steps of photosynthesis would have an evolutionary advantage over those limited to the inefficient process.

5.42   One mole of an ideal monatomic gas is compressed from 2.0 atm to 6.0 atm while being cooled from 400 K to 300 K. Calculate the values of ΔU, ΔH, and ΔS for the process.

The thermodynamic quantities can be readily calculated by breaking down the process into 2 steps: (1) isothermal compression at 400 K from 2.0 atm to 6.0 atm; (2) cooling at constant pressure (6.0 atm) from 400 K to 300 K.

Step 1

For an ideal gas undergoing an isothermal process, $\Delta U = 0$ and $\Delta H = 0$.

$$\Delta S = nR \ln \frac{V_2}{V_1} = nR \ln \frac{P_1}{P_2} = (1 \text{ mol}) \left(8.314 \text{ J K}^{-1} \text{ mol}^{-1}\right) \ln \frac{2.0 \text{ atm}}{6.0 \text{ atm}} = -9.13 \text{ J K}^{-1}$$

Step 2

$$\Delta H = C_P \Delta T = \frac{5}{2} nR \Delta T = \frac{5}{2} (1 \text{ mol}) \left(8.314 \text{ J K}^{-1} \text{ mol}^{-1}\right) (300 \text{ K} - 400 \text{ K}) = -2.079 \times 10^3 \text{ J}$$

$$\Delta U = \Delta H - nR \Delta T = -2.079 \times 10^3 \text{ J} - (1 \text{ mol}) \left(8.314 \text{ J K}^{-1} \text{ mol}^{-1}\right) (300 \text{ K} - 400 \text{ K})$$

$$= -1.248 \times 10^3 \text{ J}$$

$$\Delta S = C_P \ln \frac{T_2}{T_1} = \frac{5}{2} nR \ln \frac{T_2}{T_1} = \frac{5}{2} (1 \text{ mol}) \left(8.314 \text{ J K}^{-1} \text{ mol}^{-1}\right) \ln \frac{300 \text{ K}}{400 \text{ K}} = -5.979 \text{ J K}^{-1}$$

For the entire process,

$$\Delta H = 0 \text{ J} - 2.079 \times 10^3 \text{ J} = -2.08 \times 10^3 \text{ J}$$

$$\Delta U = 0 \text{ J} - 1.248 \times 10^3 \text{ J} = -1.25 \times 10^3 \text{ J}$$

$$\Delta S = -9.13 \text{ J K}^{-1} - 5.979 \text{ J K}^{-1} = -15.1 \text{ J K}^{-1}$$

5.43   The three laws of thermodynamics are sometimes stated colloquially as follows: First law: You cannot get something for nothing; Second law: The best you can do is get even; Third law: you cannot get even. Provide a scientific basis for each of these statements. (*Hint*: One consequence of the third law is that it is impossible to attain the absolute zero of temperature.)

The first law requires that energy is conserved. The production of work must be accompanied by either a reduction in energy of the system or an input of heat such that $\Delta U = q + w$. You cannot get something (work) for nothing (without energy).

The second law leads to the concept of thermodynamic efficiency, efficiency $= 1 - \frac{T_1}{T_2}$, which has a maximum of 1, or 100%, when either $T_1 = 0$ K or $T_2 = \infty$. The best you can do is break even (convert all the heat absorbed by the engine into work) with an efficiency of 100%.

As a consequence of the third law, a temperature of absolute zero cannot be attained. Since an infinite temperature is also impossible (from the definition of infinity, not as a consequence of thermodynamics), then the thermodynamic efficiency will never reach its maximum value of

100%. You cannot get even, some heat will always be returned to the surroundings without being converted to work.

5.44 Use the following data to determine the normal boiling point, in kelvins, of mercury. What assumptions must you make to do the calculation?

$$Hg(l): \quad \Delta_f \overline{H}^\circ = 0 \text{ (by definition)}$$

$$\overline{S}^\circ = 77.4 \text{ J K}^{-1} \text{ mol}^{-1}$$

$$Hg(g): \quad \Delta_f \overline{H}^\circ = 60.78 \text{ kJ mol}^{-1}$$

$$\overline{S}^\circ = 174.7 \text{ J K}^{-1} \text{ mol}^{-1}$$

The normal boiling point of mercury, T_b is related to $\Delta_{vap} S^\circ$ and $\Delta_{vap} H^\circ$:

$$\Delta_{vap} S^\circ = \frac{\Delta_{vap} H^\circ}{T_b}$$

For 1 mole of mercury, the entropy and enthalpy of vaporization are

$$\Delta_{vap} H^\circ = 60.78 \text{ kJ} - 0 \text{ kJ} = 60.78 \text{ kJ}$$

$$\Delta_{vap} S^\circ = 174.7 \text{ J K}^{-1} - 77.4 \text{ J K}^{-1} = 97.3 \text{ J K}^{-1}$$

Therefore,

$$T_b = \frac{\Delta_{vap} H^\circ}{\Delta_{vap} S^\circ} = \frac{60.78 \times 10^3 \text{ J}}{97.3 \text{ J K}^{-1}} = 625 \text{ K} = 352°C$$

The assumptions made in this calculation are that the values of $\Delta_f \overline{H}^\circ$ and $\overline{S}^\circ$ are temperature independent. These assumptions are quite good because the calculated boiling point of mercury is very close to the actual value of 356.6°C.

5.45 Referring to Trouton's rule, explain why the ratio is considerably smaller than $90 \text{ J K}^{-1} \text{ mol}^{-1}$ for liquid HF.

HF is hydrogen bonded to a significant extent in the gas phase, making it more ordered than the case in which the HF molecules were independent. Consequently, the increase in entropy accompanying vaporization, $\Delta_{vap} S$, is smaller than for most other substances.

5.46 Give a detailed example of each of the following, with an explanation: **(a)** a thermodynamically spontaneous process; **(b)** a process that would violate the first law of thermodynamics; **(c)** a process that would violate the second law of thermodynamics; **(d)** an irreversible process; **(e)** an equilibrium process.

(a) An ice cube melting in a glass of water at 20°C.

(b) A perpetual motion machine of the first kind, such as a rotating flywheel that drives a generator, the output of which is used to keep the flywheel rotating at a constant speed and also to lift a weight.

(c) A perfect air conditioner; it extracts heat from the room and warms the outside without using any energy to do so. (This does not violate the first law, since the energy deposited outside is exactly equal to that removed from inside.)

(d) Same as part (a), an ice cube melting in a glass of water at 20°C.

(e) Water and ice in a closed system at 0°C and 1 atm pressure.

5.47   In the reversible adiabatic expansion of an ideal gas, there are two contributions to entropy changes: the expansion of the gas and the cooling of the gas. Show that these two contributions are equal in magnitude but opposite in sign. Show also that for an irreversible adiabatic gas expansion, these two contributions are no longer equal in magnitude. Predict the sign of ΔS.

A reversible adiabatic expansion process can be broken down into 2 steps: (1) isothermal expansion at T_1 from V_1 to V_2, and (2) constant volume cooling at V_2 from T_1 to T_2. The entropy changes for these 2 steps are

$$\Delta S_1 = nR \ln \frac{V_2}{V_1}$$

$$\Delta S_2 = C_V \ln \frac{T_2}{T_1}$$

For a reversible adiabatic process,

$$\left(\frac{V_1}{V_2}\right)^{\gamma-1} = \frac{T_2}{T_1}$$

Therefore,

$$\Delta S_2 = C_V \ln \left(\frac{V_1}{V_2}\right)^{\gamma-1} = C_V (\gamma - 1) \ln \frac{V_1}{V_2} = C_V \left(\frac{C_P}{C_V} - 1\right) \ln \frac{V_1}{V_2} = (C_P - C_V) \ln \frac{V_1}{V_2}$$

$$= nR \ln \frac{V_1}{V_2} = -\Delta S_1$$

The contributions from the 2 steps are equal in magnitude but opposite in sign. As a result, the overall change in entropy is 0.

In an irreversible expansion between the same two volumes, less work is done by the system than in the reversible case. Since the expansion is adiabatic, $q = 0$ and $\Delta U = w$. Thus, the internal energy of the gas suffers a smaller decrease in the irreversible expansion. For an ideal gas, $\Delta U = C_V \Delta T$ so that the temperature difference is smaller, or $T_2^{\text{irrev}} > T_2^{\text{rev}}$. This means that in the irreversible expansion ΔS_2 is not as negative as for the reversible case, and $\Delta S > 0$.

5.48   A refrigerator set at 0°C discharges heat into the kitchen at 20°C. **(a)** how much work would be required to freeze 500 mL of water (about an ice tray's volume)? **(b)** How much heat would be discharged during this process? (The molar enthalpy of fusion of water is 6.01 kJ mol^{-1}, and the refrigerator operates at 35% efficency.)

(a) The work required is calculated from the heat extracted from the cold reservoir and the coefficient of performance of the refrigerator. The heat extracted depends on the amount of water to be frozen, which is calculated by assuming a density of 1 g mL^{-1}.

$$\text{Number of moles of water} = \frac{(500 \text{ mL}) \left(1 \text{ g mL}^{-1}\right)}{18.02 \text{ g mol}^{-1}} = 27.75 \text{ mol}$$

$$\text{Heat extracted from the cold reservoir} = (27.75 \text{ mol}) \left(6.01 \text{ kJ mol}^{-1}\right) = 166.8 \text{ kJ}$$

$$\text{COP} = \frac{T_1}{T_2 - T_1} \text{ (efficiency)} = \frac{273.2 \text{ K}}{293.2 \text{ K} - 273.2 \text{ K}} (0.35) = 4.78$$

$$\text{Work required} = \frac{166.8 \text{ kJ}}{4.78} = 35 \text{ kJ}$$

(b) Heat discharged = 35 kJ + 166.8 kJ = 202 kJ

5.49 Superheated water is water heated above 100°C without boiling. As for supercooled water (see Example 5.7), superheated water is thermodynamically unstable. Calculate the values of ΔS_{sys}, ΔS_{surr}, and ΔS_{univ} when 1.5 moles of superheated water at 110°C and 1.0 atm are converted to steam at the same temperature and pressure. (The molar enthalpy of vaporization of water is 40.79 kJ mol^{-1}, and the molar heat capacities of water and steam in the temperature range 100-110°C are 75.5 J K^{-1} mol^{-1} and 34.4 J K^{-1}mol^{-1}, respectively.)

The entire process can be broken down into 3 steps: (1) cooling of H$_2$O(l) from 110°C to 100°C; (2) supplying heat to system to effect the phase transformation of H$_2$O from liquid to gas at 100°C; and (3) heating H$_2$O(g) from 100°C to 110°C.

Step 1

$$\text{H}_2\text{O}(l, 110°\text{C}) \rightarrow \text{H}_2\text{O}(l, 100°\text{C})$$

$$\Delta S_1 = C_P \ln \frac{373 \text{ K}}{383 \text{ K}} = (1.5 \text{ mol}) \left(75.5 \text{ J K}^{-1} \text{ mol}^{-1}\right) \ln \frac{373 \text{ K}}{383 \text{ K}} = -3.00 \text{ J K}^{-1}$$

$$q_{surr,1} = -q_1 = -C_P \Delta T = -(1.5 \text{ mol}) \left(75.5 \text{ J K}^{-1} \text{ mol}^{-1}\right) (373 \text{ K} - 383 \text{ K}) = 1.13 \times 10^3 \text{J}$$

Step 2

$$\text{H}_2\text{O}(l, 100°\text{C}) \rightarrow \text{H}_2\text{O}(g, 100°\text{C})$$

$$\Delta S_2 = \frac{\Delta H_{vap}}{T_b} = \frac{(1.5 \text{ mol}) \left(40.79 \times 10^3 \text{ J mol}^{-1}\right)}{373 \text{ K}} = 164 \text{ J K}^{-1}$$

$$q_{surr,2} = -q_2 = -\Delta H_{vap} = -(1.5 \text{ mol}) \left(40.79 \times 10^3 \text{ J mol}^{-1}\right) = -6.12 \times 10^4 \text{ J}$$

Step 3

$$H_2O(g, 100°C) \rightarrow H_2O(g, 110°C)$$

$$\Delta S_3 = C_P \ln \frac{383 \text{ K}}{373 \text{ K}} = (1.5 \text{ mol}) \left(34.4 \text{ J K}^{-1} \text{ mol}^{-1}\right) \ln \frac{383 \text{ K}}{373 \text{ K}} = 1.37 \text{ J K}^{-1}$$

$$q_{\text{surr},3} = -q_3 = -C_P \Delta T = -(1.5 \text{ mol}) \left(34.4 \text{ J K}^{-1} \text{ mol}^{-1}\right)(383 \text{ K} - 373 \text{ K}) = -516 \text{ J}$$

Therefore, for the entire process,

$$\Delta S = -3.00 \text{ J K}^{-1} + 164 \text{ J K}^{-1} + 1.37 \text{ J K}^{-1} = 162 \text{ J K}^{-1}$$

$$q_{\text{surr}} = 1.13 \times 10^3 \text{J} - 6.12 \times 10^4 \text{ J} - 516 \text{ J} = -6.06 \times 10^4 \text{ J}$$

$$\Delta S_{\text{surr}} = \frac{q_{\text{surr}}}{T_{\text{surr}}} = \frac{-6.06 \times 10^4 \text{ J}}{383 \text{ K}} = -158 \text{ J K}^{-1}$$

$$\Delta S_{\text{univ}} = 162 \text{ J K}^{-1} - 158 \text{ J K}^{-1} = 4 \text{ J K}^{-1}$$

$\Delta S_{\text{univ}} > 0$ confirms the expectation that superheated water vaporizes spontaneously.

5.50   Toluene (C_7H_8) has a dipole moment, whereas benzene (C_6H_6) is nonpolar:

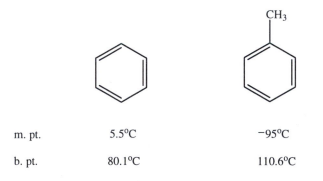

m. pt.	5.5°C	−95°C
b. pt.	80.1°C	110.6°C

Explain why, contrary to our expectation, benzene melts at a much higher temperature than toluene. Why is the boiling point of toluene higher than that of benzene?

Rearranging equation 5.18 gives $T_{\text{fus}} = \frac{\Delta_{\text{fus}}H}{\Delta_{\text{fus}}S}$ Although toluene as a polar molecule is expected to have a slightly larger $\Delta_{\text{fus}}H$ than the non-polar benzene, benzene is a highly symmetric molecule so that it has a much smaller $\Delta_{\text{fus}}S$ than toluene. The entropy effect outweighs the enthalpy effect giving benzene a melting point considerably higher than that of toluene.

For boiling, the situation is different. The entropy of the vapor phase is significantly greater than that of the liquid phase for both molecules. Thus, $\Delta_{\text{vap}}S = S_{\text{vap}} - S_{\text{liq}} \approx S_{\text{vap}}$ is roughly the same for both species. Equation 5.19, $T_{\text{vap}} = \frac{\Delta_{\text{vap}}H}{\Delta_{\text{vap}}S}$, then indicates that the boiling point is determined primarily by $\Delta_{\text{vap}}H$ which is greater for toluene than benzene so that the polar toluene has the higher boiling point.

CHAPTER 6

Gibbs and Helmholtz Energies and Their Applications

PROBLEMS AND SOLUTIONS

6.1 A quantity of 0.35 mole of an ideal gas initially at 15.6°C is expanded from 1.2 L to 7.4 L. Calculate the values of w, q, ΔU, ΔS, and ΔG if the process is carried out **(a)** isothermally and reversibly, and **(b)** isothermally and irreversibly against an external pressure of 1.0 atm.

(a) For an isothermal process of an ideal gas, $\Delta U = 0$ and $\Delta H = 0$.

$$w = -nRT \ln \frac{V_2}{V_1} = -(0.35 \text{ mol}) \left(8.314 \text{ J K}^{-1} \text{ mol}^{-1}\right) (288.8 \text{ K}) \ln \frac{7.4 \text{ L}}{1.2 \text{ L}}$$

$$= -1.53 \times 10^3 \text{ J} = -1.5 \times 10^3 \text{ J}$$

$$q = \Delta U - w = 1.53 \times 10^3 \text{ J} = 1.5 \times 10^3 \text{ J}$$

$$\Delta S = \frac{q_{\text{rev}}}{T} = \frac{1.53 \times 10^3 \text{ J}}{288.8 \text{ K}} = 5.30 \text{ J K}^{-1} = 5.3 \text{ J K}^{-1}$$

$$\Delta G = \Delta H - T \Delta S = 0 - (288.8 \text{ K}) \left(5.30 \text{ J K}^{-1}\right) = -1.5 \times 10^3 \text{ J}$$

(b) Since U, S, and G are state functions, ΔU, ΔS, ΔG depend only on the initial and final states and not on the path. Therefore, they are the same as above, that is

$$\Delta U = 0 \text{ J}$$

$$\Delta S = 5.3 \text{ J K}^{-1}$$

$$\Delta G = -1.5 \times 10^3 \text{ J}$$

w and q, however, are path dependent.

$$w = -P_{\text{ex}} \Delta V = -(1.0 \text{ atm}) (7.4 \text{ L} - 1.2 \text{ L}) \left(\frac{101.3 \text{ J}}{1 \text{ L atm}}\right) = -6.3 \times 10^2 \text{ J}$$

$$q = \Delta U - w = 6.3 \times 10^2 \text{ J}$$

6.2 At one time, the gas used for cooking, called "water gas," was prepared as follows:

$$H_2O(g) + C(\text{graphite}) \rightarrow CO(g) + H_2(g)$$

From the thermodynamic quantities listed in Appendix B, predict whether this reaction will occur at 298 K. If not, at what temperature will the reaction occur? Assume $\Delta_r H°$ and $\Delta_r S°$ are temperature independent.

The sign of $\Delta_r G°$ is used to determine if the reaction is spontaneous when the reactants and products are in their standard states. If $\Delta_r G°$ is calculated via $\Delta_r H°$ and $\Delta_r S°$, the dependence of this quantity on temperature can also be assessed.

$$\Delta_r H° = \Delta_f \overline{H}° [CO(g)] + \Delta_f \overline{H}° \left[H_2(g)\right] - \Delta_f \overline{H}° \left[H_2O(g)\right] - \Delta_f \overline{H}° \left[C(\text{graphite})\right]$$

$$= -110.5 \text{ kJ mol}^{-1} + 0 \text{ kJ mol}^{-1} - \left(-241.8 \text{ kJ mol}^{-1}\right) - 0 \text{ kJ mol}^{-1}$$

$$= 131.3 \text{ kJ mol}^{-1}$$

$$\Delta_r S° = \overline{S}° [CO(g)] + \overline{S}° \left[H_2(g)\right] - \overline{S}° \left[H_2O(g)\right] - \overline{S}° \left[C(\text{graphite})\right]$$

$$= 197.9 \text{ J K}^{-1} \text{mol}^{-1} + 130.6 \text{ J K}^{-1} \text{mol}^{-1} - 188.7 \text{ J K}^{-1} \text{mol}^{-1} - 5.7 \text{ J K}^{-1} \text{mol}^{-1}$$

$$= 134.1 \text{ J K}^{-1} \text{mol}^{-1}$$

$$\Delta_r G° = \Delta_r H° - T\Delta_r S° = 131.3 \text{ kJ mol}^{-1} - (298 \text{ K}) \left(134.1 \times 10^{-3} \text{ kJ K}^{-1} \text{mol}^{-1}\right) = 91.3 \text{ kJ mol}^{-1}$$

Since $\Delta_r G° > 0$, the reaction is not spontaneous when all the reactants and products are in their standard states and at 298 K. The reaction is spontaneous when $\Delta_r G° < 0$, that is, when

$$\Delta_r H° - T\Delta_r S° < 0$$

$$\Delta_r H° < T\Delta_r S°$$

$$T > \frac{\Delta_r H°}{\Delta_r S°} = \frac{131.3 \text{ kJ mol}^{-1}}{134.1 \times 10^{-3} \text{ kJ K}^{-1} \text{mol}^{-1}} = 979.1 \text{ K}$$

The lower limit of temperature, 979.1 K, is an estimate because $\Delta_r H°$ and $\Delta_r S°$ are not truly temperature independent, as assumed here.

6.3 Use the values listed in Appendix B to calculate $\Delta_r G°$ for the following alcohol fermentation:

$$\alpha\text{-D-glucose}(aq) \rightarrow 2C_2H_5OH(l) + 2CO_2(g)$$

$(\Delta_f G°[\alpha\text{-D-glucose}(aq)] = -914.5 \text{ kJ mol}^{-1})$

$$\Delta_r G° = 2\Delta_f \overline{G}° \left[C_2H_5OH(l)\right] + 2\Delta_f \overline{G}° \left[CO_2(g)\right] - \Delta_f \overline{G}° \left[\alpha\text{-D-glucose}(aq)\right]$$

$$= 2\left(-174.2 \text{ kJ mol}^{-1}\right) + 2\left(-394.4 \text{ kJ mol}^{-1}\right) - \left(-914.5 \text{ kJ mol}^{-1}\right)$$

$$= -222.7 \text{ kJ mol}^{-1}$$

6.4 Without referring to Appendix B, calculate the quantity $(\Delta_r G^\circ - \Delta_r A^\circ)$ for the following reaction at 298 K:

$$C(s) + CO_2(g) \rightarrow 2CO(g)$$

Assume ideal gas behavior.

The definitions of $\Delta_r G^\circ$ and $\Delta_r A^\circ$ give

$$\Delta_r G^\circ = \Delta_r H^\circ - T\Delta_r S^\circ$$

$$\Delta_r A^\circ = \Delta_r U^\circ - T\Delta_r S^\circ$$

Thus,

$$\Delta_r G^\circ - \Delta_r A^\circ = \Delta_r H^\circ - \Delta_r U^\circ$$

$$= \Delta(PV)$$

$$= \Delta(nRT)$$

$$= RT\Delta n$$

$$= \left(8.314 \text{ J K}^{-1} \text{ mol}^{-1}\right)(298 \text{ K})(1)$$

$$= 2.48 \times 10^3 \text{ J mol}^{-1}$$

$$= 2.48 \text{ kJ mol}^{-1}$$

6.5 As an approximation, we can assume that proteins exist either in the native (or physiologically functioning) state and the denatured state. The standard molar enthalpy and entropy of the denaturation of a certain protein are 512 kJ mol^{-1} and 1.60 kJ K^{-1} mol^{-1}, respectively. Comment on the signs and magnitudes of these quantities, and calculate the temperature at which the denaturation becomes spontaneous.

The enthalpy change for the denaturation of the protein is positive. It is an endothermic process, requiring energy to break up the hydrogen bonds and van der Waals interactions present in the native form. Some energy is returned due to additional protein-solvent (H_2O) intermolecular interactions in the denatured state. Nevertheless, the overall process is endothermic. Typical hydrogen bond enthalpies are 10 - 15 kJ mol^{-1}, so the value of 512 kJ mol^{-1} for $\Delta \overline{H}^\circ$ indicates that approximately 40 more hydrogen bonds were broken than were formed. Since the denaturation process takes the specific, ordered native form to a random form, $\Delta \overline{S}^\circ$ is expected to be positive, although solvent effects may also make a contribution.

The denaturation is spontaneous when $\Delta G = \Delta H - T\Delta S < 0$, or $T > \frac{\Delta H}{\Delta S}$. Since the values given are standard values, they apply only to a solution that is both 1 M in native form and 1 M in denatured form. For such a solution, the native form will spontaneously denature to reach a position of equilibrium when

$$T > \frac{\Delta H^\circ}{\Delta S^\circ}$$

$$> \frac{512 \text{ kJ mol}^{-1}}{1.60 \text{ kJ mol}^{-1}}$$

$$> 320 \text{ K} = 47^\circ C$$

6.6 Certain bacteria in the soil obtain the necessary energy for growth by oxidizing nitrite to nitrate:

$$2NO_2^-(aq) + O_2(g) \rightarrow 2NO_3^-(aq)$$

Given that the standard Gibbs energies of formation of NO_2^- and NO_3^- are $-34.6 \text{ kJ mol}^{-1}$ and $-110.5 \text{ kJ mol}^{-1}$, respectively, calculate the amount of Gibbs energy released when 1 mole of NO_2^- is oxidized to 1 mole of NO_3^-.

According to the chemical equation $2NO_2^-(aq) + O_2(g) \rightarrow 2NO_3^-(aq)$,

$$\Delta_r G^\circ = 2\Delta_f \overline{G}^\circ \left[NO_3^-(aq) \right] - 2\Delta_f \overline{G}^\circ \left[NO_2^-(aq) \right] - \Delta_f \overline{G}^\circ \left[O_2(g) \right]$$

$$= 2\left(-110.5 \text{ kJ mol}^{-1} \right) - 2\left(-34.6 \text{ kJ mol}^{-1} \right) - 0 \text{ kJ mol}^{-1}$$

$$= -151.8 \text{ kJ mol}^{-1}$$

When 1 mole of NO_2^- is oxidized to 1 mole of NO_3^-,

$$\Delta_r G^\circ = \frac{-151.8 \text{ kJ mol}^{-1}}{2} = -75.9 \text{ kJ mol}^{-1}$$

6.7 Consider the synthesis of urea according to the equation

$$CO_2(g) + 2NH_3(g) \rightarrow (NH_2)_2CO(s) + H_2O(l)$$

From the data listed in Appendix B, calculate the value of $\Delta_r G^\circ$ for the reaction at 298 K. Assuming ideal gas behavior, calculate the value of $\Delta_r G$ for the reaction at a pressure of 10.0 bar. The $\Delta_f \overline{G}^\circ$ of urea is $-197.15 \text{ kJ mol}^{-1}$.

$$\Delta_r G^\circ = \Delta_f \overline{G}^\circ \left[(NH_2)_2CO(s) \right] + \Delta_f \overline{G}^\circ \left[H_2O(l) \right] - \Delta_f \overline{G}^\circ \left[CO_2(g) \right] - 2\Delta_f \overline{G}^\circ \left[NH_3(g) \right]$$

$$= -197.15 \text{ kJ mol}^{-1} + \left(-237.2 \text{ kJ mol}^{-1} \right) - \left(-394.4 \text{ kJ mol}^{-1} \right) - 2\left(-16.6 \text{ kJ mol}^{-1} \right)$$

$$= -6.75 \text{ kJ mol}^{-1} = -6.8 \text{ kJ mol}^{-1}$$

At a pressure of 10.0 bar, the Gibbs energy of a substance is

$$\overline{G} = \overline{G}^\circ + \int_{1 \text{ bar}}^{10 \text{ bar}} \overline{V} dP$$

Therefore, $\Delta_r G$ for the reaction at a pressure of 10.0 bar is

$$\Delta_r G = \overline{G} \left[(NH_2)_2CO(s) \right] + \overline{G} \left[H_2O(l) \right] - \overline{G} \left[CO_2(g) \right] - 2\overline{G} \left[NH_3(g) \right]$$

$$= \left[\overline{G}^\circ \left[(NH_2)_2CO(s) \right] + \int_{1 \text{ bar}}^{10 \text{ bar}} \overline{V} \left[(NH_2)_2CO(s) \right] dP \right] + \left[\overline{G}^\circ \left[H_2O(l) \right] + \int_{1 \text{ bar}}^{10 \text{ bar}} \overline{V} \left[H_2O(l) \right] dP \right]$$

$$- \left[\overline{G}^\circ \left[CO_2(g) \right] + \int_{1 \text{ bar}}^{10 \text{ bar}} \overline{V} \left[CO_2(g) \right] dP \right] - 2 \left[\overline{G}^\circ \left[NH_3(g) \right] + \int_{1 \text{ bar}}^{10 \text{ bar}} \overline{V} \left[NH_3(g) \right] dP \right]$$

The above expression can be simplified by recognizing that (1) the $\overline{G}^\circ$ terms combine to give ΔG° and (2) the molar volumes of solids and liquids are negligible compared with gases.

$$\Delta_r G = \Delta_r G^\circ - \int_{1\,\text{bar}}^{10\,\text{bar}} \overline{V}\left[CO_2(g)\right]dP - 2\int_{1\,\text{bar}}^{10\,\text{bar}} \overline{V}\left[NH_3(g)\right]dP$$

Assume both CO_2 and NH_3 behave ideally, then

$$\Delta_r G = \Delta_r G^\circ - \int_{1\,\text{bar}}^{10\,\text{bar}} \frac{RT}{P} - 2\int_{1\,\text{bar}}^{10\,\text{bar}} \frac{RT}{P}$$

$$= \Delta_r G^\circ - RT \ln \frac{10.0\,\text{bar}}{1\,\text{bar}} - 2RT \ln \frac{10.0\,\text{bar}}{1\,\text{bar}}$$

$$= -6.75 \times 10^3\,\text{J mol}^{-1} - \left(8.314\,\text{J K}^{-1}\,\text{mol}^{-1}\right)(298\,\text{K}) \ln 10$$

$$- 2\left(8.314\,\text{J K}^{-1}\,\text{mol}^{-1}\right)(298\,\text{K}) \ln 10$$

$$= -2.39 \times 10^4\,\text{J mol}^{-1} = -23.9\,\text{kJ mol}^{-1}$$

6.8 This problem involves the synthesis of diamond from graphite

$$C(\text{graphite}) \rightarrow C(\text{diamond})$$

(a) Calculate the values of $\Delta_r H^\circ$ and $\Delta_r S^\circ$ for the reaction. Will the conversion occur spontaneously at 25°C or any other temperature? **(b)** From density measurements, the molar volume of graphite is found to be 2.1 cm^3 greater than that of diamond. Can the conversion of graphite to diamond be brought about at 25°C by applying pressure on graphite? If so, estimate the pressure at which the process becomes spontaneous. [*Hint:* Starting from Equation 6.16, derive the equation $\Delta_r G_2 - \Delta_r G_1 = \left(\overline{V}_{\text{diamond}} - \overline{V}_{\text{graphite}}\right)\Delta P$] for a constant-temperature process. Next, calculate the ΔP that would lead to the necessary decrease in Gibbs energy.]

(a)

$$\Delta_r H^\circ = \Delta_f \overline{H}^\circ \left[C(\text{diamond})\right] - \Delta_f \overline{H}^\circ \left[C(\text{graphite})\right]$$

$$= 1.90\,\text{kJ mol}^{-1} - 0\,\text{kJ mol}^{-1}$$

$$= 1.90\,\text{kJ mol}^{-1}$$

$$\Delta_r S^\circ = \overline{S}^\circ \left[C(\text{diamond})\right] - \overline{S}^\circ \left[C(\text{graphite})\right]$$

$$= 2.4\,\text{J K}^{-1}\,\text{mol}^{-1} - 5.7\,\text{J K}^{-1}\,\text{mol}^{-1}$$

$$= -3.3\,\text{J K}^{-1}\,\text{mol}^{-1}$$

At 25°C,

$$\Delta_r G^\circ = \Delta_r H^\circ - (298\,\text{K})\,\Delta_r S^\circ$$

$$= 1.90\,\text{kJ mol}^{-1} - (298\,\text{K})\left(-3.3 \times 10^{-3}\,\text{kJ K}^{-1}\,\text{mol}^{-1}\right)$$

$$= 2.883\,\text{kJ mol}^{-1} = 2.88\,\text{kJ mol}^{-1}$$

Therefore, the conversion from graphite to diamond is not spontaneous at 25°C and when both are in their standard states. In fact, because $\Delta_r H°$ is positive and $T\Delta_r S°$ is negative, $\Delta_r G°$ can never be negative. Thus, the conversion will not be spontaneous at any other temperature.

(b) The integration of Equation 6.16

$$\left(\frac{\partial G}{\partial P}\right)_T = V$$

gives

$$G_2 = G_1 + V\left(P_2 - P_1\right)$$

where G_1 and G_2 are the Gibbs energies at P_1 and P_2, respectively. Using molar quantities and $\Delta P = P_2 - P_1$, the equation becomes

$$\overline{G}_2 = \overline{G}_1 + \overline{V}\Delta P$$

Apply this equation to graphite and diamond,

$$\overline{G}_2\left[\text{graphite}\right] = \overline{G}_1\left[\text{graphite}\right] + \overline{V}\left[\text{graphite}\right]\Delta P$$

$$\overline{G}_2\left[\text{diamond}\right] = \overline{G}_1\left[\text{diamond}\right] + \overline{V}\left[\text{diamond}\right]\Delta P$$

These two equations are combined to relate the values of $\Delta_r G$ for the conversion of graphite to diamond at two different pressures:

$$\Delta_r G_2 = \overline{G}_2\left[\text{diamond}\right] - \overline{G}_2\left[\text{graphite}\right]$$

$$= \left\{\overline{G}_1\left[\text{diamond}\right] + \overline{V}\left[\text{diamond}\right]\Delta P\right\} - \left\{\overline{G}_1\left[\text{graphite}\right] + \overline{V}\left[\text{graphite}\right]\Delta P\right\}$$

$$= \Delta_r G_1 + \left\{\overline{V}\left[\text{diamond}\right] - \overline{V}\left[\text{graphite}\right]\right\}\Delta P$$

If $P_1 = 1$ bar, then $\Delta_r G_1 = \Delta_r G° = 2.883$ kJ mol^{-1} at 25°C. $\Delta_r G_2$ becomes

$$\Delta_r G_2 = 2.883 \text{ kJ mol}^{-1} + \left(-2.1 \text{ cm}^3 \text{ mol}^{-1}\right)\Delta P\left(\frac{1\text{ L}}{1000\text{ cm}^3}\right)\left(\frac{1\text{ atm}}{1.013\text{ bar}}\right)\left(\frac{101.3\text{ J}}{1\text{ L atm}}\right)\left(\frac{1\text{ kJ}}{1000\text{ J}}\right)$$

$$= 2.883 \text{ kJ mol}^{-1} - 2.1 \times 10^{-4}\Delta P \text{ kJ bar}^{-1}\text{ mol}^{-1}$$

If the process is spontaneous, then

$$\Delta_r G_2 = 2.883 \text{ kJ mol}^{-1} - 2.1 \times 10^{-4}\Delta P \text{ kJ bar}^{-1}\text{ mol}^{-1} < 0$$

$$2.883 \text{ kJ mol}^{-1} < 2.1 \times 10^{-4}\Delta P \text{ kJ bar}^{-1}\text{ mol}^{-1}$$

$$\Delta P > \frac{2.883 \text{ kJ mol}^{-1}}{2.1 \times 10^{-4} \text{ kJ bar}^{-1}\text{ mol}^{-1}} = 1.4 \times 10^4 \text{ bar}$$

Therefore, at a very high pressure, $P_2 = \Delta P + P_1 = 1.4 \times 10^4$ bar $+ 1$ bar $= 1.4 \times 10^4$ bar, the conversion from graphite to diamond is spontaneous.

6.9 How do the requirements that T and V are constant enter the derivation for $\Delta A_{\text{sys}} < 0$ for a spontaneous process?

By definition, $A = U - TS$. At constant T,

$$dA = dU - TdS$$

At constant V, $dw = 0$. Therefore, $dU = dq$ and

$$dA = dq - TdS$$

A process is spontaneous when

$$dS_{univ} = dS_{sys} - \frac{dq_{sys}}{T} > 0$$

or

$$TdS_{sys} - dq_{sys} > 0$$

The left hand side of the inequality is $-dA$ at constant T and V. Therefore, the inequality becomes

$$-dA_{sys} > 0$$

$$dA_{sys} < 0$$

6.10   From the standard molar enthalpy of combustion of benzene at 298 K, calculate the value of $\Delta_r A^\circ$ for the process. Compare the value of $\Delta_r A^\circ$ with that of $\Delta_r H^\circ$. Comment on the difference.

The chemical equation corresponding to the combustion of benzene is

$$C_6H_6(l) + \tfrac{15}{2}O_2(g) \rightarrow 6CO_2(g) + 3H_2O(l)$$

At constant temperature, $\Delta_r A^\circ$ can be calculated directly from $\Delta_r U^\circ$ and $\Delta_r S^\circ$. $\Delta_r U^\circ$ in turn can be calculated from $\Delta_r H^\circ$.

$$\Delta_r H^\circ = 6\Delta_f \overline{H}^\circ \left[CO_2(g)\right] + 3\Delta_f \overline{H}^\circ \left[H_2O(l)\right] - \Delta_f \overline{H}^\circ \left[C_6H_6(l)\right] - \frac{15}{2}\Delta_f \overline{H}^\circ \left[O_2(g)\right]$$

$$= 6\left(-393.5 \text{ kJ mol}^{-1}\right) + 3\left(-285.8 \text{ kJ mol}^{-1}\right) - 49.04 \text{ kJ mol}^{-1} - \frac{15}{2}\left(0 \text{ kJ mol}^{-1}\right)$$

$$= -3267.4 \text{ kJ mol}^{-1}$$

$$\Delta_r U^\circ = \Delta_r H^\circ - RT\,\Delta n = -3267.4 \text{ kJ mol}^{-1} - \left(8.314 \text{ J K}^{-1} \text{ mol}^{-1}\right)(298 \text{ K})(-1.5)\left(\frac{1 \text{ kJ}}{1000 \text{ J}}\right)$$

$$= -3263.68 \text{ kJ mol}^{-1}$$

$$\Delta_r S^\circ = 6\overline{S}^\circ \left[CO_2(g)\right] + 3\overline{S}^\circ \left[H_2O(l)\right] - \overline{S}^\circ \left[C_6H_6(l)\right] - \frac{15}{2}\overline{S}^\circ \left[O_2(g)\right]$$

$$= 6\left(213.6 \text{ J K}^{-1} \text{ mol}^{-1}\right) + 3\left(69.9 \text{ J K}^{-1} \text{ mol}^{-1}\right) - 172.8 \text{ J K}^{-1} \text{ mol}^{-1} - \frac{15}{2}\left(205.0 \text{ J K}^{-1} \text{ mol}^{-1}\right)$$

$$= -219.0 \text{ J K}^{-1} \text{ mol}^{-1}$$

$$\Delta_r A^\circ = \Delta_r U^\circ - T\Delta_r S^\circ = -3263.68 \text{ kJ mol}^{-1} - (298 \text{ K})\left(-219 \text{ J K}^{-1} \text{ mol}^{-1}\right)\left(\frac{1 \text{ kJ}}{1000 \text{ J}}\right)$$

$$= -3198.4 \text{ kJ mol}^{-1}$$

The maximum work ($\Delta_r A$) that can be obtained from this combustion is not as great as the heat released at constant pressure ($\Delta_r H$). This is because there is an entropy decrease in the system.

6.11 A student placed 1 g each of three compounds A, B, and C in a container and found that no change had occurred after one week. Offer possible explanations for the lack of reaction. Assume that A, B, and C are totally miscible liquids.

There are three possibilities, depending on the sign of $\Delta_r G$ for the expected changes.

(1) No reactions are possible, $\Delta_r G > 0$.

(2) Reactions are possible thermodynamically, $\Delta_r G < 0$, but the rates are extremely slow.

(3) The three components are already at equilibrium, $\Delta_r G = 0$.

6.12 Predict the signs of ΔH, ΔS, and ΔG of the system for the following processes at 1 atm: **(a)** ammonia melts at $-60°C$, **(b)** ammonia melts at $-77.7°C$, **(c)** ammonia melts at $-100°C$. (The normal melting point of ammonia is $-77.7°C$.)

Since melting is an endothermic process, ΔH is positive in all three cases. A substance in the liquid phase is more disordered than in the solid phase. Therefore, ΔS is positive in all three cases.

When the temperature is above the melting point, the melting process is spontaneous, that is, ΔG is negative. When the temperature is at the melting point, the melting process is at equilibrium, that is, ΔG is zero. When the temperature is below the melting point, the melting process is not spontaneous, that is, ΔG is negative.

In summary,

(a) ΔH: +, ΔS: +, ΔG: − **(b)** ΔH: +, ΔS: +, ΔG: 0 **(c)** ΔH: +, ΔS: +, ΔG: +

6.13 Crystallization of sodium acetate from a supersaturated solution occurs spontaneously. What can you deduce about the signs of ΔS and ΔH.

Since this is a spontaneous process, ΔG is negative. The solid phase of a substance is more ordered than the aqueous phase. Thus, ΔS is negative. Since $\Delta H = \Delta G + T\Delta S$, it must be negative. That is, this process is exothermic.

6.14 A student looked up the $\Delta_f \overline{G}°$, $\Delta_f \overline{H}°$, and $\overline{S}°$ values for CO_2 in Appendix B. Plugging these values into Equation 6.3, he found that $\Delta_f \overline{G}° \neq \Delta_f \overline{H}° - T\overline{S}°$ at 298 K. What is wrong with his approach?

The equation $\Delta G° = \Delta H° - T\Delta S°$ applies to a process. If the process is a formation reaction, such as the formation of CO_2 [$C(graphite) + O_2(g) \rightarrow CO_2(g)$], then

$$\Delta_r G^\circ = \Delta_f G^\circ = -394.4 \text{ kJ mol}^{-1}$$

$$\Delta_r H^\circ = \Delta_f H^\circ = -393.5 \text{ kJ mol}^{-1}$$

$$\Delta_r S^\circ = \overline{S}^\circ \left[CO_2(g) \right] - \overline{S}^\circ \left[C(\text{graphite}) \right] - \overline{S}^\circ \left[O_2(g) \right]$$

$$= 213.6 \text{ J K}^{-1} \text{mol}^{-1} - 5.7 \text{ J K}^{-1} \text{mol}^{-1} - 205.0 \text{ J K}^{-1} \text{mol}^{-1}$$

$$= 2.9 \text{ J K}^{-1} \text{mol}^{-1}$$

Note that $\Delta_r S^\circ$ is not the same as $\overline{S}^\circ \left[CO_2(g) \right]$ as stated in the question. At 298 K,

$$\Delta_r H^\circ - T\Delta_r S^\circ = -393.5 \text{ kJ mol}^{-1} - (298 \text{ K}) \left(2.9 \times 10^{-3} \text{ kJ K}^{-1} \text{mol}^{-1} \right) = -394.4 \text{ kJ mol}^{-1}$$

which is the same as $\Delta_r G^\circ$.

6.15 A certain reaction is spontaneous at 72°C. If the enthalpy change for the reaction is 19 kJ, what is the *minimum* value of $\Delta_r S$ (in joules per kelvin) for the reaction?

$$\Delta_r G = \Delta_r H - T\Delta_r S < 0$$

$$19 \text{ kJ} - (345 \text{ K}) \Delta_r S < 0$$

$$\Delta_r S > \frac{19 \times 10^3 \text{ kJ}}{345 \text{ K}} = 55 \text{ J K}^{-1}$$

Therefore, the minimum $\Delta_r S$ must be 55 J K^{-1}. At this value of $\Delta_r S$, the reaction is at equilibrium. At greater values of $\Delta_r S$, the reaction is spontaneous.

6.16 A certain reaction is known to have a $\Delta_r G^\circ$ value of -122 kJ. Will the reaction necessarily occur if the reactants are mixed together?

No, for two reasons. First, the reaction rate can be very slow. Secondly, even if the rate is rapid enough for reaction to occur, the value of $\Delta_r G$ for the reaction depends on the concentrations of reactants and products and will be different from the value $\Delta_r G^\circ$ that is appropriate when all species are present in their standard pressure or concentration.

6.17 The vapor pressure of mercury at various temperatures has been determined as follows:

T/K	P/mmHg
323	0.0127
353	0.0888
393.5	0.7457
413	1.845
433	4.189

Calculate the value of $\Delta_{vap}\overline{H}$ for mercury.

A plot of $\ln P$ vs $1/T$ has a slope of $-\Delta_{vap}\overline{H}/R$, from which $-\Delta_{vap}\overline{H}$ is obtained.

10^3 K$/T$	$\ln P$
3.10	-4.3362
2.83	-2.4214
2.54	-0.29343
2.42	0.61248
2.31	1.4325

The slope of the best fit straight line to the data is -7300 K, where the value has been rounded to reflect the regression estimate of uncertainty in the slope. Thus,

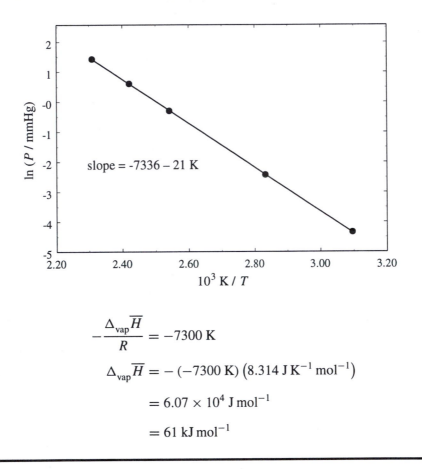

$$-\frac{\Delta_{vap}\overline{H}}{R} = -7300 \text{ K}$$

$$\Delta_{vap}\overline{H} = -(-7300 \text{ K})\left(8.314 \text{ J K}^{-1}\text{ mol}^{-1}\right)$$

$$= 6.07 \times 10^4 \text{ J mol}^{-1}$$

$$= 61 \text{ kJ mol}^{-1}$$

6.18 The pressure exerted on ice by a 60.0-kg skater is about 300 atm. Calculate the depression in freezing point. The molar volumes are $\overline{V}_L = 0.0180 \text{ L mol}^{-1}$ and $\overline{V}_S = 0.0196 \text{ L mol}^{-1}$.

The depression of freezing point can be obtained from the slope of the S–L curve in the phase diagram.

$$\frac{dP}{dT} = \frac{\Delta \overline{H}}{T \Delta \overline{V}} = \frac{\left(6.01 \times 10^3 \text{ J mol}^{-1}\right) \left(\frac{1 \text{ L atm}}{101.3 \text{ J}}\right)}{(273.2 \text{ K}) \left(0.0180 \text{ L mol}^{-1} - 0.0196 \text{ L mol}^{-1}\right)}$$

$$= -135.7 \text{ atm K}^{-1}$$

$$dP = -135.7 \text{ atm K}^{-1} dT$$

$$\int_{1 \text{ atm}}^{300 \text{ atm}} dP = -135.7 \text{ atm K}^{-1} \int_{273.15 \text{ K}}^{T \text{ K}} dT$$

$$299 \text{ atm} = -135.7 \text{ atm K}^{-1} \Delta T$$

$$\Delta T = \frac{299 \text{ atm}}{-135.7 \text{ atm K}^{-1}} = -2.20 \text{ K}$$

Therefore, the freezing point is depressed by 2.20 K when 300 atm is applied to ice. In other words, the freezing point at this pressure is $273.15 \text{ K} - 2.20 \text{ K} = 270.95 \text{ K}$ or $-2.20 \,^\circ\text{C}$.

6.19   Use the phase diagram of water (Figure 6.5) to predict the direction for the following changes: **(a)** at the triple point of water, temperature is lowered at constant pressure, and **(b)** somewhere along the S–L curve of water, pressure is increased at constant temperature.

(a) Referring to Figure 6.5, this change will cause ice to form from the liquid and vapor present at the triple point.

(b) Because of the negative slope of the S-L curve in the phase diagram for water, this change will cause some ice to melt.

6.20   Use the phase diagram of water (Figure 6.5) to predict the dependence of the freezing and boiling point of water on pressure.

The freezing point decreases with increasing pressure (because the S–L curve has a negative slope) whereas the boiling point increases with increasing pressure (because the L–V curve has a positive slope).

6.21   Consider the following system at equilibrium

$$CaCO_3(s) \rightleftharpoons CaO(s) + CO_2(g)$$

How many phases are present?

Three phases: 2 solid phases and 1 gas phase.

6.22   Below is a rough sketch of the phase diagram of carbon. **(a)** How many triple points are there, and what are the phases that can coexist at each triple point? **(b)** Which has a higher density, graphite or diamond? **(c)** Synthetic diamond can be made from graphite. Using the phase diagram, how would you go about making diamond?

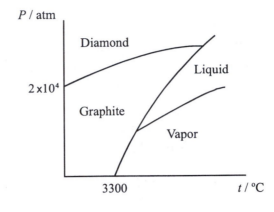

(a) There are two triple points. At one triple point, diamond, graphite and liquid carbon coexist. At the other, graphite, liquid carbon, and gaseous carbon coexist.

(b) According to the Clapeyron equation, $\frac{dP}{dT} = \frac{\Delta \overline{H}}{T \Delta \overline{V}}$. From the phase diagram, the graphite-diamond curve has positive slope for most of its length. From Problem 6.8, the standard molar enthalpy change for C(graphite) → C(diamond), $\Delta \overline{H}^\circ$, is also positive. This implies that $\Delta \overline{V}$ is positive, or that diamond has a greater molar volume than graphite. Consequently, one would conclude that diamond is less dense than graphite. In fact this conclusion is in error, and diamond is the denser phase. The problem is that by using $\Delta \overline{H}^\circ$, the assumption is made that ΔH is independent of pressure. Indeed, the evidence suggests that it changes sign and is negative at those pressures where graphite and diamond are in equilibrium.

(c) The phase diagram indicates that high pressures are required to convert graphite to diamond.

6.23   What is wrong with the following phase diagram for a one-component system?

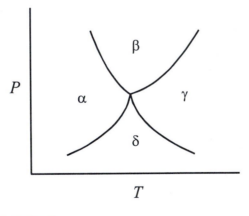

This is a one-component system ($c = 1$). At the quadruple point, $p = 4$. Therefore, $f = c - p + 2 = 1 - 4 + 2 = -1$. Since the degrees of freedom cannot be negative, this phase diagram is nonphysical.

6.24   The plot in Figure 6.4 is no longer linear at high temperatures. Explain.

The plot is no longer linear at high temperatures means that the slope is not constant over a large temperature range. This is because $\Delta_{\text{vap}}\overline{H}$, which is related to the slope, remains constant only over a limited temperature range.

6.25 Pike's Peak in Colorado is approximately 4,300 m above sea level (0°C). What is the boiling point of water at the summit? (*Hint*: See Problem 2.56. The molar mass of air is 29.0 g mol^{-1} and $\Delta_{\text{vap}}\overline{H}$ for water is 40.79 kJ mol^{-1}.)

The pressure at 4,300 m above sea level, P, is related to that at the sea level, P_0. Once P is evaluated, the boiling point of water is calculated using the Clausius-Clapeyron equation.

$$P = P_0 \exp\left(\frac{-g\mathcal{M}h}{RT}\right) = P_0 \exp\left[-\frac{\left(9.81 \text{ m s}^{-2}\right)\left(29.0 \times 10^{-3} \text{ kg mol}^{-1}\right)(4300 \text{ m})}{\left(8.314 \text{ J K}^{-1} \text{ mol}^{-1}\right)(273.2 \text{ K})}\right] = 0.5836 P_0$$

Let T_0 and T be the boiling points of water at the sea level and at 4,300 m above the sea level.

$$\ln\frac{P}{P_0} = -\frac{\Delta_{\text{vap}}\overline{H}}{R}\left(\frac{1}{T} - \frac{1}{T_0}\right)$$

$$\ln\frac{0.5836 P_0}{P_0} = -\frac{40.79 \times 10^3 \text{ J mol}^{-1}}{8.314 \text{ J K}^{-1} \text{ mol}^{-1}}\left(\frac{1}{T} - \frac{1}{373.2 \text{ K}}\right)$$

$$-0.5385 = -4906.2 \text{ K}\left(\frac{1}{T} - \frac{1}{373.2 \text{ K}}\right)$$

$$\frac{1}{T} = \frac{-0.5385}{-4906.2 \text{ K}} + \frac{1}{373.2 \text{ K}} = 2.789 \times 10^{-3} \text{ K}^{-1}$$

$$T = 358 \text{ K} = 85°\text{C}$$

6.26 The normal boiling point of ethanol is 78.3°C and its molar enthalpy of vaporization is 39.3 kJ mol^{-1}. What is its vapor pressure at 30°C?

$$\ln\frac{P_2}{P_1} = -\frac{\Delta_{\text{vap}}\overline{H}}{R}\left(\frac{1}{T_2} - \frac{1}{T_1}\right)$$

$$\ln\frac{P_2}{1 \text{ atm}} = -\frac{39.3 \times 10^3 \text{ J mol}^{-1}}{8.314 \text{ J K}^{-1} \text{ mol}^{-1}}\left(\frac{1}{303.2 \text{ K}} - \frac{1}{351.5 \text{ K}}\right) = -2.142$$

$$P_2 = (1 \text{ atm})\, e^{-2.142} = 0.117 \text{ atm} = 88.9 \text{ torr}$$

6.27 Calculate the number of components present in each of the following situations:

(a) Water, including auto-dissociation into H^+ and OH^- ions.

(b) Consider the following reaction in a closed container:

$$2NH_3(g) \rightleftharpoons N_2(g) + 3H_2(g)$$

(i) All three gases were present initially in arbitrary amounts, but the temperature is too low for the reaction to occur.

(ii) Same as **(i)** but now the temperature is raised sufficiently to allow for the equilibrium to be established.

(iii) Initially only NH_3 was present. The system is then allowed to reach equilibrium.

(a) Referring to Appendix 6.2, there are 3 constituents, H_2O, H^+, and OH^-, or $s = 3$. At the same time there are 2 algebraic relationships, $K_w = [H^+][OH^-]$ and $[H^+] = [OH^-]$, or $r = 2$. Thus, $c = s - r = 1$. This is a one-component system

(b-i) There are 3 constituents, but with no reaction there are no algebraic relationships. Thus $s = 3$ and $r = 0$, so $c = s - r = 3$. This is a three-component system.

(b-ii) Now there is one algebraic relationship, the equilibrium constant, $K_P = P_{N_2} P_{H_2}^3 / P_{NH_3}^2$. Thus, $c = s - r = 3 - 1 = 2$. This is a two-component system.

(b-iii) This adds a second relationship in addition to K_P, namely, $P_{N_2} = \frac{1}{3} P_{H_2}$. With $r = 2$ and $s = 3$, $c = s - r = 1$, and this is a one-component system.

6.28   Give the conditions under which each of the following equations may be applied. **(a)** $dA \leq 0$ (for equilibrium and spontaneity), **(b)** $dG \leq 0$ (for equilibrium and spontaneity), **(c)** $\ln \frac{P_2}{P_1} = \frac{\Delta \bar{H}}{R} \frac{(T_2 - T_1)}{T_1 T_2}$, **(d)** $\Delta G = nRT \ln \frac{P_2}{P_1}$.

(a) Constant volume and temperature, **(b)** constant pressure and temperature, **(c)** $\Delta_{vap} \bar{H}$ independent of temperature, $\bar{V}_{vap} \gg \bar{V}_{liquid}$ (or $\bar{V}_{solid}$), ideal gas behavior, **(d)** constant temperature, ideal gas behavior.

6.29   When ammonium nitrate is dissolved in water, the solution becomes colder. What conclusion can you draw about ΔS° for the process?

The reaction is spontaneous, therefore, ΔG is negative. Furthermore, the solution becomes colder, which denotes an endothermic process, that is, ΔH is positive. Since

$$\Delta S = \frac{\Delta H - \Delta G}{T}$$

ΔS must be positive (which is expected because the aqueous ammonium nitrate is more disordered than the solid form).

6.30   Protein molecules are polypeptide chains made up of amino acids. In their physiologically functioning or native state, these chains fold up in a unique manner such that the nonpolar groups of the amino acids are usually buried in the interior region of the proteins, where there is little or no contact with water. When a protein denatures, the chain unfolds so that these nonpolar groups are exposed to water. A useful estimate of the changes of the thermodynamic quantities as a result of denaturation is to consider the transfer of a hydrocarbon such as methane (a nonpolar substance) from an inert solvent (such as benzene or carbon tetrachloride) to the aqueous environment:

(a) $CH_4(\text{inert solvent}) \rightarrow CH_4(g)$ **(b)** $CH_4(g) \rightarrow CH_4(aq)$

If the values of ΔH° and ΔG° are approximately 2.0 kJ mol^{-1} and -14.5 kJ mol^{-1}, respectively, for **(a)**, and -13.5 kJ mol^{-1} and 26.5 kJ mol^{-1}, respectively, for **(b)**, calculate the values of ΔH° and ΔG° for the transfer of 1 mole of CH_4 according to the equation

$$CH_4(\text{inert solvent}) \rightarrow CH_4(aq)$$

Comment on your results. Assume $T = 298$ K.

For the process $CH_4(\text{inert solvent}) \rightarrow CH_4(aq)$

$$\Delta H^\circ = 2.0 \text{ kJ mol}^{-1} - 13.5 \text{ kJ mol}^{-1} = -11.5 \text{ kJ mol}^{-1}$$

$$\Delta G^\circ = -14.5 \text{ kJ mol}^{-1} + 26.5 \text{ kJ mol}^{-1} = 12.0 \text{ kJ mol}^{-1}$$

Since $\Delta G^\circ > 0$, the process is not spontaneous when the reactant and product are in their standard states. Furthermore,

$$\Delta S^\circ = \frac{\Delta H^\circ - \Delta G^\circ}{T} = \frac{-11.5 \text{ kJ mol}^{-1} - 12.0 \text{ kJ mol}^{-1}}{298 \text{ K}} \left(\frac{1000 \text{ J}}{1 \text{ kJ}} \right) = -78.9 \text{ J K}^{-1} \text{ mol}^{-1}$$

This negative value of ΔS° indicates that there is a increase in order when CH_4 is dissolved in H_2O. This is a result of order imposed on the solvent (water) molecules, due to the special arrangement of water molecules around each CH_4 molecule (see Section 16.6).

6.31   Find a rubber band that is about 0.5 cm wide. Quickly stretch the rubber band and then press it against your lips. You will feel a slight warming effect. Next reverse the process. Stretch a rubber band and hold it in position for a few seconds. Then quickly release the tension and press the rubber band against your lips. This time you will feel a slight cooling effect. Use Equation 6.3 to present a thermodynamic analysis for this behavior

---

Stretching the rubber band

Since this is not a spontaneous process, $\Delta G > 0$. The warming effect indicates that $\Delta H < 0$. Therefore,

$$\Delta S = \frac{\Delta H - \Delta G}{T} < 0$$

The entropy decreases as the rubber band is stretched. (See Figure 6.9 in text.)

Releasing the rubber band

This is a spontaneous process. Thus, $\Delta G < 0$. The cooling effect indicates that $\Delta H > 0$. Therefore,

$$\Delta S = \frac{\Delta H - \Delta G}{T} > 0$$

The entropy increases as the rubber band is released from a stretched state. (Again, see Figure 6.9 in text.)

6.32 A rubber band under tension will contract when heated. Explain.

When the rubber band is heated, ΔH is positive. If a process occurs, ΔG must be negative.

$$\Delta G = \Delta H - T \Delta S$$

$$\Delta S = \frac{\Delta H - \Delta G}{T}$$

The expression on the right hand side of the last expression is positive, that is, ΔS must also be positive for a process to occur. The process that can effect such a ΔS is the contraction of the rubber band.

6.33 Hydrogenation reactions are facilitated by the use of a transition metal catalyst, such as Ni or Pt. Predict the signs of $\Delta_r H$, $\Delta_r S$, and $\Delta_r G$ when hydrogen gas is adsorbed onto the surface of nickel metal.

$\Delta_r G$: $-$ because the process is spontaneous (constant temperature and pressure assumed).

$\Delta_r S$: $-$ because the adsorbed hydrogen on the surface is in a more ordered state than the gas.

$\Delta_r H$: $-$ because $\Delta_r H = \Delta_r G + T \Delta_r S$, and both $\Delta_r G < 0$ and $\Delta_r S < 0$. The process is exothermic.

6.34 A sample of supercooled water freezes at $-10°C$. What are the signs of ΔH, ΔS, and ΔG for this process? All the changes refer to the system.

ΔG: $-$ because the process is spontaneous (constant temperature given and constant pressure assumed).

ΔH: $-$ because freezing is an exothermic process.

ΔS: $-$ because ice is more ordered than liquid water.

6.35 The boiling point of benzene is $80.1°C$. Estimate (**a**) its $\Delta_{vap} H$ value and (**b**) its vapor pressure at $74°C$. (*Hint*: Use Trouton's rule on p. 144.)

(**a**) For 1 mole of benzene, $\Delta_{vap} S = 88 \ \mathrm{J \ K^{-1}}$ according to Trouton's rule.

$$\Delta_{vap} S = \frac{\Delta_{vap} H}{T_b}$$

$$\Delta_{vap} H = \Delta_{vap} S \, T_b = \left(88 \ \mathrm{J \ K^{-1}} \right) (353.3 \ \mathrm{K}) = 3.11 \times 10^4 \ \mathrm{J} = 3.1 \times 10^4 \ \mathrm{J}$$

(b) At the normal boiling point ($T_1 = 353.3$ K), the vapor pressure is 1 atm (P_1). The vapor pressure at $T_2 = 74°C = 347.2$ K, P_2, is calculated from the Clausius-Clapeyron equation:

$$\ln \frac{P_2}{P_1} = -\frac{\Delta_{vap}H}{R}\left(\frac{1}{T_2} - \frac{1}{T_1}\right)$$

$$\ln \frac{P_2}{1\text{ atm}} = -\frac{3.11 \times 10^4 \text{ J mol}^{-1}}{8.314 \text{ J K}^{-1}\text{ mol}^{-1}}\left(\frac{1}{347.2\text{ K}} - \frac{1}{353.3\text{ K}}\right) = -0.186$$

$$P_2 = (1\text{ atm})\, e^{-0.186} = 0.83\text{ atm} = 6.3 \times 10^2 \text{ torr}$$

6.36 A chemist has synthesized a hydrocarbon compound (C_xH_y). Briefly describe what measurements are needed to determine the values of $\Delta_f\overline{H}°$, $\overline{S}°$, and $\Delta_f\overline{G}°$.

A determination of the enthalpy of combustion [$C_xH_y + (x + \frac{y}{4})O_2(g) \rightarrow xCO_2(g) + \frac{y}{2}H_2O(l)$] using a calorimeter will enable a calculation of $\Delta_f\overline{H}°$ from this measurement and the known $\Delta_f\overline{H}°$'s for O_2, CO_2 and H_2O.

$\overline{S}°$ may be found via a determination of the third-law entropy from 0 K to 298 K. This assumes no residual entropy at 0 K.

Once $\overline{S}°$ is known for the compound, it is used together with the known values of $\overline{S}°$ for C(graphite) and $H_2(g)$ to calculate $\Delta_r S°$ for the reaction $xC(\text{graphite}) + \frac{y}{2}H_2(g) \rightarrow C_xH_y$. $\Delta_f\overline{G}°$ for the compound is determined via $\Delta_f\overline{G}° = \Delta_f\overline{H}° - T\Delta_r S°$.

6.37 A closed, 7.8-L flask contains 1.0 g of water. At what temperature will half of the water be in the vapor phase. (*Hint:* Look up the vapor pressures of water in the inside back matter.)

Half of the water is 0.50 g, or $\dfrac{0.50\text{ g}}{18.02\text{ g mol}^{-1}} = 0.0277$ mol. Find the temperature at which the pressure exerted by this much water in 7.8 L is equal to the vapor pressure of water. (The volume of the remaining 0.50 g of water is assumed negligible.) Assuming ideal gas behavior,

$$P = \frac{nRT}{V}$$

$$= \frac{(0.0277\text{ mol})\left(0.08206\text{ L atm K}^{-1}\text{ mol}^{-1}\right)T}{7.8\text{ L}}\left(\frac{760\text{ mmHg}}{1\text{ atm}}\right)$$

$$= \left(0.221\text{ mmHg K}^{-1}\right)T$$

$$= \left(0.221\text{ mmHg °C}^{-1}\right)(t + 273°C)$$

$$= \left(0.221\text{ mmHg °C}^{-1}\right)t + 60.3\text{ mmHg}$$

Now refer to the inside back matter of the text to find a temperature and vapor pressure that satisfy this equation.

$t/°C$	$P_{H_2O}/mmHg$	$[(0.221 \text{ mmHg}°C^{-1})t + 60.3 \text{ mmHg}]/mmHg$
40	55.32	69.1
45	71.88	70.2
50	92.51	71.4

The closest agreement is for $t = 45°C$. At this temperature, half of the water will be in the vapor phase.

6.38   A person heated water in a closed bottle in a microwave oven for tea. After removing the bottle from the oven, she added a tea bag to the hot water. To her surprise, the water started to boil violently. Explain what happened.

The water was superheated. The closed bottle allowed the pressure to become greater than 1 atm during the heating. Although thermodynamically unstable, superheated water will not boil even when exposed to a pressure of about 1 atm. Any mechanical disturbance, such as shaking, will cause it to boil. Adding the tea bag acts like adding boiling chips, which facilitates the boiling action.

6.39   Consider the reversible, isothermal compression of 0.45 mole of helium gas from 0.50 atm and 22 L to 1.0 atm at 25°C. **(a)** Calculate the values of w, ΔU, ΔH, ΔS, and ΔG for the process. **(b)** Can you use the sign of ΔG to predict whether the process is spontaneous? Explain. **(c)** What is the maximum work that can be done for the compression process? Assume ideal-gas behavior.

(a) For an isothermal process of an ideal gas, $\Delta U = 0$ and $\Delta H = 0$.

$$w = -nRT \ln \frac{P_1}{P_2} = -(0.45 \text{ mol})(8.314 \text{ J K}^{-1}\text{ mol}^{-1})(298.2 \text{ K}) \ln \frac{0.50 \text{ atm}}{1.0 \text{ atm}}$$

$$= 7.73 \times 10^2 \text{ J} = 7.7 \times 10^2 \text{ J}$$

$$q = q_{rev} = \Delta U - w = -w = -7.73 \times 10^2 \text{ J}$$

$$\Delta S = \frac{q_{rev}}{T} = \frac{-7.73 \times 10^2 \text{ J}}{298.2 \text{ K}} = -2.59 \text{ J K}^{-1} = -2.6 \text{ J K}^{-1}$$

$$\Delta G = \Delta H - T\Delta S = 0 - (298.2 \text{ K})(-2.59 \text{ J K}^{-1}) = 7.7 \times 10^2 \text{ J}$$

(b) The sign of ΔG cannot be used to predict whether the process is spontaneous because the pressure is not kept constant. $\Delta G < 0$ is a criterion for spontaneity only if both T and P are kept constant.

(c) The maximum work done on the system is obtained by using a maximum pressure, that is, an external pressure the same as the final pressure of the system ($P_f = 1.0$ atm), at all times.

$$w_{max} = -P_f(V_f - V_i) = -P_f\left(\frac{nRT}{P_f} - \frac{nRT}{P_i}\right) = -nRTP_f\left(\frac{1}{P_f} - \frac{1}{P_i}\right)$$

$$= -(0.45 \text{ mol})(8.314 \text{ J K}^{-1}\text{ mol}^{-1})(298.2 \text{ K})(1 \text{ atm})\left(\frac{1}{1.0 \text{ atm}} - \frac{1}{0.50 \text{ atm}}\right)$$

$$= 1.1 \times 10^3 \text{ J}$$

6.40 The molar entropy of argon (Ar) is given by

$$\overline{S}^{\,\circ} = (36.4 + 20.8 \ln T) \ \text{J K}^{-1} \, \text{mol}^{-1}$$

Calculate the change in Gibbs energy when 1.0 mole of Ar is heated at constant pressure from 20°C to 60°C. (*Hint*: Use the relation $\int \ln x \, dx = x \ln x - x$.)

ΔG° can be evaluated by using the relation

$$\left(\frac{\partial G}{\partial T} \right)_P = -S$$

At constant pressure,

$$dG = -SdT$$

$$\int_{G_1}^{G_2} dG = - \int_{T_1}^{T_2} SdT$$

$$\Delta G = - \int_{293.2 \, \text{K}}^{333.2 \, \text{K}} (1 \ \text{mol}) \left[(36.4 + 20.8 \ln T) \ \text{J K}^{-1} \, \text{mol}^{-1} \right] dT$$

$$= - [36.4T + 20.8 \, (T \ln T - T)]_{293.2}^{333.2} \ \text{J}$$

$$= - \{ [36.4 \, (333.2) + 20.8 \, (333.2 \ln 333.2 - 333.2)]$$

$$- [36.4 \, (293.2) + 20.8 \, (293.2 \ln 293.2 - 293.2)] \} \ \text{J}$$

$$= -6.24 \times 10^3 \ \text{J}$$

6.41 Derive the thermodynamic equation of state

$$\left(\frac{\partial U}{\partial V} \right)_T = -P + T \left(\frac{\partial P}{\partial T} \right)_V$$

Apply the equation to **(a)** an ideal gas and **(b)** a van der Waals gas. Comment on your results. (*Hint*: See Appendix 6.1 for thermodynamic relationships.)

To obtain an expression for $(\partial U/\partial V)_T$, differentiate $dU = -PdV + TdS$ (Equation 1 in Appendix 6.1) with respect to V while keeping T constant:

$$\left(\frac{\partial U}{\partial V} \right)_T = -P \left(\frac{\partial V}{\partial V} \right)_T + T \left(\frac{\partial S}{\partial V} \right)_T = -P + T \left(\frac{\partial S}{\partial V} \right)_T$$

According to Equation 19 in Appendix 6.1,

$$\left(\frac{\partial S}{\partial V} \right)_T = \left(\frac{\partial P}{\partial T} \right)_V$$

Substitute this expression into the $(\partial U/\partial V)_T$ expression:

$$\left(\frac{\partial U}{\partial V}\right)_T = -P + T\left(\frac{\partial P}{\partial T}\right)_V$$

(a) For an ideal gas,

$$P = \frac{nRT}{V}$$

$$\left(\frac{\partial P}{\partial T}\right)_V = \frac{nR}{V}$$

$$\left(\frac{\partial U}{\partial V}\right)_T = -P + T\frac{nR}{V} = -\frac{nRT}{V} + \frac{nRT}{V} = 0$$

This says that as an ideal gas expands (or is compressed) at constant temperature, the internal energy remains constant. Since there are no intermolecular forces for an ideal gas, this is as expected.

(b) For a van der Waals gas,

$$P = \frac{nRT}{V - nb} - \frac{an^2}{V^2}$$

$$\left(\frac{\partial P}{\partial T}\right)_V = \frac{nR}{V - nb}$$

$$\left(\frac{\partial U}{\partial V}\right)_T = -P + T\frac{nR}{V - nb} = -\left(\frac{nRT}{V - nb} - \frac{an^2}{V^2}\right) + \frac{nRT}{V - nb} = \frac{an^2}{V^2}$$

For the van der Waals gas $\left(\frac{\partial U}{\partial V}\right)_T > 0$, since $a > 0$. Thus, a graph of U versus V has positive slope, and the internal energy increases as the volume increases. The presence of the V^2 term in the denominator of $\frac{an^2}{V^2}$, however, means that the rate of increase becomes smaller as the volume grows. This all makes sense because as the gas expands, molecules have to break away from attractive intermolecular forces, which requires an increase in potential energy. Since the partial derivative is determined at constant T, the average kinetic energy of the gas remains constant. Consequently, the total internal energy must increase upon expansion of a van der Waals gas if a constant temperature is to be maintained. Those gases with stronger attractive forces have a larger value for the constant a, and thus a greater increase in energy. On the other hand, as the volume becomes larger, the average intermolecular separation grows, and the effect of the attractive forces is diminished. This results in a smaller increase in potential energy.

CHAPTER 7

Nonelectrolyte Solutions

PROBLEMS AND SOLUTIONS

7.1 How many grams of water must be added to 20.0 g of urea to prepare a 5.00% aqueous urea solution by weight?

$$\frac{\text{mass of urea}}{\text{mass of urea} + \text{mass of water}} = 5.00\% = 0.0500$$

$$\text{mass of urea} = 0.0500 \, (\text{mass of urea} + \text{mass of water})$$

$$0.0500 \, (\text{mass of water}) = 0.9500 \, (\text{mass of urea})$$

$$\text{mass of water} = 19.00 \, (\text{mass of urea}) = 19.00 \, (20.0 \text{ g}) = 380 \text{ g}$$

7.2 What is the molarity of a 2.12 mol kg^{-1} aqueous sulfuric acid solution? The density of this solution is 1.30 g cm^{-3}.

To find the molarity of the solution, the number of moles of solute and the volume of solution in a sample must be determined. Assume 1 kg of water is present in the solution.

$$\text{Number of moles of H}_2\text{SO}_4 = 2.12 \text{ mol}$$

$$\text{Mass of solution} = \text{mass of water} + \text{mass of H}_2\text{SO}_4$$

$$= 1000 \text{ g} + \left(2.12 \text{ mol H}_2\text{SO}_4\right) \left(\frac{98.09 \text{ g}}{1 \text{ mol H}_2\text{SO}_4}\right) = 1208.0 \text{ g}$$

$$\text{Volume of solution} = \frac{1208.0 \text{ g}}{\left(1.30 \text{ g cm}^{-3}\right) \left(\frac{1000 \text{ cm}^3}{1 \text{ L}}\right)} = 0.9292 \text{ L}$$

$$\text{Molarity of solution} = \frac{2.12 \text{ mol}}{0.9292 \text{ L}} = 2.28 \, M$$

7.3   Calculate the molality of a 1.50 M aqueous ethanol solution. The density of the solution is 0.980 g cm^{-3}.

To find the molality of the solution, the number of moles of solute and the mass of the solvent in a sample must be determined. Assume 1 L of solution is present.

$$\text{Number of moles of ethanol} = 1.50 \text{ mol}$$

$$\text{Mass of water} = \text{mass of solution} - \text{ mass of ethanol}$$

$$= (1000 \text{ cm}^3) (0.980 \text{ g cm}^{-3}) - (1.50 \text{ mol ethanol}) \left(\frac{46.07 \text{ g}}{1 \text{ mol ethanol}} \right)$$

$$= 910.9 \text{ g} = 0.9109 \text{ kg}$$

$$\text{Molality of solution} = \frac{1.50 \text{ mol}}{0.9109 \text{ kg}} = 1.65 \text{ } m$$

7.4   The concentrated sulfuric acid we use in the laboratory is 98.0% sulfuric acid by weight. Calculate the molality and molarity of concentrated sulfuric acid if the density of the solution is 1.83 g cm^{-3}.

Assume 100 g of solution is present. The solution contains 98.0 g H_2SO_4 and 2.0 g H_2O.

$$\text{Number of moles of } H_2SO_4 = (98.0 \text{ g}) \left(\frac{1 \text{ mol}}{98.09 \text{ g}} \right) = 0.9991 \text{ mol}$$

The molality of the solution is the ratio between the number of moles of solute and the mass of solvent:

$$\text{Molality of solution} = \frac{0.9991 \text{ mol}}{2.0 \times 10^{-3} \text{ kg}} = 5.0 \times 10^2 \text{ } m$$

The molarity of the solution is the ratio between the number of moles of solute and the volume of the solution.

$$\text{Volume of solution} = \frac{100 \text{ g}}{(1.83 \text{ g cm}^{-3}) \left(\frac{1000 \text{ cm}^3}{1 \text{ L}} \right)} = 5.464 \times 10^{-2} \text{ L}$$

$$\text{Molarity of solution} = \frac{0.9991 \text{ mol}}{5.464 \times 10^{-2} \text{ L}} = 18.3 \text{ } M$$

7.5   Convert a 0.25 mol kg^{-1} sucrose solution into percent by weight. The density of the solution is 1.2 g cm^{-3}.

Assume the solution contains 1 kg of water. This solution also contains 0.25 mole of sucrose.

$$\text{Mass of sucrose} = (0.25 \text{ mol}) \left(342.30 \text{ g mol}^{-1}\right) = 85.6 \text{ g}$$

$$\text{Mass of solution} = \text{mass of water} + \text{mass of sucrose} = 1000 \text{ g} + 85.6 \text{ g} = 1085.6 \text{ g}$$

$$\% \text{ by weight} = \frac{85.6 \text{ g}}{1085.6 \text{ g}} \times 100\% = 7.9\%$$

7.6   For dilute aqueous solutions in which the density of the solution is roughly equal to that of the pure solvent, the molarity of the solution is equal to its molality. Show that this statement is correct for a 0.010 M aqueous urea $[(NH_2)_2CO]$ solution.

To convert molarity to molality, the volume of the solution has to be converted to the mass of solvent. Assume 1 L of solution is present.

$$\text{Number of moles of urea} = 0.010 \text{ mol}$$

$$\text{Mass of solvent} = \text{mass of solution} - \text{ mass of solute}$$

$$= \left(1000 \text{ cm}^3\right) \left(1.00 \text{ g cm}^{-3}\right) - (0.010 \text{ mol urea}) \left(\frac{60.06 \text{ g}}{1 \text{ mol urea}}\right)$$

$$= 999.4 \text{ g} = 0.9994 \text{ kg}$$

$$\text{Molality of solution} = \frac{0.010 \text{ mol}}{0.9994 \text{ kg}} = 0.010 \text{ } m$$

Therefore, for a dilute aqueous solution, such as 0.010 M urea, its molality is numerically the same as its molarity.

7.7   The blood sugar (glucose) level of a diabetic patient is approximately 0.140 g of glucose/100 mL of blood. Every time the patient ingests 40 g of glucose, her blood glucose level rises to approximately 0.240 g/100 mL of blood. Calculate the number of moles of glucose per milliliter of blood and the total number of moles and grams of glucose in the blood before and after consumption of glucose. (Assume that the total volume of blood in her body is 5.0 L.)

Before consumption of glucose

$$\text{Number of moles of glucose/mL of blood} = \frac{(0.140 \text{ g}) \left(\frac{1 \text{ mol}}{180.16 \text{ g}}\right)}{100 \text{ mL}}$$

$$= 7.771 \times 10^{-6} \text{ mol mL}^{-1} = 7.77 \times 10^{-6} \text{ mol mL}^{-1}$$

$$\text{Total number of moles of glucose in blood} = \left(7.771 \times 10^{-6} \text{ mol mL}^{-1}\right) \left(5.0 \times 10^3 \text{ mL}\right)$$

$$= 3.9 \times 10^{-2} \text{ mol}$$

$$\text{Total number of grams of glucose in blood} = \left(\frac{0.140 \text{ g}}{100 \text{ mL}}\right) \left(5.0 \times 10^3 \text{ mL}\right)$$

$$= 7.0 \text{ g}$$

After consumption of glucose

$$\text{Number of moles of glucose/mL of blood} = \frac{(0.240 \text{ g}) \left(\frac{1 \text{ mol}}{180.16 \text{ g}} \right)}{100 \text{ mL}}$$

$$= 1.332 \times 10^{-5} \text{ mol mL}^{-1} = 1.33 \times 10^{-5} \text{ mol mL}^{-1}$$

$$\text{Total number of moles of glucose in blood} = \left(1.33 \times 10^{-5} \text{ mol mL}^{-1} \right) \left(5.0 \times 10^3 \text{ mL} \right)$$

$$= 6.7 \times 10^{-2} \text{ mol}$$

$$\text{Total number of grams of glucose in blood} = \left(\frac{0.240 \text{ g}}{100 \text{ mL}} \right) \left(5.0 \times 10^3 \text{ mL} \right)$$

$$= 12 \text{ g}$$

7.8 The strength of alcoholic beverages is usually described in terms of "proof," which is defined as twice the percentage by volume of ethanol. Calculate the number of grams of alcohol in 2 quarts of 75-proof gin. What is the molality of the gin? (The density of ethanol is 0.80 g cm^{-3}; 1 quart = 0.946 L.)

Since mass of a liquid is proportional to its volume, the % by weight of ethanol is the same as % by volume of ethanol, which is half the proof, or 37.5%. To find the molality of the gin, the number of moles of ethanol and the mass of water in a quantity of gin have to be calculated. In 2 quarts of gin,

$$\text{Volume of ethanol} = (37.5\%) (2 \text{ quarts}) \left(\frac{0.946 \text{ L}}{1 \text{ quart}} \right) = 0.7095 \text{ L}$$

$$\text{Mass of ethanol} = \left(0.7095 \times 10^3 \text{ cm}^3 \right) \left(0.80 \text{ g cm}^{-3} \right) = 568 \text{ g} = 5.7 \times 10^2 \text{ g}$$

$$\text{Number of moles of ethanol} = (568 \text{ g}) \left(\frac{1 \text{ mol}}{46.07 \text{ g}} \right) = 12.3 \text{ mol}$$

To evaluate the mass of water in 2 quarts of gin, two assumptions have to be made: (1) the volumes of ethanol and water are additive, and (2) the density of water is 1 g cm^{-3}. The second assumption is not bad, but the first is not particularly good. Indeed, there is a significant nonideality in water-ethanol solutions.

$$\text{Volume of water in 2 quarts of gin} = \text{volume of gin} - \text{volume of ethanol}$$

$$= (2 \text{ quarts}) \left(\frac{0.946 \text{ L}}{1 \text{ quart}} \right) - 0.7095 \text{ L} = 1.1825 \text{ L}$$

$$\text{Mass of water in 2 quarts of gin} = \left(1.1825 \times 10^3 \text{ cm}^3 \right) \left(1 \text{ g cm}^{-3} \right)$$

$$= 1.1825 \times 10^3 \text{ g} = 1.1825 \text{ kg}$$

Therefore,

$$\text{Molality of the gin} = \frac{12.3 \text{ mol}}{1.1825 \text{ kg}} = 10 \ m$$

7.9 Liquids A and B form a nonideal solution. Provide a molecular interpretation for each of the following situations: $\Delta_{mix}H > 0$, $\Delta_{mix}H < 0$, $\Delta_{mix}V > 0$, $\Delta_{mix}V < 0$.

If the intermolecular forces between molecules A and B are stronger (more attractive) than those between A molecules and those between B molecules, then $\Delta_{mix}H < 0$. If, on the other hand, the forces between A and B molecules are weaker (less attractive) than those between like molecules, then $\Delta_{mix}H > 0$.

If A molecules fit between B molecules (or vice versa) better than they do among themselves, then $\Delta_{mix}V < 0$, otherwise $\Delta_{mix}V > 0$.

7.10 Calculate the changes in entropy for the following processes: **(a)** mixing of 1 mole of nitrogen and 1 mole of oxygen, and **(b)** mixing of 2 moles of argon, 1 mole of helium, and 3 moles of hydrogen. Both **(a)** and **(b)** are carried out under conditions of constant temperature (298 K) and constant pressure. Assume ideal behavior.

(a)

$$\Delta_{mix}S = -nR\left(x_{N_2}\ln x_{N_2} + x_{O_2}\ln x_{O_2}\right)$$

$$= -(2\text{ mol})\left(8.314\text{ J K}^{-1}\text{ mol}^{-1}\right)\left(\frac{1}{2}\ln\frac{1}{2} + \frac{1}{2}\ln\frac{1}{2}\right) = 11.53\text{ J K}^{-1}$$

(b)

$$\Delta_{mix}S = -nR\left(x_{Ar}\ln x_{Ar} + x_{He}\ln x_{He} + x_{H_2}\ln x_{H_2}\right)$$

$$= -(6\text{ mol})\left(8.314\text{ J K}^{-1}\text{ mol}^{-1}\right)\left(\frac{2}{6}\ln\frac{2}{6} + \frac{1}{6}\ln\frac{1}{6} + \frac{3}{6}\ln\frac{3}{6}\right) = 50.45\text{ J K}^{-1}$$

7.11 At 25°C and 1 atm pressure, the absolute third-law entropies of methane and ethane are 186.19 J K^{-1} mol^{-1} and 229.49 J K^{-1} mol^{-1}, respectively in the gas phase. Calculate the absolute third-law entropy of a "solution" containing 1 mole of each gas. Assume ideal behavior.

The absolute third-law entropy of the gas solution, S_f, can be calculated from the third-law entropy of the gases before they are mixed, S_i, and from the entropy of mixing, $\Delta_{mix}S$.

$$S_i = S_{CH_4} + S_{C_2H_5}$$

$$= (1\text{ mol})\left(186.19\text{ J K}^{-1}\text{ mol}^{-1}\right) + (1\text{ mol})\left(229.49\text{ J K}^{-1}\text{ mol}^{-1}\right)$$

$$= 415.68\text{ J K}^{-1}$$

$$\Delta_{mix}S = -nR\left(x_{CH_4}\ln x_{CH_4} + x_{C_2H_5}\ln x_{C_2H_5}\right)$$

$$= -(2\text{ mol})\left(8.314\text{ J K}^{-1}\text{ mol}^{-1}\right)\left(\frac{1}{2}\ln\frac{1}{2} + \frac{1}{2}\ln\frac{1}{2}\right)$$

$$= 11.526\text{ J K}^{-1}$$

$$S_f = S_i + \Delta_{mix}S = 415.68\text{ J K}^{-1} + 11.526\text{ J K}^{-1} = 427.21\text{ J K}^{-1}$$

7.12   Prove the statement that an alternative way to express Henry's law of gas solubility is to say that the volume of gas that dissolves in a fixed volume of solution is independent of pressure at a given temperature.

When Henry's law, $P_2 = K x_2$, is expressed in terms of the number of moles of gas dissolved in the solution, n_2, the result is

$$P_2 = K \frac{n_2}{n_T}$$

where n_T is the total number of moles.

The volume that would be occupied by this number of moles of gas were it in the gas phase at pressure P_2, is taken to satisfy the ideal gas law,

$$P_2 = \frac{n_2 RT}{V}$$

These two expressions for P_2 must agree,

$$K \frac{n_2}{n_T} = \frac{n_2 RT}{V}$$

$$V = \frac{n_T RT}{K}$$

Thus, V is independent of the pressure, although the amount of gas (number of moles) that occupies this volume is, of course, pressure dependent.

7.13   A miner working 900 ft below the surface had a soft drink beverage during the lunch break. To his surprise, the drink seemed very flat (that is, not much effervescence was observed upon removing the cap). Shortly after lunch, he took the elevator to the surface. During the trip up, he felt a great urge to belch. Explain.

At 900 ft below sea level, the total pressure, and likewise the partial pressure of CO_2 is greater than it is at or slightly above sea level, where presumably the soft drink was bottled. Thus, the solubility of CO_2 is increased, resulting in a higher concentration in the solution. Upon returning to the surface, the partial pressure returns to normal (*i.e.* decreases), the solubility of CO_2 is reduced, and gaseous CO_2 begins to come out of solution in the miner's stomach, which leads to a natural urge to let this excess gas escape.

7.14   The Henry's law constant of oxygen in water at 25°C is 773 atm mol^{-1} kg of water. Calculate the molality of oxygen in water under a partial pressure of 0.20 atm. Assuming that the solubility of oxygen in blood at 37°C is roughly the same as that in water at 25°C, comment on the prospect for our survival without hemoglobin molecules. (The total volume of blood in the human body is about 5 L.)

$$P_2 = k'm$$

$$m = \frac{P_2}{k'} = \frac{0.20 \text{ atm}}{773 \text{ atm mol}^{-1} \text{ kg}} = 2.6 \times 10^{-4} \text{ mol kg}^{-1}$$

In a dilute aqueous solution, molality and molarity are numerically the same (see Problem 7.6). Therefore,

$$\text{Number of moles of } O_2 \text{ in blood} = (2.6 \times 10^{-4} \text{ } M)(5 \text{ L}) = 1.3 \times 10^{-3} \text{ mol}$$

$$\text{Mass of } O_2 \text{ in blood} = (1.3 \times 10^{-3} \text{ mol})(32.00 \text{ g mol}^{-1}) = 0.042 \text{ g}$$

This amount of O_2 is too small to sustain metabolic processes. This is the reason why we need hemoglobin molecules to transport O_2.

7.15   The solubility of N_2 in blood at 37°C and a partial pressure of 0.80 atm is 5.6×10^{-4} mol L^{-1}. A deep-sea diver breathes compressed air with a partial pressure of N_2 equal to 4.0 atm. Assume that the total volume of blood in the body is 5.0 L. Calculate the amount of N_2 gas released (in liters) when the diver returns to the surface of water, where the partial pressure of N_2 is 0.80 atm.

Assuming N_2 is an ideal gas, the volume of N_2 released can be readily calculated from the number of moles of N_2 released, which in turn is related to the number of moles of N_2 dissolved in blood when the partial pressures of N_2 are 4.0 atm and 0.80 atm, respectively.

According to Henry's law, the solubility of a substance is proportional to the applied pressure. Therefore,

$$\frac{\text{Solubility of } N_2 \text{ when } P_{N_2} \text{ is 4.0 atm}}{\text{Solubility of } N_2 \text{ when } P_{N_2} \text{ is 0.80 atm}} = \frac{4.0 \text{ atm}}{0.80 \text{ atm}} = 5.0$$

$$\text{Solubility of } N_2 \text{ when } P_{N_2} \text{ is 4.0 atm} = 5.0 \, (5.6 \times 10^{-4} \text{ mol L}^{-1}) = 2.80 \times 10^{-3} \text{ mol}$$

When $P_{N_2} = 0.80$ atm,

$$\text{Number of moles of } N_2 \text{ in blood} = (5.6 \times 10^{-4} \text{ mol L}^{-1})(5.0 \text{ L}) = 2.80 \times 10^{-3} \text{ mol}$$

When $P_{N_2} = 4.0$ atm,

$$\text{Number of moles of } N_2 \text{ in blood} = (2.80 \times 10^{-3} \text{ mol L}^{-1})(5.0 \text{ L}) = 1.40 \times 10^{-2} \text{ mol}$$

When the diver returns to the surface of water,

$$\text{Number of moles of } N_2 \text{ released} = 1.40 \times 10^{-2} \text{ mol} - 2.80 \times 10^{-3} \text{ mol} = 1.12 \times 10^{-2} \text{ mol}$$

and therefore,

$$\text{Volume of } N_2 \text{ released} = \frac{(1.12 \times 10^{-2} \text{ mol})(0.08206 \text{ L atm K}^{-1} \text{ mol}^{-1})(310 \text{ K})}{0.80 \text{ atm}} = 0.36 \text{ L}$$

7.16   Which of the following has a higher chemical potential? If neither, answer "same." **(a)** $H_2O(s)$ or $H_2O(l)$ at water's normal melting point, **(b)** $H_2O(s)$ at $-5°C$ and 1 bar or $H_2O(l)$ at $-5°C$ and 1 bar. **(c)** Benzene at 25°C and 1 bar or benzene in a 0.1 M toluene solution in benzene at 25°C and 1 bar.

The less stable species of a pair has a higher chemical potential. When both substances in a pair are at equilibrium, then they have the same chemical potential.

(a) Same

(b) $H_2O(l)$ at $-5°C$ and 1 bar

(c) Benzene at 25°C and 1 bar. This is because $x_{benzene} < 0$ in the following relation:

$$\mu_{benzene}(l) = \mu^*_{benzene}(l) + RT \ln x_{benzene}$$

Therefore,

$$\mu_{benzene}(l) < \mu^*_{benzene}(l)$$

7.17   A solution of ethanol and n-propanol behaves ideally. Calculate the chemical potential of ethanol in solution relative to that of pure ethanol when its mole fraction is 0.40 at its boiling point (78.3°C).

$$\mu_{ethanol}(l) = \mu^*_{ethanol}(l) + RT \ln x_{ethanol}$$

$$= \mu^*_{ethanol}(l) + \left(8.314 \, J\,K^{-1}\,mol^{-1}\right)(351.5 \, K) \ln 0.40$$

$$= \mu^*_{ethanol}(l) - 2.7 \times 10^3 \, J\,mol^{-1}$$

The chemical potential of ethanol in solution is lower than the chemical potential of pure ethanol by $2.7 \times 10^3 \, J\,mol^{-1}$.

7.18   Derive the phase rule (Equation 6.23) in terms of chemical potentials.

The total number of independent intensive variables for a system containing c components and p phases is found as in Appendix 6.2 to be $p(c-1) + 2$. At equilibrium, the chemical potential of each component is the same in all phases. Taking the phases to be α, β, γ, etc., this means that for a component present in two phases,

$$\mu_\alpha = \mu_\beta$$

which gives one equation. For a component present in three phases,

$$\mu_\alpha = \mu_\beta$$

$$\mu_\beta = \mu_\gamma$$

or two equations. Extending this process is seen to result in $(p - 1)$ independent equations relating the chemical potentials for the p phases present. Since there are c components, this gives a total of $c(p - 1)$ known variables restricting the total number of degrees of freedom, which is found via the difference

$$f = [p(c - 1) + 2] - c(p - 1)$$

$$= c - p + 2$$

7.19 The following data give the pressures for carbon disulfide-acetone solutions at 35.2°C. Calculate the activity coefficients of both components based on deviations from Raoult's law and Henry's law. (*Hint*: First determine Henry's law constants graphically.)

x_{CS_2}	0	0.20	0.45	0.67	0.83	1.00
P_{CS_2}/torr	0	272	390	438	465	512
$P_{C_3H_6O}$/torr	344	291	250	217	180	0

From a graph of the vapor pressure data, Henry's law constants may be obtained from the limiting slopes of the curves as they approach their respective infinite dilution (mole fraction equals zero) limits. (Note that the graph pictured is scaled to show the $x = 1$ intercept of each Henry's law line. In making the actual determination of the limiting slopes it is better to use an expanded scale that shows the limiting behavior in detail.) The Henry's law constants obtained from these data are $K_{CS_2} = 1570$ torr and $K_{C_3H_6O} = 1250$ torr. The activity coefficient of the ith component accounts for deviations from Henry's law as

$$P_i = K_i \gamma_i x_i$$

or

$$\gamma_i = \frac{P_i}{K_i x_i}$$

The necessary data to find the Henry's law activity coefficients, γ_i (H) are in the table. (Note that $x_{C_3H_6O} = 1 - x_{CS_2}$.)

x_{CS_2}	0	0.20	0.45	0.67	0.83	1.00
P_{CS_2}/torr	0	272	390	438	465	512
$P_{C_3H_6O}$/torr	344	291	250	217	180	0
γ_{CS_2} (H)	1.00	0.87	0.55	0.42	0.36	0.33
$\gamma_{C_3H_6O}$(H)	0.28	0.29	0.36	0.53	0.85	1.00

The Raoult's law activity is found through deviations from Raoult's law via

$$P_i = \gamma_i x_i P_i^*$$

or

$$\gamma_i = \frac{P_i}{x_i P_i^*}$$

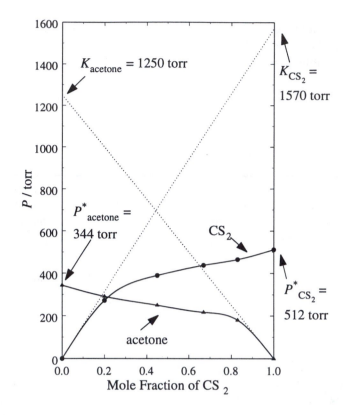

The tabulated data are used once again, noting that the vapor pressures of the pure components are included, namely $P^*_{CS_2} = 512$ torr and $P^*_{C_3H_6O} = 344$ torr. The Raoult's law activity is denoted $\gamma_i(R)$, and is not defined at zero mole fraction.

x_{CS_2}	0	0.20	0.45	0.67	0.83	1.00
P_{CS_2}/torr	0	272	390	438	465	512
$P_{C_3H_6O}$/torr	344	291	250	217	180	0
$\gamma_{CS_2}(R)$	—	2.66	1.69	1.28	1.09	1.00
$\gamma_{C_3H_6O}(R)$	1.00	1.06	1.32	1.91	3.08	—

The activities determined from the two laws differ because the standard state is chosen differently in the two cases.

7.20 A solution is made up by dissolving 73 g of glucose ($C_6H_{12}O_6$; molar mass 180.2 g) in 966 g of water. Calculate the activity coefficient of glucose in this solution if the solution freezes at −0.66°C.

If the solution is ideal, the freezing point depression ΔT is $K_f m_2$. However, if it is nonideal, then the molality of the glucose solution should be replaced by the activity of glucose. The activity coefficient of glucose can be calculated by comparing the activity with molality.

The freezing point depression of the solution is $0°C - (-0.66°C) = 0.66°C = 0.66$ K

$$a = \frac{0.66 \text{ K}}{K_f} = \frac{0.66 \text{ K}}{1.86 \text{ K mol}^{-1} \text{ kg}} = 0.355 \text{ mol kg}^{-1}$$

The molality of the solution is

$$m = \frac{(73 \text{ g}) \left(\frac{1 \text{ mol}}{180.16 \text{ g}} \right)}{966 \times 10^{-3} \text{ kg}} = 0.419 \text{ mol kg}^{-1}$$

The activity coefficient is

$$\gamma = \frac{a}{m} = \frac{0.355 \text{ mol kg}^{-1}}{0.419 \text{ mol kg}^{-1}} = 0.85$$

7.21   A certain dilute solution has an osmotic pressure of 12.2 atm at 20°C. Calculate the difference between the chemical potential of the solvent in the solution and that of pure water. Assume that the density is the same as that of water. (*Hint*: Express the chemical potential in terms of mole fraction, x_1, and rewrite the osmotic pressure equation as $\pi V = n_2 RT$, where n_2 is the number of moles of the solute and $V = 1$ L.)

Let the chemical potential of pure water be $\mu_1^* \, (l)$ and the chemical potential of the solvent be $\mu_1 \, (l)$. These two chemical potential are related by

$$\mu_1 \, (l) = \mu_1^* \, (l) + RT \ln x_1$$

The mole fraction of water, x_1, is then calculated from the number of moles of solute, n_2, and the number of moles of water, n_1, in 1 L of solution. The number of moles of solute is related to the osmotic pressure of the solution.

$$\pi = MRT = \frac{n_2}{V} RT$$

$$n_2 = \frac{\pi V}{RT} = \frac{(12.2 \text{ atm}) \, (1 \text{ L})}{(0.08206 \text{ L atm K}^{-1} \text{ mol}^{-1}) \, (293.2 \text{ K})} = 0.5071 \text{ mol}$$

Since the solution is dilute, the volume of water is assumed to be the same as the volume of the solution and the density of water is assumed to be 1 g cm^{-3}. Then

$$n_1 = \frac{(1000 \text{ cm}^3) \, (1 \text{ g cm}^{-3})}{18.02 \text{ g mol}^{-1}} = 55.494 \text{ mol}$$

$$\frac{(\text{vol}) \, (\text{density})}{\text{molar mas}} = \frac{\text{mass}}{\text{molar}} = m$$

The mole fraction of water is

$$x_1 = \frac{n_1}{n_1 + n_2} = \frac{55.494 \text{ mol}}{55.494 \text{ mol} + 0.5071 \text{ mol}} = 0.9909$$

Substituting x_1 into the chemical potential expression above,

$$\mu_1 \, (l) = \mu_1^* \, (l) + (8.314 \text{ J K}^{-1} \text{ mol}^{-1}) \, (293.2 \text{ K}) \ln (0.9909) = \mu_1^* \, (l) - 22.3 \text{ J mol}^{-1}$$

The chemical potential of the solution is lower than that of water by 22.3 J mol^{-1}.

7.22   At 45°C, the vapor pressure of water for a glucose solution in which the mole fraction of glucose is 0.080 is 65.76 mmHg. Calculate the activity and activity coefficient of the water in the solution. The vapor pressure of pure water at 45°C is 71.88 mmHg.

The activity is calculated by comparing the vapor pressure of the solution and that of pure water. This together with the mole fraction of water gives the activity coefficient.

$$a_{H_2O} = \frac{P_{H_2O}}{P^*_{H_2O}} = \frac{65.76 \text{ mmHg}}{71.88 \text{ mmHg}} = 0.9149$$

$$\gamma_{H_2O} = \frac{a_{H_2O}}{x_{H_2O}} = \frac{a_{H_2O}}{1 - x_{glucose}} = \frac{0.9149}{1 - 0.080} = 0.994$$

7.23   Consider a binary liquid mixture A and B, where A is volatile and B is nonvolatile. The composition of the solution in terms of mole fraction is $x_A = 0.045$ and $x_B = 0.955$. The vapor pressure of A from the mixture is 5.60 mmHg, and that of pure A is 196.4 mmHg at the same temperature. Calculate the activity coefficient of A at this concentration.

The activity of A is

$$a_A = \frac{P_A}{P^*_A} = \frac{5.60 \text{ mmHg}}{196.4 \text{ mmHg}} = 2.851 \times 10^{-2}$$

The activity coefficient of A is

$$\gamma_A = \frac{a_A}{x_A} = \frac{2.851 \times 10^{-2}}{0.045} = 0.63$$

7.24   List the important assumptions in the derivation of Equation 7.39.

The derivation makes use of several assumptions. The solute is assumed to be nonvolatile. The solution is assumed to be dilute (so that $\Delta_{vap}\overline{H}$ is unchanged from the pure solvent value). The solution is assumed to be dilute-ideal (so that equation 7.34 is valid). The solute is assumed to not be an electrolyte, and the boiling point elevation is assumed to be small.

7.25   Liquids A (bp = T_A°) and B (bp = T_B°) form an ideal solution. Predict the range of boiling points of solutions formed by mixing different amounts of A and B.

The range of boiling points will be between T_A° and T_B°.

7.26   A mixture of ethanol and *n*-propanol behaves ideally at 36.4°C. **(a)** Determine graphically the mole fraction of *n*-propanol in a mixture of ethanol and *n*-propanol that boils at 36.4°C and 72 mmHg. **(b)** What is the total vapor pressure over the mixture at 36.4°C when the mole fraction of *n*-propanol is 0.60? **(c)** Calculate the composition of the vapor in **(b)**. (The equilibrium vapor pressures of ethanol and *n*-propanol at 36.4°C are 108 mmHg and 40.0 mmHg, respectively.)

(a) The vapor pressure of ethanol, P_{ethanol}, the vapor pressure of propanol, P_{propanol} and the total vapor pressure, P_{total} can each be expressed as a function of the mole fraction of ethanol, x_{ethanol}:

$$P_{\text{ethanol}} = x_{\text{ethanol}} P^*_{\text{ethanol}} = x_{\text{ethanol}} (108 \text{ mmHg}) = \left(1 - x_{\text{propanol}}\right)(108 \text{ mmHg})$$

$$P_{\text{propanol}} = x_{\text{propanol}} P^*_{\text{propanol}} = x_{\text{propanol}} (40.0 \text{ mmHg})$$

$$P_{\text{total}} = P_{\text{ethanol}} + P_{\text{propanol}}$$

equilibrium VP

These relations are plotted below.

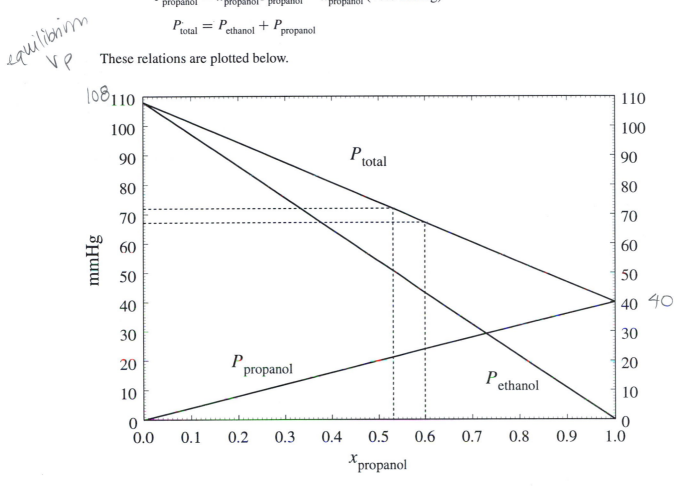

The vapor pressure of the mixture is the same as the external pressure when it boils. According to the graph, when the total vapor pressure is 72 mmHg, the mole fraction of propanol is 0.53 and the mole fraction of ethanol is $1 - 0.53 = 0.47$.

(b) According to the graph, when the mole fraction of propanol is 0.60, the total vapor pressure is 67.2 mmHg.

(c) The vapor pressure of each component can be written in terms of (1) its mole fraction in the vapor and the total pressure or (2) its mole fraction in the solution and the vapor pressure of the pure component.

$$P_{\text{ethanol}} = x^v_{\text{ethanol}} P_{\text{total}} = x_{\text{ethanol}} P^*_{\text{ethanol}}$$

$$P_{\text{propanol}} = x^v_{\text{propanol}} P_{\text{total}} = x_{\text{propanol}} P^*_{\text{propanol}}$$

Take the ratio between $P_{ethanol}$ and $P_{propanol}$,

$$\frac{P_{ethanol}}{P_{propanol}} = \frac{x_{ethanol}^{v} P_{total}}{x_{propanol}^{v} P_{total}} = \frac{x_{ethanol} P_{ethanol}^{*}}{x_{propanol} P_{propanol}^{*}}$$

$$\frac{x_{ethanol}^{v}}{x_{propanol}^{v}} = \frac{(0.40)\,(108 \text{ mmHg})}{(0.60)\,(40.0 \text{ mmHg})} = 1.80$$

$$x_{ethanol}^{v} = 1.80 x_{propanol}^{v} = 1.80\left(1 - x_{ethanol}^{v}\right)$$

$$2.80 x_{ethanol}^{v} = 1.80$$

$$x_{ethanol}^{v} = 0.64$$

$$x_{propanol}^{v} = 1 - 0.64 = 0.36$$

7.27 Two beakers, 1 and 2, containing 50 mL of 0.10 M urea and 50 mL of 0.20 M urea, respectively, are placed under a tightly sealed bell jar at 298 K. Calculate the mole fraction of urea in the solutions at equilibrium. Assume ideal behavior. (*Hint*: Use Raoult's law and note that at equilibrium, the mole fraction of urea is the same in both solutions.)

First find the number of moles of urea in each beaker.

$$\# \text{ moles urea} = (0.050 \text{ L})\,(0.10M) = 0.00500 \text{ mol}$$

and the second

$$\# \text{ moles urea} = (0.050 \text{ L})\,(0.20M) = 0.0100 \text{ mol}$$

Finding the mole fraction requires the number of moles of water in each beaker as well. Assume that the volume of each solution is equal to the volume of water, which is to say $V_{H_2O} = 50$ mL. Thus,

$$\# \text{ moles water} = \frac{(50 \text{ mL})\left(1 \text{ g mL}^{-1}\right)}{18.01 \text{ g mol}^{-1}}$$

$$= 2.78 \text{ mol}$$

The mole fraction of urea in each beaker is then

$$x_1 = \frac{0.00500 \text{ mol}}{0.00500 \text{ mol} + 2.78 \text{ mol}} = 1.80 \times 10^{-3}$$

$$x_2 = \frac{0.0100 \text{ mol}}{0.0100 \text{ mol} + 2.78 \text{ mol}} = 3.58 \times 10^{-3}$$

Equilibrium is attained by the transfer of water (via water vapor) from the less concentrated solution to the more concentrated one until the mole fractions of urea are equal. At this point, the mole fractions of water in each beaker are also equal and Raoult's law implies that the vapor pressures of the water over each beaker are the same. Thus, there is no more net transfer of solvent between beakers. Let y be the number of moles of water transferred to reach equilibrium.

$$\frac{0.00500 \text{ mol}}{0.00500 \text{ mol} + 2.78 \text{ mol} - y} = \frac{0.0100 \text{ mol}}{0.0100 \text{ mol} + 2.78 \text{ mol} + y}$$

$$y = 0.927 \text{ mol}$$

and the mole fraction of urea is

$$\frac{0.0100 \text{ mol}}{0.0100 \text{ mol} + 2.78 \text{ mol} + 0.927 \text{ mol}} = 2.7 \times 10^{-3}$$

This solution to the problem assumes that the volume of water left in the bell jar as vapor is negligible compared to the volumes of the solutions. It is interesting to note that at equilibrium 16.7 mL of water has been transferred from one beaker to the other.

7.28   At 298 K, the vapor pressure of pure water is 23.76 mmHg and that of seawater is 22.98 mmHg. Assuming that seawater contains only NaCl, estimate its concentration. (*Hint*: Sodium chloride is a strong electrolyte.)

The mole fraction of the solute is related to the vapor pressure lowering of the solvent.

$$\Delta P = x_2 P_1^*$$

$$x_2 = \frac{\Delta P}{P_1^*} = \frac{23.76 \text{ mmHg} - 22.98 \text{ mmHg}}{23.76 \text{ mmHg}} = 3.2828 \times 10^{-2}$$

Since NaCl is a strong electrolyte, it dissociates virtually completely into ions in solutions. Thus, x_2 calculated above represents the mole fraction of Na^+ and Cl^-. The mole fraction of NaCl, x_{NaCl}, is half of the calculated amount, or 1.6414×10^{-2}, and the mole fraction of water is $1 - 1.6414 \times 10^{-2} = 0.983586$. In other words, for every mole of molecules in the solution there are 1.6414×10^{-2} mole of NaCl and 0.983586 mole of water. Assuming the density of water is 1 kg L^{-1}, the volume of water is

$$V = \frac{(0.983586 \text{ mol}) \left(\frac{18.02 \times 10^{-3} \text{ kg}}{1 \text{ mol}} \right)}{1 \text{ kg L}^{-1}} = 0.017724 \text{ L}$$

The concentration of NaCl is

$$M = \frac{1.6414 \times 10^{-2} \text{ mol}}{0.017724 \text{ L}} = 0.9261 \text{ mol L}^{-1}$$

7.29   Trees in cold climates may be subjected to temperatures as low as $-60°C$. Estimate the concentration of an aqueous solution in the body of the tree that would remain unfrozen at this temperature. Is this a reasonable concentration? Comment on your result.

The freezing point of water is depressed by 60°C or 60 K. The concentration of solute giving rise to this depression is

$$m_2 = \frac{\Delta T}{K_f} = \frac{60\ \text{K}}{1.86\ \text{K}\,\text{mol}^{-1}\,\text{kg}} = 32\ \text{mol}\,\text{kg}^{-1}$$

This is a very high concentration. Indeed it would be impossible for any species with a molar mass greater than approximately 31 g mol^{-1} to be present in a solution at this concentration. (In such a case, 32 moles of the solute would have a mass greater than 1 kg.) Indeed, some of the species present are certainly electrolytes, which would lower the concentration required to attain such a freezing point depression. Additionally, the core of the tree is insulated by the bark, so that the full 60 K depression may not be required. Finally, living systems that are adapted to such harsh conditions may produce proteins that inhibit the formation of ice crystals. Such "antifreeze" proteins are well-studied in fish, whether cold-adapted trees possess similar proteins is an open question.

7.30   Explain why jams can be stored under atmospheric conditions for long periods of time without spoilage.

The high concentration of sugar in a jam results in a hypertonic solution. Any bacteria landing in the jam will lose its intracellular water through osmosis to the more concentrated solution (the jam). The dessicated bacteria are no longer viable.

7.31   Provide a molecular interpretation for the positive and negative deviations in the boiling-point curves and the formation of azeotropes.

If the intermolecular forces between unlike molecules are greater (more attractive) than those between like molecules, the solution will show a positive deviation in the boiling point curve. On the other hand, if the intermolecular forces are weaker between unlike molecules than between like molecules there will be a negative deviation. At a particular concentration the intermolecular forces between unlike molecules may be equal to those between like molecules. When this happens the solution will boil as if it were a pure liquid.

7.32   The freezing-point-depression measurement of benzoic acid in acetone yields a molar mass of 122 g; the same measurement in benzene gives a value of 242 g. Account for this discrepancy. *Hint*: Consider solvent-solute and solute-solute interactions.

In benzene, benzoic acid exists largely in a dimeric form while in acetone it exists as a monomer. The dimerization is not complete so that the apparent molar mass in benzene, 242 g, is not quite twice 122 g.

7.33   A common antifreeze for car radiators is ethylene glycol, $CH_2(OH)CH_2(OH)$. How many milliliters of this substance would you add to 6.5 L of water in the radiator if the coldest day in winter is $-20°C$? Would you keep this substance in the radiator in the summer to prevent the water from boiling? (The density and boiling point of ethylene glycol are 1.11 g cm^{-3} and 470 K, respectively.)

The molality of ethylene glycol that can depress the freezing point by 20°C or 20 K is

$$m_2 = \frac{\Delta T}{K_f} = \frac{20 \text{ K}}{1.86 \text{ K mol}^{-1} \text{ kg}} = 10.75 \text{ mol kg}^{-1}$$

Assuming the density of water is 1 kg L^{-1}, the mass of water in the radiator is 6.5 kg, which, together with the molality of ethylene glycol, gives the number of moles of ethylene glycol.

$$\text{Number of moles of ethylene glycol} = \left(10.75 \text{ mol kg}^{-1}\right)(6.5 \text{ kg}) = 69.88 \text{ mol}$$

which corresponds to a volume of

$$V = \frac{(69.88 \text{ mol})\left(62.07 \text{ g mol}^{-1}\right)}{1.11 \text{ g cm}^{-3}} = 3.91 \times 10^3 \text{ cm}^3 = 3.91 \times 10^3 \text{ mL}$$

Since it has a higher boiling point, ethylene glycol can also elevate the boiling point of water. With a 10.75 m solution, the boiling point of water will be elevated by

$$\Delta T = K_b m_2 = \left(0.51 \text{ K mol}^{-1} \text{ kg}\right)\left(10.75 \text{ mol kg}^{-1}\right) = 5.5 \text{ K}$$

Therefore, keeping ethylene glycol in the radiator in the summer will increase the boiling point by 5.5°C.

7.34 For intravenous injections, great care is taken to ensure that the concentration of solutions to be injected is comparable to that of blood plasma. Why?

If the injected solution were either hypertonic or hypotonic, the resulting osmotic pressure would cause water to be transferred across cell (such as red blood cell) membranes leading to either hemolysis or crenation.

7.35 The tallest trees known are the redwoods in California. Assuming the height of a redwood to be 105 m (about 350 ft), estimate the osmotic pressure required to push water up from the roots to the treetop.

Assume that the density of water is $1 \times 10^3 \text{ kg m}^{-3}$.

$$\pi = h\rho g \quad \text{(See Example 7.5 in text.)}$$

$$= (105 \text{ } m)\left(1 \times 10^3 \text{ kg m}^{-3}\right)\left(9.81 \text{ m s}^{-2}\right)$$

$$= \left(1.030 \times 10^6 \text{ N m}^{-2}\right)\left(\frac{1 \text{ atm}}{1.013 \times 10^5 \text{ N m}^{-2}}\right) = 10.2 \text{ atm}$$

7.36 A mixture of liquids A and B exhibits ideal behavior. At 84°C, the total vapor pressure of a solution containing 1.2 moles of A and 2.3 moles of B is 331 mmHg. Upon the addition of another mole of B to the solution, the vapor pressure increases to 347 mmHg. Calculate the vapor pressures of pure A and B at 84°C.

The total vapor pressure depends on the vapor pressures of A and B in a mixture, which in turn depends on the vapor pressures of pure A and B. With the total vapor pressure of the two mixtures known, a pair of simultaneous equations can be written in terms of the vapor pressures of pure A and B.

For the solution containing 1.2 moles of A and 2.3 moles of B,

$$x_A = \frac{1.2}{3.5} = 0.343$$

$$x_B = 1 - 0.343 = 0.657$$

$$P_{total} = P_A + P_B = x_A P_A^* + x_B P_B^*$$

$$331 \text{ mmHg} = 0.343 P_A^* + 0.657 P_B^*$$

Solve the last equation for P_A^*:

$$P_A^* = \frac{331 \text{ mmHg} - 0.657 P_B^*}{0.343} = 965 \text{ mmHg} - 1.92 P_B^* \qquad (7.36.1)$$

Now consider the solution with the additional mole of B.

$$x_A = \frac{1.2}{4.5} = 0.267$$

$$x_B = 1 - 0.267 = 0.733$$

$$P_{total} = P_A + P_B = x_A P_A^* + x_B P_B^*$$

$$347 \text{ mmHg} = 0.267 P_A^* + 0.733 P_B^* \qquad (7.36.2)$$

Substitute Eq. 7.36.1 into Eq. 7.36.2:

$$347 \text{ mmHg} = 0.267 \left(965 \text{ mmHg} - 1.92 P_B^*\right) + 0.733 P_B^*$$

$$0.220 P_B^* = 89.3 \text{ mmHg}$$

$$P_B^* = 406 \text{ mmHg} = 4.1 \times 10^2 \text{ mmHg}$$

Substitute the value of P_B^* into Eq. 7.36.1:

$$P_A^* = 965 \text{ mmHg} - 1.92 (406 \text{ mmHg}) = 1.9 \times 10^2 \text{ mmHg}$$

7.37 Fish breathe the dissolved air in water through their gills. Assuming the partial pressures of oxygen and nitrogen in air to be 0.20 atm and 0.80 atm, respectively, calculate the mole fractions of oxygen and nitrogen in water at 298 K. Comment on your results.

The mole fractions of oxygen and nitrogen can be calculated using Henry's law (See Table 7.1).

$$x_{O_2} = \frac{P_{O_2}}{K_{O_2}} = \left(\frac{0.20 \text{ atm}}{3.27 \times 10^7 \text{ torr}}\right)\left(\frac{760 \text{ torr}}{1 \text{ atm}}\right) = 4.6 \times 10^{-6}$$

$$x_{N_2} = \frac{P_{N_2}}{K_{N_2}} = \left(\frac{0.80 \text{ atm}}{6.80 \times 10^7 \text{ torr}}\right)\left(\frac{760 \text{ torr}}{1 \text{ atm}}\right) = 8.9 \times 10^{-6}$$

In water, the concentration of N_2 is just twice that of O_2. In the atmosphere, $\frac{1}{5}$ of the gas is O_2, but in water $\frac{1}{3}$ of the dissolved gas is O_2.

7.38 Liquids A (molar mass 100 g mol^{-1}) and B (molar mass 110 g mol^{-1}) form an ideal solution. At 55°C, A has a vapor pressure of 95 mmHg and B a vapor pressure of 42 mmHg. A solution is prepared by mixing equal weights of A and B. **(a)** Calculate the mole fraction of each component in the solution. **(b)** Calculate the partial pressures of A and B over the solution at 55°C. **(c)** Suppose that some of the vapor described in **(b)** is condensed to a liquid. Calculate the mole fraction of each component in this liquid and the vapor pressure of each component above this liquid at 55°C.

(a) Let m g be the mass of A and therefore the mass of B.

$$\text{Number of moles of A} = \frac{m \text{ g}}{100 \text{ g mol}^{-1}}$$

$$\text{Number of moles of B} = \frac{m \text{ g}}{110 \text{ g mol}^{-1}}$$

The mole fractions are

$$x_A = \frac{\frac{m \text{ g}}{100 \text{ g mol}^{-1}}}{\frac{m \text{ g}}{100 \text{ g mol}^{-1}} + \frac{m \text{ g}}{110 \text{ g mol}^{-1}}} = \frac{\frac{1}{100}}{\frac{1}{100} + \frac{1}{110}} = 0.5238 = 0.524$$

$$x_B = 1 - 0.5238 = 0.4762 = 0.476$$

(b)

$$P_A = x_A P_A^* = (0.5238)(95 \text{ mmHg}) = 49.8 \text{ mmHg} = 50 \text{ mmHg}$$

$$P_B = x_B P_B^* = (0.4762)(42 \text{ mmHg}) = 20.0 \text{ mmHg} = 20 \text{ mmHg}$$

(c) The composition of the liquid is the same as the composition of the vapor from which it is condensed. Using Dalton's law of partial pressure,

$$x_A = x_A^v = \frac{P_A}{P_A + P_B} = \frac{49.8 \text{ mmHg}}{49.8 \text{ mmHg} + 20.0 \text{ mmHg}} = 0.713 = 0.71$$

$$x_B = x_B^v = 1 - 0.713 = 0.287 = 0.29$$

Note that the vapor, and therefore, the condensed liquid, contains a higher mole fraction of A, the more volatile component, than the original mixture.

The vapor pressure of each component above this liquid is

$$P_A = x_A P_A^* = (0.713)(95 \text{ mmHg}) = 68 \text{ mmHg}$$

$$P_B = x_B P_B^* = (0.287)(42 \text{ mmHg}) = 12 \text{ mmHg}$$

7.39 Lysozyme extracted from chicken egg white has a molar mass of 13,930 g mol^{-1}. Exactly 0.1 g of this protein is dissolved in 50 g of water at 298 K. Calculate the vapor pressure lowering, the depression in freezing point, the elevation of boiling point, and the osmotic pressure of this solution. The vapor pressure of pure water at 298 K is 23.76 mmHg.

First calculate the mole fraction, molality, and molarity of lysozyme.

$$\text{Number of moles of lysozyme} = \frac{0.1 \text{ g}}{13930 \text{ g mol}^{-1}} = 7.17875 \times 10^{-6} \text{ mol}$$

$$\text{Number of moles of water} = \frac{50 \text{ g}}{18.02 \text{ g mol}^{-1}} = 2.7747 \text{ mol}$$

$$x_{\text{lysozyme}} = \frac{7.17875 \times 10^{-6} \text{ mol}}{7.17875 \times 10^{-6} \text{ mol} + 2.7747 \text{ mol}} = 2.5872 \times 10^{-6}$$

$$m = \frac{7.17875 \times 10^{-6} \text{ mol}}{50 \times 10^{-3} \text{ kg}} = 1.43575 \times 10^{-4} \text{ mol kg}^{-1}$$

For a dilute aqueous solution, the molality and molarity are numerically the same (see Problem 7.6). Therefore, $M = 1.43575 \times 10^{-4} \text{ mol L}^{-1}$.

Vapor pressure lowering:

$$\Delta P = x_{\text{lysozyme}} P_{\text{H}_2\text{O}}^* = (2.5872 \times 10^{-6})(23.76 \text{ mmHg}) = 6.147 \times 10^{-5} \text{ mmHg}$$

Depression in freezing point:

$$\Delta T = K_f m = (1.86 \text{ K m}^{-1})(1.43575 \times 10^{-4} m) = 2.67 \times 10^{-4} \text{ K}$$

Elevation of boiling point:

$$\Delta T = K_b m = (0.51 \text{ K m}^{-1})(1.43575 \times 10^{-4} m) = 7.3 \times 10^{-5} \text{ K}$$

Osmotic pressure:

$$\pi = MRT = (1.43575 \times 10^{-4} M)(0.08206 \text{ L atm K}^{-1} \text{ mol}^{-1})(298 \text{ K})$$

$$= 3.51 \times 10^{-3} \text{ atm} \left(\frac{760 \text{ torr}}{1 \text{ atm}}\right) = 2.67 \text{ torr}$$

Note that the only property that is readily measurable is the osmotic pressure.

7.40 The following argument is frequently used to explain the fact that the vapor pressure of the solvent is lower over a solution than over the pure solvent and that lowering is proportional to the concentration. A dynamic equilibrium exists in both cases, so that the rate at which molecules of solvent evaporate from the liquid is always equal to that at which they condense. The rate of condensation is proportional to the partial pressure of the vapor, whereas that of evaporation is unimpaired in the pure solvent but is impaired by solute molecules in the surface of the solution. Hence the rate of escape is reduced in proportion to the concentration of the solute, and maintenance of equilibrium requires a corresponding lowering of the rate of condensation and therefore of the partial pressure of the vapor phase. Explain why this argument is incorrect. [*Source*: K. J. Mysels, *J. Chem. Educ.* **32,** 179 (1955).]

Two reasons suggest themselves from general principles.

1) If the explanation given were correct, then the nature of the solute molecules would have a significant effect on the vapor pressure lowering. Namely, larger molecules would be more efficient at blocking the surface and would reduce the vapor pressure to a greater extent than would smaller molecules at the same concentration. On the contrary, Raoult's law shows the same vapor pressure lowering as a function of concentration regardless of molecular identity.

2) The solute molecules would impede the incorporation of solvent molecules back into solution to the same extent as they would the escape into the vapor phase. This is a requirement of the principle of microscopic reversibility to be discussed in Chapter 12.

The following reasons rely on more specific observations.

3) Surfactant molecules are known to accumulate at the surface of a solution for dilute concentrations, but do not affect the vapor pressure lowering. (At higher concentrations, micelles are formed, which is clearly a non-ideal effect. The resulting change in the chemical potential causes significant deviations in the vapor pressure curve from ideal behavior. It should be noted, however, that this is *not* a surface effect.)

4) Certain insoluble materials, which are observed to have no effect on the equilibrium vapor pressure of the solution, are known to form tightly packed monolayers on the surface of the solution (cetyl alcohol in water is an example) that greatly inhibit the rate of evaporation. This argues in support of the second reason given above that the rate of condensation must likewise be slowed.

7.41 A compound weighing 0.458 g is dissolved in 30.0 g of acetic acid. The freezing point of the solution is found to be 1.50 K below that of the pure solvent. Calculate the molar mass of the compound.

The molality of the solution is related to the depression in freezing point.

$$m = \frac{\Delta T}{K_f} = \frac{1.50 \text{ K}}{3.90 \text{ K mol}^{-1} \text{ kg}} = 0.3846 \text{ mol kg}^{-1}$$

The number of moles of the compound is

$$n = \left(0.3846 \text{ mol kg}^{-1}\right)\left(30.0 \times 10^{-3} \text{ kg}\right) = 0.01154 \text{ mol}$$

Therefore,

$$\text{Molar mass of the compound} = \frac{0.458 \text{ g}}{0.01154 \text{ mol}} = 39.7 \text{ g mol}^{-1}$$

7.42   Two aqueous urea solutions have osmotic pressures of 2.4 atm and 4.6 atm, respectively, at a certain temperature. What is the osmotic pressure of a solution prepared by mixing equal volumes of these two solutions at the same temperature?

First calculate the number of moles of urea in a given volume, V L, of each solution. The solutions with osmotic pressures of 2.4 atm and 4.6 atm are denoted Solutions 1 and 2, respectively.

$$\pi_1 = M_1 RT = \frac{n_1}{V} RT$$

$$n_1 = \frac{\pi_1 V}{RT}$$

Similarly,

$$n_2 = \frac{\pi_2 V}{RT}$$

If V L of Solution 1 is mixed with V L of Solution 2, then the total number of moles is

$$n_{mix} = n_1 + n_2 = \frac{(\pi_1 + \pi_2) V}{RT}$$

and the molarity of the solution is

$$M_{mix} = \frac{\frac{(\pi_1 + \pi_2) V}{RT}}{2V} = \frac{\pi_1 + \pi_2}{2RT}$$

The osmotic pressure of the mixture is

$$\pi_{mix} = M_{mix} RT = \frac{\pi_1 + \pi_2}{2RT} RT = \frac{\pi_1 + \pi_2}{2} = \frac{2.4 \text{ atm} + 4.6 \text{ atm}}{2} = 3.5 \text{ atm}$$

This simple result comes about because the osmotic pressure is proportional to the molarity of the solute. When equal volumes of the two solutions are mixed, the molarity will just be the mean of the molarities of the two solutions (assuming additive volume). Since the osmotic pressure is proportional to the molarity, the osmotic pressure of the solution will be the average of the osmotic pressures of the two solutions.

7.43   A forensic chemist is given a white powder for analysis. She dissolves 0.50 g of the substance in 8.0 g of benzene. The solution freezes at 3.9°C. Can the chemist conclude that the compound is cocaine ($C_{17}H_{21}NO_4$)? What assumptions are made in the analysis? The freezing point of benzene is 5.5°C.

The depression in freezing point of benzene ($5.5°C - 3.9\,°C = 1.6°C = 1.6$ K) furnishes the molar mass of the white powder, which is then compared with the molar mass of cocaine.

The molality of the compound is

$$m = \frac{\Delta T}{K_f} = \frac{1.6\ \text{K}}{5.12\ \text{K mol}^{-1}\,\text{kg}} = 0.313\ \text{mol kg}^{-1}$$

The number of moles of the compound is

$$n = \left(0.313\ \text{mol kg}^{-1}\right)\left(8.0 \times 10^{-3}\ \text{kg}\right) = 2.50 \times 10^{-3}\ \text{mol}$$

Therefore, the molar mass of the substance is

$$\mathcal{M} = \frac{0.50\ \text{g}}{2.50 \times 10^{-3}\ \text{mol}} = 2.0 \times 10^2\ \text{g mol}^{-1}$$

The molar mass of cocaine is $303.35\ \text{g mol}^{-1}$. Thus, the compound is not likely to be cocaine. The assumptions implicit in this analysis are that the compound is pure, that it is monomeric and does not either associate or dissociate in benzene, and that it is not an electrolyte.

7.44 "Time-release" drugs have the advantage of releasing the drug to the body at a constant rate so that the drug concentration at any time is not high enough to have harmful side effects or so low as to be ineffective. A schematic diagram of a pill that works on this basis is shown below. Explain how it works.

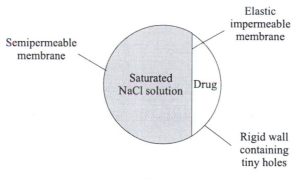

Schematic of time-release pill

When swallowed, the pill will be in a hypotonic solution and the resulting osmotic pressure will cause water to move across the semipermeable membrane into the pill. This increases the volume of the salt solution and pushes the elastic membrane to the right, which causes the drug to exit through the small holes at a constant rate.

7.45 A nonvolatile organic compound, Z, was used to make up two solutions. Solution A contains 5.00 g of Z dissolved in 100 g of water, and solution B contains 2.31 g of Z dissolved in 100 g of benzene. Solution A has a vapor pressure of 754.5 mmHg at the normal boiling point of water, and solution B has the same vapor pressure at the normal boiling point of benzene. Calculate the molar mass of Z in solutions A and B and account for the difference.

Since Z causes the same amount of vapor pressure lowering in Solutions A and B, the mole fraction of Z, x_Z, must be the same in both solutions.

$$x_Z = \frac{\Delta P}{P^*} = \frac{760 \text{ mmHg} - 754.5 \text{ mmHg}}{760 \text{ mmHg}} = 7.24 \times 10^{-3}$$

The molar mass of Z in each solution can then be calculated from the number of moles of solvent, the mole fraction of Z, and the mass of Z.

Solution A

$$n_{H_2O} = \frac{100 \text{ g}}{18.02 \text{ g mol}^{-1}} = 5.549 \text{ mol}$$

$$x_Z = 7.24 \times 10^{-3} = \frac{n_Z}{n_Z + n_{H_2O}}$$

$$n_Z = 7.24 \times 10^{-3} \left(n_Z + n_{H_2O} \right)$$

$$0.99276 n_Z = 7.24 \times 10^{-3} n_{H_2O} = 0.0402$$

$$n_Z = 4.05 \times 10^{-2} \text{ mol}$$

$$\text{Molar mass of Z} = \frac{5.00 \text{ g}}{4.05 \times 10^{-2} \text{ mol}} = 1.2 \times 10^2 \text{ g mol}^{-1}$$

Solution B

$$n_{benzene} = \frac{100 \text{ g}}{78.11 \text{ g mol}^{-1}} = 1.280 \text{ mol}$$

$$x_Z = 7.24 \times 10^{-3} = \frac{n_Z}{n_Z + n_{benzene}}$$

$$n_Z = 7.24 \times 10^{-3} \left(n_Z + n_{benzene} \right)$$

$$0.99276 n_Z = 7.24 \times 10^{-3} n_{benzene} = 0.00927$$

$$n_Z = 9.34 \times 10^{-3} \text{ mol}$$

$$\text{Molar mass of Z} = \frac{2.31 \text{ g}}{9.34 \times 10^{-3} \text{ mol}} = 2.5 \times 10^2 \text{ g mol}^{-1}$$

The molar mass of Z in benzene is about twice that in water. This suggests some sort of dimerization is occurring in a nonpolar solvent such as benzene.

7.46 Acetic acid is a polar molecule that can form hydrogen bonds with water molecules. Therefore, it has a high solubility in water. Yet acetic acid is also soluble in benzene (C_6H_6), a nonpolar solvent that lacks the ability to form hydrogen bonds. A solution of 3.8 g of CH_3COOH in 80 g C_6H_6 has a freezing point of 3.5°C. Calculate the molar mass of the solute, and suggest what its structure might be. (*Hint*: Acetic acid molecules can form hydrogen bonds among themselves.)

The acetic acid depresses the freezing point of benzene by

$$\Delta T = 5.5°C - 3.5°C = 2.0°C = 2.0 \text{ K}$$

The molality of acetic acid is

$$m_2 = \frac{2.0 \text{ K}}{5.12 \text{ K mol}^{-1} \text{ kg}} = 0.391 \text{ mol kg}^{-1}$$

The number of moles of acetic acid is

$$n_2 = (0.391 \text{ mol kg}^{-1})(80 \times 10^{-3} \text{ kg}) = 0.0313 \text{ mol}$$

which gives a molar mass of

$$\mathcal{M} = \frac{3.8 \text{ g}}{0.0313 \text{ mol}} = 1.2 \times 10^2 \text{ g mol}^{-1}$$

The calculated molar mass is twice that of the molar mass of acetic acid. This suggests that acetic acid, when placed in benzene, exists as a dimer held together by hydrogen bonds.

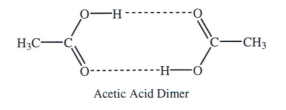

Acetic Acid Dimer

7.47   At 85°C, the vapor pressure of A is 566 torr and that of B is 250 torr. Calculate the composition of a mixture of A and B that boils at 85°C when the pressure is 0.60 atm. Also calculate the composition of the vapor mixture. Assume ideal behavior.

The total vapor pressure is the same as the external pressure, 0.60 atm, when the mixture boils. The total vapor pressure is the sum of the vapor pressures of A and B from the mixture, which depend on the mole fractions of A and B in the mixture.

$$P_{\text{total}} = (0.60 \text{ atm})\left(\frac{760 \text{ torr}}{1 \text{ atm}}\right) = P_A + P_B = x_A P_A^* + x_B P_B^*$$

$$456 \text{ torr} = x_A (566 \text{ torr}) + x_B (250 \text{ torr})$$

Since $x_B = 1 - x_A$, the above equation depends on only one variable, x_A.

$$456 \text{ torr} = x_A (566 \text{ torr}) + (1 - x_A)(250 \text{ torr})$$

$$(316 \text{ torr}) x_A = 206 \text{ torr}$$

$$x_A = 0.652 = 0.65$$

Therefore,

$$x_B = 1 - x_A = 0.348 = 0.35$$

The mole fractions of A and B are 0.65 and 0.35, respectively, in the mixture. The composition in the vapor can be evaluated from the vapor pressures of A and B.

$$P_A = x_A P_A^* = (0.652)(566 \text{ torr}) = 369 \text{ torr}$$

$$P_B = P_{total} - P_A = 456 \text{ torr} - 369 \text{ torr} = 87 \text{ torr}$$

$$\frac{P_A}{P_B} = \frac{369 \text{ torr}}{87 \text{ torr}} = \frac{x_A^v P_{total}}{x_B^v P_{total}} = \frac{x_A^v}{x_B^v}$$

$$\frac{x_A^v}{x_B^v} = 4.24$$

$$x_A^v = 4.24 x_B^v = 4.24 \left(1 - x_A^v\right)$$

$$5.24 x_A^v = 4.24$$

$$x_A^v = 0.81$$

$$x_B^v = 0.19$$

The vapor of the mixture is richer in A, the more volatile component.

7.48 Comment on whether each of the following statements is true or false, and briefly explain your answer: **(a)** If one component of a solution obeys Raoult's law, then the other component must also obey the same law. **(b)** Intermolecular forces are small in ideal solutions. **(c)** When 15.0 mL of an aqueous 3.0 M ethanol solution is mixed with 55.0 mL of an aqueous 3.0 M ethanol solution, the total volume is 70.0 mL.

---

a) False, consider the case of a dilute solution where the solvent obeys Raoult's law and the solute obeys Henry's law.

b) False, the intermolecular forces need only be the same between like and unlike molecules regardless of their magnitude.

c) True, the solutions are identical, all having the same density.

7.49 Liquids A and B form an ideal solution at a certain temperature. The vapor pressures of pure A and B are 450 torr and 732 torr, respectively, at this temperature. **(a)** A sample of the solution's vapor is condensed. Given that the original solution contains 3.3 moles of A and 8.7 moles of B, calculate the composition of this condensed liquid in mole fractions. **(b)** Suggest a method for measuring the partial pressures of A and B at equilibrium.

(a) First calculate the vapor pressures of A and B using Raoult's law.

$$x_A = \frac{3.3 \text{ mol}}{3.3 \text{ mol} + 8.7 \text{ mol}} = 0.275$$

$$x_B = 1 - 0.275 = 0.725$$

$$P_A = x_A P_A^* = (0.275)(450 \text{ torr}) = 124 \text{ torr}$$

$$P_B = x_B P_B^* = (0.725)(732 \text{ torr}) = 531 \text{ torr}$$

The composition of the condensed liquid is the same as the composition of the vapor before condensation, the latter can be calculated from P_A and P_B.

$$\frac{P_A}{P_B} = \frac{124 \text{ torr}}{531 \text{ torr}} = \frac{x_A^v P_{total}}{x_B^v P_{total}} = \frac{x_A^v}{x_B^v}$$

$$\frac{x_A^v}{x_B^v} = 0.234$$

$$x_A^v = 0.234 x_B^v = 0.234 \left(1 - x_A^v\right)$$

$$1.234 x_A^v = 0.234$$

$$x_A^v = 0.19$$

$$x_B^v = 0.81$$

(b) Using Dalton's Law, knowledge of the mole fractions of A and B in the vapor phase will allow the determination of their partial pressures via $P_i = x_i P_{total}$. Thus, mass spectrometric analysis of a sample of the vapor in equilibrium with the solution to determine the mole fraction of each plus a measurement of the total pressure will provide the needed information.

7.50 Nonideal solutions are the result of unequal intermolecular forces between components. Based on this knowledge, comment on whether a racemic mixture of a liquid compound would behave as an ideal solution.

The solution will not behave as an ideal solution. The two enantiomers interact differently with each other than they do with themselves. That is, although the interactions between the members of a $(+)(+)$ pair are the same as those in a $(-)(-)$ pair, these will be different from the interactions in a $(+)(-)$ pair.

7.51 Calculate the molal boiling-point elevation constant (K_b) for water. The molar enthalpy of vaporization of water is 40.79 kJ mol^{-1} at 100°C

$$K_b = \frac{R T_0^2 \mathcal{M}_1}{\Delta \overline{H}_{vap}}$$

$$= \frac{\left(8.314 \text{ J K}^{-1} \text{ mol}^{-1}\right) (373.15 \text{ K})^2 \left(18.02 \times 10^{-3} \text{ kg mol}^{-1}\right)}{40.79 \times 10^3 \text{ J mol}^{-1}}$$

$$= 0.5114 \text{ K kg mol}^{-1} = 0.5114 \text{ K m}^{-1}$$

7.52 Explain the following phenomena. **(a)** A cucumber placed in concentrated brine (saltwater) shrivels into a pickle. **(b)** A carrot placed in fresh water swells in volume.

(a) The cucumber is in a hypertonic solution, it loses water through osmosis to the concentrated brine.

(b) The carrot is in a hypotonic solution. There is a higher concentration of salts, etc. in the cells in the carrot, and water enters into the cells.

7.53   Calculate the change in the Gibbs energy at 37°C when the human kidneys secrete 0.275 mole of urea per kilogram of water from blood plasma to urine if the concentrations of urea in blood plasma and urine are 0.005 mol kg^{-1} and 0.326 mol kg^{-1}, respectively.

The chemical potentials of urea in urine and blood are

$$\left(\mu_{\text{urea}}\right)_{\text{urine}} = \mu_{\text{urea}}^* + RT \ln \left(x_{\text{urea}}\right)_{\text{urine}}$$

$$\left(\mu_{\text{urea}}\right)_{\text{blood}} = \mu_{\text{urea}}^* + RT \ln \left(x_{\text{urea}}\right)_{\text{blood}}$$

The change in chemical potential for the kidneys to secret water from blood plasma to urine is

$$\left(\mu_{\text{urea}}\right)_{\text{urine}} - \left(\mu_{\text{urea}}\right)_{\text{blood}} = \left[\mu_{\text{urea}}^* + RT \ln \left(x_{\text{urea}}\right)_{\text{urine}}\right] - \left[\mu_{\text{urea}}^* + RT \ln \left(x_{\text{urea}}\right)_{\text{blood}}\right]$$

$$= RT \ln \frac{\left(x_{\text{urea}}\right)_{\text{urine}}}{\left(x_{\text{urea}}\right)_{\text{blood}}}$$

Since mole fraction is proportional to molality, the above equation can be rewritten as

$$\left(\mu_{\text{urea}}\right)_{\text{urine}} - \left(\mu_{\text{urea}}\right)_{\text{blood}} = RT \ln \frac{\left(m_{\text{urea}}\right)_{\text{urine}}}{\left(m_{\text{urea}}\right)_{\text{blood}}}$$

$$= \left(8.314 \text{ J K}^{-1} \text{ mol}^{-1}\right) (310 \text{ K}) \ln \frac{0.326 \, m}{0.005 \, m}$$

$$= 1.1 \times 10^4 \text{ J mol}^{-1}$$

The change in the Gibbs energy for this secretion process is

$$\Delta G = n \left[\left(\mu_{\text{urea}}\right)_{\text{urine}} - \left(\mu_{\text{urea}}\right)_{\text{blood}}\right]$$

$$= (0.275 \text{ mol}) \left(1.1 \times 10^4 \text{ J mol}^{-1}\right) = 3 \times 10^3 \text{ J}$$

7.54   **(a)** Which of the following expressions is incorrect as a representation of the partial molar volume of component A in a two-component solution? Why? How would you correct it?

$$\left(\frac{\partial V_m}{\partial n_A}\right)_{T,P,n_B} \qquad\qquad \left(\frac{\partial V_m}{\partial x_A}\right)_{T,P,x_B}$$

(b) Given that the molar volume of this mixture (V_m) is given by

$$V_m = 0.34 + 3.6 x_A x_B + 0.4 x_B \left(1 - x_A\right) \text{ L mol}^{-1}$$

derive an expression for the partial molar volume for A at $x_A = 0.20$.

(a) The second expression is incorrect, since it is impossible to vary x_A while holding x_B constant. The correct expression is $\left(\frac{\partial V_m}{\partial x_A}\right)_{T,P}$

(b) Write V_m in terms of x_A.

$$V_m = \left[0.34 + 3.6x_A\left(1 - x_A\right) + 0.4\left(1 - x_A\right)\left(1 - x_A\right)\right] \text{ L mol}^{-1}$$

$$= \left[0.34 + 3.6x_A - 3.6x_A^2 + 0.4 - 0.8x_A + 0.4x_A^2\right] \text{ L mol}^{-1}$$

$$= \left(0.38 + 2.8x_A - 3.2x_A^2\right) \text{ L mol}^{-1}$$

Differentiate V_m with respect to x_A.

$$\left(\frac{\partial V_m}{\partial x_A}\right)_{P,T} = \left(2.8 - 6.4x_A\right) \text{ L mol}^{-1}$$

At $x_A = 0.20$,

$$\left(\frac{\partial V_m}{\partial x_A}\right)_{P,T} = \left[2.8 - 6.4\left(0.20\right)\right] \text{ L mol}^{-1} = 1.5 \text{ L mol}^{-1}$$

7.55   The partial molar volumes for a benzene-carbon tetrachloride solution at 25°C at a mole fraction of 0.5 are: $\overline{V}_b = 0.106 \text{ L mol}^{-1}$ and $\overline{V}_c = 0.100 \text{ L mol}^{-1}$, respectively, where the subscripts b and c denote C_6H_6 and CCl_4. **(a)** What is the volume of a solution made up of one mole of each? **(b)** Given that the molar volumes are: $C_6H_6 = 0.089 \text{ L mol}^{-1}$ and $CCl_4 = 0.097 \text{ L mol}^{-1}$, what is the change in volume on mixing 1 mole each of C_6H_6 and CCl_4? **(c)** What can you deduce about the nature of intermolecular forces between C_6H_6 and CCl_4?

(a)

$$V = n_b\overline{V}_b + n_c\overline{V}_c = \left(1 \text{ mol}\right)\left(0.106 \text{ L mol}^{-1}\right) + \left(1 \text{ mol}\right)\left(0.100 \text{ L mol}^{-1}\right) = 0.206 \text{ L}$$

(b) Before mixing, the volume of the two components are

$$V = \left(1 \text{ mol}\right)\left(0.089 \text{ L mol}^{-1}\right) + \left(1 \text{ mol}\right)\left(0.097 \text{ L mol}^{-1}\right) = 0.186 \text{ L}$$

After mixing, the volume is 0.206 L, as determined in (a). Therefore, the change in volume is

$$\Delta V = 0.206 \text{ L} - 0.186 \text{ L} = 0.020 \text{ L}$$

(c) The volume of the solution is larger than the sum of the volumes of the individual components. This indicates that C_6H_6 and CCl_4 repel each other and on average are farther apart from each other than they are from themselves.

7.56   The osmotic pressure of poly(methylmethacrylate) in toluene has been measured at a series of concentrations at 298 K. Determine graphically the molar mass of the polymer.

π/atm	8.40×10^{-4}	1.72×10^{-3}	2.52×10^{-3}	3.23×10^{-3}	7.75×10^{-3}
c/g·L^{-1}	8.10	12.31	15.00	18.17	28.05

In the limit of a dilute solution, where all virial coefficients except the second may be ignored, the osmotic pressure π is given by

$$\frac{\pi}{c} = RT \left(\frac{1}{\mathcal{M}} + Bc \right)$$

Thus, a graph of π/c vs. c will extrapolate to a y intercept of $\frac{RT}{\mathcal{M}}$, where $\mathcal{M}$ is the molar mass of the solute.

The data for the graph is given in the table below.

(π/c)/atm·L·g^{-1}	1.037×10^{-4}	1.397×10^{-4}	1.680×10^{-4}	1.778×10^{-4}	2.763×10^{-4}
c/g·L^{-1}	8.10	12.31	15.00	18.17	28.05

As seen in the graph, the extrapolation to zero concentration gives an intercept of 3.4×10^{-5} atm L g^{-1}.

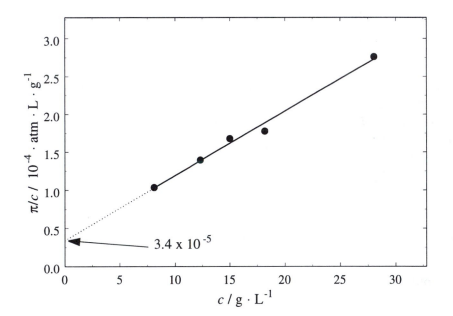

Thus,

$$3.4 \times 10^{-5} \text{atm L g}^{-1} = \frac{RT}{\mathcal{M}}$$

$$= \frac{\left(0.08206 \text{ L atm K}^{-1} \text{ mol}^{-1}\right)(298 \text{ K})}{\mathcal{M}}$$

$$\mathcal{M} = 7.2 \times 10^5 \text{ g mol}^{-1}$$

7.57   Benzene and toluene form an ideal solution. Prove that to achieve the maximum entropy of mixing, the mole fraction of each component must be 0.5.

The entropy of mixing is

$$\Delta_{mix}S = -nR\left(x_A \ln x_A + x_B \ln x_B\right) = -nR\left[x_A \ln x_A + \left(1 - x_A\right) \ln \left(1 - x_A\right)\right]$$

The equation is written in terms of only one variable, x_A. To achieve the maximum entropy of mixing, $d\Delta_{mix}S/dx_A$ must be 0.

$$\frac{d\Delta_{mix}S}{dx_A} = -nR\left[\ln x_A + x_A \frac{1}{x_A} - \ln\left(1 - x_A\right) - \left(1 - x_A\right)\frac{1}{1 - x_A}\right] = 0$$

$$-nR\ln\frac{x_A}{1 - x_A} = 0$$

$$\frac{x_A}{1 - x_A} = 1$$

$$x_A = 0.5$$

Therefore, the maximum entropy of mixing is achieved when the mixture is made up of equal number of moles of each component. To check that this is a maximum, note that

$$\frac{d^2\Delta_{mix}S}{dx_A^2} = -nR\left[\frac{1}{x_A\left(1 - x_A\right)}\right] < 0$$

7.58 Suppose 2.6 moles of He at 0.80 atm and 25°C are mixed with 4.1 moles of Ne at 2.7 atm and 25°C. Calculate the Gibbs energy change for the process. Assume ideal behavior.

The expression $\Delta_{mix}G = nRT\left(x_1 \ln x_1 + x_2 \ln x_2\right)$ is valid only for mixing two gases originally at the same pressure and is not directly applicable here. G is a state function, however, so consider a path in which the two gases are first individually brought to the final pressure of the mixture and then mixed. The overall ΔG is the sum of the ΔG's for the steps. The process is assumed to take place with the gases originally in two containers connected by a valve which is opened. The final volume of the mixture is then the sum of the original volumes. Since $V_{He} = \frac{n_{He}RT}{P_{He}}$, with an similar expression for Ne,

$$P_f = \frac{n_{total}RT}{V_f}$$

$$= \frac{\left(n_{He} + n_{Ne}\right)RT}{\frac{n_{He}RT}{P_{He}} + \frac{n_{Ne}RT}{P_{Ne}}}$$

$$= \frac{n_{He} + n_{Ne}}{\frac{n_{He}}{P_{He}} + \frac{n_{Ne}}{P_{Ne}}}$$

$$= \frac{2.6\ \text{mol} + 4.1\ \text{mol}}{\frac{2.6\ \text{mol}}{0.80\ \text{atm}} + \frac{4.1\ \text{mol}}{2.7\ \text{atm}}}$$

$$= 1.41\ \text{atm}$$

For the compression (expansion) of an ideal gas, $\Delta G = nRT \ln \frac{P_f}{P_i}$, so that

$$\Delta G_{He} = (2.6 \text{ mol}) \left(8.314 \text{ J K}^{-1} \text{ mol}^{-1}\right) (298 \text{ K}) \ln \frac{1.41 \text{ atm}}{0.80 \text{ atm}} = 3.65 \times 10^3 \text{ J}$$

$$\Delta G_{Ne} = (4.1 \text{ mol}) \left(8.314 \text{ J K}^{-1} \text{ mol}^{-1}\right) (298 \text{ K}) \ln \frac{1.41 \text{ atm}}{2.7 \text{ atm}} = -6.60 \times 10^3 \text{ J}$$

$$\Delta_{mix} G = n_{total} RT \left(x_{He} \ln x_{He} + x_{Ne} \ln x_{Ne}\right)$$

$$= (2.6 \text{ mol} + 4.1 \text{ mol}) \left(8.314 \text{ J K}^{-1} \text{ mol}^{-1}\right) (298 \text{ K})$$

$$\times \left[\frac{2.6 \text{ mol}}{2.6 \text{ mol} + 4.1 \text{mol}} \ln \left(\frac{2.6 \text{ mol}}{2.6 \text{ mol} + 4.1 \text{mol}} \right) \right.$$

$$\left. + \frac{4.1 \text{ mol}}{2.6 \text{ mol} + 4.1 \text{mol}} \ln \left(\frac{4.1 \text{ mol}}{2.6 \text{ mol} + 4.1 \text{mol}} \right) \right]$$

$$= -1.11 \times 10^4 \text{ J}$$

$$\Delta G_{total} = \Delta G_{He} + \Delta G_{Ne} + \Delta_{mix} G = -1.4 \times 10^4 \text{ J} = -14 \text{ kJ}$$

7.59   Two beakers are placed in a closed container. Beaker A initially contains 0.15 mole of naphthalene ($C_{10}H_8$) in 100 g of benzene (C_6H_6) and beaker B initially contains 31 g of an unknown compound dissolved in 100 g of benzene. At equilibrium, beaker A is found to have lost 7.0 g. Assuming ideal behavior, calculate the molar mass of the unknown compound. State any assumptions made.

This problem is very similar to problem 7.27. At equilibrium, the vapor pressure of benzene over each beaker must be the same. Assuming ideal solutions, this means that the mole fraction of benzene in each beaker must be identical at equilibrium. Consequently, the mole fraction of solute is also the same in each beaker, even though the solutes are different in the two. Assuming the solutes to be non-volatile, equilibrium is reached by the transfer of benzene, via the vapor phase, from beaker A to beaker B.

The mole fraction of naphthalene in beaker A at equilibrium can be determined from the data given. The number of moles of naphthalene is given, and that for benzene is calculated from the molar mass knowing that 100 g − 7.0 g = 93.0 g of benzene is left in the beaker.

$$x_{C_{10}H_8} = \frac{0.15 \text{ mol}}{0.15 \text{ mol} + \frac{93.0 \text{ g}}{78.11 \text{ g mol}^{-1}}} = 0.112$$

Now let the number of moles of unknown compound be n. Assuming all the benzene lost from beaker A is transferred to beaker B, there are 100 g + 7.0 g = 107 g of benzene in the beaker. Thus, the mole fraction of the unknown compound is

$$x_{unknown} = \frac{n}{n + \frac{107 \text{ g}}{78.11 \text{ g mol}^{-1}}}$$

$$= \frac{n}{n + 1.370 \text{ mol}}$$

$$= 0.112$$

Solving for n gives $n = 0.172$ mol. The molar mass is

$$\frac{31 \text{ g}}{0.172 \text{ mol}} = 1.8 \times 10^2 \text{ g mol}^{-1}$$

In addition to those assumptions mentioned above, the temperature is assumed constant.

Electrolyte Solutions

PROBLEMS AND SOLUTIONS

8.1 The resistance of a 0.010 M NaCl solution is 172 Ω. If the equivalent conductance of the solution is 153 Ω^{-1} equiv^{-1} cm^2, what is the cell constant?

Resistance is related to specific conductance by

$$\frac{1}{R} = \kappa \frac{A}{l}$$

and specific conductance is related to molar (or equivalent) conductance by

$$\Lambda = \frac{\kappa}{c}$$

Combining these two equations, we have an equation relating resistance and molar (or equivalent) conductance:

$$\frac{1}{R} = \Lambda c \frac{A}{l}$$

Rearranging this equation, we have an expression for the cell constant:

$$\frac{l}{A} = \Lambda c R$$

$$= \left(153 \ \Omega^{-1} \text{equiv}^{-1} \text{cm}^2\right) \left(\frac{1 \text{ equiv}}{1 \text{ mol}}\right) \left(0.010 \text{ mol L}^{-1}\right) (172 \ \Omega) \left(\frac{1 \text{ L}}{1000 \text{ cm}^3}\right)$$

$$= 0.263 \text{ cm}^{-1} = 0.26 \text{ cm}^{-1}$$

8.2 Using the cell described in Problem 8.1, a student determined the resistance of a 0.086 M KCl solution to be 20.4 Ω. Calculate the equivalent conductance of this solution.

Rearranging the relation between resistance and molar (or equivalent) conductance

$$\frac{1}{R} = \Lambda c \frac{A}{l}$$

obtained in Problem 8.1, the equivalent conductance is

$$
\begin{aligned}
\Lambda &= \frac{1}{Rc}\frac{l}{A} \\
&= \frac{1}{(20.4\ \Omega)\,(0.086\ \text{mol L}^{-1})}\,(0.263\ \text{cm}^{-1})\left(\frac{1\ \text{mol}}{1\ \text{equiv}}\right)\left(\frac{1000\ \text{cm}^3}{1\ \text{L}}\right) \\
&= 1.5 \times 10^2\ \Omega^{-1}\,\text{equiv}^{-1}\,\text{cm}^2
\end{aligned}
$$

8.3   The cell constant (l/A) of a conductance cell is 388.1 m^{-1}. At 25°C the resistance of a 4.8×10^{-4} mol dm^{-3} aqueous solution of sodium chloride is $6.4 \times 10^4\ \Omega$ and that of a sample of water is $7.4 \times 10^6\ \Omega$. Calculate the molar conductivity of the NaCl in solution at this concentration.

The conductance of the NaCl solution is a sum of the conductance of NaCl and that of water. Thus, the conductance of NaCl, C, is

$$
C = C(\text{NaCl solution}) - C(\text{water}) = \frac{1}{R(\text{NaCl solution})} - \frac{1}{R(\text{water})}
$$

$$
= \frac{1}{6.4 \times 10^4\ \Omega} - \frac{1}{7.4 \times 10^6\ \Omega} = 1.55 \times 10^{-5}\ \Omega^{-1}
$$

The specific conductance of NaCl, κ, is

$$
\kappa = C\frac{l}{A} = \left(1.55 \times 10^{-5}\ \Omega^{-1}\right)\left(388.1\ \text{m}^{-1}\right) = 6.02 \times 10^{-3}\ \Omega^{-1}\,\text{m}^{-1}
$$

The molar conductivity of NaCl is

$$
\Lambda = \frac{\kappa}{c} = \left(\frac{6.02 \times 10^{-3}\ \Omega^{-1}\,\text{m}^{-1}}{4.8 \times 10^{-4}\ \text{mol dm}^{-3}}\right)\left(\frac{1\ \text{m}}{10\ \text{dm}}\right)^3 = 1.3 \times 10^{-2}\ \Omega^{-1}\,\text{mol}^{-1}\,\text{m}^2
$$

8.4   Given that the measurement of Λ_0 for weak electrolytes is generally difficult to obtain, how would you deduce the value of Λ_0 for CH$_3$COOH from the data listed in Table 8.1? (*Hint:* Consider CH$_3$COONa, HCl, and NaCl.)

At infinite dilution, any electrolyte is completely dissociated. Therefore, Λ_0 for CH$_3$COOH can be obtained from Λ_0 for CH$_3$COO$^-$ and Λ_0 for H$^+$, which in turn can be derived by combining Λ_0's for CH$_3$COONa and HCl and subtracting Λ_0 for NaCl.

$$
\Lambda_0(\text{CH}_3\text{COOH}) = \Lambda_0(\text{CH}_3\text{COONa}) + \Lambda_0(\text{HCl}) - \Lambda_0(\text{NaCl})
$$

$$
= (91.00 + 426.16 - 126.45)\ \Omega^{-1}\,\text{equiv}^{-1}\,\text{cm}^2
$$

$$
= 390.71\ \Omega^{-1}\,\text{equiv}^{-1}\,\text{cm}^2
$$

8.5   A simple way to determine the salinity of water is to measure its conductivity and assume that the conductivity is entirely due to sodium chloride. In a particular experiment, the resistance of a sample solution is found to be 254 Ω. The resistance of a 0.050 M KCl solution measured in the same cell is 467 Ω. Estimate the concentration of NaCl in the solution. (*Hint*: First derive an equation relating R to Λ and c and then use Λ_0 values for Λ.)

According to the solution of Problem 8.1, a relation between the cell constant, c, Λ, and R is

$$\frac{1}{R} = \Lambda c \frac{A}{l}$$

The cell constant is the same for both measurements. Therefore,

$$\frac{R(\text{KCl})}{R(\text{NaCl})} = \frac{\Lambda(\text{NaCl})\,c(\text{NaCl})}{\Lambda(\text{KCl})\,c(\text{KCl})}$$

$$c(\text{NaCl}) = \frac{R(\text{KCl})}{R(\text{NaCl})}\,\frac{\Lambda(\text{KCl})\,c(\text{KCl})}{\Lambda(\text{NaCl})}$$

Since the KCl solution is dilute, $\Lambda_0 \approx \Lambda$. The NaCl solution is more conductive (less resistive) than the KCl solution, but not too much so. Thus, Λ_0 approximates Λ for the NaCl solution also.

$$c(\text{NaCl}) = \left(\frac{467\ \Omega}{254\ \Omega}\right)\left[\frac{(149.85\ \Omega^{-1}\ \text{equiv}^{-1}\ \text{cm}^2)\,(0.050\ M)}{126.45\ \Omega^{-1}\ \text{equiv}^{-1}\ \text{cm}^2}\right]$$

$$= 0.11\ M$$

8.6   A conductance cell consists of two electrodes, each with an area of 4.2×10^{-4} m^2, separated by 0.020 m. The resistance of the cell when filled with a 6.3×10^{-4} M KNO$_3$ solution is 26.7 Ω. What is the molar conductivity of the solution?

Rearranging the relation between resistance and molar conductivity

$$\frac{1}{R} = \Lambda c \frac{A}{l}$$

obtained in Problem 8.1, the molar conductivity of the solution is

$$\Lambda = \frac{1}{Rc}\frac{l}{A} = \left[\frac{1}{(26.7\ \Omega)\,(6.3 \times 10^{-4}\ \text{mol L}^{-1})}\right]\left(\frac{0.020\ \text{m}}{4.2 \times 10^{-4}\ \text{m}^2}\right)\left(\frac{1\ \text{m}^3}{1000\ \text{L}}\right)$$

$$= 2.8\ \Omega^{-1}\ \text{mol}^{-1}\ \text{m}^2$$

8.7   Referring to Figure 8.4, explain why the slope of conductance versus volume of NaOH added rises right at the start if the acid employed in the titration is weak.

Before the equivalence point in a titration of a weak acid, for example acetic acid, the neutralization reaction has the effect of replacing a weak electrolyte, CH_3COOH, with a strong electrolyte, CH_3COONa. The salt formed by the titration dissociates completely and increases the number of ions present which leads to an increase in the conductance. Contrast this to the situation with the titration of a strong acid where the net effect is to replace highly mobile H^+ ions with less mobile Na^+ ions.

8.8   Calculate the solubility of $BaSO_4$ (in $g\,L^{-1}$) in **(a)** water and **(b)** a $6.5 \times 10^{-5}\,M\,MgSO_4$ solution. The solubility product of $BaSO_4$ is 1.1×10^{-10}. Assume ideal behavior.

The equation for the dissolution of $BaSO_4$ is

$$BaSO_4(s) \rightleftharpoons Ba^{2+}(aq) + SO_4^{2-}(aq)$$

(a) If the solubility of $BaSO_4$ is $x\,M$, then there are $x\,M$ of Ba^{2+} and $x\,M\,SO_4^{2-}$ in the solution.

$$K_{sp} = 1.1 \times 10^{-10} = [Ba^{2+}][SO_4^{2-}] = x \cdot x$$

$$x = 1.05 \times 10^{-5}$$

Therefore,

$$\text{The solubility of } BaSO_4 = \left(1.05 \times 10^{-5}\,mol\,L^{-1}\right)\left(\frac{233.4\,g}{1\,mol}\right) = 2.5 \times 10^{-3}\,g\,L^{-1}$$

(b) The $MgSO_4$ solution contains $6.5 \times 10^{-5}\,M$ of SO_4^{2-}. If the solubility of $BaSO_4$ is $x\,M$, then there are $x\,M$ of Ba^{2+} and $x + 6.5 \times 10^{-5}\,M\,SO_4^{2-}$ in the solution.

$$K_{sp} = 1.1 \times 10^{-10} = [Ba^{2+}][SO_4^{2-}] = x\left(x + 6.5 \times 10^{-5}\right)$$

$$x^2 + 6.5 \times 10^{-5}x - 1.1 \times 10^{-10} = 0$$

$$x = 1.65 \times 10^{-6} \quad \text{or} \quad x = -6.67 \times 10^{-5} \text{ (nonphysical)}$$

Therefore,

$$\text{The solubility of } BaSO_4 = \left(1.65 \times 10^{-6}\,mol\,L^{-1}\right)\left(\frac{233.4\,g}{1\,mol}\right) = 3.9 \times 10^{-4}\,g\,L^{-1}$$

8.9   The thermodynamic solubility product of $AgCl$ is 1.6×10^{-10}. What is $[Ag^+]$ in **(a)** a $0.020\,M$ KNO_3 solution and **(b)** a $0.020\,M\,KCl$ solution?

(a) First calculate the ionic strength and the mean ionic activity coefficient of the KNO_3 solution. Since this solution is dilute, its molality has the same numerical value as its molarity. For this solution, $z_+ = 1$, $z_- = -1$, $m_+ = m_- = 0.020\,m$.

$$I = \frac{1}{2}\left[(0.020m)\,(1)^2 + (0.020m)\,(-1)^2\right] = 0.020\,m$$

$$\log \gamma_\pm = -0.509\,|(1)\,(-1)|\,\sqrt{0.020} = -7.20 \times 10^{-2}$$

$$\gamma_\pm = 0.847$$

The molalities of Ag^+ (m_+) and Cl^- (m_-) are the same. m_+ is calculated from K_{sp}^o and $\gamma_{\pm}$:

$$K_{sp}^o = 1.6 \times 10^{-10} = a_+ a_- = \gamma_+ m_+ \gamma_- m_-$$

$$= \gamma_{\pm}^2 m_+^2$$

$$= (0.847)^2 m_+^2$$

$$m_+ = 1.5 \times 10^{-5}$$

Since the concentration of Ag is very small, its molarity has the same numerical value as its molality. Thus, $[Ag^+] = 1.5 \times 10^{-5} M$

(b) The ionic strength of the 0.020 M KCl ($\approx$ 0.020 m) solution is the same as the 0.020 M KNO_3 solution in part (a). However, when AgCl dissolves in this solution, the molalities of Ag^+ (m_+) and Cl^- (m_-) are no longer the same, as KCl contributes extra Cl^-. In fact, $m_- = m_+ + 0.020$ m.

$$K_{sp}^o = 1.6 \times 10^{-10} = a_+ a_- = \gamma_+ m_+ \gamma_- m_-$$

$$= \gamma_{\pm}^2 m_+ m_-$$

$$= (0.847)^2 m_+ (m_+ + 0.020)$$

$$m_+ (m_+ + 0.020) = 2.23 \times 10^{-10}$$

Since m_+ is expected to be much smaller than 0.020, $m_+ + 0.020 \approx 0.020$. Therefore,

$$m_+ (0.020) = 2.23 \times 10^{-10}$$

$$m_+ = 1.1 \times 10^{-8}$$

Indeed, m_+ is insignificant compared with 0.020.

Since the concentration of Ag^+ is very small, its molarity has the same numerical value as its molality. Thus, $[Ag^+] = 1.1 \times 10^{-8} M$

8.10 Referring to Problem 8.9, calculate ΔG^o for the process

$$AgCl(s) \rightleftharpoons Ag^+(aq) + Cl^-(aq)$$

to yield a saturated solution at 298 K. (*Hint*: Use the well-known equation $\Delta G^o = -RT \ln K$.)

The value of K for this reaction is the thermodynamic solubility product, K_{sp}^o, for AgCl from Problem 8.9.

$$\Delta G^o = -RT \ln K = -\left(8.314 \, J \, K^{-1} \, mol^{-1}\right) (298 \, K) \ln 1.6 \times 10^{-10} = 5.6 \times 10^4 \, J \, mol^{-1}$$

8.11 The apparent solubility products of CdS and CaF_2 at 25°C are 3.8×10^{-29} and 4.0×10^{-11}, respectively. Calculate the solubility (g/100 g of solution) of these compounds.

CdS

The reaction corresponding to the dissolution of CdS in water is

$$CdS(s) \rightleftharpoons Cd^{2+}(aq) + S^{2-}(aq)$$

Let the solubility of CdS be x M. The concentrations of Cd^{2+} and S^{2-} are therefore also x M.

$$K_{sp} = 3.8 \times 10^{-29} = [Cd^{2+}][S^{2-}] = x \cdot x$$

$$x = 6.16 \times 10^{-15}$$

Assuming the density of the solution to be 1 kg L^{-1},

$$\text{Solubility of CdS} = \left(6.16 \times 10^{-15} \text{ mol L}^{-1}\right) \left(\frac{144.5 \text{ g}}{1 \text{ mol}}\right) \left(\frac{1 \text{ L}}{1000 \text{ g}}\right)$$

$$= 8.90 \times 10^{-13} \text{ g/1000 g of solution}$$

$$= 8.9 \times 10^{-14} \text{ g/100 g of solution}$$

CaF$_2$

The reaction corresponding to the dissolution of CaF_2 in water is

$$CaF_2(s) \rightleftharpoons Ca^{2+}(aq) + 2F^-(aq)$$

Let the solubility of CaF_2 be x M. The concentrations of Ca^{2+} and F^- are therefore x M and $2x$ M, respectively.

$$K_{sp} = 4.0 \times 10^{-11} = [Ca^{2+}][F^-]^2 = x\,(2x)^2 = 4x^3$$

$$x = 2.15 \times 10^{-4}$$

Assuming the density of the solution to be 1 kg L^{-1},

$$\text{Solubility of CaF}_2 = \left(2.15 \times 10^{-4} \text{ mol L}^{-1}\right) \left(\frac{78.08 \text{ g}}{1 \text{ mol}}\right) \left(\frac{1 \text{ L}}{1000 \text{ g}}\right)$$

$$= 1.68 \times 10^{-2} \text{ g/1000 g of solution}$$

$$= 1.7 \times 10^{-3} \text{ g/100 g of solution}$$

8.12 Oxalic acid, $(COOH)_2$, is a poisonous compound present in many plants and vegetables, including spinach. Calcium oxalate is only slightly soluble in water ($K_{sp} = 3.0 \times 10^{-9}$ at 25°C) and its ingestion can result in kidney stones. Calculate **(a)** the apparent and thermodynamic solubility of calcium oxalate in water, and **(b)** the concentrations of calcium and oxalate ions in a 0.010 M $Ca(NO_3)_2$ solution. Assume ideal behavior in **(b)**.

(a) The reaction corresponding to the dissolution of calcium oxalate, CaOxa, in water is

$$CaOxa(s) \rightleftharpoons Ca^{2+}(aq) + Oxa^{2-}(aq)$$

The apparent solubility is readily calculated from the given (apparent) solubility product and the stoichiometry of the dissociation. Let the solubility of CaOxa be x M, then $[Ca^{2+}] = [Oxa^{2-}] = x$.

$$K_{sp} = [Ca^{2+}][Oxa^{2-}] = x^2 = 3.0 \times 10^{-9}$$

$$x = 5.48 \times 10^{-5} M = 5.5 \times 10^{-5} \, M$$

The thermodynamic solubility is determined from the thermodynamic solubility product, $K_{sp}^{\circ} = \gamma_{\pm}^2 K_{sp}$, which is calculated from the apparent concentrations and the mean ionic activity obtained from the Debye–Hückel limiting law.

Since this is a very dilute solution, the molarities of the ionic species are numerically equal to the molalities required in determining the ionic strength.

$$I = \frac{1}{2} \sum_i m_i z_i^2$$

$$= \frac{1}{2} \left[(5.48 \times 10^{-5} \, m) \, (2)^2 + (5.48 \times 10^{-5} \, m) \, (2)^2 \right]$$

$$= 2.19 \times 10^{-4} \, m$$

This is then used to determine the mean ionic activity

$$\log \gamma_{\pm} = -0.509 \, |z_+ z_-| \sqrt{I}$$

$$= -0.509 \, |(2)(-2)| \sqrt{(2.19 \times 10^{-4})}$$

$$= -3.01 \times 10^{-2}$$

$$\gamma_{\pm} = 0.933$$

Thus, the thermodynamic solubility constant and thermodynamic solubility, x°, are

$$K_{sp}^{\circ} = (0.933)^2 \, (3.0 \times 10^{-9}) = 2.61 \times 10^{-9} = 2.6 \times 10^{-9} = (x^{\circ})^2$$

$$x^{\circ} = \sqrt{2.61 \times 10^{-9}} = 5.1 \times 10^{-5} \, M$$

(b) The solubility in the $Ca(NO_3)_2$ solution is decreased due to the presence of Ca^{2+} ions. If the concentration of dissolved calcium oxalate is taken as x, then in the solution $[Ca^{2+}] = 0.010 + x$ and $[Oxa^{2-}] = x$.

$$K_{sp} = [Ca^{2+}][Oxa^{2-}] = 3.0 \times 10^{-9}$$

$$(0.010 + x) \, x = 3.0 \times 10^{-9}$$

$$x = 3.0 \times 10^{-7} \, M = [Oxa^{2-}]$$

$$[Ca^{2+}] = 0.010 M + 3.0 \times 10^{-7} \, M = 0.010 \, M$$

8.13 Express mean activity, mean activity coefficient, and mean molality in terms of the individual ionic quantities (a_+, a_-, γ_+, γ_-, m_+, and m_-) for the following electrolytes: KI, $SrSO_4$, $CaCl_2$, Li_2CO_3, $K_3Fe(CN)_6$ and $K_4Fe(CN)_6$.

	ν_+	ν_-	ν	$a_\pm$	$\gamma_\pm$	$m_\pm$
KI	1	1	2	$(a_+a_-)^{1/2}$	$(\gamma_+\gamma_-)^{1/2}$	$(m_+m_-)^{1/2}$
$SrSO_4$	1	1	2	$(a_+a_-)^{1/2}$	$(\gamma_+\gamma_-)^{1/2}$	$(m_+m_-)^{1/2}$
$CaCl_2$	1	2	3	$(a_+a_-^2)^{1/3}$	$(\gamma_+\gamma_-^2)^{1/3}$	$(m_+m_-^2)^{1/3}$
Li_2CO_3	2	1	3	$(a_+^2a_-)^{1/3}$	$(\gamma_+^2\gamma_-)^{1/3}$	$(m_+^2m_-)^{1/3}$
$K_3Fe(CN)_6$	3	1	4	$(a_+^3a_-)^{1/4}$	$(\gamma_+^3\gamma_-)^{1/4}$	$(m_+^3m_-)^{1/4}$
$K_4Fe(CN)_6$	4	1	5	$(a_+^4a_-)^{1/5}$	$(\gamma_+^4\gamma_-)^{1/5}$	$(m_+^4m_-)^{1/5}$

8.14 Calculate the ionic strength and the mean activity coefficient for the following solutions at 298 K: **(a)** 0.10 m NaCl, **(b)** 0.010 m $MgCl_2$, and **(c)** 0.10 m $K_4Fe(CN)_6$.

The ionic strength can be obtained from the equation

$$I = \frac{1}{2} \sum_i m_i z_i^2$$

and subsequently the mean activity from the Debye–Hückel limiting law

$$\log \gamma_\pm = -0.509 \left| z_+ z_- \right| \sqrt{I}$$

(a) 0.10 m NaCl: $z_+ = 1$, $z_- = -1$, $m_+ = 0.10\ m$, $m_- = 0.10\ m$

$$I = \frac{1}{2}\left[(0.10\ m)(1)^2 + (0.10\ m)(-1)^2\right] = 0.10\ m$$

$$\log \gamma_\pm = -0.509 \left|(1)(-1)\right| \sqrt{0.10} = -0.161$$

$$\gamma_\pm = 0.69$$

(b) 0.010 m $MgCl_2$: $z_+ = 2$, $z_- = -1$, $m_+ = 0.010\ m$, $m_- = 0.020\ m$

$$I = \frac{1}{2}\left[(0.010\ m)(2)^2 + (0.020\ m)(-1)^2\right] = 0.030\ m$$

$$\log \gamma_\pm = -0.509 \left|(2)(-1)\right| \sqrt{0.030} = -0.176$$

$$\gamma_\pm = 0.67$$

(c) 0.10 m $K_4Fe(CN)_6$: $z_+ = 1$, $z_- = -4$, $m_+ = 0.40\ m$, $m_- = 0.10\ m$

$$I = \frac{1}{2}\left[(0.40\ m)(1)^2 + (0.10\ m)(-4)^2\right] = 1.0\ m$$

$$\log \gamma_\pm = -0.509 \left|(1)(-4)\right| \sqrt{1.0} = -2.04$$

$$\gamma_\pm = 9.1 \times 10^{-3}$$

8.15   The mean activity coefficient of a 0.010 m H_2SO_4 solution is 0.544. What is its mean ionic activity?

First calculate the mean molality. For the H_2SO_4 solution, $v_+ = 2$, $v_- = 1$, $v = 3$, $m_+ = 0.020\ m$, $m_- = 0.010\ m$.

$$m_\pm = \left[m_+^{v_+} m_-^{v_-} \right]^{1/v} = \left[(0.020\ m)^2\ (0.010\ m) \right]^{1/3} = 1.59 \times 10^{-2}\ m$$

Therefore, the mean ionic activity is

$$a_\pm = \gamma_\pm m_\pm = (0.544)\left(1.59 \times 10^{-2} \right) = 8.6 \times 10^{-3}$$

8.16   A 0.20 m $Mg(NO_3)_2$ solution has a mean ionic activity coefficient of 0.13 at 25°C. Calculate the mean molality, the mean ionic activity, and the activity of the compound.

For the $Mg(NO_3)_2$ solution, $v_+ = 1$, $v_- = 2$, $v = 3$, $m_+ = 0.20\ m$, $m_- = 0.40\ m$. The mean molality is

$$m_\pm = \left[m_+^{v_+} m_-^{v_-} \right]^{1/v} = \left[(0.20\ m)\ (0.40\ m)^2 \right]^{1/3} = 0.317\ m = 0.32\ m$$

The mean ionic activity is

$$a_\pm = \gamma_\pm m_\pm = (0.13)\ (0.317) = 0.0412 = 0.041$$

The activity is

$$a = a_\pm^v = (0.0412)^3 = 7.0 \times 10^{-5}$$

8.17   The Debye–Hückel limiting law is more reliable for 1:1 electrolytes than for 2:2 electrolytes. Explain.

The Debye–Hückel limiting law assumes a dilute solution in which the ions are on average far apart from each other and the influence of electrostatic forces is small. For ions carrying a higher charge, these forces are more significant.

8.18   In theory, the size of the ionic atmosphere is $1/\kappa$, called the Debye radius, and κ is given by:

$$\kappa = \left(\frac{e^2 N_A}{\epsilon_0 \epsilon k_B T} \right)^{1/2} \sqrt{I}$$

where e is the electronic charge, N_A Avogadro's constant, ϵ_0 the permittivity of vacuum $(8.854 \times 10^{-12}\ C^2\ N^{-1}\ m^{-2})$, ϵ the dielectric constant of the solvent, k_B the Boltzmann constant, T the absolute temperature, and I the ionic strength (see the physical chemistry texts listed in Chapter 1). Calculate the Debye radius in a $0.010\ m$ aqueous Na_2SO_4 solution at $25°C$.

The ionic strength of a $0.010\ m\ Na_2SO_4$ solution is

$$I = \frac{1}{2}\sum_i m_i z_i^2$$

$$= \frac{1}{2}\left[(0.020\ m)\,(1)^2 + (0.010\ m)\,(-2)^2\right]$$

$$= 0.030\ m$$

Since this is a dilute solution, the ionic strength in $mol\ L^{-1}$ can be taken as numerically equal to that in $mol\ kg^{-1}$, or $0.030\ M$.

For water, $\epsilon = 78.54$. Thus,

$$\kappa = \left(\frac{e^2 N_A}{\epsilon_0 \epsilon k_B T}\right)^{1/2}\sqrt{I}$$

$$= \left[\frac{\left(1.602 \times 10^{-19}\ C\right)^2 \left(6.022 \times 10^{23}\ mol^{-1}\right)}{\left(8.854 \times 10^{-12}\ C^2\ N^{-1}\ m^{-2}\right)(78.54)\left(1.381 \times 10^{-23}\ J\ K^{-1}\right)(298\ K)}\right]^{1/2}$$

$$\times \sqrt{\left(0.030\ mol\ L^{-1}\right)\left(\frac{1000\ L}{1\ m^3}\right)}$$

$$= 4.025 \times 10^8\ m^{-1}$$

$$\frac{1}{\kappa} = 2.48 \times 10^{-9}\ m = 24.8\ Å$$

8.19 Explain why it is preferable to take the geometric mean rather than the arithmetic mean when defining mean activity, mean molality, and mean activity coefficient.

In taking the logarithm of the geometric mean, defined as the square root of the product of two quantities (or more generally as the nth root of the product of n quantities), the individual terms separate out according to the laws of logarithms. Note that the logarithm of a geometric mean is the arithmetic mean of the logarithms.

$$\log\left(\prod_{i=1}^{n} x_i\right)^{1/n} = \frac{1}{n}\sum_{i=1}^{n}\log x_i$$

This would not be possible if the arithmetic mean were used initially.

8.20 The freezing-point depression of a $0.010\ m$ acetic acid solution is $0.0193\ K$. Calculate the degree of dissociation for acetic acid at this concentration.

The degree of dissociation is related to the van't Hoff factor, which can be obtained from the freezing point depression.

$$\Delta T = K_f \left(i m_2 \right)$$

$$i = \frac{\Delta T}{K_f m_2} = \frac{0.0193 \text{ K}}{\left(1.86 \text{ K mol}^{-1} \text{ kg} \right) \left(0.010 \text{ mol kg}^{-1} \right)} = 1.04$$

The degree of dissociation of the acetic acid is

$$\alpha = \frac{i - 1}{\nu - 1} = \frac{1.04 - 1}{2 - 1} = 0.04 = 4\%$$

8.21 A 0.010 m aqueous solution of the ionic compound $Co(NH_3)_5Cl_3$ has a freezing point depression of 0.0558 K. What can you conclude about its structure? Assume the compound is a strong electrolyte.

The van't Hoff factor is

$$i = \frac{\Delta T}{K_f m_2} = \frac{0.0558 \text{ K}}{\left(1.86 \text{ K mol}^{-1} \text{ kg} \right) \left(0.010 \text{ mol kg}^{-1} \right)} = 3.0$$

There are 3 particles in the solution per 1 particle before dissociation. Thus, the compound is probably $[Co(NH_3)_5Cl]Cl_2$, which on dissolution, dissociates into $[Co(NH_3)_5Cl]^{2+}$ and $2Cl^-$.

8.22 The osmotic pressure of blood plasma is about 7.5 atm at 37°C. Estimate the total concentration of dissolved species and the freezing point of blood plasma.

The total concentration of dissolved species is

$$c = \frac{\pi}{RT} = \frac{7.5 \text{ atm}}{\left(0.08206 \text{ L atm K}^{-1} \text{ mol}^{-1} \right) \left(310 \text{ K} \right)} = 0.295 \, M = 0.30 \, M$$

Assuming that molality can be approximated by molarity, the freezing point depression caused by the dissolved species is

$$\Delta T = K_f m_2 = \left(1.86 \text{ K mol}^{-1} \text{ kg} \right) \left(0.295 \text{ mol kg}^{-1} \right) = 0.55 \text{ K}$$

Therefore, blood plasma freezes at −0.55°C or 272.60 K.

8.23 Calculate the ionic strength of a 0.0020 m aqueous solution of $MgCl_2$ at 298 K. Use the Debye–Hückel limiting law to estimate **(a)** the activity coefficients of the Mg^{2+} and Cl^- ions in this solution and **(b)** the mean ionic activity coefficient of these ions.

The ionic strength of the solution is

$$I = \frac{1}{2} \sum_i m_i z_i^2 = \frac{1}{2} \left[(0.0020 \ m) \ (2)^2 + (0.0040 \ m) \ (-1)^2 \right] = 0.0060 \ m$$

(a) The activity coefficients of Mg^{2+} and Cl^- can be evaluated using

$$\log \gamma_i = -0.509 z_i^2 \sqrt{I}$$

For Mg^{2+},

$$\log \gamma_+ = -0.509 \ (2)^2 \ \sqrt{0.0060} = -0.158$$

$$\gamma_+ = 0.695 = 0.70$$

For Cl^-,

$$\log \gamma_- = -0.509 \ (-1)^2 \ \sqrt{0.0060} = -3.94 \times 10^{-2}$$

$$\gamma_- = 0.913 = 0.91$$

(b) The mean ionic activity coefficient is

$$\gamma_\pm = \left(\gamma_+^{v_+} \gamma_-^{v_-} \right)^{1/v} = \left[(0.695) \ (0.913)^2 \right]^{1/3} = 0.83$$

8.24   Referring to Figure 8.13, calculate the osmotic pressure for the following cases at 298 K:
(a) The left compartment contains 200 g of hemoglobin in 1 liter of solution; the right compartment contains pure water. (b) The left compartment contains the same hemoglobin solution as in part (a); the right compartment initially contains 6.0 g of NaCl in 1 liter of solution. Assume that the pH of the solution is such that the hemoglobin molecules are in the Na^+Hb^- form. (The molar mass of hemoglobin is 65,000 g mol^{-1}).

(a) To maintain electrical neutrality, all the Na^+ ions remain in the left compartment with the Hb^- anionic form of the protein. The total concentration is twice the hemoglobin concentration.

$$c = 2 \left[\frac{200 \ g \left(\frac{1 \ mol}{65000 \ g \ mol^{-1}} \right)}{1 \ L} \right] = 2 \left(3.077 \times 10^{-3} \ M \right) = 6.154 \times 10^{-3} \ M$$

From the osmotic pressure equation,

$$\pi = cRT$$

$$= \left(6.154 \times 10^{-3} \ M \right) \left(0.08206 \ L \ atm \ K^{-1} \ mol^{-1} \right) (298 \ K)$$

$$= 0.150 \ atm$$

(b) Na^+ and Cl^- will diffuse through the membrane from right to left, maintaining electrical neutrality, until the chemical potentials of NaCl on both sides of the membrane are equal.

According to Equation 8.35, the concentration of the NaCl, x, that diffuses from right to left depends on the concentration of NaCl initially in the right compartment, b, and the concentration of the nondiffusible ion, c.

$$b = \frac{\frac{6.0\ g}{58.44\ g\,mol^{-1}}}{1\ L} = 0.103\ M$$

$$c = 3.077 \times 10^{-3}\ M \text{ (from above)}$$

$$x = \frac{b^2}{c + 2b} = \frac{(0.103\ M)^2}{3.077 \times 10^{-3}M + 2\,(0.103\ M)} = 0.0507\ M$$

The osmotic pressure is determined by the difference between the number of particles in the left compartment and that in the right compartment.

$$c = \left([Hb^-] + [Na^+] + [Cl^-]\right)_L - \left([Na^+] + [Cl^-]\right)_R$$

$$= \{(3.077 \times 10^{-3}\ M + 3.077 \times 10^{-3}\ M + 0.0507\ M + 0.0507\ M)$$

$$- [(0.103\ M - 0.0507\ M) + (0.103\ M - 0.0507\ M)]\}$$

$$= 2.95 \times 10^{-3}\ M$$

Using the osmotic pressure equation,

$$\pi = cRT$$

$$= (2.95 \times 10^{-3}\ M)\,(0.08206\ L\,atm\,K^{-1}\ mol^{-1})\,(298\ K)$$

$$= 0.072\ atm$$

8.25 From the following data, calculate the heat of solution for KI:

	NaCl	NaI	KCl	KI
Lattice energy/kJ·mol^{-1}	787	700	716	643
Heat of solution/kJ·mol^{-1}	3.8	−5.1	17.1	?

Begin by using $\Delta_{soln}H = U_0 + \Delta_{hydr}H$, where U_0 is the lattice energy.

$$Na^+(g) + Cl^-(g) \longrightarrow Na^+(aq) + Cl^-(aq)$$
$$\Delta_{hydr}H = 3.8\ kJ\,mol^{-1} - 787\ kJ\,mol^{-1} = -783.2\ kJ\,mol^{-1} \quad (8.25.1)$$

$$Na^+(g) + I^-(g) \longrightarrow Na^+(aq) + I^-(aq)$$
$$\Delta_{hydr}H = -5.1\ kJ\,mol^{-1} - 700\ kJ\,mol^{-1} = -705.1\ kJ\,mol^{-1} \quad (8.25.2)$$

$$K^+(g) + Cl^-(g) \longrightarrow K^+(aq) + Cl^-(aq)$$
$$\Delta_{hydr}H = 17.1\ kJ\,mol^{-1} - 716\ kJ\,mol^{-1} = -698.9\ kJ\,mol^{-1} \quad (8.25.3)$$

Taking equations 8.25.2 plus 8.25.3 minus 8.25.1 results in

$$K^+(g) + I^-(g) \longrightarrow K^+(aq) + I^-(aq)$$

$$\Delta_{hydr}H = -705.1\ kJ\,mol^{-1} - 698.9\ kJ\,mol^{-1} + 783.2\ kJ\,mol^{-1} = -620.8\ kJ\,mol^{-1}$$

Combine this last result with the given value of the lattice energy to arrive at the desired heat of solution.

$$\Delta_{soln}H = U_0 + \Delta_{hydr}H$$

$$= 643 \text{ kJ mol}^{-1} - 620.8 \text{ kJ mol}^{-1}$$

$$= 22.2 \text{ kJ mol}^{-1}$$

8.26 The concentrations of K^+ and Na^+ ions in the intracellular fluid of a nerve cell are approximately 400 mM and 50 mM, respectively, but in the extracellular fluid the K^+ and Na^+ concentrations are 20 mM and 440 mM, respectively. Given that the electric potential inside the cell is −70 mV relative to the outside, calculate the Gibbs energy change for the transfer of 1 mole of each type of ion against the concentration gradient at 37°C.

In each case the Gibbs energy change going against the concentration gradient is found using

$$\Delta G = RT \ln \frac{[K^+]_{higher}}{[K^+]_{lower}} + zF\Delta V.$$

Note that $1 \text{ V} = 1 \text{ J C}^{-1}$.

For K^+:

$$\Delta G = RT \ln \frac{[K^+]_{in}}{[K^+]_{out}} + zF\Delta V$$

$$= (8.314 \text{ J K}^{-1} \text{ mol}^{-1}) (310 \text{ K}) \ln \frac{400 \text{ mM}}{20 \text{ mM}} + (1) (96500 \text{ C mol}^{-1}) (-0.070 \text{ V})$$

$$= 966 \text{ J mol}^{-1} = 0.97 \text{ kJ mol}^{-1}$$

For Na^+:

$$\Delta G = RT \ln \frac{[Na^+]_{out}}{[Na^+]_{in}} + zF\Delta V$$

$$= (8.314 \text{ J K}^{-1} \text{ mol}^{-1}) (310 \text{ K}) \ln \frac{440 \text{ mM}}{50 \text{ mM}} + (1) (96500 \text{ C mol}^{-1}) (+0.070 \text{ V})$$

$$= 1.24 \times 10^4 \text{ J mol}^{-1} = 12 \text{ kJ mol}^{-1}$$

Although the concentration gradients are similar, it is easier to transport K^+ ions into the cell, since they are going towards a negative electric potential, than to transport Na^+ ions outside the cell towards a positive potential.

8.27 In this chapter (see Figures 8.2, and 8.11) and in Chapter 7 (see π measurements in Figure 7.19) we extrapolated concentration-dependent values to zero solute concentration. Explain what these extrapolated values mean physically and why they differ from the value obtained for the pure solvent.

Extrapolating a concentration-dependent quantity to zero solute concentration corresponds to the value the quantity would take in the absence of solute-solute interactions. This infinitely dilute solution is not the same as pure solvent, since the solute molecules are still there. The physical quantities determined in such extrapolations correspond to those of an ideal solution.

8.28 **(a)** The root cells of plants contain a solution that is hypertonic in relation to water in the soil. Thus water can move into the roots by osmosis. Explain why salts (NaCl and $CaCl_2$) spread on roads to melt ice can be harmful to nearby trees. **(b)** Just before urine leaves the human body, the collecting ducts in the kidney (which contain the urine) pass through a fluid whose salt concentration is considerably greater than is found in the blood and tissues. Explain how this action helps to conserve water in the body.

(a) When the road salts get into the soil water, the concentration there becomes greater than the concentration in the plant root cells, making the solution in the cells hypotonic. Thus the osmotic pressure difference will be reversed, and water will flow out of the plant roots into the soil. This action is harmful and potentially fatal to the plant. Even if the effect is not as severe, there will be a reduction in the osmotic pressure, which will limit the height to which the water can rise in the plant.

(b) The high-salt fluid is hypertonic relative to urine. Thus, some of the water in the urine flows into the fluid by osmosis. This action concentrates the waste products in the urine and helps to conserve water in the body.

8.29 A very long pipe is capped at one end with a semipermeable membrane. How deep (in meters) must the pipe be immersed into the sea for fresh water to begin passing through the membrane? Assume seawater is at 20°C and treat it as a 0.70 M NaCl solution. The density of seawater is 1.03 g cm^{-3}.

The desired process is for (fresh) water to move from a more concentrated solution (seawater) to pure solvent. This is an example of reverse osmosis, and external pressure must be provided to overcome the osmotic pressure of the sea water. The source of the pressure here is the water pressure, which increases with increasing depth. The osmotic pressure of the sea water is

$$\pi = cRT$$

$$= (0.70\ M) \left(0.08206\ \text{L atm K}^{-1}\ \text{mol}^{-1}\right) (293\ \text{K})$$

$$= 16.8\ \text{atm}$$

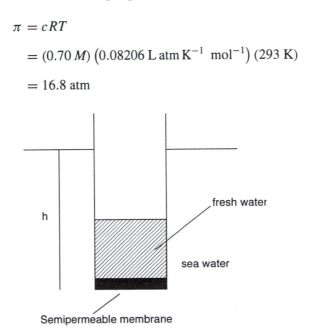

The water pressure at the membrane depends on the height of the sea above it, *i.e.* the depth, $P = \rho g h$, and fresh water will begin to pass through the membrane when $P = \pi$.

$$h = \frac{\pi}{g\rho}$$

$$= \frac{16.8 \text{ atm}}{\left(9.81 \text{ m s}^{-1}\right)\left(1.03 \text{ g cm}^{-3}\right)} \left(\frac{1 \text{ m}^3}{100 \text{ cm}^3}\right)^3 \left(\frac{1000 \text{ g}}{1 \text{ kg}}\right) \left(\frac{1.01325 \times 10^5 \text{ Pa}}{1 \text{ atm}}\right)$$

$$= 168 \text{ m}$$

8.30 **(a)** Using the Debye–Hückel limiting law, calculate the value of $\gamma_\pm$ for a 2.0×10^{-3} m Na_3PO_4 solution at 25°C. **(b)** Calculate the values of γ_+ and γ_- for the Na_3PO_4 solution, and show that they give the same value for $\gamma_\pm$ as that obtained in **(a)**.

For the Na_3PO_4 solution, $z_+ = 1$, $z_- = -3$, $m_+ = 6.0 \times 10^{-3}$ m, and $m_- = 2.0 \times 10^{-3}$ m.

The ionic strength of the solution is

$$I = \frac{1}{2}\sum_i m_i z_i^2 = \frac{1}{2}\left[\left(6.0 \times 10^{-3}m\right)(1)^2 + \left(2.0 \times 10^{-3}m\right)(-3)^2\right] = 1.2 \times 10^{-2} \, m$$

(a)

$$\log \gamma_\pm = -0.509\left|z_+ z_-\right|\sqrt{I} = -0.509\left|(1)(-3)\right|\sqrt{1.2 \times 10^{-2}} = -0.167$$

$$\gamma_\pm = 0.681 = 0.68$$

(b) The activity coefficients of Na^+ and PO_4^{3-} can be evaluated using

$$\log \gamma_i = -0.509 z_i^2 \sqrt{I}$$

For Na^+,

$$\log \gamma_+ = -0.509 (1)^2 \sqrt{1.2 \times 10^{-2}} = -5.58 \times 10^{-2}$$

$$\gamma_+ = 0.879 = 0.88$$

For PO_4^{3-},

$$\log \gamma_- = -0.509 (-3)^2 \sqrt{1.2 \times 10^{-2}} = -0.502$$

$$\gamma_- = 0.315 = 0.32$$

The mean ionic activity coefficient is

$$\gamma_\pm = \left(\gamma_+^{\nu_+} \gamma_-^{\nu_-}\right)^{1/\nu} = \left[(0.879)^3 (0.315)\right]^{1/4} = 0.680$$

which is the same as that obtained in **(a)**.

CHAPTER 9

Chemical Equilibrium

PROBLEMS AND SOLUTIONS

9.1 Equilibrium constants of gaseous reactions can be expressed in terms of pressures only (K_P), concentrations only (K_c), or mole fractions only (K_x). For the hypothetical reaction

$$aA(g) \rightleftharpoons bB(g)$$

derive the following relationships: **(a)** $K_P = K_c (RT)^{\Delta n} (P^\circ)^{-\Delta n}$ and **(b)** $K_P = K_x P^{\Delta n} (P^\circ)^{-\Delta n}$, where Δn is the difference in the number of moles of products and reactants, and P is the total pressure of the system. Assume ideal-gas behavior.

(a) The expression for K_P is

$$K_P = \frac{(P_B/P^\circ)^b}{(P_A/P^\circ)^a} = \frac{P_B^b}{P_A^a} (P^\circ)^{-\Delta n}$$

where $\Delta n = b - a$. Since A and B are ideal gases,

$$P_A = \frac{n_A}{V} RT = [A] RT$$

$$P_B = \frac{n_B}{V} RT = [B] RT$$

Substitute these results into the K_P expression.

$$K_P = \frac{([B] RT)^b}{([A] RT)^a} (P^\circ)^{-\Delta n}$$

$$= \frac{[B]^b}{[A]^a} (RT)^{\Delta n} (P^\circ)^{-\Delta n}$$

$$= K_c (RT)^{\Delta n} (P^\circ)^{-\Delta n}$$

(b) Substitute mole fractions for pressures in the K_P expression, using

$$P_A = x_A P$$

$$P_B = x_B P$$

Thus,

$$K_P = \frac{(x_B P)^b}{(x_A P)^a}(P^\circ)^{-\Delta n}$$

$$= \frac{x_B^b}{x_A^a}(P)^{\Delta n}(P^\circ)^{-\Delta n}$$

$$= K_x (P)^{\Delta n}(P^\circ)^{-\Delta n}$$

9.2 At 1024°C, the pressure of oxygen gas from the decomposition of copper (II) oxide (CuO) is 0.49 bar:

$$4CuO(s) \rightleftharpoons 2Cu_2O(s) + O_2(g)$$

(a) What is the value of K_P for the reaction? **(b)** Calculate the fraction of CuO that will decompose if 0.16 mole of it is placed in a 2.0-L flask at 1024°C. **(c)** What would the fraction be if a 1.0-mole sample of CuO were used? **(d)** What is the smallest amount of CuO (in moles) that would establish the equilibrium?

(a)

$$K_P = \frac{P_{O_2}}{P^\circ} = \frac{0.49\ \text{bar}}{1\ \text{bar}} = 0.49$$

(b) First calculate the number of moles of O_2 formed by the reaction, from which the number of moles of CuO decomposed is determined. Assume O_2 behaves ideally.

$$\text{Number of moles of } O_2 \text{ formed} = \frac{PV}{RT} = \frac{(0.49\ \text{bar})\left(\frac{1\ \text{atm}}{1.013\ \text{bar}}\right)(2.0\ \text{L})}{(0.08206\ \text{L atm K}^{-1}\ \text{mol}^{-1})(1297\ \text{K})}$$

$$= 9.09 \times 10^{-3}\ \text{mol}$$

$$\text{Number of moles of CuO decomposed} = (9.09 \times 10^{-3}\ \text{mol O}_2)\left(\frac{4\ \text{mol CuO}}{1\ \text{mol O}_2}\right) = 0.0364\ \text{mol}$$

Therefore,

$$\text{Fraction of CuO decomposed} = \frac{0.0364\ \text{mol}}{0.16\ \text{mol}} = 0.23$$

(c) If a 1.0-mole sample of CuO were used, the pressure of O_2 would still be the same (0.49 bar) and it would be due to the same quantity of O_2. (A pure solid does not affect the equilibrium position, as long as it is in excess at equilibrium.) The number of moles of CuO lost would still be 0.0364 mol. Therefore,

$$\text{Fraction of CuO decomposed} = \frac{0.0364\ \text{mol}}{1.0\ \text{mol}} = 0.036$$

(d) If the number of moles of CuO were less than 0.036 mole, the equilibrium could not be established because the pressure of O_2 would be less than 0.49 bar. Therefore, the smallest number of moles of CuO needed to establish equilibrium must be slightly greater than 0.036 mole.

9.3   Gaseous nitrogen dioxide is actually a mixture of nitrogen dioxide (NO_2) and dinitrogen tetroxide (N_2O_4). If the density of such a mixture is 2.3 g L^{-1} at 74°C and 1.3 atm, calculate the partial pressures of the gases and the value of K_P for the dissociation of N_2O_4.

First calculate the average molar mass of the mixture, $\mathcal{M}_{mix}$, assuming the mixture behaves ideally. Since

$$P = \frac{n_{mix}RT}{V} = \frac{m_{mix}RT}{\mathcal{M}_{mix}V} = \frac{m_{mix}}{V}\frac{RT}{\mathcal{M}_{mix}} = \frac{\rho_{mix}RT}{\mathcal{M}_{mix}},$$

the average molar mass is

$$\mathcal{M}_{mix} = \frac{\rho_{mix}RT}{P} = \frac{\left(2.3 \text{ g L}^{-1}\right)\left(0.08206 \text{ L atm K}^{-1}\text{ mol}^{-1}\right)(347 \text{ K})}{1.3 \text{ atm}} = 50.4 \text{ g mol}^{-1}$$

The average molar mass is also related to the mole fractions of the components and their molar masses:

$$x_{NO_2}\mathcal{M}_{NO_2} + x_{N_2O_4}\mathcal{M}_{N_2O_4} = 50.4 \text{ g mol}^{-1}$$

$$x_{NO_2}\left(46.01 \text{ g mol}^{-1}\right) + x_{N_2O_4}\left(92.02 \text{ g mol}^{-1}\right) = 50.4 \text{ g mol}^{-1}$$

Since $x_{N_2O_4} = 1 - x_{NO_2}$, the above equation contains only 1 unknown and can be readily solved.

$$x_{NO_2}\left(46.01 \text{ g mol}^{-1}\right) + \left(1 - x_{NO_2}\right)\left(92.02 \text{ g mol}^{-1}\right) = 50.4 \text{ g mol}^{-1}$$

$$46.01x_{NO_2} + 92.02 - 92.02x_{NO_2} = 50.4$$

$$46.01x_{NO_2} = 41.62$$

$$x_{NO_2} = 0.905$$

$$x_{N_2O_4} = 1 - 0.905 = 0.095$$

The partial pressures of the gases are

$$P_{NO_2} = x_{NO_2}P = (0.905)(1.3 \text{ atm}) = 1.18 \text{ atm} = 1.2 \text{ atm}$$

$$P_{N_2O_4} = P - P_{NO_2} = 1.3 \text{ atm} - 1.18 \text{ atm} = 0.12 \text{ atm} = 0.1 \text{ atm}$$

The equilibrium constant for the dissociation reaction, $N_2O_4(g) \rightleftharpoons 2NO_2(g)$, is

$$K_P = \frac{\left(P_{NO_2}/P^\circ\right)^2}{P_{N_2O_4}/P^\circ} = \frac{\left[(1.18 \text{ atm})\left(\frac{1.013 \text{ bar}}{1 \text{ atm}}\right)/1 \text{ bar}\right]^2}{(0.12 \text{ atm})\left(\frac{1.013 \text{ bar}}{1 \text{ atm}}\right)/1 \text{ bar}} = 12$$

9.4 About 75% of the hydrogen produced for industrial use is produced by the *steam-reforming* process. This process is carried out in two stages called primary and secondary reforming. In the primary stage, a mixture of steam and methane at about 30 atm is heated over a nickel catalyst at 800°C to give hydrogen and carbon monoxide:

$$CH_4(g) + H_2O(g) \rightleftharpoons CO(g) + 3\,H_2(g) \qquad \Delta_r H^\circ = 206 \text{ kJ mol}^{-1}$$

The secondary stage is carried out at about 1000°C, in the presence of air, to convert the remaining methane to hydrogen:

$$CH_4(g) + \tfrac{1}{2}O_2(g) \rightleftharpoons CO(g) + 2H_2(g) \qquad \Delta_r H^\circ = 35.7 \text{ kJ mol}^{-1}$$

(a) What conditions of temperature and pressure would favor the formation of products in both the primary and secondary stages? **(b)** The equilibrium constant, K_c, for the primary stage is 18 at 800°C. **(i)** Calculate the value of K_p for the reaction. **(ii)** If the partial pressures of methane and steam were both 15 atm at the start, what would the pressures of all the gases be at equilibrium?

(a) Since both reactions are endothermic ($\Delta_r H^\circ$ is positive for each), according to Le Chatelier's principle the products would be favored at high temperatures. Indeed, the steam-reforming process is carried out at very high temperatures (between 800°C and 1000°C). It is interesting to note that in a plant that uses natural gas (methane) for both hydrogen generation and heating, about one-third of the gas is burned to maintain the high temperatures.

In each reaction there are more moles of products than reactants; therefore, products are favored at low pressures. In reality, the reactions are carried out at high pressures. The reason is that when the hydrogen gas produced is used captively (usually in the synthesis of ammonia), high pressure leads to higher yields of ammonia.

(b) (i) According to Problem 9.1,

$$K_P = K_c\,(RT)^{\Delta n}\left(P^\circ\right)^{-\Delta n}$$

In this equation, if P° is in bar, R has to be in L bar K^{-1} mol^{-1}:

$$R = \left(0.08206 \text{ L atm K}^{-1} \text{ mol}^{-1}\right)\left(\frac{1.01325 \text{ bar}}{1 \text{ atm}}\right) = 0.08315 \text{ L bar K}^{-1} \text{ mol}^{-1}$$

and $\Delta n = (1+3) - (1+1) = 2$. Therefore,

$$K_P = (18)\left[(0.08315)\,(1073)\right]^2 (1)^{-2} = 1.43 \times 10^5 = 1.4 \times 10^5$$

(b)(ii) The pressures need to be converted to bars in order to use the K_P expression where $P^\circ = 1$ bar. The partial pressures of the reactants are

$$(15 \text{ atm})\left(\frac{1.013 \text{ bar}}{1 \text{ atm}}\right) = 15.2 \text{ bar}$$

The initial and equilibrium pressures of all species are shown in the following.

	CH_4	$+$	H_2O	$\rightleftharpoons$	CO	$+$	$3H_2$	
Initial	15.2		15.2					bar
At equilibrium	$15.2 - x$		$15.2 - x$		x		$3x$	bar

$$K_P = \frac{(P_{CO}/P^\circ)\left(P_{H_2}/P^\circ\right)^3}{\left(P_{CH_4}/P^\circ\right)\left(P_{H_2O}/P^\circ\right)}$$

$$1.43 \times 10^5 = \frac{x\,(3x)^3}{(15.2 - x)^2} = \frac{27x^4}{(15.2 - x)^2}$$

Take the square root of K_P and the last term in the above expression.

$$378 = \frac{5.20x^2}{15.2 - x}$$

$$5.20x^2 + 378x - 5745.6 = 0$$

$$x = 12.9 \qquad \text{or} \qquad -85.6\ \text{(nonphysical)}$$

Therefore, at equilibrium,

$$P_{CH_4} = (15.2 - x)\ \text{bar} = 2\ \text{bar} = 2\ \text{atm}$$

$$P_{H_2O} = (15.2 - x)\ \text{bar} = 2\ \text{bar} = 2\ \text{atm}$$

$$P_{CO} = x\ \text{bar} = 13\ \text{bar} = 13\ \text{atm}$$

$$P_{H_2} = 3x\ \text{bar} = 39\ \text{bar} = 38\ \text{atm}$$

9.5 Consider the reaction

$$PCl_5(g) \rightleftharpoons PCl_3(g) + Cl_2(g)$$

for which $K_P = 1.05$ at 250°C. A quantity of 2.50 g of PCl_5 is placed in an evacuated flask of volume 0.500 L and heated to 250°C. **(a)** Calculate the pressure of PCl_5 if it did not dissociate. **(b)** Calculate the partial pressure of PCl_5 at equilibrium. **(c)** What is the total pressure at equilibrium? **(d)** What is the degree of dissociation of PCl_5? (The degree of dissociation is given by the fraction of PCl_5 that has undergone dissociation.)

(a) Assume PCl_5 behaves ideally. If it did not dissociate, its pressure would be

$$P_{PCl_5} = \frac{nRT}{V} = \frac{\left(\dfrac{2.50\ \text{g}}{208.22\ \text{g mol}^{-1}}\right)(0.08206\ \text{L atm K}^{-1}\ \text{mol}^{-1})(523.2\ \text{K})}{0.500\ \text{L}}$$

$$= 1.031\ \text{atm} = 1.03\ \text{atm}$$

(b) The partial pressure of PCl_5 at equilibrium can be determined using the K_P expression. Since $P^\circ = 1$ bar, the pressure of PCl_5 needs to be converted to bars:

$$(1.031\ \text{atm})\left(\frac{1.013\ \text{bar}}{1\ \text{atm}}\right) = 1.044\ \text{bar}$$

The initial and equilibrium pressures of all species are shown in the following.

$$PCl_5(g) \quad \rightleftharpoons \quad PCl_3(g) \quad + \quad Cl_2(g)$$

Initial	1.044			bar
At equilibrium	$1.044 - x$	x	x	bar

$$K_P = 1.05 = \frac{x^2}{1.044 - x}$$

$$x^2 + 1.05x - 1.096 = 0$$

$$x = 0.6462 \quad \text{or} \quad -1.696 \text{ (nonphysical)}$$

Therefore, at equilibrium,

$$P_{PCl_5} = (1.044 - x) \text{ bar} = 0.40 \text{ bar}$$

(c) The total pressure at equilibrium, P, is the sum of the partial pressures of the reactants and products at equilibrium.

$$P = P_{PCl_5} + P_{PCl_3} + P_{Cl_2} = [(1.044 - x) + x + x] \text{ bar} = (1.044 + x) \text{ bar} = 1.69 \text{ bar}$$

(d)

$$\text{The degree of dissociation of } PCl_5 = \frac{x \text{ bar}}{1.044 \text{ bar}} = \frac{0.6462 \text{ bar}}{1.044 \text{ bar}} = 0.619$$

9.6 The vapor pressure of mercury is 0.002 mmHg at 26°C. **(a)** Calculate the values of K_c and K_P for the process $Hg(l) \rightleftharpoons Hg(g)$. **(b)** A chemist breaks a thermometer and spills mercury onto the floor of a laboratory measuring 6.1 m long, 5.3 m wide, and 3.1 m high. Calculate the mass of mercury (in grams) vaporized at equilibrium and the concentration of mercury vapor in $mg \, m^{-3}$. Does this concentration exceed the safety limit of $0.05 \, mg \, m^{-3}$? (Ignore the volume of furniture and other objects in the laboratory.)

(a)

$$K_P = \frac{P_{Hg}}{P^\circ} = \frac{(0.002 \text{ mmHg})\left(\frac{1 \text{ atm}}{760 \text{ mmHg}}\right)\left(\frac{1.013 \text{ bar}}{1 \text{ atm}}\right)}{1 \text{ bar}} = 2.666 \times 10^{-6} = 2.67 \times 10^{-6}$$

From Problem 9.1,

$$K_c = \frac{K_P}{(RT)^{\Delta n} (P^\circ)^{-\Delta n}}$$

To use this expression, R has to be expressed in $L \, bar \, K^{-1} \, mol^{-1}$, which has been done in Problem 9.4. Δn is 1.

$$K_c = \frac{2.666 \times 10^{-6}}{[(0.08315)(299.2)]^1 (1)^{-1}} = 1.07 \times 10^{-7}$$

(b) The volume occupied by Hg vapor is the same as the volume of the room.

$$V = (6.1 \text{ m}) (5.3 \text{ m}) (3.1 \text{ m}) = (100 \text{ m}^3) \left(\frac{1000 \text{ L}}{1 \text{ m}^3} \right) = 1.00 \times 10^5 \text{ L}$$

At equilibrium, the vapor pressure of Hg is 0.002 mmHg. The number of moles of Hg vapor is calculated by using the ideal gas law.

$$n = \frac{PV}{RT} = \frac{(0.002 \text{ mmHg}) \left(\frac{1 \text{ atm}}{760 \text{ mmHg}} \right) (1.00 \times 10^5 \text{ L})}{(0.08206 \text{ L atm K}^{-1} \text{ mol}^{-1}) (299.2 \text{ K})} = 1.07 \times 10^{-2} \text{ mol}$$

Therefore, the mass of Hg vapor is

$$m = (1.07 \times 10^{-2} \text{ mol}) (200.6 \text{ g mol}^{-1}) = 2.15 \text{ g}$$

and the concentration of Hg is

$$\text{Concentration} = \frac{2.15 \text{ g}}{100 \text{ m}^3} = 2.2 \times 10^{-2} \text{ g m}^{-3} = 22 \text{ mg m}^{-3}$$

This concentration greatly exceeds the safety limit.

9.7   A quantity of 0.20 mole of carbon dioxide was heated to a certain temperature with an excess of graphite in a closed container until the following equilibrium was reached:

$$C(s) + CO_2(g) \rightleftharpoons 2CO(g)$$

Under these conditions, the average molar mass of the gases was 35 g mol^{-1}. **(a)** Calculate the mole fractions of CO and CO_2. **(b)** What is the value of K_P if the total pressure is 11 atm? (*Hint:* The average molar mass is the sum of the products of the mole fraction of each gas times its molar mass.)

(a) The average molar mass of the gases is related to the mole fractions of the gases and their molar masses.

$$x_{CO} M_{CO} + x_{CO_2} M_{CO_2} = 35 \text{ g mol}^{-1}$$

$$x_{CO} (28.01 \text{ g mol}^{-1}) + x_{CO_2} (44.01 \text{ g mol}^{-1}) = 35 \text{ g mol}^{-1}$$

Since $x_{CO_2} = 1 - x_{CO}$, the above equation contains only 1 unknown and can be solved readily.

$$x_{CO} (28.01 \text{ g mol}^{-1}) + (1 - x_{CO}) (44.01 \text{ g mol}^{-1}) = 35 \text{ g mol}^{-1}$$

$$28.01 x_{CO} + 44.01 - 44.01 x_{CO} = 35$$

$$16.00 x_{CO} = 9.0$$

$$x_{CO} = 0.56$$

$$x_{CO_2} = 1 - x_{CO} = 0.44$$

Therefore, the mole fractions of CO and CO_2 are 0.6 and 0.4, respectively.

(b) The partial pressures of CO and CO_2 have to be determined before the calculation of K_P.

$$P_{CO} = x_{CO}P = (0.56)(11 \text{ atm})\left(\frac{1.013 \text{ bar}}{1 \text{ atm}}\right) = 6.2 \text{ bar}$$

$$P_{CO_2} = x_{CO_2}P = (0.44)(11 \text{ atm})\left(\frac{1.013 \text{ bar}}{1 \text{ atm}}\right) = 4.9 \text{ bar}$$

The equilibrium constant is

$$K_P = \frac{(P_{CO}/P^\circ)^2}{P_{CO_2}/P^\circ} = \frac{6.2^2}{4.9} = 8$$

9.8 Consider the thermal decomposition of $CaCO_3$:

$$CaCO_3(s) \rightleftharpoons CaO(s) + CO_2(g)$$

The equilibrium vapor pressures of CO_2 are 22.6 mmHg at 700°C and 1829 mmHg at 950°C. Calculate the standard enthalpy of the reaction.

The van't Hoff equation,

$$\ln\frac{K_2}{K_1} = \frac{\Delta_r H^\circ}{R}\left(\frac{1}{T_1} - \frac{1}{T_2}\right),$$

is used to solve this problem. Since

$$K_P = \frac{P_{CO_2}}{P^\circ},$$

K_P is proportional to P_{CO_2}. Thus,

$$\ln\frac{K_2}{K_1} = \ln\frac{P_{CO_2,2}}{P_{CO_2,1}} = \frac{\Delta_r H^\circ}{R}\left(\frac{1}{T_1} - \frac{1}{T_2}\right)$$

$$\ln\frac{1829 \text{ mmHg}}{22.6 \text{ mmHg}} = \frac{\Delta_r H^\circ}{8.314 \text{ J K}^{-1}\text{ mol}^{-1}}\left(\frac{1}{973.2 \text{ K}} - \frac{1}{1223.2 \text{ K}}\right)$$

$$\Delta_r H^\circ = 1.74 \times 10^5 \text{ J mol}^{-1}$$

9.9 Consider the following reaction:

$$CO_2(g) + H_2(g) \rightleftharpoons CO(g) + H_2O(g)$$

The equilibrium constant is 0.534 at 960 K and 1.571 at 1260 K. What is the enthalpy of the reaction?

According to the van't Hoff equation,

$$\ln \frac{K_2}{K_1} = \frac{\Delta_r H^\circ}{R} \left(\frac{1}{T_1} - \frac{1}{T_2} \right)$$

$$\ln \frac{1.571}{0.534} = \frac{\Delta_r H^\circ}{8.314 \, \text{J K}^{-1} \, \text{mol}^{-1}} \left(\frac{1}{960 \, \text{K}} - \frac{1}{1260 \, \text{K}} \right)$$

$$\Delta_r H^\circ = 3.62 \times 10^4 \, \text{J mol}^{-1}$$

9.10 The vapor pressure of dry ice (solid CO_2) is 672.2 torr at $-80°C$ and 1486 torr at $-70°C$. Calculate the molar heat of sublimation of CO_2.

The van't Hoff equation

$$\ln \frac{K_2}{K_1} = \frac{\Delta_r H^\circ}{R} \left(\frac{1}{T_1} - \frac{1}{T_2} \right)$$

is used to solve this problem. The process is

$$CO_2(s) \rightleftharpoons CO_2(g),$$

and the equilibrium constant is

$$K_P = \frac{P_{CO_2}}{P^\circ}.$$

Since K_P is proportional to P_{CO_2}, the van't Hoff equation becomes

$$\ln \frac{K_2}{K_1} = \ln \frac{P_{CO_2,2}}{P_{CO_2,1}} = \frac{\Delta_r H^\circ}{R} \left(\frac{1}{T_1} - \frac{1}{T_2} \right)$$

$$\ln \frac{1486 \, \text{torr}}{672.2 \, \text{torr}} = \frac{\Delta_r H^\circ}{8.314 \, \text{J K}^{-1} \, \text{mol}^{-1}} \left(\frac{1}{193.2 \, \text{K}} - \frac{1}{203.2 \, \text{K}} \right)$$

$$\Delta_r H^\circ = 2.59 \times 10^4 \, \text{J mol}^{-1}$$

9.11 Nitric oxide from car exhaust is a primary air pollutant. Calculate the equilibrium constant for the reaction

$$N_2(g) + O_2(g) \rightleftharpoons 2NO(g)$$

at 25°C using the data listed in Appendix B. Assume that both $\Delta_r H^\circ$ and $\Delta_r S^\circ$ are temperature independent. Calculate the equilibrium constant at 1500°C, which is the typical temperature inside the cylinders of a car's engine after it has been running for some time.

Using the standard molar Gibbs energy of formation for the reactants and products, calculate $\Delta_r G°$ for the reaction at 25°C, from which the equilibrium constant at 25°C is obtained.

$$\Delta_r G° = 2\Delta_f \overline{G}° \left[NO(g)\right] - \Delta_f \overline{G}° \left[N_2(g)\right] - \Delta_f \overline{G}° \left[O_2(g)\right]$$

$$= 2\left(86.7 \text{ kJ mol}^{-1}\right) - 0 \text{ kJ mol}^{-1} - 0 \text{ kJ mol}^{-1}$$

$$= 173.4 \text{ kJ mol}^{-1}$$

Since $\Delta_r G° = -RT \ln K$,

$$\ln K = -\frac{\Delta_r G°}{RT} = -\frac{173.4 \times 10^3 \text{ J mol}^{-1}}{\left(8.314 \text{ J K}^{-1} \text{ mol}^{-1}\right)(298.2 \text{ K})} = -69.940$$

$$K = 4.221 \times 10^{-31} = 4.22 \times 10^{-31}$$

The equilibrium constant at 1500 K can be calculated using the van't Hoff equation once $\Delta_r H°$ for the reaction is known.

$$\Delta_r H° = 2\Delta_f \overline{H}° \left[NO(g)\right] - \Delta_f \overline{H}° \left[N_2(g)\right] - \Delta_f \overline{H}° \left[O_2(g)\right]$$

$$= 2\left(90.4 \text{ kJ mol}^{-1}\right) - 0 \text{ kJ mol}^{-1} - 0 \text{ kJ mol}^{-1}$$

$$= 180.8 \text{ kJ mol}^{-1}$$

Apply the van't Hoff equation, assuming $\Delta_r H°$ is temperature independent.

$$\ln \frac{K_2}{K_1} = \frac{\Delta_r H°}{R}\left(\frac{1}{T_1} - \frac{1}{T_2}\right)$$

$$\ln \frac{K_2}{4.221 \times 10^{-31}} = \frac{180.8 \times 10^3 \text{ J mol}^{-1}}{8.314 \text{ J K}^{-1} \text{ mol}^{-1}}\left(\frac{1}{298.2 \text{ K}} - \frac{1}{1773.2 \text{ K}}\right)$$

$$K_2 = 9.34 \times 10^{-5}$$

9.12   Calculate the value of $\Delta_r G°$ for each of the following equilibrium constants: 1.0×10^{-4}, 1.0×10^{-2}, 1.0, 1.0×10^2, 1.0×10^4 at 298 K.

$\Delta_r G°$ can be evaluated using $\Delta_r G° = -RT \ln K$.

K	$\Delta_r G°/\text{J·mol}^{-1}$
1.0×10^{-4}	2.3×10^4
1.0×10^{-2}	1.1×10^4
1.0	0
1.0×10^2	-1.1×10^4
1.0×10^4	-2.3×10^4

9.13 Use the data listed in Appendix B to calculate the equilibrium constant, K_P, for the synthesis of HCl at 298 K:

$$H_2(g) + Cl_2(g) \rightleftharpoons 2HCl(g)$$

What is the value of K_P if the equilibrium is expressed as

$$\tfrac{1}{2}H_2(g) + \tfrac{1}{2}Cl_2(g) \rightleftharpoons HCl(g)$$

For the reaction $H_2(g) + Cl_2(g) \rightleftharpoons 2HCl(g)$,

$$\Delta_r G^\circ = 2\Delta_f \overline{G}^\circ\,[HCl(g)] - \Delta_f \overline{G}^\circ\,[H_2(g)] - \Delta_f \overline{G}^\circ\,[Cl_2(g)]$$

$$= 2\left(-95.3\ \text{kJ mol}^{-1}\right) - 0\ \text{kJ mol}^{-1} - 0\ \text{kJ mol}^{-1}$$

$$= -190.6\ \text{kJ mol}^{-1}$$

The equilibrium constant K_P is evaluated from $\Delta_r G^\circ$.

$$\ln K_P = -\frac{\Delta_r G^\circ}{RT} = -\frac{-190.6 \times 10^3\ \text{J mol}^{-1}}{\left(8.314\ \text{J K}^{-1}\,\text{mol}^{-1}\right)(298\ \text{K})} = 76.93$$

$$K_P = 2.572 \times 10^{33} = 2.57 \times 10^{33}$$

For the reaction $\tfrac{1}{2}H_2(g) + \tfrac{1}{2}Cl_2(g) \rightleftharpoons HCl(g)$, the equilibrium constant is the square root of that calculated above, that is

$$K_P = \sqrt{2.572 \times 10^{33}} = 5.07 \times 10^{16}$$

9.14 The dissociation of N_2O_4 into NO_2 is 16.7% complete at 298 K and 1 atm:

$$N_2O_4(g) \rightleftharpoons 2NO_2(g)$$

Calculate the equilibrium constant and the standard Gibbs energy change for the reaction. [*Hint:* Let α be the degree of dissociation and show that $K_P = 4\alpha^2 P/(1 - \alpha^2)$, where P is the total pressure.]

The equilibrium constant is related to the degree of dissociation, α. The exact relation is derived as follows.

The dissociation reaction is

$$N_2O_4(g) \rightleftharpoons 2NO_2(g)$$

Let the initial number of moles of N_2O_4 be n. At equilibrium, $n\alpha$ moles of N_2O_4 dissociates, giving $2n\alpha$ moles of NO_2. The number of moles of N_2O_4 remaining is $n - n\alpha = n(1 - \alpha)$. The total number of moles of gases are $n(1 - \alpha) + 2n\alpha = n(1 + \alpha)$.

The equilibrium constant can be calculated once the partial pressures of the gases are known. The partial pressures are calculated from the mole fractions of the gases.

$$x_{N_2O_4} = \frac{n(1-\alpha)}{n(1+\alpha)} = \frac{1-\alpha}{1+\alpha}$$

$$x_{NO_2} = \frac{2n\alpha}{n(1+\alpha)} = \frac{2\alpha}{1+\alpha}$$

The partial pressures are

$$P_{N_2O_4} = x_{N_2O_4}P = \frac{1-\alpha}{1+\alpha}P$$

$$P_{NO_2} = x_{NO_2}P = \frac{2\alpha}{1+\alpha}P$$

The equilibrium constant can now be calculated using the partial pressures in bars. The total pressure is 1 atm, which is equivalent to 1.013 bar.

$$K_P = \frac{P_{NO_2}^2}{P_{N_2O_4}} = \frac{\left(\frac{2\alpha}{1+\alpha}P\right)^2}{\frac{1-\alpha}{1+\alpha}P} = \frac{4\alpha^2 P}{(1+\alpha)(1-\alpha)} = \frac{4\alpha^2}{1-\alpha^2}P = \frac{4(0.167)^2}{1-0.167^2}(1.013)$$

$$= 0.1162 = 0.116$$

The standard Gibbs energy change is

$$\Delta_r G^\circ = -RT \ln K = -(8.314\,\text{J K}^{-1}\,\text{mol}^{-1})(298\,\text{K})\ln 0.1162 = 5.33 \times 10^3\,\text{J mol}^{-1}$$

9.15   The standard Gibbs energies of formation of gaseous *cis*- and *trans*-2-butene are 67.15 kJ mol^{-1} and 64.10 kJ mol^{-1}, respectively. Calculate the ratio of equilibrium pressures of the gaseous isomers at 298 K.

For the reaction *cis*-2-butene(g) $\rightleftharpoons$ *trans*-2-butene(g),

$$\Delta_r G^\circ = \Delta_f \overline{G}^\circ\,[\text{\textit{trans}-2-butene}] - \Delta_f \overline{G}^\circ\,[\text{\textit{cis}-2-butene}]$$

$$= 64.10\,\text{kJ mol}^{-1} - 67.15\,\text{kJ mol}^{-1}$$

$$= -3.05\,\text{kJ mol}^{-1}$$

Since

$$K_P = \frac{P_{\text{\textit{trans}-2-butene}}}{P_{\text{\textit{cis}-2-butene}}}$$

The equilibrium constant gives the ratio of equilibrium pressures of the isomers. This ratio can be determined using $\Delta_r G^\circ$.

$$\ln K_P = -\frac{\Delta_r G^\circ}{RT} = -\frac{-3.05 \times 10^3 \text{ J mol}^{-1}}{\left(8.314 \text{ J K}^{-1} \text{ mol}^{-1}\right)(298 \text{ K})} = 1.231$$

$$K_P = \frac{P_{trans\text{-2-butene}}}{P_{cis\text{-2-butene}}} = 3.42$$

9.16 Consider the decomposition of calcium carbonate:

$$CaCO_3(s) \rightleftharpoons CaO(s) + CO_2(g)$$

(a) Write an equilibrium constant expression (K_P) for the reaction. **(b)** The rate of decomposition is slow until the partial pressure of carbon dioxide is equal to 1 bar. Calculate the temperature at which the decomposition becomes spontaneous. Assume that $\Delta_r H^\circ$ and $\Delta_r S^\circ$ are temperature independent. Use the data in Appendix B for your calculation.

(a)

$$K_P = \frac{P_{CO_2}}{P^\circ}$$

(b) When the reactants and products are in their standard states, the decomposition becomes spontaneous when $\Delta_r G^\circ < 0$. The temperature at which this occurs can be calculated using the relation

$$\Delta_r G^\circ = \Delta_r H^\circ - T \Delta_r S^\circ < 0$$

$\Delta_r H^\circ$ and $\Delta_r S^\circ$ are

$$\Delta_r H^\circ = \Delta_f \overline{H}^\circ [CaO(s)] + \Delta_f \overline{H}^\circ \left[CO_2(g)\right] - \Delta_f \overline{H}^\circ \left[CaCO_3(s)\right]$$

$$= -635.6 \text{ kJ mol}^{-1} + \left(-393.5 \text{ kJ mol}^{-1}\right) - \left(-1206.9 \text{ kJ mol}^{-1}\right)$$

$$= 177.8 \text{ kJ mol}^{-1}$$

$$\Delta_r S^\circ = \overline{S}^\circ [CaO(s)] + \overline{S}^\circ \left[CO_2(g)\right] - \overline{S}^\circ \left[CaCO_3(s)\right]$$

$$= 39.8 \text{ J K}^{-1} \text{ mol}^{-1} + 213.6 \text{ J K}^{-1} \text{ mol}^{-1} - 92.9 \text{ J K}^{-1} \text{ mol}^{-1}$$

$$= 160.5 \text{ J K}^{-1} \text{ mol}^{-1}$$

Therefore, for the reaction to become spontaneous,

$$\Delta_r H^\circ - T \Delta_r S^\circ = 177.8 \times 10^3 \text{ J mol}^{-1} - T \left(160.5 \text{ J K}^{-1} \text{ mol}^{-1}\right) < 0$$

$$T > \frac{177.8 \times 10^3 \text{ J mol}^{-1}}{160.5 \text{ J K}^{-1} \text{ mol}^{-1}}$$

$$T > 1108 \text{ K}$$

9.17   Use the data in Appendix B to calculate the equilibrium constant (K_P) for the following reaction at 25°C:

$$2SO_2(g) + O_2(g) \rightleftharpoons 2SO_3(g)$$

Calculate K_P for the reaction at 60°C **(a)** using the van't Hoff equation, that is, Equation 9.17; **(b)** using the Gibbs-Helmholtz equation, that is, Equation 6.15, to find $\Delta_r G°$ at 60°C and hence K_P at the same temperature; and **(c)** using $\Delta_r G° = \Delta_r H° - T\Delta_r S°$ to find $\Delta_r G°$ at 60°C and hence K_P at the same temperature. State the approximations employed in each case and compare your results. (*Hint*: From 6.15 you can derive the relationship

$$\frac{\Delta_r G_2}{T_2} - \frac{\Delta_r G_1}{T_1} = \Delta_r H \left(\frac{1}{T_2} - \frac{1}{T_1} \right)$$

At 25°C, K_P can be calculated from $\Delta_r G°$ of the reaction.

$$\Delta_r G° = 2\Delta_f \overline{G}° \left[SO_3(g) \right] - 2\Delta_f \overline{G}° \left[SO_2(g) \right] - \Delta_f \overline{G}° \left[O_2(g) \right]$$

$$= 2\left(-370.4 \text{ kJ mol}^{-1} \right) - 2\left(-300.1 \text{ kJ mol}^{-1} \right) - 0 \text{ kJ mol}^{-1}$$

$$= -140.6 \text{ kJ mol}^{-1}$$

$$\ln K = -\frac{\Delta_r G°}{RT} = -\frac{-140.6 \times 10^3 \text{ J mol}^{-1}}{\left(8.314 \text{ J K}^{-1} \text{ mol}^{-1} \right) (298.2 \text{ K})} = 56.71$$

$$K = 4.254 \times 10^{24} = 4.25 \times 10^{24}$$

(a) $\Delta_r H°$ is evaluated from the standard molar enthalpies of formation of the reactants and products.

$$\Delta_r H° = 2\Delta_f \overline{H}° \left[SO_3(g) \right] - 2\Delta_f \overline{H}° \left[SO_2(g) \right] - \Delta_f \overline{H}° \left[O_2(g) \right]$$

$$= 2\left(-395.2 \text{ kJ mol}^{-1} \right) - 2\left(-296.1 \text{ kJ mol}^{-1} \right) - 0 \text{ kJ mol}^{-1}$$

$$= -198.2 \text{ kJ mol}^{-1}$$

Assuming $\Delta_r H°$ is temperature independent, the van't Hoff equation can be used to calculate the equilibrium constant at 60°C.

$$\ln \frac{K_2}{K_1} = \frac{\Delta_r H°}{R} \left(\frac{1}{T_1} - \frac{1}{T_2} \right)$$

$$\ln \frac{K_2}{4.254 \times 10^{24}} = \frac{-198.2 \times 10^3 \text{ J mol}^{-1}}{8.314 \text{ J K}^{-1} \text{ mol}^{-1}} \left(\frac{1}{298.2 \text{ K}} - \frac{1}{333.2 \text{ K}} \right) = -8.397$$

$$K_2 = 9.59 \times 10^{20}$$

(b) The Gibbs-Helmholtz equation at the standard state is

$$\left[\frac{\partial \left(\Delta_r G°/T \right)}{\partial T} \right]_P = -\frac{\Delta_r H°}{T^2}$$

At constant P,

$$d\left(\frac{\Delta_r G^\circ}{T}\right) = -\frac{\Delta_r H^\circ}{T^2}dT$$

Integrate both sides of the equation while assuming $\Delta_r H^\circ$ is temperature independent.

$$\int_{\Delta_r G_1^\circ/T_1}^{\Delta_r G_2^\circ/T_2} d\left(\frac{\Delta_r G^\circ}{T}\right) = \int_{T_1}^{T_2} -\frac{\Delta_r H^\circ}{T^2}dT$$

$$\frac{\Delta_r G_2^\circ}{T_2} - \frac{\Delta_r G_1^\circ}{T_1} = \Delta_r H^\circ\left(\frac{1}{T_2} - \frac{1}{T_1}\right)$$

Now $\Delta_r G^\circ$ at 60°C, $\Delta_r G_2^\circ$, can be calculated.

$$\frac{\Delta_r G_2^\circ}{333.2\ \text{K}} - \frac{-140.6\ \text{kJ mol}^{-1}}{298.2\ \text{K}} = \left(-198.2\ \text{kJ mol}^{-1}\right)\left(\frac{1}{333.2\ \text{K}} - \frac{1}{298.2\ \text{K}}\right)$$

$$\Delta_r G_2^\circ = -133.8\ \text{kJ mol}^{-1}$$

The equilibrium constant at 60°C can now be calculated.

$$\ln K_2 = -\frac{\Delta_r G_2^\circ}{RT} = -\frac{-133.8 \times 10^3\ \text{J mol}^{-1}}{\left(8.314\ \text{J K}^{-1}\ \text{mol}^{-1}\right)(333.2\ \text{K})} = 48.30$$

$$K_2 = 9.47 \times 10^{20}$$

(c) The standard entropy of reaction is calculated from the molar entropies of the reactants and products.

$$\Delta_r S^\circ = 2\overline{S}^\circ\left[SO_3(g)\right] - 2\overline{S}^\circ\left[SO_2(g)\right] - \overline{S}^\circ\left[O_2(g)\right]$$

$$= 2\left(256.2\ \text{J K}^{-1}\ \text{mol}^{-1}\right) - 2\left(248.5\ \text{J K}^{-1}\ \text{mol}^{-1}\right) - 205.0\ \text{J K}^{-1}\ \text{mol}^{-1}$$

$$= -189.6\ \text{J K}^{-1}\ \text{mol}^{-1}$$

Assuming $\Delta_r H^\circ$ and $\Delta_r S^\circ$ are temperature independent, $\Delta_r G^\circ$ at 60°C, $\Delta_r G_2^\circ$, can be calculated.

$$\Delta_r G_2^\circ = \Delta_r H^\circ - T_2\Delta_r S^\circ = -198.2 \times 10^3\ \text{J mol}^{-1} - (333.2\ \text{K})\left(-189.6\ \text{J K}^{-1}\ \text{mol}^{-1}\right)$$

$$= -135.03 \times 10^3\ \text{J mol}^{-1}$$

Now K_2 can be calculated.

$$\ln K_2 = -\frac{\Delta_r G_2^\circ}{RT} = -\frac{-135.03 \times 10^3\ \text{J mol}^{-1}}{\left(8.314\ \text{J K}^{-1}\ \text{mol}^{-1}\right)(333.2\ \text{K})} = 48.74$$

$$K_2 = 1.47 \times 10^{21}$$

Methods **(a)** and **(b)** give nearly identical results, but method **(c)** is different by close to a factor of 1.5. The agreement between the first two methods is not unexpected, since the integration of the Gibbs-Helmholtz equation in part **(b)** gives

$$\frac{\Delta_r G_2^{\circ}}{T_2} - \frac{\Delta_r G_1^{\circ}}{T_1} = \frac{-RT_2 \ln K_2}{T_2} - \frac{-RT_1 \ln K_1}{T_1}$$

$$= -R \ln K_2 - R \ln K_1$$

$$= -R \ln \frac{K_2}{K_1}$$

$$= \Delta_r H^{\circ} \left(\frac{1}{T_2} - \frac{1}{T_1} \right)$$

which is completely equivalent to the van't Hoff equation of part **(a)**. The minor difference between the answers is due to differences in rounding. Both part **(a)** and **(b)** require only the one assumption that $\Delta_r H^{\circ}$ is temperature independent and equal to the value calculated at 298 K. Part **(c)** requires in addition to this same assumption, the second assumption that $\Delta_r S^{\circ}$ is temperature independent *and* equal to the value calculated at 298 K. Although the first assumption implies the temperature independence of $\Delta_r S^{\circ}$ as noted in the text, the value of $\Delta_r S^{\circ}$ it enforces may differ from the actual value at any given temperature in the range under consideration. Thus method **(c)** is not as reliable as the first two.

9.18   Consider the reaction

$$2NO_2(g) \rightleftharpoons N_2O_4(g) \qquad \Delta_r H^{\circ} = -58.04 \text{ kJ mol}^{-1}$$

Predict what happens to the system at equilibrium if **(a)** the temperature is raised, **(b)** the pressure on the system is increased, **(c)** an inert gas is added to the system at constant pressure, **(d)** an inert gas is added to the system at constant volume, and **(e)** a catalyst is added to the system.

(a) The equilibrium will shift from right to left. The equilibrium constant K_P will decrease.

(b) The equilibrium will shift from left to right. K_P remains unchanged.

(c) The volume of the system expands, and the gases are "diluted." K_P remains unchanged, but the equilibrium shifts from right to left.

(d) The pressure of the system will increase, but the partial pressures of NO_2 and N_2O_4 remain constant. K_P remains unchanged, and the position of equilibrium will not shift.

(e) The catalyst has no effect on either K_P or the position of equilibrium.

9.19   Referring to Problem 9.14, calculate the degree of dissociation of N_2O_4 if the total pressure is 10 atm. Comment on your result.

First express the total pressure in bars.

$$P = (10 \text{ atm}) \left(\frac{1.013 \text{ bar}}{1 \text{ atm}} \right) = 10.1 \text{ bar}$$

The degree of dissociation is related to K_P, which, according to the solution of Problem 9.14, is 0.1162.

$$K_P = 0.1162 = \frac{4\alpha^2 P}{1 - \alpha^2} = \frac{4\alpha^2 (10.1)}{1 - \alpha^2}$$

$$0.1162 \left(1 - \alpha^2\right) = 40.4\alpha^2$$

$$40.5\alpha^2 = 0.1162$$

$$\alpha = 5.4 \times 10^{-2}$$

As pressure increases from 1 atm (Problem 9.14) to 10 atm, the equilibrium shifts from right to left. Therefore, α decreases.

9.20   At a certain temperature, the equilibrium pressures of NO_2 and N_2O_4 are 1.6 bar and 0.58 bar, respectively. If the volume of the container is doubled at constant temperature, what would be the partial pressures of the gases when equilibrium is re-established?

The equilibrium process is

$$N_2O_4(g) \rightleftharpoons 2NO_2(g)$$

The equilibrium constant can be calculated using the equilibrium pressures.

$$K_P = \frac{1.6^2}{0.58} = 4.41$$

When the volume of the container is doubled at constant temperature, the partial pressures of the gases are halved. That is,

$$P_{NO_2} = 0.80 \text{ bar}$$

$$P_{N_2O_4} = 0.29 \text{ bar}$$

According to Le Chatelier's principle, when the volume is increased, the reaction will shift to produce more molecules. In the case at hand, more NO_2 will be produced.

Assume the equilibrium pressure of N_2O_4 to be $0.29 - x$ bars. The equilibrium pressure of NO_2 is obtained using the stoichiometric relationship between the two compounds.

$$
\begin{array}{ccc}
N_2O_4 & \rightleftharpoons & 2NO_2 \\
\text{At equilibrium} \quad 0.29 - x & & 0.80 + 2x \quad \text{bar}
\end{array}
$$

Solve for x using the equilibrium expression and discarding the non-physical root.

$$K_P = 4.41 = \frac{(0.80 + 2x)^2}{0.29 - x}$$

$$4.41 (0.29 - x) = (0.80 + 2x)^2$$

$$4x^2 + 7.61x - 0.639 = 0$$

$$x = 8.06 \times 10^{-2}$$

Therefore, when equilibrium is re-established,

$$P_{N_2O_4} = 0.29 - x = 0.21 \text{ bar}$$

$$P_{NO_2} = 0.80 + 2x = 0.96 \text{ bar}$$

9.21 Eggshells are composed mostly of calcium carbonate ($CaCO_3$) formed by the reaction

$$Ca^{2+}(aq) + CO_3^{2-}(aq) \rightleftharpoons CaCO_3(s)$$

The carbonate ions are supplied by carbon dioxide produced during metabolism. Explain why eggshells are thinner in the summer when the rate of panting by chickens is greater. Suggest a remedy for this situation.

Panting reduces the concentration of CO_2 because it is exhaled during respiration. Possible solutions include cooling the chickens' environment (to reduce panting) or supplying an additional source of carbonate, such as carbonated drinking water or by adding ground oyster shells to the chicken feed.

9.22 Photosynthesis can be represented by

$$6CO_2(g) + 6H_2O(l) \rightleftharpoons C_6H_{12}O_6(s) + 6O_2(g) \qquad \Delta_r H^\circ = 2801 \text{ kJ mol}^{-1}$$

Explain how the equilibrium would be affected by the following changes: **(a)** the partial pressure of CO_2 is increased, **(b)** O_2 is removed from the mixture, **(c)** $C_6H_{12}O_6$ (sucrose) is removed from the mixture, **(d)** more water is added, **(e)** a catalyst is added, **(f)** the temperature is decreased, and **(g)** more sunlight shines on the plants.

(a) The equilibrium would shift from left to right.

(b) The equilibrium would shift from left to right.

(c) The equilibrium would be unaffected, since $C_6H_{12}O_6$ is a solid. (As long as excess solid remains.)

(d) The equilibrium would be unaffected, since water is a liquid. (As long as excess water remains.)

(e) The catalyst has no effect on the position of equilibrium.

(f) The equilibrium would shift from right to left.

(g) Assuming constant temperature, the position of equilibrium is unaffected by the amount of sunlight shining on the plant, although the rate of attaining equilibrium is increased.

9.23 When a gas was heated at atmospheric pressure and 25°C, its color deepened. Heating above 150°C caused the color to fade, and at 550°C the color was barely detectable. At 550°C, however, the color was partially restored by increasing the pressure of the system. Which of the following scenarios best fits the above description? Justify your choice. **(a)** A mixture of hydrogen and bromine, **(b)** pure bromine, **(c)** a mixture of nitrogen dioxide and dinitrogen tetroxide. (*Hint*: Bromine is reddish, and nitrogen dioxide is brown. The other gases are colorless.)

Since neither H_2 nor Br_2 dissociate appreciably upon heating to 150°C, the only possible reaction for choice **(a)** is $H_2 + Br_2 \longrightarrow 2HBr$, but since HBr is colorless, this would lighten rather than darken the gas.

Similarly, since Br_2 does not begin to dissociate into Br atoms at 150°C, choice **(b)** is eliminated, as the color would be expected to change.

This leaves choice **(c)**. At low temperatures colorless N_2O_4 is the predominant species present at equilibrium, but raising the temperature from 25°C to 150°C shifts the equilibrium toward the brown NO_2 (see Problem 9.18), darkening the color. Above 150°C, the NO_2 begins to dissociate according to the equilibrium $2NO_2(g) \rightleftharpoons 2NO(g) + O_2(g)$, forming colorless NO and O_2 gases. At 550°C this equilibrium lies predominantly to the right, but an increase in pressure causes a shift to the left, forming NO_2 and darkening the gas.

9.24 Industrially, sodium metal is obtained by electrolyzing molten sodium chloride. The reaction at the cathode is $Na^+ + e^- \rightarrow Na$. We might expect that potassium metal could also be prepared by electrolyzing molten potassium chloride. Potassium metal is soluble in molten potassium chloride, however, and therefore is hard to recover. Furthermore, potassium vaporizes readily at the operating temperature, creating hazardous conditions. Instead, potassium is prepared by the distillation of molten potassium chloride in the presence of sodium vapor at 892°C:

$$Na(g) + KCl(l) \rightleftharpoons NaCl(l) + K(g)$$

In view of the fact that potassium is a stronger reducing agent than sodium, explain why this approach works. (The boiling points of sodium and potassium are 892°C and 770°C, respectively.)

Potassium is more volatile than sodium and is removed from the system more rapidly, causing the equilibrium to shift from left to right.

9.25 People living at high altitudes have higher hemoglobin content in their red blood cells than those living near sea level. Explain.

At higher altitudes, the air is thinner with less O_2 present than at sea level. There is insufficient oxygen carried to cells and the body shifts the (simplified) equilibrium

$$Hb + O_2 \rightleftharpoons HbO_2$$

to the right by making additional Hb molecules.

9.26   Derive Equation 9.23 from 9.21.

$$Y = \frac{[L]}{[L] + K_d}$$

$$Y\left([L] + K_d\right) = [L]$$

$$Y[L] + YK_d = [L]$$

$$\frac{Y[L] + YK_d}{[L]} = 1$$

$$Y + \frac{Y}{[L]}K_d = 1$$

$$\frac{Y}{[L]}K_d = 1 - Y$$

$$\frac{Y}{[L]} = \frac{1}{K_d} - \frac{Y}{K_d}$$

9.27   The calcium ion binds to a certain protein to form a 1:1 complex. The following data were obtained in an experiment:

Total $Ca^{2+}/\mu M$	60	120	180	240	480
Ca^{2+} bound to Protein/μM	31.2	51.2	63.4	70.8	83.4

Determine graphically the dissociation constant of the Ca^{2+}—protein complex. The protein concentration was kept at 96 μM for each run. (1 $\mu M = 1 \times 10^{-6}$ M.)

The dissociation constant can be obtained by plotting $Y/[L]$ vs Y. The slope is $-1/K_d$.

L and Y are obtained using the following relations:

$$[L] = \left[Ca^{2+}\right]_{total} - \left[Ca^{2+}\right]_{bound}$$

$$Y = \frac{\left[Ca^{2+}\right]_{bound}}{96 \ \mu M}$$

According to the data given above, the values of [L], Y, and $Y/[L]$ are

$[L]/\mu M$	28.8	68.8	116.6	169.2	396.6
Y	0.3250	0.5333	0.6604	0.7375	0.8688
$(Y/[L])/10^{-3} \cdot M^{-1}$	11.28	7.751	5.664	4.359	2.191

The slope of the plot is -16.72×10^{-3}. Therefore,

$$K_d = \frac{1}{16.72 \times 10^{-3}} = 59.8$$

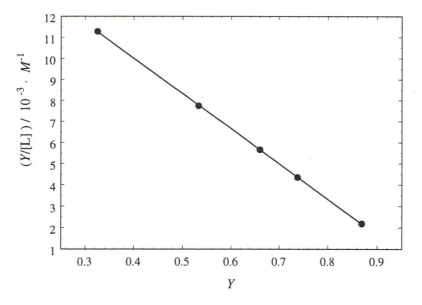

9.28   An equilibrium dialysis experiment showed that the concentrations of the free ligand, bound ligand, and protein are $1.2 \times 10^{-5}\ M$, $5.4 \times 10^{-6}\ M$, and $4.9 \times 10^{-6}\ M$, respectively. Calculate the dissociation constant for the reaction $PL \rightleftharpoons P + L$. Assume that there is one binding site per protein molecule.

Since there is one binding site per protein molecule, $[PL] = [L]_{bound}$.

$$K_d = \frac{[P][L]}{[PL]} = \frac{\left(4.9 \times 10^{-6}\right)\left(1.2 \times 10^{-5}\right)}{5.4 \times 10^{-6}} = 1.1 \times 10^{-5}$$

9.29   The reaction

$$\text{L-glutamate + pyruvate} \rightarrow \alpha\text{-ketoglutarate + L-alanine}$$

is catalyzed by the enzyme L-glutamate-pyruvate aminotransferase. At 300 K, the equilibrium constant for the reaction is 1.11. Predict whether the forward reaction (left to right) will occur spontaneously if the concentrations of the reactants and products are [L-glutamate] = $3.0 \times 10^{-5}\ M$, [pyruvate] = $3.3 \times 10^{-4}\ M$, [α-ketoglutarate] = $1.6 \times 10^{-2}\ M$, and [L-alanine] = $6.25 \times 10^{-3}\ M$.

First calculate $\Delta_r G^\circ$, which, together with the reaction quotient, gives $\Delta_r G$. The sign of $\Delta_r G$ determines if the reaction is spontaneous.

$\Delta_r G^\circ$ is related to the equilibrium constant by

$$\Delta_r G^\circ = -RT \ln K = -\left(8.314\ \text{J K}^{-1}\ \text{mol}^{-1}\right)(300\ \text{K}) \ln 1.11 = -260.3\ \text{J mol}^{-1}$$

$\Delta_r G$ can now be determined.

$$\Delta_r G = \Delta_r G^\circ + RT \ln Q = \Delta_r G^\circ + RT \ln \frac{\left[\alpha\text{-ketoglutarate}\right]\left[\text{L-alanine}\right]}{\left[\text{L-glutamate}\right]\left[\text{pyruvate}\right]}$$

$$= -260.3 \text{ J mol}^{-1} + \left(8.314 \text{ J K}^{-1} \text{ mol}^{-1}\right)(300 \text{ K}) \ln \frac{\left(1.6 \times 10^{-2}\right)\left(6.25 \times 10^{-3}\right)}{\left(3.0 \times 10^{-5}\right)\left(3.3 \times 10^{-4}\right)}$$

$$= 2.3 \times 10^4 \text{ J mol}^{-1} > 0$$

Therefore, the reaction is not spontaneous.

9.30 As mentioned in the chapter, the standard Gibbs energy for the hydrolysis of ATP to ADP at 310 K is approximately -30.5 kJ mol^{-1}. Calculate the value of $\Delta_r G^{\circ\prime}$ for the reaction in the muscle of a polar sea fish at $-1.5°$C. (*Hint*: $\Delta_r H^{\circ\prime} = -20.1$ kJ mol^{-1}.)

$\Delta_r S^{\circ\prime}$ can be determined from $\Delta_r G^{\circ\prime}$ at 310 K and $\Delta_r H^{\circ\prime}$, which is assumed to be temperature independent.

$$\Delta_r G^{\circ\prime} = \Delta_r H^{\circ\prime} - T \Delta_r S^{\circ\prime}$$

$$\Delta_r S^{\circ\prime} = \frac{\Delta_r H^{\circ\prime} - \Delta_r G^{\circ\prime}}{T} = \frac{-20.1 \text{ kJ mol}^{-1} - \left(-30.5 \text{ kJ mol}^{-1}\right)}{310 \text{ K}} = 3.355 \times 10^{-2} \text{ kJ K}^{-1} \text{ mol}^{-1}$$

Assuming that $\Delta_r S^{\circ\prime}$ is temperature independent, in addition to $\Delta_r H^{\circ\prime}$, $\Delta_r G^{\circ\prime}$ at $-1.5°$C (271.65 K) can be determined.

$$\Delta_r G^{\circ\prime} = \Delta_r H^{\circ\prime} - T \Delta_r S^{\circ\prime} = -20.1 \text{ kJ mol}^{-1} - (271.65 \text{ K})\left(3.355 \times 10^{-2} \text{ kJ K}^{-1} \text{ mol}^{-1}\right)$$

$$= -29.2 \text{ kJ mol}^{-1}$$

9.31 Under standard-state conditions one of the steps in glycolysis does not occur spontaneously:

$$\text{glucose} + \text{HPO}_4^{2-} \rightarrow \text{glucose-6-phosphate} + \text{H}_2\text{O} \qquad \Delta_r G^{\circ\prime} = 13.4 \text{ kJ mol}^{-1}$$

Can the reaction take place in the cytoplasm of a cell where the concentrations are [glucose] = 4.5×10^{-2} M, [HPO$_4^{2-}$] = 2.7×10^{-3} M, and [glucose-6-phosphate] = 1.6×10^{-4} M and the temperature is 310 K?

$$\Delta_r G = \Delta_r G^{\circ\prime} + RT \ln Q = \Delta_r G^{\circ\prime} + RT \ln \frac{\left[\text{glucose-6-phosphate}\right]}{\left[\text{glucose}\right]\left[\text{HPO}_4^{2-}\right]}$$

$$= 13.4 \times 10^3 \text{ J mol}^{-1} + \left(8.314 \text{ J K}^{-1} \text{ mol}^{-1}\right)(310 \text{ K}) \ln \frac{1.6 \times 10^{-4}}{\left(4.5 \times 10^{-2}\right)\left(2.7 \times 10^{-3}\right)}$$

$$= 14.1 \times 10^3 \text{ J mol}^{-1} > 0$$

Therefore, the reaction is not spontaneous.

9.32   The formation of a dipeptide is the first step toward the synthesis of a protein molecule. Consider the following reaction:

$$glycine + glycine \rightarrow glycylglycine + H_2O$$

Use the data in Appendix B to calculate the value of $\Delta_r G^{o\prime}$ and the equilibrium constant at 298 K, keeping in mind that the reaction is carried out in an aqueous buffer solution. Assume that the value of $\Delta_r G^{o\prime}$ is essentially the same at 310 K. What conclusion can you draw about your result?

$\Delta_r G^{o\prime}$ is calculated from the standard molar Gibbs energies of formation of the reactants and products.

$$\Delta_r G^{o\prime} = \Delta_f \overline{G}^{\,o} \text{ (glycylglycine)} + \Delta_f \overline{G}^{\,o} \text{ (H}_2\text{O)} - 2\Delta_f \overline{G}^{\,o} \text{ (glycine)}$$

$$= \left(-493.1 \text{ kJ mol}^{-1}\right) + \left(-237.2 \text{ kJ mol}^{-1}\right) - 2\left(-379.9 \text{ kJ mol}^{-1}\right)$$

$$= 29.5 \text{ kJ mol}^{-1}$$

The equilibrium constant can now be determined.

$$\ln K' = -\frac{\Delta_r G^{o\prime}}{RT} = -\frac{29.5 \times 10^3 \text{ J mol}^{-1}}{\left(8.314 \text{ J K}^{-1} \text{ mol}^{-1}\right)(298 \text{ K})} = -11.91$$

$$K' = 6.72 \times 10^{-6}$$

If $\Delta_r G^{o\prime}$ at 310 K is essentially the same as that at 298 K, then K' at 310 K is also approximately the same as that at 298 K. The small K' indicates that the formation of a dipeptide (and hence a protein molecule) is not a spontaneous process under standard-state conditions. Protein synthesis *in vivo* is carried out both under other conditions and with the aid of ATP.

9.33   From the following reactions at 25°C:

$$\text{fumarate}^{2-} + \text{NH}_4^+ \rightarrow \text{aspartate}^- \qquad\qquad \Delta_r G^{o\prime} = -36.7 \text{ kJ mol}^{-1}$$

$$\text{fumarate}^{2-} + \text{H}_2\text{O} \rightarrow \text{malate}^{2-} \qquad\qquad \Delta_r G^{o\prime} = -2.9 \text{ kJ mol}^{-1}$$

Calculate the standard Gibbs energy change and the equilibrium constant for the following reaction:

$$\text{malate}^{2-} + \text{NH}_4^+ \rightarrow \text{aspartate}^- + \text{H}_2\text{O}$$

The desired reaction can be obtained as a sum of the following reactions:

$$\text{fumarate}^{2-} + \text{NH}_4^+ \rightarrow \text{aspartate}^- \qquad\qquad \Delta_r G^{o\prime} = -36.7 \text{ kJ mol}^{-1}$$

$$\text{malate}^{2-} \rightarrow \text{fumarate}^{2-} + \text{H}_2\text{O} \qquad\qquad \Delta_r G^{o\prime} = 2.9 \text{ kJ mol}^{-1}$$

Therefore, $\Delta_r G^{o\prime}$ for

$$malate^{2-} + NH_4^+ \rightarrow aspartate^- + H_2O$$

is

$$\Delta_r G^{o\prime} = -36.7 \text{ kJ mol}^{-1} + 2.9 \text{ kJ mol}^{-1} = -33.8 \text{ kJ mol}^{-1}$$

The equilibrium constant can now be determined.

$$\ln K' = -\frac{\Delta_r G^{o\prime}}{RT} = -\frac{-33.8 \times 10^3 \text{ J mol}^{-1}}{(8.314 \text{ J K}^{-1} \text{ mol}^{-1})(298.2 \text{ K})} = 13.63$$

$$K' = 8.31 \times 10^5$$

9.34 A polypeptide can exist in either the helical or random coil forms. The equilibrium constant for equilibrium reaction of the helix to the random coil transition is 0.86 at 40°C and 0.35 at 60°C. Calculate the values of $\Delta_r H^\circ$ and $\Delta_r S^\circ$ for the reaction.

$\Delta_r H^\circ$ is calculated from the van't Hoff equation.

$$\ln \frac{K_2}{K_1} = \frac{\Delta_r H^\circ}{R}\left(\frac{1}{T_1} - \frac{1}{T_2}\right)$$

$$\ln \frac{0.35}{0.86} = \frac{\Delta_r H^\circ}{8.314 \text{ J K}^{-1} \text{ mol}^{-1}}\left(\frac{1}{313 \text{ K}} - \frac{1}{333 \text{ K}}\right)$$

$$\Delta_r H^\circ = -3.90 \times 10^4 \text{ J mol}^{-1} = -3.9 \times 10^4 \text{ J mol}^{-1}$$

To calculate $\Delta_r S^\circ$, $\Delta_r G^\circ$ at a particular temperature is needed. The following calculations are carried out using 40°C.

$$\Delta_r G^\circ = -RT \ln K = -(8.314 \text{ J K}^{-1} \text{ mol}^{-1})(313 \text{ K}) \ln 0.86 = 392 \text{ J mol}^{-1}$$

Assuming $\Delta_r H^\circ$ and $\Delta_r S^\circ$ to be independent of temperature, the latter can be determined.

$$\Delta_r G^\circ = \Delta_r H^\circ - T\Delta_r S^\circ$$

$$\Delta_r S^\circ = \frac{\Delta_r H^\circ - \Delta_r G^\circ}{T} = \frac{-3.90 \times 10^4 \text{ J mol}^{-1} - 392 \text{ J mol}^{-1}}{313 \text{ K}} = -1.3 \times 10^2 \text{ J K}^{-1} \text{ mol}^{-1}$$

9.35 List two important differences between a steady state and an equilibrium state.

A system at steady state requires that mass be continually supplied and removed. A system at steady state also displays a concentration gradient whereas a system at equilibrium is homogeneous.

9.36   At 720 K, the equilibrium partial pressures are $P_{NH_3} = 321.6$ atm, $P_{N_2} = 69.6$ atm, and $P_{H_2} = 208.8$ atm, respectively. (a) Calculate the value of K_P for the reaction described in Example 9.1. (b) Calculate the thermodynamic equilibrium constant if $\gamma_{NH_3} = 0.782$, $\gamma_{N_2} = 1.266$, and $\gamma_{H_2} = 1.243$.

The reaction is

$$N_2(g) + 3H_2(g) \rightleftharpoons 2NH_3(g)$$

(a) The partial pressures have to be expressed in bars to be used in the equilibrium expression.

$$P_{NH_3} = (321.6 \text{ atm}) \left(\frac{1.01325 \text{ bar}}{1 \text{ atm}} \right) = 325.86 \text{ bar}$$

$$P_{N_2} = (69.6 \text{ atm}) \left(\frac{1.01325 \text{ bar}}{1 \text{ atm}} \right) = 70.52 \text{ bar}$$

$$P_{H_2} = (208.8 \text{ atm}) \left(\frac{1.01325 \text{ bar}}{1 \text{ atm}} \right) = 211.57 \text{ bar}$$

The equilibrium constant is

$$K_P = \frac{P_{NH_3}^2}{P_{N_2} P_{H_2}^3} = \frac{(325.86)^2}{(70.52)(211.57)^3} = 1.590 \times 10^{-4} = 1.59 \times 10^{-4}$$

(b)

$$K_f = K_\gamma K_P = \frac{\gamma_{NH_3}^2}{\gamma_{N_2} \gamma_{H_2}^3} K_P = \frac{(0.782)^2}{(1.266)(1.243)^3} \left(1.590 \times 10^{-4} \right) = 4.00 \times 10^{-5}$$

9.37   Based on the material covered so far in the text, describe as many ways as you can for calculating the $\Delta_r G°$ value of a process.

There are at least three methods.

(a) From $\Delta_r H°$, $\Delta_r S°$ and T via $\Delta_r G° = \Delta_r H° - T\Delta_r S°$

(b) From $\Delta_f \overline{G}°$ values for the reactants and products

(c) Using the value of the equilibrium constant for the process to give $\Delta_r G° = -RT \ln K$

9.38   The solubility of n-heptane in water is 0.050 g per liter of solution at 25°C. What is the Gibbs energy change for the hypothetical process of dissolving n-heptane in water at a concentration of

2.0 g L^{-1} at the same temperature? (*Hint*: First calculate the value of $\Delta_r G^\circ$ from the equilibrium process and then the $\Delta_r G$ value using Equation 9.7.)

The solubility of *n*-heptane in water is $\dfrac{0.050 \text{ g L}^{-1}}{100.20 \text{ g mol}^{-1}} = 4.99 \times 10^{-4}$ *M*. The appropriate equilibrium is

$$n\text{-heptane}(l) \rightleftharpoons n\text{-heptane}(aq)$$

with $K_{sp} = [n\text{-heptane}] = 4.99 \times 10^{-4}$. For the saturated solution, $\Delta_r G = 0$, and

$$\Delta_r G^\circ = -RT \ln K$$

$$= -\left(8.314 \text{ J K}^{-1} \text{ mol}^{-1}\right)(298 \text{ K}) \ln\left(4.99 \times 10^{-4}\right)$$

$$= 1.88 \times 10^4 \text{ J mol}^{-1}$$

For the hypothetical process, the concentration is $c = \dfrac{2.0 \text{ g L}^{-1}}{100.20 \text{ g mol}^{-1}} = 2.00 \times 10^{-2}$ *M*, and

$$\Delta_r G = \Delta_r G^\circ + RT \ln c$$

$$= 1.88 \times 10^4 \text{ J mol}^{-1} + \left(8.314 \text{ J K}^{-1} \text{ mol}^{-1}\right)(298 \text{ K}) \ln\left(2.00 \times 10^{-2}\right)$$

$$= 9.14 \times 10^3 \text{ J mol}^{-1}$$

$$= 9.1 \text{ kJ mol}^{-1}$$

9.39   In this chapter, we introduced the quantity $\Delta_r G^{\circ\prime}$, which is the standard Gibbs energy change for a reaction in which the reactants and products are in their biochemical standard states. The discussion focused on the uptake or liberation of H$^+$ ions. The $\Delta_r G^{\circ\prime}$ can also be applied to reactions involving the uptake and liberation of gases such as O$_2$ and CO$_2$. In these cases, the biochemical standard states are $P_{O_2} = 0.2$ bar and $P_{CO_2} = 0.0003$ bar, where 0.2 bar and 0.0003 bar are the partial pressures of O$_2$ and CO$_2$ in air, respectively. Consider the reaction

$$A(aq) + B(aq) \rightarrow C(aq) + CO_2(g)$$

where A, B, and C are molecular species. Derive a relation between $\Delta_r G^\circ$ and $\Delta_r G^{\circ\prime}$ for this reaction at 310 K.

Using 1 bar and 1 *M* as standard states for gases and solutions,

$$\Delta_r G = \Delta_r G^\circ + RT \ln \frac{\left(P_{CO_2}/1 \text{ bar}\right)([C]/1M)}{([A]/1M)\,([B]/1M)}$$

Using the biochemical standard states,

$$\Delta_r G = \Delta_r G^{\circ\prime} + RT \ln \frac{\left(P_{CO_2}/0.0003 \text{ bar}\right)([C]/1M)}{([A]/1M)\,([B]/1M)}$$

Since the value of $\Delta_r G$ does not depend on the standard states chosen, the two expressions above can be equated.

$$\Delta_r G^{o\prime} + RT \ln \frac{\left(P_{CO_2}/0.0003\text{ bar}\right)\left([C]/1M\right)}{([A]/1M)\,([B]/1M)} = \Delta_r G^{\circ} + RT \ln \frac{\left(P_{CO_2}/1\text{ bar}\right)\left([C]/1M\right)}{([A]/1M)\,([B]/1M)}$$

At 310 K,

$$\Delta_r G^{o\prime} = \Delta_r G^{\circ} + \left(8.314\text{ J K}^{-1}\text{ mol}^{-1}\right)(310\text{ K})\ln\frac{0.0003\text{ bar}}{1\text{ bar}}$$

$$\Delta_r G^{o\prime} = \Delta_r G^{\circ} - 20.9\text{ kJ mol}^{-1}$$

Thus for reactions producing CO_2, $\Delta_r G^{\circ}$ is greater than $\Delta_r G^{o\prime}$ by 20.9 kJ mol^{-1} for each mole of CO_2 produced at 310 K.

9.40 The binding of oxygen to hemoglobin (Hb) is quite complex, but for our purpose we can represent the reaction as

$$Hb(aq) + O_2(g) \rightarrow HbO_2(aq)$$

If the value of $\Delta_r G^{\circ}$ for the reaction is -11.2 kJ mol^{-1} at 20°C, calculate the value of $\Delta_r G^{o\prime}$ for the reaction. (*Hint*: Refer to the result in Problem 9.39.)

Using 1 bar and 1 M as standard states for gases and solutions,

$$\Delta_r G = \Delta_r G^{\circ} + RT \ln \frac{[HbO_2]/1M}{([Hb]/1M)\left(P_{O_2}/1\text{ bar}\right)}$$

Using the biochemical standard states,

$$\Delta_r G = \Delta_r G^{o\prime} + RT \ln \frac{[HbO_2]/1M}{([Hb]/1M)\left(P_{O_2}/0.2\text{ bar}\right)}$$

Since the value of $\Delta_r G$ does not depend on the standard states chosen, the two expressions above can be equated.

$$\Delta_r G^{o\prime} + RT \ln \frac{[HbO_2]/1M}{([Hb]/1M)\left(P_{O_2}/0.2\text{ bar}\right)} = \Delta_r G^{\circ} + RT \ln \frac{[HbO_2]/1M}{([Hb]/1M)\left(P_{O_2}/1\text{ bar}\right)}$$

$$\Delta_r G^{o\prime} = \Delta_r G^{\circ} + RT \ln \frac{1\text{ bar}}{0.2\text{ bar}}$$

$$\Delta_r G^{o\prime} = -11.2 \times 10^3\text{ J mol}^{-1} + \left(8.314\text{ J K}^{-1}\text{ mol}^{-1}\right)(293.2\text{ K})\ln\frac{1}{0.2} = -7.3 \times 10^3\text{ J mol}^{-1}$$

9.41 The K_{sp} value of AgCl is 1.6×10^{-10} at 25°C. What is its value at 60°C?

The reaction is

$$AgCl(s) \rightleftharpoons Ag^+(aq) + Cl^-(aq)$$

First calculate $\Delta_r H°$ for the reaction.

$$\Delta_r H° = \Delta_f \overline{H}° \left[Ag^+(aq)\right] + \Delta_f \overline{H}° \left[Cl^-(aq)\right] - \Delta_f \overline{H}° \left[AgCl(s)\right]$$

$$= 105.9 \text{ kJ mol}^{-1} + \left(-167.2 \text{ kJ mol}^{-1}\right) - \left(-127.0 \text{ kJ mol}^{-1}\right)$$

$$= 65.7 \text{ kJ mol}^{-1}$$

The van't Hoff equation is now used to calculate K_{sp} at 60°C.

$$\ln \frac{K_2}{K_1} = \frac{\Delta_r H°}{R} \left(\frac{1}{T_1} - \frac{1}{T_2}\right)$$

$$\ln \frac{K_2}{1.6 \times 10^{-10}} = \frac{65.7 \times 10^3 \text{ J mol}^{-1}}{8.314 \text{ J K}^{-1} \text{ mol}^{-1}} \left(\frac{1}{298 \text{ K}} - \frac{1}{333 \text{ K}}\right)$$

$$K_2 = 2.6 \times 10^{-9}$$

The increase in the value of K at higher temperature indicates that the solubility of AgCl increases with temperature.

9.42 Many hydrocarbons exist as structural isomers, which are compounds that have the same molecular formula but different structures. For example, both butane and isobutane have the same molecular formula: C_4H_{10}. Calculate the mole percent of these molecules in an equilibrium mixture at 25°C, given that the standard Gibbs energy of formation of butane is $-15.9 \text{ kJ mol}^{-1}$ and that of isobutane is $-18.0 \text{ kJ mol}^{-1}$. Does your result support the notion that straight-chain hydrocarbons (that is, hydrocarbons in which the C atoms are joined along a line) are less stable than branch-chain hydrocarbons?

The isomerization process can be expressed as

$$\text{butane}(g) \rightleftharpoons \text{isobutane}(g)$$

First calculate $\Delta_r G°$, from which the equilibrium constant K_P is obtained.

$$\Delta_r G° = \Delta_f \overline{G}° \text{ [isobutane]} - \Delta_f \overline{G}° \text{ [butane]}$$

$$= -18.0 \text{ kJ mol}^{-1} - \left(-15.9 \text{ kJ mol}^{-1}\right)$$

$$= -2.1 \text{ kJ mol}^{-1}$$

$$\ln K_P = -\frac{\Delta_r G°}{RT}$$

$$= -\frac{-2.1 \times 10^3 \text{ J mol}^{-1}}{\left(8.314 \text{ J K}^{-1} \text{ mol}^{-1}\right)(298 \text{ K})} = 0.848$$

$$K_P = 2.33$$

The equilibrium constant is a ratio between the pressures of isobutane and butane. The pressure of each gas is proportional to number of moles of gas at constant T and V. Therefore,

$$K_P = \frac{P_{\text{isobutane}}}{P_{\text{butane}}} = \frac{n_{\text{isobutane}}}{n_{\text{butane}}} = 2.33$$

According to the above expression, for each mole of butane, there are 2.33 moles of isobutane.

$$\text{Mole percent of isobutane} = \frac{2.33}{3.33} \times 100\% = 70\%$$

$$\text{Mole percent of butane} = 1 - 70\% = 30\%$$

These results support the notion that straight-chain hydrocarbons like butane are less stable than branch-chain hydrocarbons like isobutane.

9.43 Consider the equilibrium system $3A \rightleftharpoons B$. Sketch the change in the concentrations of A and B with time for the following situations: **(a)** initially only A is present; **(b)** initially only B is present; and **(c)** initially both A and B are present (with A in higher concentration). In each case, assume that the concentration of B is higher than that of A at equilibrium.

(a)

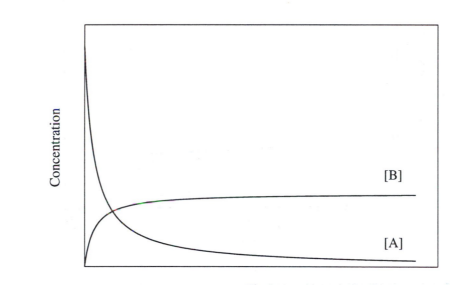

(b)

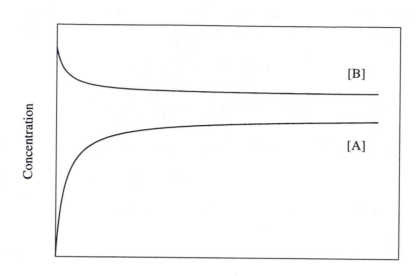

(c)

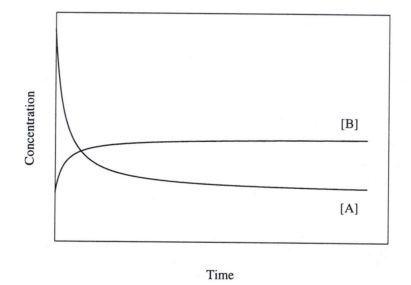

CHAPTER 10

Electrochemistry

PROBLEMS AND SOLUTIONS

10.1 Calculate the standard emf for the following reaction at 298 K:

$$Fe(s) + Tl^{3+} \rightarrow Fe^{2+} + Tl^{+}$$

The half reactions are:

Anode:	$Fe \rightarrow Fe^{2+} + 2e^{-}$
Cathode:	$Tl^{3+} + 2e^{-} \rightarrow Tl^{+}$

Therefore, E° for the reaction $Fe(s) + Tl^{3+} \rightarrow Fe^{2+} + Tl^{+}$ is

$$E^{\circ} = 1.252 \text{ V} - (-0.447 \text{ V}) = 1.699 \text{ V}$$

10.2 Calculate the emf of the Daniell cell at 298 K when the concentrations of $CuSO_4$ and $ZnSO_4$ are 0.50 M and 0.10 M, respectively. What would the emf be if activities were used instead of concentrations? (The $\gamma_{\pm}$ values for $CuSO_4$ and $ZnSO_4$ at their respective concentrations are 0.068 and 0.15, respectively.)

The half reactions for the Daniell cell are

Anode:	$Zn \rightarrow Zn^{2+} + 2e^{-}$
Cathode:	$Cu^{2+} + 2e^{-} \rightarrow Cu$

Thus, for the cell,

$$E^\circ = 0.342 \text{ V} - (-0.762 \text{ V}) = 1.104 \text{ V}$$

The emf at the specified Cu^{2+} and Zn^{2+} concentrations is

$$E = E^\circ - \frac{0.0257 \text{ V}}{\nu} \ln \frac{[Zn^{2+}]}{[Cu^{2+}]} = 1.104 \text{ V} - \frac{0.0257 \text{ V}}{2} \ln \frac{0.10}{0.50} = 1.125 \text{ V}$$

Using activities,

$$E = E^\circ - \frac{0.0257 \text{ V}}{\nu} \ln \frac{a_{Zn^{2+}}}{a_{Cu^{2+}}}$$

$$= E^\circ - \frac{0.0257 \text{ V}}{\nu} \ln \frac{\gamma_{\pm,ZnSO_4} [Zn^{2+}]}{\gamma_{\pm,CuSO_4} [Cu^{2+}]}$$

$$= 1.104 \text{ V} - \frac{0.0257 \text{ V}}{2} \ln \frac{(0.15)(0.10)}{(0.068)(0.50)}$$

$$= 1.115 \text{ V}$$

10.3 The half-reaction at an electrode is

$$Al^{3+}(aq) + 3e^- \rightarrow Al(s)$$

Calculate the number of grams of aluminum that can be produced by passing 1.00 faraday through the electrode.

1.00 faraday supplies 1 mole e^-. Therefore, since each mole of Al^{3+} requires 3 moles e^- to be completely reduced to Al,

$$\text{Number of moles of Al produced} = \frac{1}{3} \text{ mol}$$

$$\text{Mass of Al produced} = \left(\frac{1}{3} \text{ mol} \right) (26.98 \text{ g mol}^{-1}) = 8.99 \text{ g}$$

10.4 Consider a Daniell cell operating under non-standard-state conditions. Suppose that the cell's reaction is multiplied by 2. What effect does this have on each of the following quantities in the Nernst equation? **(a)** E, **(b)** E°, **(c)** Q, **(d)** $\ln Q$, and **(e)** ν

(a) None. E is an intensive property,   **(b)** none. E° is an intensive property,   **(c)** squared,
(d) doubled,   **(e)** doubled.

10.5 A student is given two beakers in the laboratory. One beaker contains a solution that is 0.15 M in Fe^{3+} and 0.45 M in Fe^{2+}, and the other beaker contains a solution that is 0.27 M in I^- and 0.050 M in I_2. A piece of platinum wire is dipped into each solution. **(a)** Calculate the potential of each electrode relative to a standard hydrogen electrode at 25°C. **(b)** Predict the chemical reaction that will occur when these two electrodes are connected and a salt bridge is used to join the two solutions together.

(a) At the Pt | Fe^{3+}, Fe^{2+} electrode, the emf for

$$Fe^{3+} + e^- \rightarrow Fe^{2+}$$

is

$$E = E^\circ - \frac{0.0257 \text{ V}}{\nu} \ln \frac{\left[Fe^{2+}\right]}{\left[Fe^{3+}\right]} = 0.771 \text{ V} - \frac{0.0257 \text{ V}}{1} \ln \frac{0.45}{0.15} = 0.743 \text{ V}$$

At the Pt | I_2, I^- electrode, the emf for

$$I_2 + 2e^- \rightarrow 2I^-$$

is

$$E = E^\circ - \frac{0.0257 \text{ V}}{\nu} \ln \frac{\left[I^-\right]^2}{\left[I_2\right]} = 0.536 \text{ V} - \frac{0.0257 \text{ V}}{2} \ln \frac{(0.27)^2}{0.050} = 0.531 \text{ V}$$

(b) The chemical reaction that would occur is one with a positive E, that is

$$2Fe^{3+} + 2I^- \rightarrow 2Fe^{2+} + I_2$$

The emf for this reaction is

$$E = 0.743 \text{ V} - 0.531 \text{ V} = 0.212 \text{ V}$$

Since this cell is not operating under standard conditions of concentration, the appropriate half cell potentials are those calculated in part **(a)** and not the E° values from Table 10.1.

10.6 From the standard reduction potentials listed in Table 10.1 for Cu^{2+} | Cu and Pt | Cu^{2+}, Cu^+, calculate the standard reduction potential for Cu^+ | Cu.

The reduction half reactions and the standard reduction potentials for Cu^{2+} | Cu and Pt | Cu^{2+}, Cu^+ are

$$Cu^{2+} + 2e^- \rightarrow Cu \qquad\qquad E_1^\circ = 0.342 \text{ V}$$

$$Cu^{2+} + e^- \rightarrow Cu^+ \qquad\qquad E_2^\circ = 0.153 \text{ V}$$

The reduction half reaction for Cu^+ | Cu is

$$Cu^+ + e^- \rightarrow Cu.$$

Let its reduction potential be denoted as E_3^o. This half reaction can be derived from the previous two half reactions. The Gibbs energy changes are related by

$$\Delta_r G_3^o = \Delta_r G_1^o - \Delta_r G_2^o$$

Since $\Delta_r G^o = -\nu F E^o$, the above relation becomes

$$\left(-\nu_3 F E_3^o\right) = \left(-\nu_1 F E_1^o\right) - \left(-\nu_2 F E_2^o\right)$$

$$\nu_3 E_3^o = \nu_1 E_1^o - \nu_2 E_2^o$$

$$E_3^o = \frac{\nu_1 E_1^o - \nu_2 E_2^o}{\nu_3}$$

$$= \frac{2\,(0.342\ \text{V}) - 1\,(0.153\ \text{V})}{1} = 0.531\ \text{V}$$

10.7 Complete the following table, indicating in the third column whether the cell reaction is spontaneous:

E	$\Delta_r G$	Cell Reaction
+		
	+	
0		

E	$\Delta_r G$	Cell Reaction
+	−	Spontaneous
−	+	Not spontaneous
0	0	At equilibrium

10.8 Calculate the values of E^o, $\Delta_r G^o$, and K for the following reactions at 25°C:

(a) $\text{Zn} + \text{Sn}^{4+} \rightleftharpoons \text{Zn}^{2+} + \text{Sn}^{2+}$

(b) $\text{Cl}_2 + 2\text{I}^- \rightleftharpoons 2\text{Cl}^- + \text{I}_2$

(c) $5\text{Fe}^{2+} + \text{MnO}_4^- + 8\text{H}^+ \rightleftharpoons \text{Mn}^{2+} + 4\text{H}_2\text{O} + 5\text{Fe}^{3+}$

(a)

Anode: $\quad\quad\quad\quad\quad\quad\quad\quad\quad \text{Zn} \rightarrow \text{Zn}^{2+} + 2\text{e}^-$

Cathode: $\quad\quad\quad\quad\quad\quad\quad \text{Sn}^{4+} + 2\text{e}^- \rightarrow \text{Sn}^{2+}$

$$E^o = 0.151\ \text{V} - (-0.762\ \text{V}) = 0.913\ \text{V}$$

$$\Delta_r G^o = -\nu F E^o = -2\left(96500\ \text{C mol}^{-1}\right)(0.913\ \text{V}) = -1.762 \times 10^5\ \text{J mol}^{-1} = -1.76 \times 10^5\ \text{J mol}^{-1}$$

$$K = \exp\left(-\frac{\Delta_r G^\circ}{RT}\right) = \exp\left[-\frac{(-1.762 \times 10^5 \text{ J mol}^{-1})}{(8.314 \text{ J K}^{-1} \text{ mol}^{-1})(298.2 \text{ K})}\right] = 7.34 \times 10^{30}$$

(b)

Anode: $\qquad 2I^- \rightarrow I_2 + 2e^-$

Cathode: $\qquad Cl_2 + 2e^- \rightarrow 2Cl^-$

$$E^\circ = 1.36 \text{ V} - 0.536 \text{ V} = 0.824 \text{ V}$$

$$\Delta_r G^\circ = -\nu F E^\circ = -2 (96500 \text{ C mol}^{-1})(0.824 \text{ V}) = -1.590 \times 10^5 \text{ J mol}^{-1} = -1.59 \times 10^5 \text{ J mol}^{-1}$$

$$K = \exp\left(-\frac{\Delta_r G^\circ}{RT}\right) = \exp\left[-\frac{(-1.590 \times 10^5 \text{ J mol}^{-1})}{(8.314 \text{ J K}^{-1} \text{ mol}^{-1})(298.2 \text{ K})}\right] = 7.12 \times 10^{27}$$

(c)

Anode: $\qquad Fe^{2+} \rightarrow Fe^{3+} + e^-$

Cathode: $\qquad MnO_4^- + 8H^+ + 5e^- \rightarrow Mn^{2+} + 4H_2O$

$$E^\circ = 1.507 \text{ V} - 0.771 \text{ V} = 0.736 \text{ V}$$

$$\Delta_r G^\circ = -\nu F E^\circ = -5 (96500 \text{ C mol}^{-1})(0.739 \text{ V}) = -3.551 \times 10^5 \text{ J mol}^{-1} = -3.55 \times 10^5 \text{ J mol}^{-1}$$

$$K = \exp\left(-\frac{\Delta_r G^\circ}{RT}\right) = \exp\left[-\frac{(-3.551 \times 10^5 \text{ J mol}^{-1})}{(8.314 \text{ J K}^{-1} \text{ mol}^{-1})(298.2 \text{ K})}\right] = 1.60 \times 10^{62}$$

10.9 The equilibrium constant for the reaction

$$Sr + Mg^{2+} \rightleftharpoons Sr^{2+} + Mg$$

is 6.56×10^{19} at 25°C. Calculate E° for a cell made up of the $Sr^{2+} \mid Sr$ and $Mg^{2+} \mid Mg$ half-cells.

$$E^\circ = \frac{RT \ln K}{\nu F} = \frac{(8.314 \text{ J K}^{-1} \text{ mol}^{-1})(298.2 \text{ K})(\ln 6.56 \times 10^{17})}{2 (96500 \text{ C mol}^{-1})} = 0.527 \text{ V}$$

10.10 Consider a concentration cell consisting of two hydrogen electrodes. At 25°C the cell emf is found to be 0.0267 V. If the pressure of hydrogen gas at the anode is 4.0 bar, what is the pressure of hydrogen gas at the cathode?

Let the pressure of hydrogen gas at the cathode be x bars. The half reactions are

Anode: $H_2 \text{ (4.0 bar)} \rightarrow 2H^+ + 2e^-$

Cathode: $2H^+ + 2e^- \rightarrow H_2 \text{ (}x \text{ bar)}$

The overall reaction is

$$H_2 \text{ (4.0 bar)} \rightarrow H_2 \text{ (}x \text{ bar)}$$

Since this is a concentration cell, $E^\circ = 0$ V. The emf for the cell is

$$E = E^\circ - \frac{0.0257 \text{ V}}{v} \ln \frac{x}{4.0}$$

$$0.0267 \text{ V} = 0 \text{ V} - \frac{0.0257 \text{ V}}{2} \ln \frac{x}{4.0}$$

$$x = 0.50 \text{ bar}$$

10.11   An electrochemical cell consists of a half-cell in which a piece of platinum wire is dipped into a solution that is 2.0 M in KBr and 0.050 M in Br_2. The other half-cell consists of magnesium metal immersed in a 0.38 M Mg^{2+} solution. **(a)** Which electrode is the anode and which is the cathode? **(b)** What is the emf of the cell? **(c)** What is the spontaneous cell reaction? **(d)** What is the equilibrium constant of the cell reaction? Assume that the temperature is 25°C.

(a) The reduction reaction for the Pt | Br_2, Br^- half cell is

$$Br_2(aq) + 2e^- \rightarrow 2Br^-(aq) \qquad E^\circ = 1.087 \text{ V}$$

The emf for this half cell is

$$E = E^\circ - \frac{0.0257 \text{ V}}{v} \ln \frac{\left[Br^-\right]^2}{\left[Br_2\right]} = 1.087 \text{ V} - \frac{0.0257 \text{ V}}{2} \ln \frac{(2.0)^2}{0.050} = 1.0307 \text{ V}$$

The reduction reaction for the Mg^{2+} | Mg half cell is

$$Mg^{2+} + 2e^- \rightarrow Mg \qquad E^\circ = -2.372 \text{ V}$$

The emf for this half cell is

$$E = E^\circ - \frac{0.0257 \text{ V}}{v} \ln \frac{1}{\left[Mg^{2+}\right]} = -2.372 \text{ V} - \frac{0.0257 \text{ V}}{2} \ln \frac{1}{0.38} = -2.3844 \text{ V}$$

The cell reaction that would occur is one with a positive E. This is possible if Mg^{2+} | Mg is the anode and Pt | Br_2, Br^- is the cathode.

(b) $E = 1.0307 \text{ V} - (-2.3844 \text{ V}) = 3.415 \text{ V}$

(c) The spontaneous cell reaction is

$$Br_2 + Mg \rightarrow 2Br^- + Mg^{2+}$$

(d) $E°$ for the cell is

$$E° = 1.087 \text{ V} - (-2.372 \text{ V}) = 3.459 \text{ V}$$

Therefore, the equilibrium constant is

$$K = \exp\left(\frac{vFE°}{RT}\right) = \exp\left[\frac{2\left(96500 \text{ C mol}^{-1}\right)(3.459 \text{ V})}{\left(8.314 \text{ J K}^{-1} \text{ mol}^{-1}\right)(298.2 \text{ K})}\right] = e^{269.3} = 9.03 \times 10^{116}$$

10.12   From the standard reduction potentials listed in Table 10.1 for Sn^{2+} | Sn and Pb^{2+} | Pb, calculate the ratio of $[Sn^{2+}]$ to $[Pb^{2+}]$ at equilibrium at 25°C and the $\Delta_r G°$ value for the reaction.

The reduction potentials for the half cells are

$$Sn^{2+} + 2e^- \rightarrow Sn \qquad\qquad E° = -0.138 \text{ V}$$

$$Pb^{2+} + 2e^- \rightarrow Pb \qquad\qquad E° = -0.126 \text{ V}$$

The standard emf for the reaction $Sn + Pb^{2+} \rightarrow Sn^{2+} + Pb$ is

$$E° = -0.126 \text{ V} - (-0.138 \text{ V}) = 0.012 \text{ V}$$

The ratio of $[Sn^{2+}]$ to $[Pb^{2+}]$ at equilibrium is directly related to the equilibrium constant, which can be calculated from $E°$.

$$K = \frac{\left[Sn^{2+}\right]}{\left[Pb^{2+}\right]} = \exp\left(\frac{vFE°}{RT}\right) = \exp\left[\frac{2\left(96500 \text{ C mol}^{-1}\right)(0.012 \text{ V})}{\left(8.314 \text{ J K}^{-1} \text{ mol}^{-1}\right)(298.2 \text{ K})}\right] = 2.55$$

The standard Gibbs energy is

$$\Delta_r G° = -vFE° = -2\left(96500 \text{ C mol}^{-1}\right)(0.012 \text{ V}) = -2.32 \times 10^3 \text{ J mol}^{-1}$$

10.13   Consider the following cell:

$$Ag(s) \mid AgCl(s) \mid NaCl(aq) \mid Hg_2Cl_2(s) \mid Hg(l)$$

(a) Write the half-cell reactions. (b) The standard emfs of the cell at several temperatures are as follows:

T/K	291	298	303	311
$E°$/mV	43.0	45.4	47.1	50.1

Calculate the values of $\Delta_r G^\circ$, $\Delta_r S^\circ$, and $\Delta_r H^\circ$ for the reaction at 298 K.

(a) The half-cell reactions are

$$\text{Anode:} \qquad\qquad\qquad \text{Ag} + \text{Cl}^- \rightarrow \text{AgCl} + \text{e}^-$$

$$\text{Cathode:} \qquad\qquad\qquad \text{Hg}_2\text{Cl}_2 + 2\text{e}^- \rightarrow 2\text{Hg} + 2\text{Cl}^-$$

(b) $\Delta_r S^\circ$ can be calculated once the slope for the plot E° vs T is determined graphically.

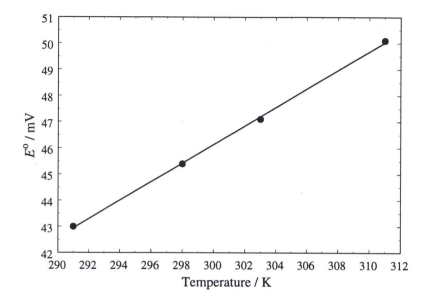

The slope of the plot is $0.3544\ \text{mV K}^{-1}$. Therefore,

$$\Delta_r S^\circ = \nu F \left(\frac{\partial E^\circ}{\partial T} \right)_P = 2\left(96500\ \text{C mol}^{-1}\right)\left(0.3544 \times 10^{-3}\ \text{V K}^{-1}\right)$$

$$= 68.40\ \text{J K}^{-1}\,\text{mol}^{-1} = 68.4\ \text{J K}^{-1}\,\text{mol}^{-1}$$

At 298 K, $E^\circ = 45.4\ \text{mV}$. Thus,

$$\Delta_r G^\circ = -\nu F E^\circ = -2\left(96500\ \text{C mol}^{-1}\right)\left(45.4 \times 10^{-3}\ \text{V}\right)$$

$$= -8.762 \times 10^3\ \text{J mol}^{-1} = -8.76 \times 10^3\ \text{J mol}^{-1}$$

$$\Delta_r H^\circ = \Delta_r G^\circ + T\Delta_r S^\circ = -8.762 \times 10^3\ \text{J mol}^{-1} + (298\ \text{K})\left(68.40\ \text{J K}^{-1}\,\text{mol}^{-1}\right)$$

$$= 1.162 \times 10^4\ \text{J mol}^{-1}$$

10.14 Calculate the emf of the following concentration cell at 298 K:

$$\text{Mg}(s)\ |\ \text{Mg}^{2+}(0.24\ M)\ \|\ \text{Mg}^{2+}(0.53\ M)\ |\ \text{Mg}(s)$$

Anode: $Mg \rightarrow Mg^{2+} (0.24 \ M) + 2e^-$

Cathode: $Mg^{2+} (0.53 \ M) + 2e^- \rightarrow Mg$

The overall reaction is

$$Mg^{2+} (0.53 \ M) \rightarrow Mg^{2+} (0.24 \ M)$$

The emf of the cell depends on the concentrations of Mg^{2+} at both the anode and the cathode.

$$E = E^\circ - \frac{0.0257 \ V}{v} \ln \frac{0.24}{0.53} = 0 \ V - \frac{0.0257 \ V}{2} \ln \frac{0.24}{0.53} = 0.010 \ V$$

10.15 An electrochemical cell consists of a silver electrode in contact with 346 mL of 0.100 M $AgNO_3$ solution and a magnesium electrode in contact with 288 mL of a 0.100 M $Mg(NO_3)_2$ solution. **(a)** Calculate the value of E for the cell at 25°C. **(b)** A current is drawn from the cell until 1.20 g of silver have been deposited at the silver electrode. Calculate the value of E for the cell at this stage of operation.

(a) The reduction reaction for the $Ag^+ \mid Ag$ half cell is

$$Ag^+ + e^- \rightarrow Ag \qquad E^\circ = 0.800 \ V$$

The emf for this half cell is

$$E = E^\circ - \frac{0.0257 \ V}{v} \ln \frac{1}{[Ag^+]} = 0.800 \ V - \frac{0.0257 \ V}{1} \ln \frac{1}{0.100} = 0.7408 \ V$$

The reduction reaction for the $Mg^{2+} \mid Mg$ half cell is

$$Mg^{2+} + 2e^- \rightarrow Mg \qquad E^\circ = -2.372 \ V$$

The emf for this half cell is

$$E = E^\circ - \frac{0.0257 \ V}{v} \ln \frac{1}{[Mg^{2+}]} = -2.372 \ V - \frac{0.0257 \ V}{2} \ln \frac{1}{0.100} = -2.4016 \ V$$

The cell reaction that would occur is one with a positive E. This is possible if $Mg^{2+} \mid Mg$ is the anode and $Ag^+ \mid Ag$ is the cathode. In other words, the overall reaction is

$$2Ag^+ + Mg \rightarrow 2Ag + Mg^{2+} \qquad E^\circ = 3.172 \ V$$

and the emf is

$$E = 0.7408 \ V - (-2.4016 \ V) = 3.1424 \ V = 3.142 \ V$$

(b) The emf of the cell depends on the concentrations of Ag^+ and Mg^{2+}, which are related to the amount of Ag deposited.

The number of moles of Ag deposited represents the number of moles of Ag^+ reduced, which is $\frac{1.20\,g}{107.9\,g\,mol^{-1}} = 1.112 \times 10^{-2}$ mol. When subtracted from the original number of moles of Ag^+ present, this gives the number of moles of Ag^+ remaining in the solution, n_{Ag^+}:

$$n_{Ag^+} = (0.100M)\,(0.346\,L) - 1.112 \times 10^{-2}\,mol = 2.348 \times 10^{-2}\,mol$$

The concentration of Ag^+ in the solution is then

$$\left[Ag^+\right] = \frac{2.348 \times 10^{-2}\,mol}{0.346\,L} = 6.786 \times 10^{-2}\,M$$

The number of moles of Mg^{2+} produced by the reaction is half of the number of moles of Ag^+ reduced. This additional amount of Mg^{2+}, when added to the original amount of Mg^{2+}, gives the total amount of Mg^{2+} present, $n_{Mg^{2+}}$:

$$n_{Mg^{2+}} = (0.100M)\,(0.288\,L) + \left(\frac{1}{2}\right)\left(1.112 \times 10^{-2}\,mol\right) = 3.436 \times 10^{-2}\,mol$$

The concentration of Mg^{2+} in the solution is therefore

$$\left[Mg^{2+}\right] = \frac{3.436 \times 10^{-2}\,mol}{0.288\,L} = 0.1193\,M$$

The emf of the cell is

$$E = E^\circ - \frac{0.0257\,V}{\nu}\,\ln\frac{\left[Mg^{2+}\right]}{\left[Ag^+\right]^2} = 3.172\,V - \frac{0.0257\,V}{2}\,\ln\frac{0.1193}{\left(6.786 \times 10^{-2}\right)^2} = 3.130\,V$$

10.16   For the reaction

$$NAD^+ + H^+ + 2e^- \rightarrow NADH$$

$E^{o\prime}$ is -0.320 V at 25°C. Calculate the value of E' at pH $= 1$. Assume that both NAD^+ and NADH are at unimolar concentration.

$$E' = E^{o\prime} - \frac{0.0257\,V}{\nu}\,\ln\frac{[NADH]}{[NAD]\left(\left[H^+\right]/10^{-7}\right)} = -0.320 - \frac{0.0257\,V}{2}\,\ln\frac{1}{(1)\left(0.1/10^{-7}\right)} = -0.142\,V$$

10.17   From the $E^{o\prime}$ value for the following reaction in Table 10.3,

$$CH_3CHO + 2H^+ + 2e^- \rightarrow C_2H_5OH$$

calculate the value of E' at pH 5.0 and 298 K, given that $[C_2H_5OH] = 5.0 \times 10^{-6}\,M$, and $[CH_3CHO] = 2.4 \times 10^{-4}\,M$.

$$E' = E^{o'} - \frac{0.0257 \text{ V}}{\nu} \ln \frac{[\text{C}_2\text{H}_5\text{OH}]}{[\text{CH}_3\text{CHO}] \left([\text{H}^+]/10^{-7}\right)^2}$$

$$= -0.197 \text{ V} - \frac{0.0257 \text{ V}}{2} \ln \frac{5.0 \times 10^{-6}}{\left(2.4 \times 10^{-4}\right) \left(10^{-5}/10^{-7}\right)^2} = -0.029 \text{ V}$$

10.18 Look up the $E^{o'}$ values in Table 10.3 for the reactions

$$\text{CH}_3\text{CHO} + 2\text{H}^+ + 2e^- \rightarrow \text{C}_2\text{H}_5\text{OH}$$

$$\text{NAD}^+ + \text{H}^+ + 2e^- \rightarrow \text{NADH}$$

Calculate the equilibrium constant for the following reaction at 298 K.

$$\text{CH}_3\text{CHO} + \text{NADH} + \text{H}^+ \rightleftharpoons \text{C}_2\text{H}_5\text{OH} + \text{NAD}^+$$

$$\text{CH}_3\text{CHO} + 2\text{H}^+ + 2e^- \rightarrow \text{C}_2\text{H}_5\text{OH} \qquad E^{o'} = -0.197 \text{ V}$$

$$\text{NAD}^+ + \text{H}^+ + 2e^- \rightarrow \text{NADH} \qquad E^{o'} = -0.320 \text{ V}$$

The standard emf for the reaction

$$\text{CH}_3\text{CHO} + \text{NADH} + \text{H}^+ \rightleftharpoons \text{C}_2\text{H}_5\text{OH} + \text{NAD}^+$$

is

$$E^{o'} = -0.197 \text{ V} - (-0.320 \text{ V}) = 0.123 \text{ V}$$

The equilibrium constant for this reaction is

$$K' = \exp\left(\frac{\nu F E^{o'}}{RT}\right) = \exp\left[\frac{2 \left(96500 \text{ C mol}^{-1}\right) (0.123 \text{ V})}{\left(8.314 \text{ J K}^{-1}\text{ mol}^{-1}\right) (298 \text{ K})}\right] = 1.45 \times 10^4$$

10.19 The following reaction, which takes place just before the citric acid cycle, is catalyzed by the enzyme lactate dehydrogenase:

$$\underset{\text{pyruvate}}{\text{CH}_3\text{COCOO}^-} + \text{NADH} + \text{H}^+ \rightleftharpoons \underset{\text{lactate}}{\text{CH}_3\text{CH(OH)COO}^-} + \text{NAD}^+$$

From the data listed in Table 10.3, calculate the value of $\Delta_r G^{o'}$ and the equilibrium constant for the reaction at 298 K.

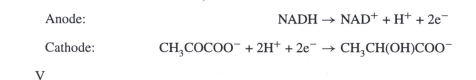

Anode: $NADH \rightarrow NAD^+ + H^+ + 2e^-$

Cathode: $CH_3COCOO^- + 2H^+ + 2e^- \rightarrow CH_3CH(OH)COO^-$

V

The standard emf of the overall reaction is

$$E^{o'} = -0.185 \text{ V} - (-0.320 \text{ V}) = 0.135 \text{ V}$$

The standard Gibbs energy and equilibrium constant for the reaction are

$$\Delta_r G^{o'} = -\nu F E^{o'} = -2 \left(96500 \text{ C mol}^{-1}\right)(0.135 \text{ V}) = -2.606 \times 10^4 \text{ J mol}^{-1} = -2.61 \times 10^4 \text{ J mol}^{-1}$$

$$K' = \exp\left(-\frac{\Delta_r G^{o'}}{RT}\right) = \exp\left[-\frac{\left(-2.606 \times 10^4 \text{ J mol}^{-1}\right)}{\left(8.314 \text{ J K}^{-1} \text{ mol}^{-1}\right)(298 \text{ K})}\right] = 3.70 \times 10^4$$

10.20   Calculate the number of moles of cytochrome c^{3+} formed from cytochrome c^{2+} with the Gibbs energy derived from the oxidation of 1 mole of glucose. ($\Delta_r G^\circ = -2879$ kJ for the degradation of 1 mole of glucose to CO_2 and H_2O.)

The standard emf for the oxidation of cyt c^{2+} to cyt c^{3+} is

$$Fe^{2+}(\text{cyt } c^{2+}) \rightarrow Fe^{3+}(\text{cyt } c^{3+}) + e^- \qquad E^{o'} = -0.254 \text{ V}$$

$\Delta_r G^\circ$ for the oxidation reaction is the same as $\Delta_r G^{o'}$, since no H^+ is involved.

$$\Delta_r G^\circ = \Delta_r G^{o'} = -\nu F E^{o'} = -\left(96500 \text{ C mol}^{-1}\right)(-0.254 \text{ V}) = 2.451 \times 10^4 \text{ J mol}^{-1}$$

The number of moles of cyt c^{3+} formed from the oxidation of 1 mole of glucose is

$$\frac{2879 \times 10^3 \text{ J}}{2.451 \times 10^4 \text{ J mol}^{-1}} = 117 \text{ mol}$$

10.21   The terminal respiratory chain involves the redox couples $NAD^+ \mid NADH$ and $FAD \mid FADH_2$. Calculate $\Delta_r G^{o'}$ value for the following reaction at 298 K:

$$NADH + FAD + H^+ \rightarrow NAD^+ + FADH_2$$

Is this Gibbs energy change sufficient to synthesize ATP from ADP and inorganic phosphate? Draw a diagram showing the experimental arrangement for measuring the emf of a cell consisting of these two couples.

The half reactions are

Anode: $NADH \rightarrow NAD^+ + H^+ + 2e^-$

Cathode: $FAD + 2H^+ + 2e^- \rightarrow FADH_2$

The standard emf for the reaction is

$$E^{o\prime} = -0.219\ \text{V} - (-0.320\ \text{V}) = 0.101\ \text{V}$$

and the standard Gibbs energy is

$$\Delta_r G^{o\prime} = -\nu F E^{o\prime} = -2\left(96500\ \text{C mol}^{-1}\right)(0.101\ \text{V}) = -1.95 \times 10^4\ \text{J mol}^{-1}$$

This Gibbs energy change is not sufficient to synthesize ATP from ADP and inorganic phosphate. An experimental arrangement for measuring the emf is shown below.

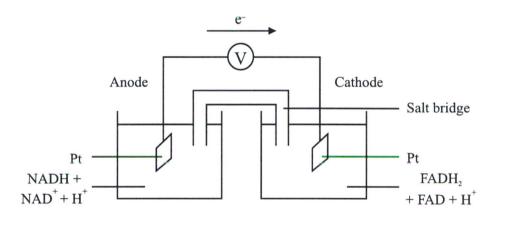

10.22   The oxidation of malate to oxaloacetate is a key reaction in the citric acid cycle:

$$\text{malate} + NAD^+ \rightarrow \text{oxaloacetate} + NADH + H^+$$

Calculate the value of $\Delta_r G^{o\prime}$ and the equilibrium constant for the reaction at pH 7 and 298 K.

The half reactions are

Anode: $\text{malate} \rightarrow \text{oxaloacetate} + 2H^+ + 2e^-$

Cathode: $NAD^+ + H^+ + 2e^- \rightarrow NADH$

The standard emf for the reaction is

$$E^{o\prime} = -0.320\ \text{V} - (-0.166\ \text{V}) = -0.154\ \text{V}$$

The standard Gibbs energy and the equilibrium constant are

$$\Delta_r G^{o\prime} = -\nu F E^{o\prime} = -2 \left(96500 \ C \ mol^{-1}\right) (-0.154 \ V) = 2.972 \times 10^4 \ J \ mol^{-1} = 2.97 \times 10^4 \ J \ mol^{-1}$$

$$K' = \exp\left(-\frac{\Delta_r G^{o\prime}}{RT}\right) = \exp\left[-\frac{2.972 \times 10^4 \ J \ mol^{-1}}{\left(8.314 \ J \ K^{-1} \ mol^{-1}\right)(298 \ K)}\right] = 6.17 \times 10^{-6}$$

10.23 Calculate the value of $\Delta_r G^{o\prime}$ for the oxidation of succinate to fumarate by cytochrome c at 298 K.

The half reactions are

Anode: succinate $\rightarrow$ fumarate $+ 2H^+ + 2e^-$

Cathode: $Fe^{3+}(cyt \ c^{3+}) + e^- \rightarrow Fe^{2+}(cyt \ c^{2+})$

The standard emf for the reaction is

$$E^{o\prime} = 0.254 \ V - 0.031 \ V = 0.223 \ V$$

The standard Gibbs energy is

$$\Delta_r G^{o\prime} = -\nu F E^{o\prime} = -2 \left(96500 \ C \ mol^{-1}\right)(0.223 \ V) = -4.30 \times 10^4 \ J \ mol^{-1}$$

10.24 Flavin adenine dinucleotide (FAD) participates in several biological redox reactions according to the half-reaction

$$FAD + 2H^+ + 2e^- \rightarrow FADH_2$$

If the value of $E^{o\prime}$ of this couple is -0.219 V at 298 K and pH 7, calculate its reduction potential at this temperature and pH when the solution contains **(a)** 85% of the oxidized form and **(b)** 15% of the oxidized form.

At pH $= 7$ and 298 K, the emf for the reduction of FAD is

$$E' = E^{o\prime} - \frac{0.0257 \ V}{\nu} \ln \frac{\left[FADH_2\right]}{[FAD]\left(\left[H^+\right]/10^{-7}\right)^2}$$

$$= -0.219 \ V - \frac{0.0257 \ V}{2} \ln \frac{\left[FADH_2\right]}{[FAD]\left(10^{-7}/10^{-7}\right)^2}$$

$$= -0.219 \ V - 0.01285 \ V \ln \frac{\left[FADH_2\right]}{[FAD]}$$

(a)
$$E' = -0.219 \text{ V} - 0.01285 \text{ V } \ln \frac{0.15}{0.85} = -0.197 \text{ V}$$

(b)
$$E' = -0.219 \text{ V} - 0.01285 \text{ V } \ln \frac{0.85}{0.15} = -0.241 \text{ V}$$

10.25 According to the chemiosmotic theory, the synthesis of 1 mole of ATP is coupled to the movement of 4 moles of H^+ ions from the low-pH side of the membrane to the high-pH side. **(a)** Derive an expression for ΔG for the movement of $4H^+$. **(b)** Calculate the change in pH across the membrane that is required at 25°C to synthesize one mole of ATP from ADP and P_i under standard-state conditions. $\Delta_r G^{o\prime} = 31.4$ kJ for the synthesis of 1 mole of ATP.

(a) The question asks for the ΔG associated with the movement of 4 moles of H^+ ions from a region of higher concentration outside the membrane to one of lower concentration inside.

$$4H^+_{outer} \rightarrow 4H^+_{inner}$$

ΔG for this process depends on the two concentrations as well as the standard state Gibbs energy change, $\Delta G^{o\prime}$.

$$\Delta G = \Delta G^{o\prime} + RT \ln \frac{\left[H^+\right]^4_{inner}}{\left[H^+\right]^4_{outer}}$$

At equilibrium, $\Delta G = 0$ and both concentrations are the same, so that the ln term is zero. Thus, $\Delta G^{o\prime} = 0$, and

$$\Delta G = RT \ln \frac{\left[H^+\right]^4_{inner}}{\left[H^+\right]^4_{outer}}$$

$$= 4RT \ln \frac{\left[H^+\right]_{inner}}{\left[H^+\right]_{outer}}$$

$$= 4(2.303) RT \left(\log \left[H^+\right]_{inner} - \log \left[H^+\right]_{outer}\right)$$

$$= -4(2.303) RT \Delta pH$$

(b) The movement of H^+ across the membrane must provide 31.4 kJ of Gibbs energy for the synthesis of one mole of ATP under the stated conditions, or $\Delta G = -31.4$ kJ mol^{-1}. From part **(a)**,

$$-31.4 \times 10^3 \text{ J mol}^{-1} = -4(2.303)\left(8.314 \text{ J K}^{-1} \text{ mol}^{-1}\right)(298 \text{ K}) \Delta pH$$

$$\Delta pH = 1.38$$

10.26 The nitrite in soil is oxidized to nitrate by the bacteria *nitrobacter agilis* in the presence of oxygen. The half reduction reactions are

$$NO_3^- + 2H^+ + 2e^- \rightarrow NO_2^- + H_2O \qquad\qquad E^{o\prime} = 0.42 \text{ V}$$

$$\tfrac{1}{2}O_2 + 2H^+ + 2e^- \rightarrow H_2O \qquad\qquad E^{o\prime} = 0.82 \text{ V}$$

Calculate the yield of ATP synthesis per mole of nitrite oxidized, assuming an efficiency of 55%. (The $\Delta_r G^{o\prime}$ for ATP synthesis from ADP and P_i is 31.4 kJ mol^{-1}.)

The standard emf when 1 mole of nitrite is oxidized is

$$E^{o\prime} = 0.82 \text{ V} - 0.42 \text{ V} = 0.40 \text{ V}$$

and the standard Gibbs energy is

$$\Delta_r G^{o\prime} = -nFE^{o\prime} = -2\left(96500 \text{ C mol}^{-1}\right)(0.40 \text{ V}) = -7.72 \times 10^4 \text{ J mol}^{-1}$$

Therefore, the yield of ATP synthesis per mole of nitrite oxidized is

$$\frac{7.72 \times 10^4 \text{ J}}{31.4 \times 10^3 \text{ J (mol ATP)}^{-1}} \times 55\% = 1.4 \text{ mol ATP}$$

10.27   Describe an experiment that would show that the nerve cell membrane is much more permeable to K^+ than Na^+.

In separate experiments with a nerve cell, add the same concentrations of Na^+ and K^+ and monitor the changes in the membrane potential. The potential will be more susceptible to changes in the K^+ concentration.

10.28   A membrane permeable only to K^+ ions is used to separate the following two solutions:

$$\alpha \quad [\text{KCl}] = 0.10 \text{ } M \qquad [\text{NaCl}] = 0.050 \text{ } M$$
$$\beta \quad [\text{KCl}] = 0.050 \text{ } M \qquad [\text{NaCl}] = 0.10 \text{ } M$$

Calculate the membrane potential at 25°C, and determine which solution has the more negative potential.

The membrane potential can only be affected by K^+ ions.

$$\Delta E_{K^+} = E_{K^+,\alpha} - E_{K^+,\beta} = \frac{0.0257 \text{ V}}{\nu} \ln \frac{[K^+]_\beta}{[K^+]_\alpha} = 0.0257 \text{ V} \ln \frac{0.050}{0.10}$$

$$= -1.8 \times 10^{-2} \text{ V} = -18 \text{ mV}$$

The α solution has the more negative potential.

10.29   Referring to Figure 10.14 b, carry the following operations: **(a)** Calculate the membrane potential due to K^+ ions at 25°C. **(b)** Given that biological membranes typically have a capacitance of approximately 1 μF cm^{-2}, calculate the charge in coulombs on a unit area (1 cm^2) of the membrane. (See Appendix 8.1 for units of capacitance.) **(c)** Convert the charge in **(b)** to number of K^+ ions. **(d)** Compare the result in **(c)** with the number of K^+ ions in 1 cm^3 of the solution in the left compartment. What can you conclude about the relative number of K^+ ions needed to establish the membrane potential?

(a)

$$E_{K^+} = \frac{0.0257 \text{ V}}{v} \ln \frac{[K^+]_r}{[K^+]_l} = 0.0257 \text{ V} \ln \frac{0.010}{0.10} = -5.918 \times 10^{-2} \text{ V} = -59.2 \text{ mV}$$

(b)

$$Q = CV = \left(1 \times 10^{-6} \, F \text{cm}^{-2}\right) \left(5.918 \times 10^{-2} \text{ V}\right)$$

$$= 5.918 \times 10^{-8} \text{ C cm}^{-2} = 5.92 \times 10^{-8} \text{ C cm}^{-2}$$

(c)

$$\text{Number of } K^+ \text{ ions} = \frac{(5.918 \times 10^{-8} \text{ C cm}^{-2})(6.022 \times 10^{23} \text{ mol}^{-1})}{96500 \text{ C mol}^{-1}} = 3.69 \times 10^{11} \text{ K}^+ \text{ ions cm}^{-2}$$

(d) The number of K^+ ions in 1 cm³ of the solution in the left compartment is

$$(0.1 \, M) \left(1 \text{ cm}^3\right) \left(\frac{1 \text{ L}}{1000 \text{ cm}^3}\right) \left(6.022 \times 10^{23} \text{ ions mol}^{-1}\right) = 6.022 \times 10^{19} \text{ ions}$$

Very few of the K^+ ions in the solution are involved in establishing the membrane potential.

10.30 Look up the values of E° for the following half-cell reactions:

$$Ag^+ + e^- \rightarrow Ag$$

$$AgBr + e^- \rightarrow Ag + Br^-$$

Describe how you would use these values to determine the solubility product (K_{sp}) of AgBr at 25°C.

$$Ag^+ + e^- \rightarrow Ag \qquad\qquad E^\circ = 0.800 \text{ V}$$

$$AgBr + e^- \rightarrow Ag + Br^- \qquad\qquad E^\circ = 0.0713 \text{ V}$$

The solubility product of AgBr is the equilibrium constant for the following reaction:

$$AgBr \rightarrow Ag^+ + Br^-$$

and the standard emf for this reaction is given by those for the half reactions above.

$$E^\circ = 0.0713 \text{ V} - 0.800 \text{ V} = -0.7287 \text{ V}$$

Therefore,

$$K_{sp} = \exp\left(\frac{vFE^\circ}{RT}\right) = \exp\left[\frac{(96500 \text{ C mol}^{-1})\,(-0.7287 \text{ V})}{(8.314 \text{ J K}^{-1} \text{ mol}^{-1})\,(298.2 \text{ K})}\right] = 4.81 \times 10^{-13}$$

10.31   A well-known organic redox system is the quinone–hydroquinone couple. In an aqueous solution at a pH below 8, we have

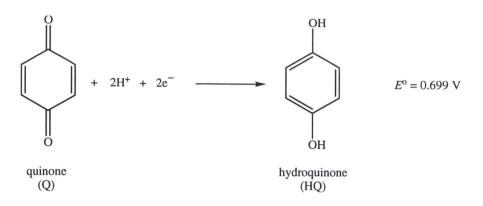

quinone
(Q)

hydroquinone
(HQ)

$E^\circ = 0.699$ V

This system can be prepared by dissolved quinhydrone, QH (a complex consisting of equimolar amounts of Q and HQ), in water. A quinhydrone electrode can be constructed by immersing a piece of platinum wire in a quinhydrone solution. **(a)** Derive an expression for the electrode potential of this couple in terms of E° and the hydrogen-ion concentration. **(b)** When the quinone–hydroquinone couple is joined to a saturated calomel electrode, the emf of the cell is found to be 0.18 V. In this arrangement the saturated calomel electrode acts as the anode. Calculate the pH of the quinhydrone solution. Assume the temperature is 25°C.

(a)

$$E_{QH} = E^\circ - \frac{0.0257 \text{ V}}{\nu} \ln \frac{[HQ]}{[Q]\left[H^+\right]^2}$$

$$= 0.699 \text{ V} - \frac{0.0257 \text{ V}}{2} \ln \frac{1}{\left[H^+\right]^2}$$

$$= 0.699 \text{ V} - \frac{0.0257 \text{ V}}{2} (2)(2.303)\left(- \log\left[H^+\right]\right)$$

$$= 0.699 \text{ V} - (0.0592 \text{ pH}) \text{ V}$$

(b) The standard reduction potential for calomel electrode is

$$Hg_2Cl_2 + 2e^- \rightarrow 2Hg + 2Cl^- \qquad E^\circ = 0.268 \text{ V}$$

The quinhydrone electrode serves as the cathode, and the calomel electrode serves as the anode. The measured emf for the cell can be expressed using the emf of these half-cells.

$$E_{QH} - 0.268 \text{ V} = 0.18 \text{ V}$$

$$0.699 \text{ V} - (0.0592 \text{ pH}) \text{ V} - 0.268 \text{ V} = 0.18 \text{ V}$$

$$\text{pH} = 4.2$$

10.32 A 25.0-mL quantity of a 0.10 M Fe^{2+} solution is titrated against a 0.10 M Ce^{4+} solution added from a buret. What is the emf after the addition of 30.0 mL of the Ce^{4+} solution? The temperature is at 298 K.

The reaction is

$$Fe^{2+} + Ce^{4+} \rightarrow Fe^{3+} + Ce^{3+}$$

The standard emf for this reaction is

$$E^{\circ} = 1.72 \text{ V} - 0.771 \text{ V} = 0.949 \text{ V}$$

The emf depends on the concentration of various ions after the addition of Ce^{4+} to the Fe^{2+} solution. These concentrations can be calculated using the initial concentrations of the solutions and the equilibrium constant.

Before the reaction, the number of moles of Fe^{2+} and Ce^{4+} are

$$n_{Fe^{2+}} = (0.10 \, M) \left(25.0 \times 10^{-3} \text{ L}\right) = 2.50 \times 10^{-3} \text{ mol}$$

$$n_{Ce^{4+}} = (0.10 \, M) \left(30.0 \times 10^{-3} \text{ L}\right) = 3.00 \times 10^{-3} \text{ mol}$$

Convert these numbers of moles to concentrations using a final volume of 55.0 mL.

$$M_{Fe^{2+}} = \frac{2.50 \times 10^{-3} \text{ mol}}{55.0 \times 10^{-3} \text{ L}} = 4.545 \times 10^{-2} \, M$$

$$M_{Ce^{4+}} = \frac{3.00 \times 10^{-3} \text{ mol}}{55.0 \times 10^{-3} \text{ L}} = 5.455 \times 10^{-2} \, M$$

The equilibrium constant is calculated using the standard emf of the reaction

$$K = \exp\left(\frac{\nu F E^{\circ}}{RT}\right) = \exp\left[\frac{(96500 \text{ C mol}^{-1})\,(0.949 \text{ V})}{(8.314 \text{ J K}^{-1}\,\text{mol}^{-1})\,(298 \text{ K})}\right] = 1.13 \times 10^{16}$$

Since the equilibrium constant is very large, the concentration of Fe^{2+} at equilibrium should be very small. Suppose at equilibrium, $[Fe^{2+}] = x \, M$, then the concentrations of the other ions are

$$[Ce^{4+}] = 5.455 \times 10^{-2} \, M - 4.545 \times 10^{-2} \, M + x \, M = \left(9.10 \times 10^{-3} + x\right) M$$

$$[Fe^{3+}] = \left(4.545 \times 10^{-2} - x\right) M$$

$$[Ce^{3+}] = \left(4.545 \times 10^{-2} - x\right) M$$

The equilibrium expression is

$$K = 1.13 \times 10^{16} = \frac{\left[Fe^{3+}\right]\left[Ce^{3+}\right]}{\left[Fe^{2+}\right]\left[Ce^{4+}\right]} = \frac{\left(4.545 \times 10^{-2} - x\right)\left(4.545 \times 10^{-2} - x\right)}{x\left(9.10 \times 10^{-3} + x\right)}$$

Since x is very small, the above expression can be approximated by

$$1.13 \times 10^{16} = \frac{\left(4.545 \times 10^{-2}\right)^2}{x \left(9.10 \times 10^{-3}\right)}$$

$$x = 2.01 \times 10^{-17} \, M$$

Therefore, at equilibrium,

$$[Ce^{4+}] = \left(9.10 \times 10^{-3} + x\right) M = 9.10 \times 10^{-3} \, M$$

$$[Fe^{3+}] = \left(4.545 \times 10^{-2} - x\right) M = 4.545 \times 10^{-2} \, M$$

$$[Ce^{3+}] = \left(4.545 \times 10^{-2} - x\right) M = 4.545 \times 10^{-2} \, M$$

The emf depends only upon the $Ce^{4+} \mid Ce^{3+}$ couple. The standard reduction emf for this couple is 1.72 V.

$$E = 1.72 \text{ V} - 0.0257 \text{ V} \ln \frac{[Ce^{3+}]}{[Ce^{4+}]} = 1.72 \text{ V} - 0.0257 \text{ V} \ln \frac{4.545 \times 10^{-2}}{9.10 \times 10^{-3}} = 1.68 \text{ V}$$

10.33 One way to prevent a buried iron pipe from rusting is to connect it with a piece of wire to a magnesium or zinc rod. What is the electrochemical principle for this action?

Both $Mg^{2+} \mid Mg$ ($E^\circ = -2.372$ V) and $Zn^{2+} \mid Zn$ ($E^\circ = -0.762$ V) are more electropositive than $Fe^{2+} \mid Fe$ ($E^\circ = -0.447$ V). The more electropositive metals will be preferentially oxidized, protecting the iron pipe. These protective electrodes are often called sacrificial anodes.

10.34 Aluminum has a more negative standard reduction potential than iron. Yet aluminum does not form rust or corrode as easily as iron. Explain.

In fact aluminum does rust to form Al_2O_3. Al_2O_3, however, forms a thin, tough layer over the metallic aluminum underneath, isolating it from the environment and protecting it from further corrosion.

10.35 Given that the $\Delta_r S^\circ$ value for the Daniell cell is $-21.7 \text{ J K}^{-1} \text{ mol}^{-1}$, calculate the temperature coefficient $\left(\partial E^\circ / \partial T\right)_P$ of the cell and the emf of the cell at 80°C.

The temperature coefficient is directly related to $\Delta_r S^\circ$.

$$\Delta_r S^\circ = \nu F \left(\frac{\partial E^\circ}{\partial T}\right)_P$$

$$\left(\frac{\partial E^\circ}{\partial T}\right)_P = \frac{\Delta_r S^\circ}{\nu F} = \frac{-21.7 \text{ J K}^{-1} \text{ mol}^{-1}}{2 \left(96500 \text{ C mol}^{-1}\right)}$$

$$= -1.124 \times 10^{-4} \text{ V K}^{-1} = -1.12 \times 10^{-4} \text{ V K}^{-1}$$

The standard emf of the cell at 80°C can be calculated using the temperature coefficient and the standard emf (1.104 V) at 25°C.

$$\left(\frac{\partial E^{\circ}}{\partial T}\right)_P = \frac{E^{\circ}_{353.2\,K} - E^{\circ}_{298.2\,K}}{353.2\,K - 298.2\,K} = \frac{E^{\circ}_{353.2\,K} - 1.104\,V}{55.0\,K} = 1.124 \times 10^{-4}\,V\,K^{-1}$$

$$E^{\circ}_{353.2\,K} = 1.098\,V$$

Note that because of the small temperature coefficient, E° changes only slightly with temperature.

10.36   For years it was not clear whether mercury(I) ions existed in solution as Hg^+ or as Hg_2^{2+}. To distinguish between these two possibilities, we could set up the following system:

$$Hg(l) \mid soln\ A \parallel soln\ B \mid Hg(l)$$

where solution A contained 0.263 g mercury(I) nitrate per liter and solution B contained 2.63 g mercury(I) nitrate per liter. If the measured emf of such a cell is 0.0289 V at 18°C, what can you deduce about the nature of the mercury ions?

The cell described in the question is a concentration cell (that is, $E^{\circ} = 0$ V) and the emf depends on the nature of the mercury ions. Note that

$$\frac{[\text{mercury ions}]_A}{[\text{mercury ions}]_B} = 0.100$$

Assuming mercury(I) ions exist in solution as Hg^+, the half reactions are

Anode: $Hg \rightarrow Hg^+ (Soln\ A) + e^-$

Cathode: $Hg^+ (Soln\ B) + e^- \rightarrow Hg$

The overall reaction is

$$Hg^+ (Soln\ B) \rightarrow Hg^+ (Soln\ A)$$

and the cell emf is

$$E = E^{\circ} - \frac{RT}{\nu F}\ln\frac{[Hg^+]_A}{[Hg^+]_B} = 0\,V - \frac{(8.314\,J\,K^{-1}\,mol^{-1})\,(291.2\,K)}{96500\,C\,mol^{-1}}\ln 0.100 = 0.0578\,V$$

This emf does not match the measured emf.

Assuming mercury(I) ions exist in solution as Hg_2^{2+}, the half reactions are

Anode: $Hg \rightarrow Hg_2^{2+} (Soln\ A) + 2e^-$

Cathode: $Hg_2^{2+} (Soln\ B) + 2e^- \rightarrow Hg$

The overall reaction is

$$Hg_2^{2+} (Soln\ B) \rightarrow Hg_2^{2+} (Soln\ A)$$

and the cell emf is

$$E = E^\circ - \frac{RT}{\nu F} \ln \frac{[Hg^+]_A}{[Hg^+]_B} = 0 \text{ V} - \frac{(8.314 \text{ J K}^{-1} \text{ mol}^{-1}) (291.2 \text{ K})}{2 (96500 \text{ C mol}^{-1})} \ln 0.1 = 0.0289 \text{ V}$$

This emf matches the measured emf. Thus, mercury(I) ions exist as Hg_2^{2+}.

10.37 Given the following standard reduction potentials, calculate the ion-product K_w value ($[H^+][OH^-]$) at 25°C:

$$2H^+(aq) + 2e^- \rightarrow H_2(g) \qquad\qquad\qquad E^\circ = 0.00 \text{ V}$$

$$2H_2O(l) + 2e^- \rightarrow H_2(g) + 2OH^-(aq) \qquad\qquad E^\circ = -0.828 \text{ V}$$

The ion-product is the equilibrium constant for the following process.

$$H_2O \rightleftharpoons H^+ + OH^-$$

The standard emf for this reaction can be calculated using those for the half-cells described in the question.

$$E^\circ = -0.828 \text{ V} - 0 \text{ V} = -0.828 \text{ V}$$

The ion-product is calculated from E°.

$$K_w = \exp\left(\frac{\nu F E^\circ}{RT}\right) = \exp\left[\frac{(96500 \text{ C mol}^{-1}) (-0.828 \text{ V})}{(8.314 \text{ J K}^{-1} \text{ mol}^{-1}) (298.2 \text{ K})}\right] = 1.01 \times 10^{-14}$$

10.38 Given that

$$2Hg^{2+}(aq) + 2e^- \rightarrow Hg_2^{2+}(aq) \qquad\qquad E^\circ = 0.920 \text{ V}$$

$$Hg_2^{2+}(aq) + 2e^- \rightarrow 2Hg(l) \qquad\qquad E^\circ = 0.797 \text{ V}$$

Calculate the values of $\Delta_r G^\circ$ and K for the following process at 25°C:

$$Hg_2^{2+}(aq) \rightarrow Hg^{2+}(aq) + Hg(l)$$

(The above reaction is an example of a *disproportionation reaction*, in which an element in one oxidation state is both oxidized and reduced.)

The sum of the two half-reactions

$$Hg_2^{2+}(aq) \rightarrow 2Hg^{2+}(aq) + 2e^-$$

$$Hg_2^{2+}(aq) + 2e^- \rightarrow 2Hg(l)$$

gives

$$2Hg_2^{2+}(aq) \rightarrow 2Hg^{2+}(aq) + 2Hg(l)$$

Therefore, the emf for the reaction is

$$E° = 0.797 \text{ V} - 0.920 \text{ V} = -0.123 \text{ V}$$

Since emf is an intensive property, this is also the emf for

$$Hg_2^{2+}(aq) \rightarrow Hg^{2+}(aq) + Hg(l)$$

In this particular reaction, only 1 mole of electrons is transferred. The standard Gibbs energy and equilibrium constant are

$$\Delta_r G° = -\nu F E° = - \left(96500 \text{ C mol}^{-1}\right)(-0.123 \text{ V}) = 1.187 \times 10^4 \text{ J mol}^{-1} = 1.19 \times 10^4 \text{ J mol}^{-1}$$

$$K = \exp\left(-\frac{\Delta_r G°}{RT}\right) = \exp\left[-\frac{1.187 \times 10^4 \text{ J mol}^{-1}}{\left(8.314 \text{ J K}^{-1} \text{ mol}^{-1}\right)(298 \text{ K})}\right] = 8.30 \times 10^{-3}$$

10.39 The magnitudes of the standard electrode potentials of two metals, X and Y, are

$$X^{2+} + 2e^- \rightarrow X \qquad\qquad |E°| = 0.25 \text{ V}$$

$$Y^{2+} + 2e^- \rightarrow Y \qquad\qquad |E°| = 0.34 \text{ V}$$

where the || notation denotes that only the magnitude (but *not* the sign) of the $E°$ value is shown. When the half-cells of X and Y are connected, electrons flow from X to Y. When X is connected to a SHE, electrons flow from X to SHE. **(a)** Which value of $E°$ is positive and which is negative? **(b)** What is the standard emf of a cell made up of X and Y?

(a) Since electrons flow from X to SHE, the anode reaction involves X^{2+} | X. The standard emf for a SHE is 0 V. Therefore,

$$E° = E°_{\text{cathode}} - E°_{\text{anode}} = 0 - E°_{\text{anode}}$$

must be positive. In other words, the standard emf for

$$X^{2+} + 2e^- \rightarrow X$$

is −0.25 V.

Since electrons flow from X to Y, the anode reaction involves X^{2+} | X and the cathode reaction involves Y^{2+} | Y:

Anode:	$X \rightarrow X^{2+} + 2e^-$	$E° = 0.25 \text{ V}$
Cathode:	$Y^{2+} + 2e^- \rightarrow Y$	$\| E° \| = 0.34 \text{ V}$

The emf for the reaction must be positive. Thus, the reduction potential for the Y^{2+} | Y pair is positive, that is, $E° = 0.34$ V.

(b) The standard emf of a cell made up of X and Y is

$$E^\circ = 0.34 \text{ V} - (-0.25 \text{ V}) = 0.59 \text{ V}$$

10.40 An electrochemical cell is constructed as follows. One half-cell consists of a platinum wire immersed in a solution containing $1.0 \, M \, Sn^{2+}$ and $1.0 \, M \, Sn^{4+}$, and the other half-cell has a thallium rod immersed in a solution of $1.0 \, M \, Tl^+$. **(a)** Write the half-cell reactions and the overall reaction. **(b)** What is the equilibrium constant at 25°C? **(c)** What is the cell voltage if the Tl^+ concentration is increased tenfold?

(a) The reduction potentials of the relevant half-reactions are

$$Sn^{4+} + 2e- \rightarrow Sn^{2+} \qquad\qquad E^\circ = 0.151 \text{ V}$$

$$Tl^+ + e^- \rightarrow Tl \qquad\qquad E^\circ = -0.336 \text{ V}$$

To obtain a positive standard emf, the overall reaction must be

$$Sn^{4+} + 2Tl \rightarrow Sn^{2+} + 2Tl^+ \qquad E^\circ = 0.151 \text{ V} - (-0.336 \text{ V}) = 0.487 \text{ V}$$

(b)

$$K = \exp\left(\frac{\nu F E^\circ}{RT}\right) = \exp\left[\frac{2\left(96500 \text{ C mol}^{-1}\right)(0.487 \text{ V})}{\left(8.314 \text{ J K}^{-1} \text{ mol}^{-1}\right)(298.2 \text{ K})}\right] = 2.92 \times 10^{16}$$

(c) $[Tl^+] = 10.0 \, M$

$$E = E^\circ - \frac{0.0257 \text{ V}}{\nu} \ln \frac{\left[Sn^{2+}\right]\left[Tl^+\right]^2}{\left[Sn^{4+}\right]} = 0.487 \text{ V} - \frac{0.0257 \text{ V}}{2} \ln \frac{(1.0)(10.0)^2}{1.0} = 0.428 \text{ V}$$

10.41 Given the standard reduction potential for Au^{3+} in Table 10.1 and

$$Au^+(aq) + e^- \rightarrow Au(s) \qquad E^\circ = 1.69 \text{ V}$$

answer the following questions. **(a)** Why does gold not tarnish in air? **(b)** Will the following disproportionation occur spontaneously?

$$3Au^+(aq) \rightarrow Au^{3+}(aq) + 2Au(s)$$

(c) Predict the reaction between gold and fluorine gas.

(a) Gold does not tarnish (oxidize) in air because the reduction potential for oxygen is insufficient to oxidize gold (either to Au^+ as given or to Au^{3+} with $E^\circ = 1.498$ V).

$$O_2 + 4H^+ + 4e^- \rightarrow 2H_2O \qquad E^\circ = 1.229 \text{ V}$$

(b) The disproportionation can be considered as the sum of the two half reactions,

$$3 \left(Au^+ + e^- \rightarrow Au \right)$$

$$Au \rightarrow Au^{3+} + 3e^-$$

$$-1.498V$$

so that the standard emf for the reaction is

$$E^\circ = 1.692 \text{ V} - 1.498 \text{ V} = 0.194 \text{ V}.$$

Since E° for the reaction is positive, the disproportionation will be spontaneous under standard state conditions.

(c) With $E^\circ = 2.87$ V, the half reaction

$$F_2 + 2e^- \rightarrow 2F^-$$

is able to oxidize gold completely to Au^{3+} via

$$3F_2 + 2Au \rightarrow 2AuF_3$$

10.42   Consider the Daniell cell shown in Figure 10.1. In the diagram, the anode appears to be negative and the cathode positive (electrons are flowing from the anode to the cathode). Yet anions in solution are moving toward the anode, which must therefore seem positive to the anions. Because the anode cannot simultaneously be negative and positive, give an explanation for this apparently contradictory situation.

At the Zn electrode (anode)/solution interface, Zn atoms are being oxidized to form Zn^{2+} ions and electrons. The electrons enter the bulk metal of the electrode, making it negative and driving electrons towards the cathode. The cations in the solution immediately surrounding the electrode give this region of the solution a net positive charge, and it is to this that the anions are attracted.

10.43   Calculate the pressure of H_2 (in bar) required to maintain equilibrium with respect to the following reaction at 25°C

$$Pb(s) + 2H^+(aq) \rightleftharpoons Pb^{2+}(aq) + H_2(g)$$

given that $[Pb^{2+}] = 0.035 \ M$ and the solution is buffered at pH 1.60.

The cell reaction is

Anode:	$Pb \rightarrow Pb^{2+} + 2e^-$
Cathode:	$2H^+ + 2e^- \rightarrow H_2$

The standard emf for the reaction

$$Pb(s) + 2H^+(aq) \rightleftharpoons Pb^{2+}(aq) + H_2(g)$$

is

$$E^\circ = 0 \text{ V} - (-0.126 \text{ V}) = 0.126 \text{ V}$$

The emf of the cell is

$$E = E^\circ - \frac{0.0257 \text{ V}}{\nu} \ln \frac{[Pb^{2+}] P_{H_2}}{[H^+]^2}$$

From the pH, $[H^+] = 10^{-1.60} = 2.51 \times 10^{-2}$ M. At equilibrium, $E = 0$. Therefore, the above equation becomes

$$0 = 0.126 - \frac{0.0257 \text{ V}}{2} \ln \frac{(0.035) P_{H_2}}{(2.51 \times 10^{-2})^2}$$

$$P_{H_2} = 3.3 \times 10^2 \text{ bar}$$

10.44   Use the data in Appendix B and the convention that $\Delta_f \overline{G}^\circ [H^+(aq)] = 0$ to determine the standard reduction potentials for sodium and fluorine. (Like sodium, fluorine also reacts violently with water.)

To determine the standard reduction potential for sodium, consider the following reaction

$$Na^+(aq) + \tfrac{1}{2} H_2(g) \rightarrow Na(s) + H^+(aq)$$

The standard Gibbs energy for this reaction is

$$\Delta_r G^\circ = \Delta_f \overline{G}^\circ [Na(s)] + \Delta_f \overline{G}^\circ [H^+(aq)] - \Delta_f \overline{G}^\circ [Na^+(aq)] - \frac{1}{2} \Delta_f \overline{G}^\circ [H_2(g)]$$

$$= 0 \text{ kJ mol}^{-1} + 0 \text{ kJ mol}^{-1} - (-261.9 \text{ kJ mol}^{-1}) - \frac{1}{2} (0 \text{ kJ mol}^{-1})$$

$$= 261.9 \text{ kJ mol}^{-1}$$

The standard emf for the reaction is

$$E^\circ = -\frac{\Delta_r G^\circ}{\nu F} = -\frac{261.9 \times 10^3 \text{ J mol}^{-1}}{96500 \text{ C mol}^{-1}} = -2.71 \text{ V}$$

Since $E^\circ = E^\circ_{cathode} - E^\circ_{anode}$, where the half-reactions are

Anode: $H_2 \rightarrow 2H^+ + e^-$

Cathode: $Na^+ + e^- \rightarrow Na$

the reduction potential of Na is -2.71 V.

To determine the standard reduction potential for fluorine, consider the following reaction

$$\tfrac{1}{2} F_2(g) + \tfrac{1}{2} H_2(g) \rightarrow F^-(aq) + H^+ (aq)$$

The standard Gibbs energy for this reaction is

$$\Delta_r G^\circ = \Delta_f \overline{G}^\circ \left[F^-(aq) \right] + \Delta_f \overline{G}^\circ \left[H^+(aq) \right] - \frac{1}{2} \Delta_f \overline{G}^\circ \left[F_2(g) \right] - \frac{1}{2} \Delta_f \overline{G}^\circ \left[H_2(g) \right]$$

$$= -276.5 \text{ kJ mol}^{-1} + 0 \text{ kJ mol}^{-1} - \frac{1}{2} \left(0 \text{ kJ mol}^{-1} \right) - \frac{1}{2} \left(0 \text{ kJ mol}^{-1} \right)$$

$$= -276.5 \text{ kJ mol}^{-1}$$

The standard emf for the reaction is

$$E^\circ = -\frac{\Delta_r G^\circ}{\nu F} = -\frac{-276.5 \times 10^3 \text{ J mol}^{-1}}{96500 \text{ C mol}^{-1}} = 2.87 \text{ V}$$

Since $E^\circ = E^\circ_{\text{cathode}} - E^\circ_{\text{anode}}$, where the half-reactions are

Anode: $\qquad\qquad\qquad\qquad\qquad\qquad H_2 \rightarrow 2H^+ + e^-$

Cathode: $\qquad\qquad\qquad\qquad\qquad\quad F_2 + 2e^- \rightarrow 2F^-$

the reduction potential of F_2 is 2.87 V

10.45   Use the data in Table 10.1 to determine the value of $\Delta_f \overline{G}^\circ$ for $Fe^{2+}(aq)$.

The reduction potential for the $Fe^{2+} \mid Fe$ couple is

$$Fe^{2+} + 2e^- \rightarrow Fe \qquad E^\circ = -0.447 \text{ V}$$

Consider a cell constructed using the following half-cells:

Anode: $\qquad\qquad\qquad\qquad\qquad Fe(s) \rightarrow Fe^{2+}(aq) + 2e^-$

Cathode: $\qquad\qquad\qquad\qquad\quad 2H^+(aq) + 2e^- \rightarrow H_2(g)$

The overall equation and standard emf are

$$Fe(s) + 2H^+(aq) \rightarrow Fe^{2+}(aq) + H_2(g) \qquad E^\circ = 0.447 \text{ V}$$

The standard Gibbs energy for this reaction is related to both the standard Gibbs energy of formation of the reactants and products and the standard emf of the reaction.

$$\Delta_r G^\circ = -\nu F E^\circ = \Delta_f \overline{G}^\circ \left[Fe^{2+}(aq) \right] + \Delta_f \overline{G}^\circ \left[H_2(g) \right] - \Delta_f \overline{G}^\circ \left[Fe(s) \right]$$

$$- 2\Delta_f \overline{G}^\circ \left[H^+(aq) \right]$$

$$-2 \left(96500 \text{ C mol}^{-1} \right) (0.447 \text{ V}) = \Delta_f \overline{G}^\circ \left[Fe^{2+}(aq) \right] + 0 \text{ kJ mol}^{-1} - 0 \text{ kJ mol}^{-1} - 2 \left(0 \text{ kJ mol}^{-1} \right)$$

$$\Delta_f \overline{G}^\circ \left[Fe^{2+}(aq) \right] = -8.63 \times 10^4 \text{ J mol}^{-1}$$

10.46   Consider the following cell

$$\text{Pt} \mid \text{H}_2(1 \text{ bar}) \mid \text{HCl}(m) \mid \text{AgCl}(s) \mid \text{Ag}$$

At 25°C, the emf values at various molalities are given by

$m/\text{mol·kg}^{-1}$	0.124	0.0539	0.0256	0.0134	0.00914	0.00562	0.00322
E/V	0.342	0.382	0.418	0.450	0.469	0.493	0.521

(a) Determine the value of $E°$ graphically. Compare your value of $E°$ with that listed in Table 10.1. **(b)** Calculate the mean activity coefficient ($\gamma_\pm$) for HCl at 0.124 m.

(a) Referring to the discussion in Section 10.6, prepare a graph of $y = E + 0.0514 \text{ V} \ln m$ versus m.

$m/\text{mol·kg}^{-1}$	0.124	0.0539	0.0256	0.0134	0.00914	0.00562	0.00322
y/V	0.2347	0.2319	0.2296	0.2283	0.2277	0.2267	0.2260

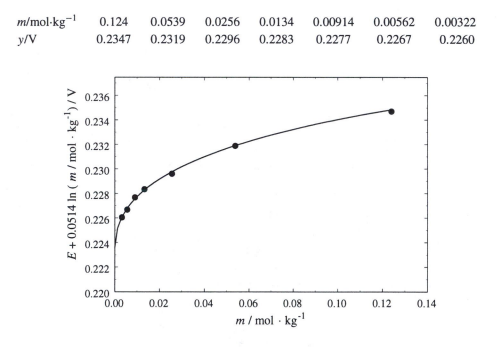

Extrapolating to zero molality gives $E° = 0.223$ V, although there is considerable uncertainty in determining $E°$ by this method. A reasonable estimate would be $E° = 0.223 \pm 0.002$ V. This does compare well, however, with the value from Table 10.1, $E° = 0.222$ V.

(b) At $m = 0.124 \text{ mol kg}^{-1}$, $E = 0.342$ V. From Section 10.6,

$$E + (0.0514 \text{ V}) \ln m = E° - (0.0514 \text{ V}) \ln \gamma_\pm$$

$$0.342 \text{ V} + (0.0514 \text{ V}) \ln 0.124 = 0.222 \text{ V} - (0.0514 \text{ V}) \ln \gamma_\pm$$

$$\ln \gamma_\pm = -0.247$$

$$\gamma_\pm = 0.781$$

where the more precisely determined value from Table 10.1 is used for $E°$. [Using the value determined in part **(a)** results in $\gamma_\pm = 0.796$.] These values compare well with the data shown in Figure 8.8.

CHAPTER 11

Acids and Bases

PROBLEMS AND SOLUTIONS

11.1 Classify each of the following species as a Brønsted acid or base, or both: **(a)** H_2O, **(b)** OH^-, **(c)** H_3O^+, **(d)** NH_3, **(e)** NH_4^+, **(f)** NH_2^-, **(g)** NO_3^-, **(h)** CO_3^{2-}, **(i)** HBr, **(j)** HCN, **(k)** HCO_3^-.

The following table classifies each species as a Brønsted acid or base, or both. In addition, the conjugate base or acid (or both) is listed.

		Brønsted acid?	Conjugate base	Brønsted base?	Conjugate acid
(a)	H_2O	Yes	OH^-	Yes	H_3O^+
(b)	OH^-	No	–	Yes	H_2O
(c)	H_3O^+	Yes	H_2O	No	–
(d)	NH_3	Yes	NH_2^{-*}	Yes	NH_4^+
(e)	NH_4^+	Yes	NH_3	No	–
(f)	NH_2^-	No	–	Yes	NH_3
(g)	NO_3^-	No	–	Yes	HNO_3
(h)	CO_3^{2-}	No	–	Yes	HCO_3^-
(i)	HBr	Yes	Br^-	No	–
(j)	HCN	Yes	CN^-	No	–
(k)	HCO_3^-	Yes	CO_3^{2-}	Yes	H_2CO_3

* In pure liquid form NH_3 can act as a Brønsted acid, but in aqueous form it can only act as a Brønsted base.

11.2 Write the formulas for the conjugate bases of the following acids: **(a)** HI, **(b)** H_2SO_4, **(c)** H_2S, **(d)** HCN, **(e)** HCOOH (formic acid).

(a) I^- **(b)** HSO_4^- **(c)** HS^- **(d)** CN^- **(e)** $HCOO^-$

11.3 Classify each of the following species as a weak or strong acid: **(a)** HNO_3, **(b)** HF, **(c)** H_2SO_4, **(d)** HSO_4^-, **(e)** H_2CO_3, **(f)** HCO_3^-, **(g)** HCl, **(h)** HCN, **(i)** HNO_2.

(a) Strong, (b) weak, (c) strong (first stage dissociation), (d) weak, (e) weak, (f) weak, (g) strong, (h) weak, (i) weak.

11.4   Classify each of the following species as a weak or strong base: (a) LiOH, (b) CN^-, (c) H_2O, (d) ClO_4^-, (e) NH_2^-.

(a) Actually LiOH is not a Brønsted base, but it dissociates into the ions Li^+ and OH^- in aqueous solution. OH^- is a strong base (the strongest base that can exist in aqueous solution). (b) Weak, (c) weak, (d) weak, (e) strong.

11.5   Calculate the pH of the following solutions: (a) 1.0 M HCl, (b) 0.10 M HCl, (c) 1.0×10^{-2} M HCl, (d) 1.0×10^{-2} M NaOH, (e) 1.0×10^{-2} M Ba(OH)$_2$. Assume ideal behavior.

Each of the compounds is a strong electrolyte and dissociates completely in aqueous solution.

(a) $[H^+] = [HCl]_0 = 1.0$ M. Thus, pH $= -\log(1.0) = 0.00$

(b) $[H^+] = [HCl]_0 = 0.10$ M. Thus, pH $= -\log(0.10) = 1.00$

(c) $[H^+] = [HCl]_0 = 1.0 \times 10^{-2}$ M. Thus, pH $= -\log(1.0 \times 10^{-2}) = 2.00$

(d) $[OH^-] = [NaOH]_0 = 1.0 \times 10^{-2}$ M. Thus, pOH $= -\log(1.0 \times 10^{-2}) = 2.00$ and pH $= 14.00 - $ pOH $= 12.00$

(e) $[OH^-] = 2[Ba(OH)_2]_0 = 2.0 \times 10^{-2}$ M. Thus, pOH $= -\log(2.0 \times 10^{-2}) = 1.699$ and pH $= 14.00 - $ pOH $= 12.30$

11.6   A 0.040 M solution of a monoprotic acid is 13.5% dissociated. What is the dissociation constant of the acid?

The dissociation of the acid can be represented by

$$HA \rightleftharpoons H^+ + A^-$$

At equilibrium,

$$[H^+] = [A^-] = (0.135)(0.040\ M) = 5.40 \times 10^{-3}\ M$$

$$[HA] = (1 - 0.135)(0.040\ M) = 3.46 \times 10^{-2}\ M$$

The dissociation constant, K_a, of the acid is therefore

$$K_a = \frac{[H^+][A^-]}{[HA]} = \frac{(5.40 \times 10^{-3})^2}{3.46 \times 10^{-2}} = 8.4 \times 10^{-4}$$

11.7   Write the equation relating K_a for a weak acid and K_b for its conjugate base. Use NH_3 and its conjugate acid NH_4^+ to derive the relationship between K_a and K_b.

The dissociation constant for the ionization of NH_4^+ ($NH_4^+ \rightleftharpoons NH_3 + H^+$) is expressed as

$$K_a = \frac{[NH_3][H^+]}{[NH_4^+]}$$

whereas the dissociation constant for the hydrolysis of NH_3, ($NH_3 + H_2O \rightleftharpoons NH_4^+ + OH^-$) is

$$K_b = \frac{[NH_4^+][OH^-]}{[NH_3]}$$

The product of K_a and K_b is seen to be

$$K_a K_b = \left(\frac{[NH_3][H^+]}{[NH_4^+]}\right)\left(\frac{[NH_4^+][OH^-]}{[NH_3]}\right) = [H^+][OH^-] = K_w$$

11.8 The dissociation constant of a monoprotic acid at 298 K is 1.47×10^{-3}. Calculate the degree of dissociation by **(a)** assuming ideal behavior and **(b)** using a mean activity coefficient $\gamma_\pm = 0.93$. The concentration of the acid is $0.010\ M$.

Let α be the degree of dissociation of the monoprotic acid. The corresponding concentrations of all species are

	HA	$\rightleftharpoons$	H^+	+	A^-	
Initial	0.010		0		0	M
At equilibrium	$0.010\,(1-\alpha)$		0.010α		0.010α	M

(a)

$$K_a = 1.47 \times 10^{-3} = \frac{[H^+][A^-]}{[HA]} = \frac{(0.010\alpha)^2}{0.010\,(1-\alpha)}$$

$$1.0 \times 10^{-4}\alpha^2 + 1.47 \times 10^{-5}\alpha - 1.47 \times 10^{-5} = 0$$

$$\alpha = 0.32$$

Therefore, assuming ideal behavior, the acid is 32% dissociated.

(b)

$$K_a = \frac{a_{H^+} a_{A^-}}{a_{HA}} = \frac{[H^+]\gamma_+[A^-]\gamma_-}{[HA]\gamma_{HA}}$$

Since HA is an uncharged species and the solution is dilute, γ_{HA} is approximately 1. Furthermore, $\gamma_+\gamma_- = \gamma_\pm^2$. The K_a expression becomes

$$K_a = 1.47 \times 10^{-3} = \frac{[H^+][A^-]\gamma_\pm^2}{[HA]} = \frac{(0.010\alpha)^2(0.93)^2}{0.010\,(1-\alpha)}$$

$$8.65 \times 10^{-5}\alpha^2 + 1.47 \times 10^{-5}\alpha - 1.47 \times 10^{-5} = 0$$

$$\alpha = 0.34$$

Therefore, accounting for non-ideality, the acid is 34% dissociated.

11.9 The ion product of D_2O is 1.35×10^{-15} at 25°C. **(a)** Calculate the value of pD for pure D_2O where $pD = -\log[D^+]$. **(b)** For what values of pD will a solution be acidic in D_2O? **(c)** Derive a relation between pD and pOD.

(a) The autoionization reaction for D_2O is

$$D_2O \rightleftharpoons D^+ + OD^-$$

and the expression for the equilibrium constant is

$$K = 1.35 \times 10^{-15} = [D^+][OD^-]$$

Since $[D^+] = [OD^-]$, the above expression becomes

$$[D^+]^2 = 1.35 \times 10^{-15}$$

$$[D^+] = 3.674 \times 10^{-8}$$

$$pD = -\log(3.674 \times 10^{-8}) = 7.435$$

(b) When $[D^+] > [OD^-]$, that is, when $pD < 7.435$, the solution will be acidic in D_2O.

(c) Starting with $K = [D^+][OD^-]$, the following expression is obtained.

$$K = [D^+][OD^-]$$

$$-\log K = -\log([D^+][OD^-])$$

$$= -\log[D^+] - \log[OD^-]$$

$$pK = pD + pOD = 14.870$$

11.10 HF is a weak acid, but its strength increases with concentration. Explain. (*Hint*: F^- reacts with HF to form HF_2^-. The equilibrium constant for this reaction is 5.2 at 25°C.)

In addition to the dissociation of HF,

$$HF \rightleftharpoons H^+ + F^-$$

there is also the reaction

$$F^- + HF \rightleftharpoons HF_2^-$$

The second equilibrium has a fairly large equilibrium constant ($K = 5.2$) so that much of the F^- produced in the first reaction is removed in the second step. Applying Le Chatelier's principle, more HF must dissociate to compensate for the removal of F^-, at the same time producing more H^+. With increasing HF concentration, Le Chatelier's principle applied to the second equation requires that even more F^- be removed and thus further dissociation of HF.

11.11 When the concentration of a strong acid is not substantially higher than 1.0×10^{-7} M, the ionization of water must be taken into account in the calculation of the solution's pH. **(a)** Derive an expression for the pH of a strong acid solution, including the contribution to $[H^+]$ from H_2O. **(b)** Calculate the pH of a 1.0×10^{-7} M HCl solution.

(a) The following equilibria have to be taken into account when the concentration of a strong acid is relatively low.

$$HA \rightleftharpoons H^+ + A^-$$

$$H_2O \rightleftharpoons H^+ + OH^-$$

Charge balance leads to the following condition:

$$[H^+] = [A^-] + [OH^-]$$

Since the concentration of OH^- is related to H^+ by

$$[OH^-] = \frac{K_w}{[H^+]},$$

the charge balance equation becomes

$$[H^+] = [A^-] + \frac{K_w}{[H^+]}$$

$$[H^+]^2 - [A^-][H^+] - K_w = 0$$

$$[H^+] = \frac{[A^-] + \sqrt{[A^-]^2 + 4K_w}}{2}$$

The nonphysical root, with a minus sign preceding the square root, has been discarded. The pH is then

$$pH = -\log\left\{ \frac{[A^-] + \sqrt{[A^-]^2 + 4K_w}}{2} \right\}$$

(b) The result from part **(a)** is used with Cl^- as A^-, and $[Cl^-] = 1.0 \times 10^{-7}$ M. The pH is

$$pH = -\log\left\{ \frac{1.0 \times 10^{-7} + \sqrt{(1.0 \times 10^{-7})^2 + 4(1.0 \times 10^{-14})}}{2} \right\} = 6.79$$

11.12 What are the concentrations of HSO_4^-, SO_4^{2-}, and H^+ in a 0.20 M $KHSO_4$ solution? (*Hint:* H_2SO_4 is a strong acid; K_a for $HSO_4^- = 1.3 \times 10^{-2}$.)

$KHSO_4$ is a strong electrolyte and dissociates completely into K^+ and HSO_4^-. K^+ is neither a Brønsted acid nor a Brønsted base and it does not hydrolyze. HSO_4^- functions as a Brønsted acid and a Brønsted base. As the Brønsted base of a strong acid, H_2SO_4, HSO_4^- does not hydrolyze to any significant extent to give the acid. As a Brønsted acid, it dissociates and makes the solution acidic. Assume that x M of HSO_4^- dissociates in the solution, then the equilibrium concentrations of various species are

$$HSO_4^- \quad \rightleftharpoons \quad H^+ \quad + \quad SO_4^{2-}$$

At equilibrium $0.20 - x$ $\quad$ x $\quad$ x M

x is determined using the equilibrium expression

$$K_a = 1.3 \times 10^{-2} = \frac{[H^+][SO_4^{2-}]}{[HSO_4^-]} = \frac{x^2}{0.2 - x}$$

$$x^2 + 1.3 \times 10^{-2}x - 2.6 \times 10^{-3} = 0$$

$$x = 4.49 \times 10^{-2}$$

Therefore,

$$[HSO_4^-] = (0.20 - x) \ M = 0.16 \ M$$

$$[H^+] = x \ M = 4.5 \times 10^{-2} \ M$$

$$[SO_4^{2-}] = x \ M = 4.5 \times 10^{-2} \ M$$

11.13   Calculate the concentrations of H^+, HCO_3^-, and CO_3^{2-} in a 0.025 M H_2CO_3 solution.

For the first-stage ionization of H_2CO_3, the concentrations of various species are

$$H_2CO_3 \quad \rightleftharpoons \quad H^+ \quad + \quad HCO_3^-$$

Initial $\quad$ 0.025 $\quad$ 0 $\quad$ 0 M

At equilibrium $0.025 - x$ $\quad$ x $\quad$ x M

$$K_a' = 4.2 \times 10^{-7} = \frac{[H^+][HCO_3^-]}{[H_2CO_3]} = \frac{x^2}{0.025 - x}$$

Since K_a' is very small, $0.025 - x$ can be assumed to be approximately 0.025, and the expression above simplifies to

$$4.2 \times 10^{-7} = \frac{x^2}{0.025}$$

$$x = 1.02 \times 10^{-4}$$

Checking the assumption,

$$\frac{x}{0.025} \times 100\% = 0.41\% < 5\%$$

Thus, the assumption is valid. After the first stage of ionization,

$$[H_2CO_3] = (0.025 - x) \ M = 0.025 \ M$$

$$[H^+] = x \ M = 1.02 \times 10^{-4} \ M$$

$$[HCO_3^-] = x \ M = 1.02 \times 10^{-4} \ M$$

The second-stage ionization involves HCO_3^-. The concentrations of various species are

$$
\begin{array}{lcccc}
 & HCO_3^- & \rightleftharpoons & H^+ & + & CO_3^{2-} \\
\text{Initial} & 1.02 \times 10^{-4} & & 1.02 \times 10^{-4} & & 0 & M \\
\text{At equilibrium} & 1.02 \times 10^{-4} - y & & 1.02 \times 10^{-4} + y & & y & M
\end{array}
$$

$$
K_a'' = 4.8 \times 10^{-11} = \frac{[H^+][CO_3^{2-}]}{[HCO_3^-]} = \frac{(1.02 \times 10^{-4} + y)\,y}{1.02 \times 10^{-4} - y}
$$

Since K_a'' is very small, both $1.02 \times 10^{-4} - y$ and $1.02 \times 10^{-4} + y$ can be approximated by 1.02×10^{-4}. The expression above simplifies to

$$
4.8 \times 10^{-11} = \frac{(1.02 \times 10^{-4})\,y}{1.02 \times 10^{-4}}
$$

$$
y = 4.8 \times 10^{-11}
$$

Checking the assumptions,

$$
\frac{y}{1.02 \times 10^{-4}} \times 100\% = 4.7 \times 10^{-5}\% < 5\%
$$

Thus, the assumption is valid.

After both stages of ionization,

$$
[H_2CO_3] = 0.025\ M
$$

$$
[H^+] = (1.02 \times 10^{-4} + y)\ M = 1.0 \times 10^{-4}\ M
$$

$$
[HCO_3^-] = (1.02 \times 10^{-4} - y)\ M = 1.0 \times 10^{-4}\ M
$$

$$
[CO_3^{2-}] = y\ M = 4.8 \times 10^{-11}\ M
$$

11.14   To which of the following would the addition of an equal volume of $0.60\ M$ NaOH lead to a solution having a lower pH? (a) Water, (b) $0.30\ M$ HCl, (c) $0.70\ M$ KOH, (d) $0.40\ M$ NaNO$_3$.

(a) The original solution has pH $= 7.00$, assuming it to be pure water. After the addition of an equal volume of $0.60\ M$ NaOH, the solution is $0.30\ M$ NaOH with pH $= 13.48$, which is higher.

(b) The original solution has pH $= -0.52$. After the addition of an equal volume of $0.60\ M$ NaOH, the solution is $0.15\ M$ NaOH (plus NaCl) with pH $= 13.18$, which is higher.

(c) The original solution has pH $= 13.85$. After the addition of an equal volume of $0.60\ M$ NaOH, the solution is $0.65\ M$ in OH$^-$ with pH $= 13.81$, which is lower.

(d) Since neither ion hydrolyzes to any extent, the original pH $= 7.00$, and the final pH is the same as in part (a), namely 13.48, which is higher.

Only the addition to solution (c) leads to a solution with lower pH.

11.15 A solution contains a weak monoprotic acid, HA, and its sodium salt, NaA, both at 0.1 M concentration. Show that $[OH^-] = K_w/K_a$.

The equilibrium reaction and the concentrations of various species are

$$
\begin{array}{lcccc}
 & HA & \rightleftharpoons & H^+ & + & A^- \\
\text{Initial} & 0.1 & & 0 & & 0.1 & M \\
\text{At equilibrium} & 0.1 - x & & x & & 0.1 + x & M
\end{array}
$$

$$K_a = \frac{[H^+][A^-]}{[HA]} = \frac{x\,(0.1 + x)}{0.1 - x}$$

x should be negligible compared with 0.1 M since the presence of A^- suppresses the ionization of HA and the presence of HA suppresses the hydrolysis of A^-. Therefore, the above expression becomes

$$K_a = \frac{x\,(0.1)}{0.1} = x = [H^+] = \frac{K_w}{[OH^-]}$$

$$[OH^-] = \frac{K_w}{K_a}$$

The assumption will be valid as long as $x < (0.05)(0.1)$, or $K_a < 5 \times 10^{-3}$.

11.16 A solution of methylamine (CH_3NH_2) has a pH of 10.64. How many grams of methylamine are in 100.0 mL of the solution?

Methylamine is a base. It ionizes to give equal concentrations of $CH_3NH_3^+$ and OH^-, which can be calculated from the pH of the solution.

$$pOH = 14.00 - pH = 14.00 - 10.64 = 3.36$$

$$[OH^-] = 10^{-3.36} = 4.37 \times 10^{-4}\ M$$

Let x M of methylamine be present initially. The equilibrium concentration is then $x - 4.37 \times 10^{-4}\ M$. The concentrations of various species are written below the chemical equation.

$$
\begin{array}{lccccc}
 & CH_3NH_2 & + & H_2O & \rightleftharpoons & CH_3NH_3^+ & + & OH^- \\
\text{Initial} & x & & & & 0 & & 0 & M \\
\text{At equilibrium} & x - 4.37 \times 10^{-4} & & & & 4.37 \times 10^{-4} & & 4.37 \times 10^{-4} & M
\end{array}
$$

$$K_b = 4.38 \times 10^{-4} = \frac{[CH_3NH_3^+][OH^-]}{[CH_3NH_2]} = \frac{(4.37 \times 10^{-4})^2}{x - 4.37 \times 10^{-4}}$$

$$x = 8.73 \times 10^{-4}$$

where x represents the initial concentration of CH_3NH_2 in solution.

The mass of methylamine can now be calculated.

$$\text{Number of moles of methylamine} = (8.73 \times 10^{-4}\ M)\,(0.100\ L) = 8.73 \times 10^{-5}\ \text{mol}$$

$$\text{Mass of methylamine} = (8.73 \times 10^{-5}\ \text{mol})\,(31.06\ \text{g mol}^{-1}) = 2.7 \times 10^{-3}\ \text{g}$$

11.17 Hydrocyanic acid (HCN) is a weak acid and a deadly poisonous compound that is used in gas chambers in the gaseous form (hydrogen cyanide). Why is it dangerous to treat sodium cyanide with acids (such as HCl) without proper ventilation?

As the conjugate base of a weak acid, CN^- from the sodium cyanide will associate with H^+ from the acid to form HCN as a major species in solution. The HCN has a tendency to escape into the gas phase, and without proper ventilation, this highly poisonous compound would pose a significant danger.

11.18 Novocaine, used as a local anesthetic by dentists, is a weak base ($K_b = 8.91 \times 10^{-6}$). What is the ratio of the concentration of the base to that of its acid in the blood plasma (pH = 7.40) of a patient?

The equilibrium between novocaine, NOV, and its acid, $HNOV^+$, can be expressed as

$$NOV + H_2O \rightleftharpoons HNOV^+ + OH^-$$

The concentration of OH^- is related to the pH of the blood plasma.

$$pOH = 14.00 - pH = 14.00 - 7.40 = 6.60$$

$$[OH^-] = 10^{-6.60} = 2.51 \times 10^{-7} \ M$$

The equilibrium expression involving novocaine and its acid is

$$K_b = \frac{[HNOV^+][OH^-]}{[NOV]}$$

Therefore,

$$\frac{[NOV]}{[HNOV^+]} = \frac{[OH^-]}{K_b} = \frac{2.51 \times 10^{-7}}{8.91 \times 10^{-6}} = 2.8 \times 10^{-2}$$

11.19 Calculate the percent dissociation of HF at the following concentrations: **(a)** 0.50 M and **(b)** 0.050 M. Comment on your results.

(a)

	HF	$\rightleftharpoons$	H^+	+	F^-	
Initial	0.50		0		0	M
At equilibrium	0.50 − x		x		x	M

$$K_a = 7.1 \times 10^{-4} = \frac{[H^+][F^-]}{[HF]} = \frac{x^2}{0.5 - x}$$

$$x^2 + 7.1 \times 10^{-4}x - 3.55 \times 10^{-4} = 0$$

$$x = 1.85 \times 10^{-2}$$

Therefore,

$$\% \text{ dissociation} = \frac{1.85 \times 10^{-2} \, M}{0.50 \, M} \times 100\% = 3.7\%$$

(b)

	HF	$\rightleftharpoons$	H$^+$	+	F$^-$	
Initial	0.050		0		0	M
At equilibrium	$0.050 - x$		x		x	M

$$K_a = 7.1 \times 10^{-4} = \frac{[\text{H}^+][\text{F}^-]}{[\text{HF}]} = \frac{x^2}{0.050 - x}$$

$$x^2 + 7.1 \times 10^{-4}x - 3.55 \times 10^{-5} = 0$$

$$x = 5.61 \times 10^{-3}$$

Therefore,

$$\% \text{ dissociation} = \frac{5.61 \times 10^{-3} \, M}{0.050 \, M} \times 100\% = 11\%$$

At more dilute concentrations the extent of dissociation increases in accord with Le Chatelier's principle.

11.20 Explain why phenol is a stronger acid than methanol:

phenol   methanol

The two conjugate bases are C$_6$H$_5$O$^-$ from phenol and CH$_3$O$^-$ from methanol. The C$_6$H$_5$O$^-$ is stabilized by resonance:

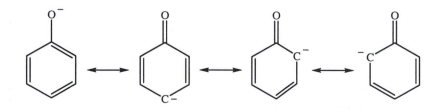

There is no such resonance stabilization for CH$_3$O$^-$. A more stable conjugate base means an increase in the strength of the acid.

11.21   Calculate the concentrations of all species in a $0.100\ M\ H_3PO_4$ solution.

The first-stage ionization of H_3PO_4 and the concentrations of various species are

	H_3PO_4	$\rightleftharpoons$	H^+	$+$	$H_2PO_4^-$	
Initial	0.100		0		0	M
At equilibrium	$0.100 - x$		x		x	M

$$K_a' = 7.5 \times 10^{-3} = \frac{[H^+][H_2PO_4^-]}{[H_3PO_4]} = \frac{x^2}{0.100 - x}$$

$$x^2 + 7.5 \times 10^{-3}x - 7.5 \times 10^{-4} = 0$$

$$x = 2.39 \times 10^{-2}$$

After the first stage of ionization,

$$[H_3PO_4] = (0.100 - x)\ M = 0.076\ M$$

$$[H^+] = x\ M = 2.39 \times 10^{-2}\ M$$

$$[H_2PO_4^-] = x\ M = 2.39 \times 10^{-2}\ M$$

The second-stage ionization involves $H_2PO_4^-$. The concentrations of various species are

	$H_2PO_4^-$	$\rightleftharpoons$	H^+	$+$	HPO_4^{2-}	
Initial	2.39×10^{-2}		2.39×10^{-2}		0	M
At equilibrium	$2.39 \times 10^{-2} - y$		$2.39 \times 10^{-2} + y$		y	M

$$K_a'' = 6.2 \times 10^{-8} = \frac{[H^+][HPO_4^{2-}]}{[H_2PO_4^-]} = \frac{(2.39 \times 10^{-2} + y)\,y}{2.39 \times 10^{-2} - y}$$

Since K_a'' is very small, both $2.39 \times 10^{-2} - y$ and $2.39 \times 10^{-2} + y$ can be approximated as 2.39×10^{-2}, and the expression above simplifies to

$$6.2 \times 10^{-8} = \frac{(2.39 \times 10^{-2})\,y}{2.39 \times 10^{-2}}$$

$$y = 6.2 \times 10^{-8}$$

Checking the assumptions,

$$\frac{y}{2.39 \times 10^{-2}} \times 100\% = 2.6 \times 10^{-4}\% < 5\%$$

Thus, the assumptions are valid.

After two stages of ionization,

$$[H_3PO_4] = 0.076\ M$$

$$[H^+] = (2.39 \times 10^{-2} + y)\ M = 2.39 \times 10^{-2}\ M$$

$$[H_2PO_4^-] = (2.39 \times 10^{-2} - y)\ M = 2.39 \times 10^{-2}\ M$$

$$[HPO_4^{2-}] = y\ M = 6.2 \times 10^{-8}\ M$$

The third-stage ionization involves HPO_4^{2-}. The concentrations of various species are

	HPO_4^{2-}	$\rightleftharpoons$	H^+	$+$	PO_4^{3-}	
Initial	6.2×10^{-8}		2.39×10^{-2}		0	M
At equilibrium	$6.2 \times 10^{-8} - z$		$2.39 \times 10^{-2} + z$		z	M

$$K_a''' = 4.8 \times 10^{-13} = \frac{[H^+][PO_4^{3-}]}{[HPO_4^{2-}]} = \frac{(2.39 \times 10^{-2} + z)z}{6.2 \times 10^{-8} - z}$$

Since K_a'' is very small, $6.2 \times 10^{-8} - z$ can be approximated as 6.2×10^{-8} and $2.39 \times 10^{-2} + z$ can be approximated as 2.39×10^{-2}. The expression above simplifies to

$$4.8 \times 10^{-13} = \frac{(2.39 \times 10^{-2})z}{6.2 \times 10^{-8}}$$

$$z = 1.2 \times 10^{-18}$$

Checking the assumptions,

$$\frac{z}{6.2 \times 10^{-8}} \times 100\% = 1.9 \times 10^{-9}\% < 5\%$$

$$\frac{z}{2.39 \times 10^{-2}} \times 100\% = 5.0 \times 10^{-15}\% < 5\%$$

Thus, the assumptions are valid.

After all stages of ionization,

$$[H_3PO_4] = 0.076\,M$$

$$[H_2PO_4^-] = 2.4 \times 10^{-2}\,M$$

$$[HPO_4^{2-}] = (6.2 \times 10^{-8} - z)\,M = 6.2 \times 10^{-8}\,M$$

$$[PO_4^{3-}] = z\,M = 1.2 \times 10^{-18}\,M$$

$$[H^+] = (2.39 \times 10^{-2} + z)\,M = 2.4 \times 10^{-2}\,M$$

11.22   The disagreeable odor of fish is mainly due to organic compounds (RNH_2) containing an amino group, $-NH_2$, where R is the rest of the molecule. Amines are bases just like ammonia. Explain why putting some lemon juice on fish can greatly reduce the odor.

Although primary amines ($R-NH_2$) interact through dipole-dipole attractions and may also participate in hydrogen bonding, they are relatively volatile. When protonated to form a salt, $R-NH_3^+$, the formal electrostatic intermolecular attractions result in much lower vapor pressures and consequently less odor. Lemon juice contains acids which react with the primary amines to form ammonium salts.

11.23 Specify which of the following salts will undergo hydrolysis: KF, NaNO$_3$, NH$_4$NO$_2$, MgSO$_4$, KCN, C$_6$H$_5$COONa, RbI, Na$_2$CO$_3$, CaCl$_2$, HCOOK.

The salts that will undergo hydrolysis are

KF (F$^-$ + H$_2$O $\rightleftharpoons$ HF + OH$^-$)

NH$_4$NO$_2$ (NH$_4^+$ + H$_2$O $\rightleftharpoons$ NH$_3$ + H$_3$O$^+$; NO$_2^-$ + H$_2$O $\rightleftharpoons$ HNO$_2$ + OH$^-$)

MgSO$_4$ (SO$_4^{2-}$ + H$_2$O $\rightleftharpoons$ HSO$_4^-$ + OH$^-$)

KCN (CN$^-$ + H$_2$O $\rightleftharpoons$ HCN + OH$^-$)

C$_6$H$_5$COONa (C$_6$H$_5$COO$^-$ + H$_2$O $\rightleftharpoons$ C$_6$H$_5$COOH + OH$^-$)

Na$_2$CO$_3$ (CO$_3^{2-}$ + H$_2$O $\rightleftharpoons$ HCO$_3^-$ + OH$^-$)

HCOOK (HCOO$^-$ + H$_2$O $\rightleftharpoons$ HCOOH + OH$^-$)

11.24 Calculate the pH of a 0.10 M NH$_4$Cl solution.

NH$_4^+$ hydrolyzes in water as described by the following equation:

	NH$_4^+$	+	H$_2$O	$\rightleftharpoons$	NH$_3$	+	H$_3$O$^+$	
Initial	0.10				0		0	M
At equilibrium	0.10 $- x$				x		x	M

$$K_a = \frac{K_w}{K_b\,(\text{NH}_3)} = \frac{1.0 \times 10^{-14}}{1.8 \times 10^{-5}} = \frac{[\text{NH}_3]\,[\text{H}_3\text{O}^+]}{[\text{NH}_4^+]}$$

$$5.56 \times 10^{-10} = \frac{x^2}{0.10 - x}$$

Since K_a is very small, $0.10 - x$ can be approximated as 0.10, and the expression above simplifies to

$$5.56 \times 10^{-10} = \frac{x^2}{0.10}$$

$$x = 7.46 \times 10^{-6}$$

Checking the assumption,

$$\frac{x}{0.10} \times 100\% = 7.5 \times 10^{-3}\% < 5\%$$

Thus, the assumption is valid. The pH of the solution is

$$\text{pH} = -\log\left(7.46 \times 10^{-6}\right) = 5.13$$

11.25 Calculate the pH and percent hydrolysis of a 0.36 M CH$_3$COONa solution.

CH_3COO^- hydrolyzes in water as described by the following equation:

$$CH_3COO^- \;+\; H_2O \;\rightleftharpoons\; CH_3COOH \;+\; OH^-$$

	CH_3COO^-		H_2O		CH_3COOH		OH^-	
Initial	0.36				0		0	M
At equilibrium	$0.36 - x$				x		x	M

$$K_b = \frac{K_w}{K_a\,(CH_3COOH)} = \frac{1.00 \times 10^{-14}}{1.75 \times 10^{-5}} = \frac{[CH_3COOH]\,[OH^-]}{[CH_3COO^-]}$$

$$5.714 \times 10^{-10} = \frac{x^2}{0.36 - x}$$

Since K_b is very small, $0.36 - x$ can be approximated by 0.36, and the expression above simplifies to

$$5.714 \times 10^{-10} = \frac{x^2}{0.36}$$

$$x = 1.43 \times 10^{-5}$$

The percent hydrolysis is

$$\frac{1.43 \times 10^{-5}}{0.36} \times 100\% = 4.0 \times 10^{-3}\%$$

Since the degree of dissociation is less than 5%, the assumption above is valid. The pH of the solution is related to $[OH^-]$:

$$pOH = -\log 1.43 \times 10^{-5} = 4.845$$

$$pH = 14.00 - 4.845 = 9.16$$

11.26 A student added NaOH solution from a buret to an Erlenmeyer flask containing HCl solution and used phenolphthalein as indicator. At the equivalence point of the titration, she observed a faint reddish-pink color. However, after a few minutes, the solution gradually turned colorless. What do you suppose happened?

CO_2 in the air is absorbed by the solution where it is converted to carbonic acid,

$$CO_2 + H_2O \rightleftharpoons H_2CO_3$$

The carbonic acid neutralizes the excess NaOH, lowering the pH sufficiently to render the phenolphthalein colorless.

11.27 The ionization constant, K_a, of an indicator, HIn, is 1.0×10^{-6}. The color of the nonionized form is red and that of the ionized form is yellow. What is the color of this indicator in a solution whose pH is 4.00?

The color of the indicator depends on [HIn]/[In⁻], which can be calculated using the Henderson–Hasselbalch equation:

$$pH = pK_a + \log \frac{[\text{In}^-]}{[\text{HIn}]}$$

$$4.00 = -\log\left(1.0 \times 10^{-6}\right) + \log \frac{[\text{In}^-]}{[\text{HIn}]}$$

$$\frac{[\text{In}^-]}{[\text{HIn}]} = 10^{-2}$$

$$\frac{[\text{HIn}]}{[\text{In}^-]} = 100$$

Therefore, the indicator should assume the acid color, that is, red.

11.28   The K_a of a certain indicator is 2.0×10^{-6}. The color of HIn is green, and that of In⁻ is red. A few drops of the indicator are added to a HCl solution, which is then titrated against a NaOH solution. At what pH will the indicator change color?

When [HIn] = [In⁻], the indicator color is a mixture of the colors of HIn and In⁻. In other words, the indicator changes color at this point, and the pH can be calculated using the Henderson–Hasselbalch equation:

$$pH = pK_a + \log \frac{[\text{In}^-]}{[\text{HIn}]} = -\log\left(2.0 \times 10^{-6}\right) = 5.70$$

11.29   The pK_a of the indicator methyl orange is 3.46. Over what pH range does this indicator change from 90% HIn to 90% In⁻?

When the indicator exists as 90% HIn,

$$pH = pK_a + \log \frac{[\text{In}^-]}{[\text{HIn}]} = 3.46 + \log \frac{0.10}{0.90} = 2.51$$

When the indicator exists as 90% In⁻,

$$pH = pK_a + \log \frac{[\text{In}^-]}{[\text{HIn}]} = 3.46 + \log \frac{0.90}{0.10} = 4.41$$

Therefore, the indicator changes from 90% HIn to 90% In⁻ over the pH range of 2.51 – 4.41.

11.30   A 200-mL volume of NaOH solution was added to 400 mL of a 2.00 M HNO₂. The pH of the mixed solution was 1.50 units greater than that of the original acid solution. Calculate the molarity of the NaOH solution.

First calculate the pH of the original solution.

$$
\begin{array}{ccccc}
 & HNO_2 & \rightleftharpoons & H^+ & + & NO_2^- \\
\text{Initial} & 2.00 & & 0 & & 0 & M \\
\text{At equilibrium} & 2.00 - x & & x & & x & M
\end{array}
$$

$$K_a = \frac{[H^+][NO_2^-]}{[HNO_2]}$$

$$4.5 \times 10^{-4} = \frac{x^2}{2.00 - x}$$

$$x^2 + 4.5 \times 10^{-4}x - 9.0 \times 10^{-4} = 0$$

$$x = 2.98 \times 10^{-2}$$

Therefore, the pH of the original solution is $-\log x = 1.526$. The pH of the mixed solution is $1.526 + 1.50 = 3.026$.

Let the molarity of the NaOH solution be $y\ M$. The concentrations of NaOH and HNO_2 in the mixture before the neutralization reaction are

$$[NaOH] = \frac{(y\ M)\,(200\ mL)}{600\ mL} = 0.3333y\ M$$

$$[HNO_2] = \frac{(2.00\ M)\,(400\ mL)}{600\ mL} = 1.333\ M$$

NaOH and HNO_2 react essentially to completion:

$$OH^- + HNO_2 \rightarrow H_2O + NO_2^-$$

After the reaction,

$$[HNO_2] = (1.333 - 0.3333y)\ M$$

$$[NO_2^-] = 0.3333y\ M$$

The value of y can be obtained using the Henderson–Hasselbalch equation:

$$pH = pK_a + \log \frac{[NO_2^-]}{[HNO_2]}$$

$$3.026 = 3.35 + \log \frac{0.3333y}{1.333 - 0.3333y}$$

$$\frac{0.3333y}{1.333 - 0.3333y} = 10^{-0.324} = 0.474$$

$$y = 1.3\ M$$

Therefore, the concentration of the NaOH solution is $1.3\ M$.

11.31   A volume of 25.0 mL of 0.100 M HCl is titrated with a 0.100 M CH_3NH_2 solution. Calculate the pH values of the solution (a) after 10.0 mL of CH_3NH_2 solution have been added, (b) after 25.0 mL of CH_3NH_2 solution have been added, and (c) after 35.0 mL of CH_3NH_2 solution have been added.

The reaction between HCl and CH_3NH_2 is

$$H^+ + CH_3NH_2 \rightarrow CH_3NH_3^+$$

(a) The concentrations of HCl and CH_3NH_2 in the reaction mixture before the neutralization reaction are

$$[HCl] = \frac{(0.100\ M)\ (25.0\ mL)}{35.0\ mL} = 7.143 \times 10^{-2}\ M$$

$$[CH_3NH_2] = \frac{(0.100\ M)\ (10.0\ mL)}{35.0\ mL} = 2.857 \times 10^{-2}\ M$$

Since $[HCl] > [CH_3NH_2]$, only part of the HCl is neutralized. The remaining concentration of HCl, and therefore, the pH are

$$[HCl] = 7.143 \times 10^{-2}\ M - 2.857 \times 10^{-2}\ M = 4.286 \times 10^{-2}\ M$$

$$pH = -\log\left(4.286 \times 10^{-2}\right) = 1.368$$

(b) The concentration and volume of HCl are the same as those of CH_3NH_2, that is, the equivalence point is reached. HCl is completely neutralized, and the amount of $CH_3NH_3^+$ formed is

$$\text{Number of moles of } CH_3NH_3^+ = \text{Number of moles of HCl}$$

$$= (0.100\ M)\left(25.0 \times 10^{-3}\ L\right)$$

$$= 2.50 \times 10^{-3}\ mol$$

So that

$$[CH_3NH_3^+] = \frac{2.50 \times 10^{-3}\ mol}{50.0 \times 10^{-3}\ L}$$

$$= 5.00 \times 10^{-2}\ M$$

$CH_3NH_3^+$ undergoes hydrolysis:

	$CH_3NH_3^+$	+	H_2O	$\rightleftharpoons$	CH_3NH_2	+	H_3O^+	
Initial	5.00×10^{-2}				0		0	M
At equilibrium	$5.00 \times 10^{-2} - x$				x		x	M

The dissociation constant for the above reaction is

$$K_a = \frac{K_w}{K_b} = \frac{1.00 \times 10^{-14}}{4.38 \times 10^{-4}} = 2.283 \times 10^{-11}$$

Using the equilibrium expression and the value of K_a, x can be calculated.

$$K_a = \frac{[CH_3NH_2]\,[H^+]}{[CH_3NH_3^+]}$$

$$2.283 \times 10^{-11} = \frac{x^2}{5.00 \times 10^{-2} - x}$$

Since K_a is very small, $5.00 \times 10^{-2} - x$ can be approximated by 5.00×10^{-2}. The expression above simplifies to

$$x^2 = 1.142 \times 10^{-12}$$

$$x = 1.069 \times 10^{-6}$$

Checking the assumption,

$$\frac{x}{5.00 \times 10^{-2}} \times 100\% = 2.1 \times 10^{-3}\% < 5\%$$

Thus, the assumption is valid. The pH of the solution after 25.0 mL of CH_3NH_2 solution is added is $-\log x$, or 5.971.

(c) The addition of 25.0 mL of CH_3NH_2 solution produces 2.50×10^{-3} mol of $CH_3NH_3^+$. Further addition of 10.0 mL of CH_3NH_2 will give

$$(0.100\ M)\,(10.0 \times 10^{-3}\ L) = 1.00 \times 10^{-3} \text{mol}$$

excess CH_3NH_2. The pH of the solution is determined by the relative concentration of $CH_3NH_3^+$ and CH_3NH_2 in the solution. Since both species are in the same solution, the concentration ratio is the same as the mole ratio.

$$pH = pK_a + \log \frac{[CH_3NH_2]}{[CH_3NH_3^+]}$$

$$= 10.64 + \log \frac{1.00 \times 10^{-3}}{2.50 \times 10^{-3}} = 10.24$$

11.32   Phenolphthalein is the common indicator for the titration of a strong acid with a strong base. (a) If the pK_a of phenolphthalein is 9.10, what is the ratio of the nonionized form of the indicator (colorless) to the ionized form (reddish pink) at pH 8.00? (b) If 2 drops of 0.060 M phenolphthalein are used in a titration involving a 50.0-mL volume, what is the concentration of the ionized form at pH 8.00? (Assume that 1 drop = 0.050 mL.)

(a) The ratio of the nonionized form of phenolphthalein to the ionized form can be calculated using the Henderson–Hasselbalch equation.

$$pH = pK_a + \log \frac{[\text{conjugate base}]}{[\text{acid}]}$$

$$8.00 = 9.10 + \log \frac{[\text{ionized}]}{[\text{nonionized}]}$$

$$\log \frac{[\text{nonionized}]}{[\text{ionized}]} = 1.10$$

$$\frac{[\text{nonionized}]}{[\text{ionized}]} = 12.59 = 12.6$$

(b) The total concentration of the indicator is the sum of the total concentrations of the nonionized and ionized forms.

$$[\text{indicator}] = [\text{nonionized}] + [\text{ionized}] = \frac{(2 \text{ drops}) \left(\frac{0.050 \text{ mL}}{1 \text{ drop}}\right) (0.060 \text{ } M)}{50.0 \text{ mL}} = 1.2 \times 10^{-4} \text{ } M$$

The concentration of the ionized form of the indicator can be found using the result of part (a).

$$\frac{[\text{nonionized}]}{[\text{ionized}]} = 12.6 = \frac{[\text{indicator}] - [\text{ionized}]}{[\text{ionized}]} = \frac{1.2 \times 10^{-4} \text{ } M - [\text{ionized}]}{[\text{ionized}]}$$

$$13.6 \, [\text{ionized}] = 1.2 \times 10^{-4} \text{ } M$$

$$[\text{ionized}] = 8.8 \times 10^{-6} \text{ } M$$

11.33   Shown below is a titration curve for carbonic acid versus sodium hydroxide. Fill in the missing species and the pH and pK_a values.

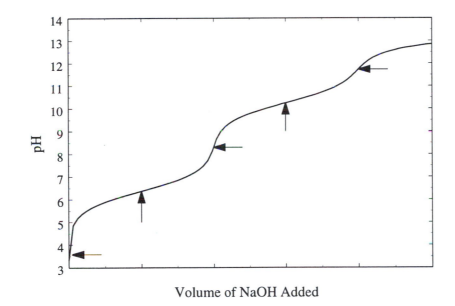

Volume of NaOH Added

At each half-equivalence point, pH $=$ pK_a and the predominant species are a mixture of the acid and conjugate base. At the beginning, the major species is H_2CO_3. At the first equivalence point, the major species is $NaHCO_3$ (completely dissociated into ions), and at the second equivalence point, Na_2CO_3. The pH's at the start and at the equivalence points depend on the actual concentrations used, which are not given.

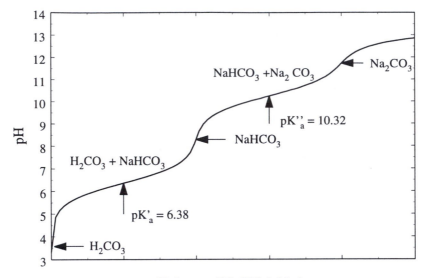

Volume of NaOH Added

11.34 Specify which of the following systems can be classified as a buffer system: **(a)** KCl/HCl, **(b)** NH_3/NH_4NO_3, **(c)** Na_2HPO_4/NaH_2PO_4, **(d)** KNO_2/HNO_2, **(e)** $KHSO_4/H_2SO_4$, **(f)** HCOOK/HCOOH.

A buffer is composed of a weak acid and a weak base. **(b)**, **(c)**, **(d)**, and **(f)** can be classified as buffer systems.

11.35 Derive the Henderson–Hasselbalch equation for the buffer system NH_4^+/NH_3.

$$NH_4^+ \rightleftharpoons H^+ + NH_3$$

$$K_a = \frac{[H^+][NH_3]}{[NH_4^+]}$$

Taking $-$ log on both sides of the equation and rearranging,

$$-\log K_a = -\log\left(\frac{[H^+][NH_3]}{[NH_4^+]}\right) = -\log[H^+] - \log\left(\frac{[NH_3]}{[NH_4^+]}\right)$$

$$pK_a = pH - \log\left(\frac{[NH_3]}{[NH_4^+]}\right)$$

$$pH = pK_a + \log\left(\frac{[NH_3]}{[NH_4^+]}\right)$$

11.36 Calculate the pH of the 0.20 M NH_3/0.20 M NH_4Cl buffer. What is the pH of the buffer after the addition of 10.0 mL of 0.10 M HCl to 65.0 mL of the buffer?

The pH of the buffer can be calculated using the Henderson–Hasselbalch equation.

$$pH = pK_a + \log\left(\frac{[NH_3]}{[NH_4^+]}\right)$$

$$= 9.25 + \log\left(\frac{0.20}{0.20}\right) = 9.25$$

When HCl is added, it will react with NH_3 to near completion.

$$\text{Initial [HCl] in the reaction mixture} = \frac{(0.10\,M)\,(10.0\,\text{mL})}{75.0\,\text{mL}} = 1.33 \times 10^{-2}\,M$$

$$\text{Initial [NH}_3\text{] in the reaction mixture} = \frac{(0.20\,M)\,(65.0\,\text{mL})}{75.0\,\text{mL}} = 0.173\,M$$

$$\text{Initial [NH}_4^+\text{] in the reaction mixture} = \frac{(0.20\,M)\,(65.0\,\text{mL})}{75.0\,\text{mL}} = 0.173\,M$$

	H^+	+	NH_3	$\rightleftharpoons$	NH_4^+	
Before reaction	1.33×10^{-2}		0.173		0.173	M
After reaction	0		0.160		0.186	M

The new pH can now be calculated using the Henderson–Hasselbalch equation.

$$pH = pK_a + \log\left(\frac{[NH_3]}{[NH_4^+]}\right)$$

$$= 9.25 + \log\left(\frac{0.160}{0.186}\right) = 9.18$$

11.37 Calculate the pH of 1.00 L of the buffer 1.00 M CH_3COONa/1.00 M CH_3COOH before and after the addition of **(a)** 0.080 mol NaOH and **(b)** 0.12 mol HCl. (Assume that there is no change in volume.)

The pH of the buffer can be calculated using the Henderson–Hasselbalch equation.

$$pH = pK_a + \log\left(\frac{[CH_3COO^-]}{[CH_3COOH]}\right)$$

$$= 4.76 + \log\left(\frac{1.00}{1.00}\right) = 4.76$$

(a) When NaOH is added, it will react with CH_3COOH to near completion.

$$\text{Initial [NaOH] in the reaction mixture} = \frac{0.080\,\text{mol}}{1.00\,\text{L}} = 0.080\,M$$

$$\text{Initial [CH}_3\text{COOH] in the reaction mixture} = 1.00\,M$$

$$\text{Initial [CH}_3\text{COO}^-\text{] in the reaction mixture} = 1.00\,M$$

	OH$^-$	+	CH$_3$COOH	$\rightleftharpoons$	CH$_3$COO$^-$	+	H$_2$O	
Before reaction	0.080		1.00		1.00			M
After reaction	0		0.92		1.08			M

The new pH can now be calculated using the Henderson–Hasselbalch equation.

$$pH = pK_a + \log\left(\frac{[CH_3COO^-]}{[CH_3COOH]}\right)$$

$$= 4.76 + \log\left(\frac{1.08}{0.92}\right) = 4.83$$

(b) When HCl is added, it will react with CH$_3$COO$^-$ to near completion.

$$\text{Initial [HCl] in the reaction mixture} = \frac{0.12 \text{ mol}}{1.00 \text{ L}} = 0.12 \, M$$

$$\text{Initial [CH}_3\text{COOH] in the reaction mixture} = 1.00 \, M$$

$$\text{Initial [CH}_3\text{COO}^-\text{] in the reaction mixture} = 1.00 \, M$$

	H$^+$	+	CH$_3$COO$^-$	$\rightleftharpoons$	CH$_3$COOH	
Before reaction	0.12		1.00		1.00	M
After reaction	0		0.88		1.12	M

The new pH can now be calculated using the Henderson–Hasselbalch equation.

$$pH = pK_a + \log\left(\frac{[CH_3COO^-]}{[CH_3COOH]}\right)$$

$$= 4.76 + \log\left(\frac{0.88}{1.12}\right) = 4.66$$

11.38 A quantity of 26.4 mL of a 0.45 M acetic acid solution is added to 31.9 mL of a 0.37 M sodium hydroxide solution. What is the pH of the final solution?

$$\text{Number of moles of NaOH added} = (0.37 \, M)\left(31.9 \times 10^{-3} \text{ L}\right) = 1.18 \times 10^{-2} \text{ mol}$$

$$\text{Number of moles of CH}_3\text{COOH initially} = (0.45 \, M)\left(26.4 \times 10^{-3} \text{ L}\right) = 1.19 \times 10^{-2} \text{ mol}$$

Since OH$^-$ reacts with CH$_3$COOH in an equimolar manner, nearly all the CH$_3$COOH will be converted to CH$_3$COO$^-$ which has a concentration of

$$[CH_3COO^-] = \frac{1.18 \times 10^{-2} \text{ mol}}{26.4 \times 10^{-3} \text{ L} + 31.9 \times 10^{-3} \text{ L}} = 0.202 \, M$$

The concentration of the remaining CH_3COOH is

$$[CH_3COOH] = \frac{0.01 \times 10^{-2}\ mol}{26.4 \times 10^{-3}\ L + 31.9 \times 10^{-3}\ L}$$

$$= 2 \times 10^{-3}\ M$$

The equilibrium between CH_3COO^- and CH_3COOH is

	CH_3COO^-	$+$	H_2O	$\rightleftharpoons$	CH_3COOH	$+$	OH^-	
Initial	0.202				2×10^{-3}		0	M
At equilibrium	$0.202 - x$				$2 \times 10^{-3} + x$		x	M

$$K_b = \frac{K_w}{K_a\ (CH_3COOH)} = \frac{1.0 \times 10^{-14}}{1.75 \times 10^{-5}} = \frac{[CH_3COOH]\,[OH^-]}{[CH_3COO^-]}$$

$$5.714 \times 10^{-10} = \frac{x\,(2 \times 10^{-3} + x)}{0.202 - x}$$

Since K_b is very small, $0.202 - x$ can be approximated by 0.202, and $2 \times 10^{-3} + x$ by 2×10^{-3}. The expression above simplifies to

$$5.714 \times 10^{-10} = \frac{x\,(2 \times 10^{-3})}{0.202}$$

$$x = 6 \times 10^{-8}$$

Checking the assumption,

$$\frac{x}{0.202} \times 100\% = 3 \times 10^{-5}\% < 5\%$$

$$\frac{x}{2 \times 10^{-3}} \times 100\% = 3 \times 10^{-3}\% < 5\%$$

Therefore, the assumptions are valid. The pH of the solution is related to $[OH^-]$:

$$pOH = -\log 6 \times 10^{-8} = 7.2$$

$$pH = 14.0 - 7.2 = 6.8$$

11.39 What is the pH of the buffer $0.10\ M\ Na_2HPO_4/0.10\ M\ KH_2PO_4$? Calculate the concentration of all the species in solution.

The species that exist in the buffer, in addition to HPO_4^{2-} and $H_2PO_4^-$, are H_3PO_4 (from the hydrolysis of $H_2PO_4^-$), PO_4^{3-} (from the dissociation of HPO_4^{2-}), and of course, H^+ and OH^-.

The pH is determined by HPO_4^{2-} and $H_2PO_4^-$:

$$pH = pK_a'' + \log\left(\frac{[HPO_4^{2-}]}{[H_2PO_4^-]}\right)$$

$$= 7.21 + \log\left(\frac{0.10}{0.10}\right) = 7.21$$

The concentrations of H^+ and OH^- corresponding to this pH are

$$\left[H^+\right] = 10^{-7.21} \, M = 6.17 \times 10^{-8} \, M$$

$$\left[OH^-\right] = \frac{K_w}{\left[H^+\right]} = \frac{1.00 \times 10^{-14}}{6.17 \times 10^{-8}} \, M = 1.62 \times 10^{-7} \, M$$

Besides being in equilibrium with HPO_4^{2-}, $H_2PO_4^-$ also hydrolyzes to give H_3PO_4:

$$H_2PO_4^- + H_2O \rightleftharpoons H_3PO_4 + OH^-$$

The equilibrium constant for this reaction is

$$K = \frac{K_w}{K_a'} = \frac{1.00 \times 10^{-14}}{7.5 \times 10^{-3}} = 1.33 \times 10^{-12}$$

Since the equilibrium constant is very small, $H_2PO_4^-$ does not hydrolyze to a great extent. Thus, at equilibrium,

$$\left[H_2PO_4^-\right] = 0.10 \, M$$

$$\left[OH^-\right] = 1.62 \times 10^{-7} \, M$$

and the concentration of H_3PO_4 is calculated from the equilibrium expression:

$$K = 1.33 \times 10^{-12} = \frac{\left[H_3PO_4\right]\left[OH^-\right]}{\left[H_2PO_4^-\right]} = \frac{\left[H_3PO_4\right]\left(1.62 \times 10^{-7}\right)}{(0.10)}$$

$$\left[H_3PO_4\right] = 8.2 \times 10^{-7} \, M$$

HPO_4^{2-} dissociates to give PO_4^{3-}:

$$HPO_4^{2-} \rightleftharpoons PO_4^{3-} + H^+$$

The equilibrium constant for this reaction is

$$K_a''' = 4.8 \times 10^{-13}$$

Since the equilibrium constant is very small, HPO_4^{2-} does not dissociate to a great extent. Thus, at equilibrium,

$$\left[HPO_4^{2-}\right] = 0.10 \, M$$

$$\left[H^+\right] = 6.17 \times 10^{-8} \, M$$

and the concentration of PO_4^{3-} is calculated from the equilibrium expression:

$$K_a''' = 4.8 \times 10^{-13} = \frac{\left[PO_4^{3-}\right]\left[H^+\right]}{\left[HPO_4^{2-}\right]} = \frac{\left[PO_4^{3-}\right]\left(6.17 \times 10^{-8}\right)}{(0.10)}$$

$$\left[PO_4^{3-}\right] = 7.8 \times 10^{-7} \, M$$

The concentration of all species are

$$[H_3PO_4] = 8.2 \times 10^{-7} \, M$$

$$[H_2PO_4^-] = 0.10 \, M$$

$$[HPO_4^{2-}] = 0.10 \, M$$

$$[PO_4^{3-}] = 7.8 \times 10^{-7} \, M$$

$$[H^+] = 6.2 \times 10^{-8} \, M$$

$$[OH^-] = 1.6 \times 10^{-7} \, M$$

11.40 A phosphate buffer has a pH equal to 7.30. **(a)** What is the predominant conjugate pair present in this buffer? **(b)** If the concentration of this buffer is 0.10 M, what is the new pH after the addition of 5.0 mL of a 0.10 M HCl to 20.0 mL of this buffer solution?

(a) The predominant conjugate pair will involve an acid with a pK_a value similar to the pH of the buffer. Since for H$_2$PO$_4^-$, p$K_a'' = 7.21$, H$_2$PO$_4^-$ and HPO$_4^{2-}$ are the chief components of this buffer.

(b) First the concentrations of H$_2$PO$_4^-$ and HPO$_4^{2-}$ need to be calculated. The sum of the concentrations is 0.10 M. In other words,

$$[HPO_4^{2-}] = 0.10 \, M - [H_2PO_4^-]$$

Using the Henderson–Hasselbalch equation,

$$pH = pK_a'' + \log \frac{[HPO_4^{2-}]}{[H_2PO_4^-]}$$

$$7.30 = 7.21 + \log \frac{0.10 \, M - [H_2PO_4^-]}{[H_2PO_4^-]}$$

$$\frac{0.10 \, M - [H_2PO_4^-]}{[H_2PO_4^-]} = 10^{0.09} = 1.2$$

$$[H_2PO_4^-] = 4.5 \times 10^{-2} \, M$$

$$[HPO_4^{2-}] = 0.10 \, M - 4.5 \times 10^{-2} \, M = 5.5 \times 10^{-2} \, M$$

Before the addition of HCl,

$$\text{Number of moles of } H_2PO_4^- = (4.5 \times 10^{-2} \, M)(20.0 \times 10^{-3} \, L) = 9.0 \times 10^{-4} \, mol$$

$$\text{Number of moles of } HPO_4^{2-} = (5.5 \times 10^{-2} \, M)(20.0 \times 10^{-3} \, L) = 1.1 \times 10^{-3} \, mol$$

When HCl is added, it will react with HPO_4^{2-}:

$$H^+ + HPO_4^{2-} \rightleftharpoons H_2PO_4^-$$

and the reaction will proceed to near completion.

$$\text{Moles of HCl added} = (0.10\ M)\left(5.0 \times 10^{-3}\ L\right) = 5.0 \times 10^{-4}\ mol$$

$$\text{Moles of } HPO_4^{2-} \text{ after reaction with HCl} = 1.1 \times 10^{-3}\ mol - 5.0 \times 10^{-4}\ mol = 6 \times 10^{-4}\ mol$$

$$\left[HPO_4^{2-}\right] \text{ after reaction with HCl} = \frac{6 \times 10^{-4}\ mol}{25.0 \times 10^{-3}\ L} = 0.02\ M$$

$$\text{Moles of } H_2PO_4^- \text{ after reaction with HCl} = 9.0 \times 10^{-4}\ mol + 5.0 \times 10^{-4}\ mol = 1.40 \times 10^{-3}\ mol$$

$$\left[H_2PO_4^{2-}\right] \text{ after reaction with HCl} = \frac{1.40 \times 10^{-3}\ mol}{25.0 \times 10^{-3}\ L} = 0.056\ M$$

The new pH can now be calculated using the Henderson–Hasselbalch equation.

$$pH = pK_a'' + \log \frac{\left[HPO_4^{2-}\right]}{\left[H_2PO_4^-\right]}$$

$$= 7.21 + \log \frac{0.02}{0.056}$$

$$= 6.76$$

11.41 Tris[tris(hydroxymethyl)aminomethane] is a common buffer for studying biological systems

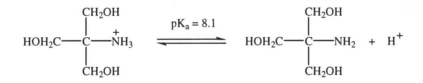

(a) Calculate the pH of the tris buffer after mixing 15.0 mL of 0.10 M HCl solution with 25.0 mL of 0.10 M tris. **(b)** This buffer was used to study an enzyme-catalyzed reaction. As a result of the reaction, 0.00015 mole of H^+ was consumed. What is the pH of the buffer at the end of the reaction? **(c)** What would be the final pH if no buffer were present?

(a) The reaction of HCl with tris can be represented by

$$H^+ + tris \rightleftharpoons tris^+$$

$$\text{Initial [HCl] in the reaction mixture} = \frac{(0.10\ M)\,(15.0\ mL)}{40.0\ mL} = 0.0375\ M$$

$$\text{Initial [tris] in the reaction mixture} = \frac{(0.10\ M)\,(25.0\ mL)}{40.0\ mL} = 0.0625\ M$$

The reaction between tris and H^+ will proceed to near completion. Therefore, after the reaction

$$[\text{tris}] = 0.0625\,M - 0.0375\,M = 0.0250\,M$$

$$\left[\text{tris}^+\right] = 0.0375\,M$$

The pH of the tris buffer can be calculated using the Henderson–Hasselbalch equation:

$$\text{pH} = \text{p}K_a + \log\frac{[\text{tris}]}{\left[\text{tris}^+\right]}$$

$$= 8.1 + \log\frac{0.0250}{0.0375}$$

$$= 7.9$$

(b) H^+ is consumed by tris to give tris^+:

$$H^+ + \text{tris} \rightleftharpoons \text{tris}^+$$

and the reaction proceeds to near completion.

$$\left[H^+\right] \text{ consumed} = \frac{0.00015\ \text{mol}}{40.0 \times 10^{-3}\ \text{L}} = 3.75 \times 10^{-3}\,M$$

$$[\text{tris}] \text{ remaining after the reaction} = 0.0250\,M - 3.75 \times 10^{-3}\,M = 0.0213\,M$$

$$\left[\text{tris}^+\right] \text{ remaining after the reaction} = 0.0375\,M + 3.75 \times 10^{-3}\,M = 0.0413\,M$$

The pH of the tris buffer can be calculated using the Henderson–Hasselbalch equation:

$$\text{pH} = \text{p}K_a + \log\frac{[\text{tris}]}{\left[\text{tris}^+\right]}$$

$$= 8.1 + \log\frac{0.0213}{0.0413}$$

$$= 7.8$$

(c) If no buffer were present, H^+ would not be consumed, and

$$\text{pH} = -\log\left(3.75 \times 10^{-3}\right) = 2.4$$

11.42   Describe the number of different ways to prepare 1 liter of a 0.050 M phosphate buffer with a pH of 7.8.

At pH = 7.8, the appropriate buffer couple is $H_2PO_4^-$/HPO_4^{2-}. There are a number of ways to prepare the buffer, but each will require the proper ratio of the two phosphate species as given by the Henderson–Hasselbalch equation,

$$pH = pK_a'' + \log \frac{[HPO_4^{2-}]}{[H_2PO_4^-]}$$

$$7.8 = 7.21 + \log \frac{[HPO_4^{2-}]}{[H_2PO_4^-]}$$

$$\frac{[HPO_4^{2-}]}{[H_2PO_4^-]} = 3.9 \approx 4$$

Thus each liter of solution would require $\frac{4}{5} \times 0.050$ mol $= 0.040$ mol of HPO_4^{2-} and $\frac{1}{5} \times 0.050$ mol $= 0.010$ mol of $H_2PO_4^-$.

The buffer could be prepared as follows. (In each case the K^+ and Na^+ counter ions are interchangeable.)

1. Mix KH_2PO_4 and K_2HPO_4 in the proper proportion

2. Titrate H_3PO_4 with NaOH to obtain the appropriate amounts of NaH_2PO_4 and Na_2HPO_4.

3. Start with NaH_2PO_4 and add NaOH to convert the proper amount to Na_2HPO_4.

4. Start with Na_2HPO_4 and add HCl to convert the appropriate amount to NaH_2PO_4.

5. Start with K_3PO_4 and convert to the correct ratio of KH_2PO_4 and K_2HPO_4 by adding HCl.

6. Mix K_3PO_4 and KH_2PO_4 in the proper proportion.

11.43 Calculate the concentration of all the species present in a solution that is 0.12 M in HCN and 0.34 M in NaCN. What is the pH of the solution? Does the solution possess buffer capacity?

The equilibrium equation between HCN and CN^- is

$$HCN \rightleftharpoons H^+ + CN^-$$

Since K_a is very small (4.9×10^{-10}), the concentrations of HCN and CN^- at equilibrium are the same as their respective initial concentrations. Therefore, the Henderson–Hasselbalch equation applies.

$$pH = pK_a + \log \frac{[CN^-]}{[HCN]}$$

$$= 9.31 + \log \frac{0.34}{0.12}$$

$$= 9.76$$

Thus,

$$[H^+] = 10^{-9.76} = 1.74 \times 10^{-10} \, M = 1.7 \times 10^{-10} \, M$$

$$[OH^-] = \frac{1.00 \times 10^{-14}}{1.74 \times 10^{-10} \, M} = 5.7 \times 10^{-5} \, M$$

$$[Na^+] = [CN^-] = 0.34 \, M$$

$$[HCN] = 0.12 \, M$$

Since the pH of this solution is within the range of $pK_a \pm 1$, the solution possesses buffer capacity. Because of its toxicity, however, great care is needed to use a cyanide buffer.

11.44 The pH of blood plasma is 7.40. Assuming the principal buffer system is HCO_3^-/H_2CO_3, calculate the ratio $[HCO_3^-]/[H_2CO_3]$. Is this buffer more effective against an added acid or an added base?

$$pH = pK_a + \log \frac{[HCO_3^-]}{[H_2CO_3]}$$

$$7.40 = 6.38 + \log \frac{[HCO_3^-]}{[H_2CO_3]}$$

$$\frac{[HCO_3^-]}{[H_2CO_3]} = 10^{1.02} = 10.5$$

The buffer should be more effective against an added acid because approximately ten times more base is present compared to acid.

11.45 A student is asked to prepare a buffer solution with pH = 8.60, using one of the following weak acids: HA ($K_a = 2.7 \times 10^{-3}$), HB ($K_a = 4.4 \times 10^{-6}$), HC ($K_a = 2.6 \times 10^{-9}$). Which acid should she choose?

Recall that to prepare a buffer solution of a desired pH, we should choose a weak acid with a pK_a value close to the desired pH. The following table lists the pK_a values for the three weak acids:

Acid	pK_a
HA	2.57
HB	5.36
HC	8.59

HC, with a pK_a of 8.59, is the best choice to prepare a buffer solution with pH = 8.60.

11.46 The buffer range is defined by the equation $pH = pK_a \pm 1$. Calculate the range of the ratio [conjugate base]/[acid] that corresponds to this equation.

When $pK_a = pH - 1$,

$$pH = pK_a + \log \frac{[\text{conjugate base}]}{[\text{acid}]}$$

$$\log \frac{[\text{conjugate base}]}{[\text{acid}]} = 1$$

$$\frac{[\text{conjugate base}]}{[\text{acid}]} = 10$$

When $pK_a = pH + 1$,

$$pH = pK_a + \log \frac{[\text{conjugate base}]}{[\text{acid}]}$$

$$\log \frac{[\text{conjugate base}]}{[\text{acid}]} = -1$$

$$\frac{[\text{conjugate base}]}{[\text{acid}]} = 0.1$$

Therefore, the range of the ratio is $0.1 < \dfrac{[\text{conjugate base}]}{[\text{acid}]} < 10$.

11.47   Describe how you would prepare 1 L of 0.20 M CH$_3$COONa/0.20 M CH$_3$COOH buffer system by **(a)** mixing a solution of CH$_3$COOH with a solution of CH$_3$COONa, **(b)** reacting a solution of CH$_3$COOH with a solution of NaOH, and **(c)** reacting a solution of CH$_3$COONa with a solution of HCl.

In each part the approximation is made that volumes are additive. None of the possibilities given below are unique.

(a) Mix 500 mL of 0.40 M CH$_3$COONa with 500 mL of 0.40 M CH$_3$COOH. The final volume is 1.00 L, and the concentrations of the two species are halved by the doubling in volume.

(b) Mix 500 mL of 0.80 M CH$_3$COOH with 500 mL of 0.40 M NaOH. There is enough base to react with half the acid, leaving a solution identical to that in part (a).

(c) Mix 500 mL of 0.80 M CH$_3$COONa with 500 mL of 0.40 M HCl. There is enough acid to react with half the salt, leaving a solution again identical to that in part (a).

11.48   How many milliliters of 1.0 M NaOH must be added to 200 mL of 0.10 M NaH$_2$PO$_4$ to make a buffer solution with a pH of 7.50?

Let x L of NaOH be the volume required. The number of moles of NaOH and NaH$_2$PO$_4$ before the reaction are

$$\text{Number of moles of NaOH} = (1.0 \, M)(x \, \text{L}) = 1.0x \text{ mol}$$

$$\text{Number of moles of NaH}_2\text{PO}_4 = (0.10 \, M)(0.200 \, \text{L}) = 0.0200 \text{ mol}$$

NaOH reacts essentially completely with $H_2PO_4^-$ according to the reaction

$$OH^- + H_2PO_4^- \rightarrow H_2O + HPO_4^{2-}$$

After the reaction,

$$\text{Number of moles of } H_2PO_4^- = (0.0200 - 1.0x) \text{ mol}$$

$$\text{Number of moles of } HPO_4^{2-} = 1.0x \text{ mol}$$

The solution now is a buffer solution. The pH of this solution is related to the concentrations of $H_2PO_4^-$ and HPO_4^{2-}:

$$pH = pK_a + \log \frac{\left[HPO_4^{2-}\right]}{\left[H_2PO_4^-\right]}$$

Since the acid/base pair is in a single solution, the concentration ratio is the same as the mole ratio:

$$pH = pK_a + \log \frac{\text{Number of moles of } HPO_4^{2-}}{\text{Number of moles of } H_2PO_4^-}$$

$$7.50 = 7.21 + \log \frac{1.0x \text{ mol}}{(0.0200 - 1.0x) \text{ mol}}$$

$$\frac{1.0x}{0.0200 - 1.0x} = 10^{0.29} = 1.95$$

$$1.0x = 1.95 (0.0200 - 1.0x) = 0.0390 - 1.95x$$

$$x = 1.3 \times 10^{-2}$$

Therefore, a volume of 1.3×10^{-2} L, or 13 mL is required to prepare the buffer.

11.49 Suggest two chemical tests that would allow you to distinguish an acid solution and a buffer solution both at pH = 3.5.

---

1. Add a quantity of base to each of the solutions. The pH of the buffer solution will increase only slightly while that of the acid will change to a greater extent.

2. Dilute each solution ten times, *e.g.* take 1 mL of each and dilute with water to 10 mL. The pH of the buffer solution will remain fairly constant while that of the acid should increase considerably. (If it is a strong acid, the pH after dilution would be 4.5. A weak acid would have a final pH between 3.5 and 4.5, depending on its K_a.)

11.50 How would you prepare a CH_3COOH/CH_3COONa buffer with a pH of 4.40 and an ionic strength of 0.050 m? Treat molarity the same as molality.

The ionic strength is a result of the dissociation of CH_3COONa into Na^+ and CH_3COO^-. CH_3COOH is too weak an acid to contribute to the ionic strength. The molality of Na^+, m_+, is the same as that of CH_3COO^-, m_-.

$$I = \frac{1}{2}\left(m_+ z_+^2 + m_- z_-^2\right)$$

$$0.050\, m = \frac{1}{2}\left[m_+\,(1)^2 + m_-\,(1)^2\right] = m_-$$

Therefore, assuming molarity is the same as molarity, $[CH_3COO^-] = 0.050\ M$ in the buffer. Using this concentration and the Henderson–Hasselbalch equation, $[CH_3COOH]$ can be calculated.

$$pH = pK_a + \log\frac{[CH_3COO^-]}{[CH_3COOH]}$$

$$4.40 = 4.76 + \log\frac{0.050}{[CH_3COOH]}$$

$$\frac{0.050}{[CH_3COOH]} = 10^{-0.36} = 0.437$$

$$[CH_3COOH] = 0.11\ M$$

The buffer should contain $0.11\ M\ CH_3COOH$ and $0.050\ M\ CH_3COONa$.

11.51 The pH of a phosphate buffer is 7.10 at 25°C. What is the pH of the buffer at 37°C? The $\Delta_r H^\circ$ for the relevant dissociation step is 3.75 kJ mol^{-1}.

The van't Hoff equation is used to calculate the relationship between the pK_a values of the buffer at 25°C, pK_1 and at 37°C, pK_2:

$$\ln\frac{K_2}{K_1} = -\frac{\Delta_r H^\circ}{R}\left(\frac{1}{T_2} - \frac{1}{T_1}\right)$$

$$2.303\log\frac{K_2}{K_1} = -\frac{3.75\times10^3\ \mathrm{J\,mol^{-1}}}{8.314\ \mathrm{J\,K^{-1}\,mol^{-1}}}\left(\frac{1}{310.2\ \mathrm{K}} - \frac{1}{298.2\ \mathrm{K}}\right)$$

$$\log\frac{K_2}{K_1} = 2.541\times10^{-2}$$

$$pK_2 - pK_1 = -2.541\times10^{-2}$$

When the temperature changes, the volume of the solution changes. Therefore, the molarity of each species also changes. However, the effect of the temperature dependence of volume disappears for a ratio between 2 concentrations. In other words, [conjugate base]/[acid] does not change with temperature. According to the Henderson–Hasselbalch equation,

$$pH = pK_a + \log\frac{[\text{conjugate base}]}{[\text{acid}]}$$

the pH of a solution changes with pK_a when temperature changes:

$$pH_2 - pH_1 = pK_2 - pK_1 = -2.541\times10^{-2}$$

Thus, at 37°C, the pH of the buffer is $7.10 - 2.541\times10^{-2} = 7.07$.

11.52 Which of the amino acids listed in Table 11.4 have a buffer capacity in the physiological region of pH 7?

An amino acid must have a pK_a (pK_a', pK_a'', or pK_a''') in the range of $6-8$ to exhibit buffer capacity in the physiological region of pH 7. Histidine, with $pK_a''' = 6.00$, is the only such amino acid.

11.53 Calculate the ionic strength of a 0.035 M serine buffer at pH 9.15.

The components of the buffer are

$$S^-$$ $$S^0$$

Since pH is the same as pK_a'', the concentration of S^- and S^0 are the same. Of the serine species present in solution, only the net charged S^- contributes to the ionic strength. The buffer can be prepared by titrating isoelectric serine (S^0) at 0.070 M with NaOH to yield

$$\left[S^0\right] = 0.035 \ M$$

$$\left[S^-\right] = 0.035 \ M$$

$$\left[Na^+\right] = 0.035 \ M$$

Thus, assuming molarity is equal to molality,

$$I = \frac{1}{2}\left(m_+ z_+^2 + m_- z_-^2\right) = \frac{1}{2}\left[(0.035)(1)^2 + (0.035)(1)^2\right] = 0.035 \ m$$

11.54 From the pK_a's listed in Table 11.4, calculate the pI value for amino acids lysine and valine.

Lysine

$$\underset{H_3\overset{+}{N}(H_2C)_3H_2C-CH-COOH}{\overset{\overset{+}{N}H_3}{|}} \quad \xrightleftharpoons{pK_a' = 2.18} \quad \underset{H_3\overset{+}{N}(H_2C)_3H_2C-CH-COO^-}{\overset{\overset{+}{N}H_3}{|}}$$

$$\xrightleftharpoons{pK_a'' = 8.95} \quad \underset{H_3\overset{+}{N}(H_2C)_3H_2C-CH-COO^-}{\overset{NH_2}{|}} \quad \xrightleftharpoons{pK_a''' = 10.53}$$

$$\underset{H_2N(H_2C)_3H_2C-CH-COO^-}{\overset{NH_2}{|}}$$

$$pI = \frac{8.95 + 10.53}{2} = 9.74$$

Valine

$$pK_a' = 2.32$$

$$pK_a'' = 9.62$$

$$pI = \frac{2.32 + 9.62}{2} = 5.97$$

11.55 Sketch the titration curve for 100 mL of 0.1 M aspartic acid hydrogen chloride titrated with sodium hydroxide.

The curve is given below. Note that the first equivalence point is not well defined because the first two pK_a's are so close together ($pK_a' = 2.09$, $pK_a'' = 3.86$).

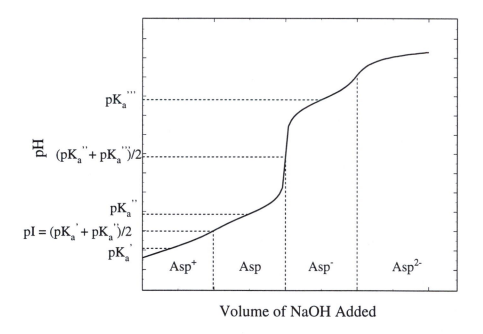

Volume of NaOH Added

11.56 At neutral pH, amino acids exist as dipolar ions. Using glycine as an example, and given that the pK_a of the carboxyl group is 2.3 and that of the ammonium group is 9.6, predict the predominant form of the molecule at pHs of 1, 7, and 12. Justify your answers using Equation 11.16.

Use the Henderson–Hasselbalch equation,

$$pH = pK_a + \log \frac{[\text{conjugate base}]}{[\text{acid}]},$$

to calculate the ratios of each acid-conjugate base pair in glycine, NH_2–CH_2–COOH.

At pH = 1,

For the –COOH group,

$$1 = 2.3 + \log \frac{[-COO^-]}{[-COOH]}$$

$$-1.3 = \log \frac{[-COO^-]}{[-COOH]}$$

$$\frac{[-COO^-]}{[-COOH]} = 5.0 \times 10^{-2}$$

For the –NH_2 group,

$$1 = 9.6 + \log \frac{[-NH_2]}{[NH_3^+]}$$

$$-8.6 = \log \frac{[-NH_2]}{[-NH_3^+]}$$

$$\frac{[-NH_2]}{[-NH_3^+]} = 2.5 \times 10^{-9}$$

Thus, the predominant species is NH_3^+–CH_2–COOH.

At pH = 7,

For the –COOH group,

$$7 = 2.3 + \log \frac{[-COO^-]}{[-COOH]}$$

$$4.7 = \log \frac{[-COO^-]}{[-COOH]}$$

$$\frac{[-COO^-]}{[-COOH]} = 5.0 \times 10^4$$

For the –NH_2 group,

$$7 = 9.6 + \log \frac{[-NH_2]}{[NH_3^+]}$$

$$-2.6 = \log \frac{[-NH_2]}{[-NH_3^+]}$$

$$\frac{[-NH_2]}{[-NH_3^+]} = 2.5 \times 10^{-3}$$

Thus, the predominant species is NH_3^+–CH_2–COO^-.

At pH = 12,

For the –COOH group,

$$12 = 2.3 + \log \frac{[-COO^-]}{[-COOH]}$$

$$9.7 = \log \frac{[-COO^-]}{[-COOH]}$$

$$\frac{[-COO^-]}{[-COOH]} = 5.0 \times 10^9$$

For the –NH_2 group,

$$12 = 9.6 + \log \frac{[-NH_2]}{[NH_3^+]}$$

$$2.4 = \log \frac{[-NH_2]}{[-NH_3^+]}$$

$$\frac{[-NH_2]}{[-NH_3^+]} = 2.5 \times 10^2$$

Thus, the predominant species is $NH_2-CH_2-COO^-$.

11.57   Describe a procedure that would allow you to compare the strength of Lewis acids.

Measure $\Delta_r H^\circ$ of the reactions of the Lewis acids with a common Lewis base. The stronger the Lewis acid, the more negative is $\Delta_r H^\circ$.

11.58   From the dependence of K_w on temperature (see p. 403), calculate the enthalpy of dissociation for water.

There are two ways to solve this problem. One uses the van't Hoff equation, and the other uses a graphical method.

<u>Method 1</u> Substitute $K_w = 0.12 \times 10^{-14}$ at 273 K and $K_w = 5.4 \times 10^{-13}$ at 373 K into the van't Hoff equation:

$$\ln \frac{K_2}{K_1} = \frac{\Delta_r H^\circ}{R} \left(\frac{1}{T_1} - \frac{1}{T_2} \right)$$

$$\ln \frac{5.4 \times 10^{-13}}{0.12 \times 10^{-14}} = \frac{\Delta_r H^\circ}{8.314 \, \text{J K}^{-1} \, \text{mol}^{-1}} \left(\frac{1}{273 \, \text{K}} - \frac{1}{373 \, \text{K}} \right)$$

$$\Delta_r H^\circ = 5.2 \times 10^4 \, \text{J mol}^{-1}$$

<u>Method 2</u> The slope of a plot of ln K_w vs $1/T$ has a value of $-\Delta_r H^\circ/R$. The following data points are used.

T/K	K_w	$10^3\ K/T$	$\ln K_w$
273	0.12×10^{-14}	3.663	-34.36
298	1.0×10^{-14}	3.356	-32.24
313	2.9×10^{-14}	3.195	-31.17
373	5.4×10^{-13}	2.681	-28.25

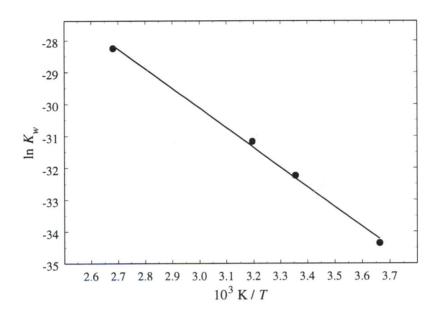

The slope of the line is -6.2×10^3 K.

$$\frac{-\Delta_r H^\circ}{R} = -6.2 \times 10^3\ \text{K}$$

$$\Delta_r H^\circ = \left(-6.2 \times 10^3\ \text{K}\right)\left(8.314\ \text{J K}^{-1}\,\text{mol}^{-1}\right)$$

$$= 5.2 \times 10^4\ \text{J mol}^{-1}$$

The two methods give the same answer, but the second method is preferred since it uses all the data and improves the statistical significance of the result. It is less likely to give a result that is skewed by a single bad data point as would happen if that point were chosen as one of the two points used in the first method.

11.59 Freshly distilled, deionized water has a pH of 7. Left standing in air, however, the water gradually becomes acidic. Calculate the pH of the "solution" at equilibrium. (*Hint*: First calculate the solubility of CO_2 in water according to Example 7.3. Assume the partial pressure of CO_2 is 0.00030 atm.)

The solubility of CO_2 in water can be calculated using Henry's law.

$$m = \frac{P_2}{K'} = \frac{3.0 \times 10^{-4}\ \text{atm}}{29.3\ \text{atm mol}^{-1}\,\text{kg H}_2\text{O}} = 1.02 \times 10^{-5}\ \text{mol (kg H}_2\text{O)}^{-1}$$

The dissolved CO_2 forms carbonic acid with a molarity of 1.02×10^{-5} M (for a dilute solution, molarity is very similar to molality), which dissociates according to

$$
\begin{array}{lcccc}
& H_2CO_3 & \rightleftharpoons & H^+ \; + & HCO_3^- \\
\text{Initial} & 1.02 \times 10^{-5} & & 0 & 0 \quad M \\
\text{At equilibrium} & 1.02 \times 10^{-5} - x & & x & x \quad M
\end{array}
$$

$$K_a = 4.2 \times 10^{-7} = \frac{x^2}{1.02 \times 10^{-5} - x}$$

$$x^2 + 4.2 \times 10^{-7}x - 4.28 \times 10^{-12} = 0$$

$$x = 1.87 \times 10^{-6}$$

Thus,

$$pH = -\log\left(1.87 \times 10^{-6}\right) = 5.73$$

11.60 Show that the acid dissociation constant, K_a, of a weak monoprotic acid in water is related to its concentration, c (mol L^{-1}), and its degree of dissociation, α, by $K_a = \alpha^2 c/(1 - \alpha)$ if the self dissociation of water is ignored. If the latter is taken into account, show that $K_a = \frac{1}{2}\alpha^2 c\left[1 + \left(1 + 4K_w\alpha^{-2}c^{-2}\right)^{1/2}\right] / (1 - \alpha)$.

If the self dissociation of water is ignored, only the following equation needs to be considered:

$$
\begin{array}{lcccc}
& HA & \rightleftharpoons & H^+ \; + & A^- \\
\text{Initial} & c & & 0 & 0 \quad M \\
\text{At equilibrium} & c\,(1 - \alpha) & & c\alpha & c\alpha \quad M
\end{array}
$$

The equilibrium expression is

$$K_a = \frac{(c\alpha)\,(c\alpha)}{c\,(1 - \alpha)} = \frac{\alpha^2 c}{1 - \alpha}$$

If the self dissociation of water is taken into account, the concentrations of HA and A^- are still the same as above at equilibrium, that is,

$$[HA] = c\,(1 - \alpha)$$

$$\left[A^-\right] = c\alpha$$

However, $[H^+]$ is derived both from HA and H_2O. It can be calculated by considering charge balance:

$$\left[H^+\right] = \left[OH^-\right] + \left[A^-\right]$$

An expression relating $[H^+]$, K_w, c, and α is obtained by solving for $[OH^-]$ and then substituting into the equilibrium expression for the self dissociation of water.

$$K_w = [H^+][OH^-] = [H^+]([H^+] - [A^-]) = [H^+]([H^+] - c\alpha)$$

$$[H^+]^2 - c\alpha[H^+] - K_w = 0$$

$$[H^+] = \frac{c\alpha \pm \sqrt{c^2\alpha^2 + 4K_w}}{2}$$

Since $\sqrt{c^2\alpha^2 + 4K_w} > c\alpha$, the only physically possible root is

$$[H^+] = \frac{c\alpha + \sqrt{c^2\alpha^2 + 4K_w}}{2}$$

The equilibrium expression for the acid dissociation is

$$K_a = \frac{[H^+][A^-]}{[HA]}$$

$$= \frac{\left(\dfrac{c\alpha + \sqrt{c^2\alpha^2 + 4K_w}}{2}\right)(c\alpha)}{c(1-\alpha)}$$

$$= \frac{\alpha}{2(1-\alpha)}\left[c\alpha + \sqrt{c^2\alpha^2\left(1 + 4K_w c^{-2}\alpha^{-2}\right)}\right]$$

$$= \frac{\alpha}{2(1-\alpha)}(c\alpha)\left[1 + \left(1 + 4K_w c^{-2}\alpha^{-2}\right)^{1/2}\right]$$

$$= \frac{\alpha^2 c}{2(1-\alpha)}\left[1 + \left(1 + 4K_w c^{-2}\alpha^{-2}\right)^{1/2}\right]$$

11.61   To correct for the effect of ionic strength, we can write the dissociation constant of an acid as

$$pK_a' = pK_a - \frac{0.509\sqrt{I}}{1 + \sqrt{I}}$$

where K_a is the acid dissociation at zero ionic strength and K_a' the corresponding value at ionic strength, I. Calculate the dissociation constant of acetic acid in a 0.15 m KCl solution at 298 K. You may neglect the ionic strength contribution due to the dissociation of the acid itself.

Ignoring the ionic contribution from the dissociation of the acid, the ionic strength of the solution is

$$I = \frac{1}{2}\left(m_+ z_+^2 + m_- z_-^2\right) = \frac{1}{2}\left[(0.15)(1)^2 + (0.15)(-1)^2\right] = 0.15\ m$$

Therefore,

$$pK_a' = pK_a - \frac{0.509\sqrt{I}}{1+\sqrt{I}} = 4.76 - \frac{0.509\sqrt{0.15}}{1+\sqrt{0.15}} = 4.618$$

$$K_a' = 2.4 \times 10^{-5}$$

This may be compared with the value of $K_a = 1.75 \times 10^{-5}$ for acetic acid at zero ionic strength.

11.62 Depending on the pH of the solution, ferric ions (Fe^{3+}) may exist in the free-ion form or form the insoluble precipitate $Fe(OH)_3$ ($K_{sp} = 1.0 \times 10^{-36}$). Calculate the pH at which 90% of the Fe^{3+} ions in a 4.5×10^{-5} M Fe^{3+} solution would be precipitated. What conclusion can you draw about the Fe^{3+} ion concentration in blood plasma whose pH is 7.40?

The concentration of Fe^{3+} remaining in the solution is

$$\left[Fe^{3+}\right] = (10\%)\left(4.5 \times 10^{-5}\ M\right) = 4.5 \times 10^{-6}\ M$$

The solubility equilibrium of $Fe(OH)_3$ is

$$Fe(OH)_3(s) \rightleftharpoons Fe^{3+}(aq) + 3OH^-(aq)$$

and $[OH^-]$ can be calculated from the equilibrium expression

$$K_{sp} = 1.0 \times 10^{-36} = \left[Fe^{3+}\right]\left[OH^-\right]^3 = \left(4.5 \times 10^{-6}\right)\left[OH^-\right]^3$$

$$\left[OH^-\right] = 6.06 \times 10^{-11} M$$

Therefore,

$$\left[H^+\right] = \frac{K_w}{\left[OH^-\right]} = \frac{1.00 \times 10^{-14}}{6.06 \times 10^{-11}} = 1.65 \times 10^{-4} M$$

$$pH = 3.783$$

In blood plasma, where pH = 7.40, and $[H^+] = 4.0 \times 10^{-8}$ M,

$$[OH^-] = \frac{K_w}{[H^+]} = \frac{2.1 \times 10^{-14}}{4.0 \times 10^{-8}} = 5.3 \times 10^{-7}\ M$$

where the value of K_w appropriate for the physiological temperature of 37°C is used. Thus, the free Fe^{3+} ion concentration in blood plasma is limited by the solubility of $Fe(OH)_3$, whose K_{sp} is assumed to be temperature independent.

$$[Fe^{3+}] \leq \frac{K_{sp}}{[OH^-]^3}$$

$$= \frac{1.0 \times 10^{-36}}{\left(5.3 \times 10^{-7}\right)^3}$$

$$= 6.7 \times 10^{-18}\ M$$

There is virtually no free Fe^{3+} in blood plasma.

11.63   A 0.020 M aqueous solution of benzoic acid has a freezing point of $-0.0392°C$. Calculate the dissociation constant of benzoic acid. Assume ideal behavior, and assume that molarity is equal to molality at this low concentration.

The freezing point depression depends on the number of particles in solution. Assuming that $0.020\ M = 0.020\ m$ at this low concentration,

$$\Delta T_f = K_f i m_2$$

$$0.0392°C = \left(1.86°C\,kg\,mol^{-1}\right) i\,\left(0.020\ mol\,kg^{-1}\right)$$

$$i = 1.05$$

The van't Hoff factor, i, is related to the degree of dissociation

$$\alpha = \frac{i-1}{\nu-1} = \frac{1.05-1}{2-1} = 0.05$$

Finally, using the result of Problem 11.60,

$$K_a = \frac{\alpha^2 c}{1-\alpha}$$

$$= \frac{(0.05)^2\,(0.020)}{1-0.05}$$

$$= 5 \times 10^{-5}$$

The value differs from that listed in Table 11.1 primarily because of the limited precision in the value for the concentration of the acid solution, although the assumption of ideality also contributes to the discrepancy.

11.64   The pH of gastric juice is about 1.00 and blood plasma is 7.40. Calculate the Gibbs energy required to secrete a mole of H^+ ions from blood plasma to the stomach at 37°C. Assume ideal behavior.

In the blood plasma, $[H^+] = 10^{-pH} = 10^{-7.40} = 4.0 \times 10^{-8}\ M$, and in the gastric juice, $[H^+] = 10^{-pH} = 10^{-1.00} = 1.0 \times 10^{-1}\ M$. The "reaction" is

$$H^+(4.0 \times 10^{-8}\ M) \rightarrow H^+(1.0 \times 10^{-1}\ M)$$

Since $\Delta G = \Delta G° + RT \ln Q$ and $\Delta G° = 0$ for this reaction,

$$\Delta G = RT \ln \frac{[H^+]_{stomach}}{[H^+]_{blood\ plasma}}$$

$$= \left(8.314\,J\,K^{-1}\,mol^{-1}\right)\left(310\ K\right)\ln\frac{1.0 \times 10^{-1}}{4.0 \times 10^{-8}}$$

$$= 3.8 \times 10^4\ J\,mol^{-1}$$

$$= 38\ kJ\,mol^{-1}$$

11.65   Chemical analysis shows that 20.0 mL of a certain sample of blood yields 12.5 mL of CO_2 gas (measured at 25°C and 1 atm) when treated with an acid. Calculate **(a)** the number of moles of CO_2 originally present in the blood, **(b)** the concentration of CO_2 and HCO_3^- at equilibrium, and **(c)** the partial pressure of CO_2 over the blood solution at equilibrium. Assume ideal behavior. The pH of blood is 7.40, and the Henry's law constant for CO_2 in blood is 29.3 atm mol^{-1} (kg H$_2$O).

(a) The number of moles of CO_2 can be calculated using the ideal gas law:

$$n = \frac{PV}{RT} = \frac{(1 \text{ atm}) \left(12.5 \times 10^{-3} \text{ L}\right)}{\left(0.08206 \text{ L atm K}^{-1} \text{ mol}^{-1}\right) (298.2 \text{ K})} = 5.108 \times 10^{-4} \text{ mol} = 5.11 \times 10^{-4} \text{ mol}$$

(b) The $[CO_2]$ (which is taken to be the same as $[H_2CO_3]$) and $[HCO_3^-]$ are related by the Henderson–Hasselbalch equation:

$$\text{pH} = \text{p}K_a' + \log \frac{\left[HCO_3^-\right]}{\left[H_2CO_3\right]}$$

$$7.40 = 6.38 + \log \frac{\left[HCO_3^-\right]}{\left[CO_2\right]}$$

$$\frac{\left[HCO_3^-\right]}{\left[CO_2\right]} = 10.5$$

The mole ratio between HCO_3^- and CO_2 is the same as the concentration ratio between HCO_3^- and CO_2 because both species are in the same solution. Therefore,

$$\frac{\left[HCO_3^-\right]}{\left[CO_2\right]} = \frac{n_{HCO_3^-}}{n_{CO_2}} = 10.5$$

HCO_3^- does not dissociate appreciably. Therefore, the number of moles of CO_2 originally present in the 20.0-mL blood sample is a sum of the number of moles of CO_2 dissolved and the number of moles of HCO_3^-:

$$n_{CO_2} \text{ originally present} = 5.108 \times 10^{-4} \text{ mol} = n_{CO_2} + n_{HCO_3^-}$$

Solve the above equation for $n_{HCO_3^-}$ and substitute it into the ratio between the number of moles of CO_2 and HCO_3^-:

$$\frac{5.108 \times 10^{-4} \text{ mol} - n_{CO_2}}{n_{CO_2}} = 10.5$$

$$n_{CO_2} = 4.44 \times 10^{-5} \text{ mol}$$

$$n_{HCO_3^-} = 5.108 \times 10^{-4} \text{ mol} - 4.44 \times 10^{-5} \text{ mol} = 4.664 \times 10^{-4} \text{ mol}$$

Thus, at equilibrium,

$$[CO_2] = \frac{4.44 \times 10^{-5} \text{ mol}}{20.0 \times 10^{-3} \text{ L}} = 2.22 \times 10^{-3} M = 2.2 \times 10^{-3} M$$

$$[HCO_3^-] = \frac{4.664 \times 10^{-4} \text{ mol}}{20.0 \times 10^{-3} \text{ L}} = 2.33 \times 10^{-2} M$$

(c) According to Henry's law, the partial pressure of CO_2 over the blood solution is related to its solubility. Since the solubility is small, the molarity and molality of CO_2 in blood are almost the same.

$$P_{CO_2} = k'm = (29.3 \text{ atm mol}^{-1} (\text{kg H}_2\text{O})) (2.22 \times 10^{-3} \text{ mol kg}^{-1}) = 6.50 \times 10^{-2} \text{ atm} = 49 \text{ mmHg}$$

11.66   Calcium oxalate is a major component of kidney stones. From the dissociation constants listed in Table 11.1 and given that the solubility product of CaC_2O_4 is 3.0×10^{-9}, predict whether the formation of kidney stones can be minimized by increasing or decreasing the pH of the fluid present in the kidneys. The pH of normal kidney fluid is about 8.2.

The solubility equilibrium of CaC_2O_4 is

$$CaC_2O_4(s) \rightleftharpoons Ca^{2+}(aq) + C_2O_4^{2-}(aq)$$

The resulting oxalate ions hydrolyze:

$$C_2O_4^{2-}(aq) + 2H_2O(l) \rightleftharpoons H_2C_2O_4(aq) + 2OH^-(aq)$$

Increasing the pH of the fluid present in the kidneys will increase $[OH^-]$, which, according to Le Chatelier's's principle, will shift the equilibrium towards $C_2O_4^{2-}$, which will in turn shift the solubility equilibrium of calcium oxalate towards CaC_2O_4. In other words, increasing the pH will enhance the formation of kidney stones. On the other hand, decreasing the pH of the fluid will decrease $[OH^-]$, which will cause the hydrolysis equilibrium to shift to the right. The removal of $C_2O_4^{2-}$ as $H_2C_2O_4$ will cause the solubility equilibrium of calcium oxalate to shift to the right. As a result, decreasing the pH of the fluid will minimize kidney stone formation.

11.67   What is the pH of a 0.050 M glycine solution at 298 K?

In solution, glycine will exist predominantly in the following three forms: $H_3N^+-CH_2-COOH$ $\equiv G^+$, $H_3N^+-CH_2-COO^- \equiv G$, and $H_2N-CH_2-COO^- \equiv G^-$. There is essentially no glycine in the non-ionic form, $H_2N-CH_2\text{-}COOH$. The concentrations of these species must satisfy the equilibria,

$$G^+ \rightleftharpoons G + H^+ \qquad \frac{[H^+][G]}{[G^+]} = K_1 = 4.57 \times 10^{-3}$$

$$G \rightleftharpoons G^- + H^+ \qquad \frac{[H^+][G^-]}{[G]} = K_2 = 2.51 \times 10^{-10},$$

as well as the charge balance condition

$$[H^+] + [G^+] = [G^-] + [OH^-].$$

Using the two equilibria to solve for the charged forms of glycine and also $[OH^-] = K_w/[H^+]$, the charge balance equation becomes,

$$[H^+] + \frac{[H^+][G]}{K_1} = \frac{K_2[G]}{[H^+]} + \frac{K_w}{[H^+]}$$

$$[H^+]\left(1 + \frac{[G]}{K_1}\right) = \frac{1}{[H^+]}\left(K_2[G] + K_w\right)$$

$$[H^+]^2 = \frac{K_1\left(K_2[G] + K_w\right)}{K_1 + [G]}$$

assuming that very little of the G has either dissociated or added an H^+, $[G] \approx 0.050\ M$

$$[H^+]^2 = \frac{\left(4.57 \times 10^{-3}\right)\left[\left(2.51 \times 10^{-10}\right)(0.050) + 1.00 \times 10^{-14}\right]}{4.57 \times 10^{-3} + 0.050}$$

$$[H^+]^2 = 1.05 \times 10^{-12}$$

$$[H^+] = 1.03 \times 10^{-6}\ M$$

$$pH = 5.99$$

At this pH, $[G^+]/[G] = [H^+]/K_1 = 2.25 \times 10^{-4}$ and $[G^-]/[G] = K_2/[H^+] = 2.44 \times 10^{-4}$, which justifies the assumption that very little of the glycine has gone to form either of the ionized forms G^+ or G^- and that $[G] \approx 0.050\ M$.

11.68 From the dissociation constant of formic acid listed in Table 11.1, calculate the Gibbs energy and the standard Gibbs energy for the dissociation of formic acid at 298 K.

The reaction is

$$HCOOH \rightleftharpoons H^+ + HCOO^-$$

At equilibrium, the Gibbs energy, $\Delta_r G = 0$. The standard Gibbs energy is related to the dissociation constant.

$$\Delta_r G^\circ = -RT \ln K_a = -\left(8.314\ J\,K^{-1}\,mol^{-1}\right)(298\ K)\ln\left(1.77 \times 10^{-4}\right) = 2.14 \times 10^4\ J\,mol^{-1}$$

11.69 (a) Calculate the percent ionization of a 0.20 M solution of the monoprotic acetylsalicylic acid (aspirin, $C_9H_8O_4$) for which $K_a = 3.0 \times 10^{-4}$. (b) The pH of gastric juice in the stomach of a certain individual is 1.00. After a few aspirin tablets have been swallowed, the concentration of acetylsalicylic acid in the stomach is 0.20 M. Calculate the percent ionization of the acid under these conditions. What effect does the nonionized acid have on the membranes lining the stomach?

(a) The ionization reaction is

$$
\begin{array}{lcccc}
 & C_9H_8O_4(aq) & \rightleftharpoons & H^+(aq) & + & C_9H_7O_4^-(aq) \\
\text{Initial} & 0.20 & & 0 & & 0 & M \\
\text{At equilibrium} & 0.20 - x & & x & & x & M
\end{array}
$$

$$K_a = 3.0 \times 10^{-4} = \frac{x^2}{0.20 - x}$$

$$x^2 + 3.0 \times 10^{-4}x - 6.0 \times 10^{-5} = 0$$

$$x = 7.60 \times 10^{-3}$$

Therefore, the percent ionization is

$$\frac{x}{0.20} \times 100\% = 3.8\%$$

(b) At pH 1.00 the concentration of H^+ is 0.10 M. According to Le Chatelier's principle, this will suppress the ionization of acetylsalicyclic acid. The H^+ contribution from the ionization of the acetylsalicyclic acid is negligible compared with the H^+ concentration in gastric juice, and to two-significant-figure accuracy, this contribution is ignored. The percent ionization of the acid can be obtained using the equilibrium expression:

$$K_a = \frac{[H^+]\left[C_9H_7O_4^-\right]}{[C_9H_8O_4]}$$

$$\% \text{ ionization} = \frac{\left[C_9H_7O_4^-\right]}{[C_9H_8O_4]} \times 100\% = \frac{K_a}{[H^+]} = \frac{3.0 \times 10^{-4}}{0.1} \times 100\% = 0.30\%$$

Although the acetylsalicyclic acid has negligible effect on the pH of the gastric juices, the high acidity of the gastric juices appears to enhance the rate of absorption of nonionized aspirin molecules through the stomach lining. In some cases this can irritate these tissues and cause bleeding.

11.70 A 0.400 M formic (HCOOH) solution freezes at $-0.758°C$. Calculate the value of K_a at that temperature. (*Hint*: Assume that molarity is equal to molality.)

The dissociation reaction is

$$
\begin{array}{lcccc}
 & HCOOH(aq) & \rightleftharpoons & H^+(aq) & + & HCOO^-(aq) \\
\text{Initial} & 0.400 & & 0 & & 0 & M \\
\text{At equilibrium} & 0.400 - x & & x & & x & M
\end{array}
$$

Assuming molarities are equal to molalities, the total molality of all species is

$$(0.400 - x + x + x)\ m = (0.400 + x)\ m$$

This molality is related to the freezing point of the solution.

$$\Delta T_f = K_f m$$

$$0.758 \text{ K} = \left(1.86 \text{ K mol}^{-1} \text{kg}\right)(0.400 + x) \text{ mol kg}^{-1}$$

$$x = 0.0075 \, m = 0.0075 \, M$$

Knowing the value of x, the dissociation constant can be calculated.

$$K_a = \frac{x^2}{0.400 - x} = 1 \times 10^{-4}$$

The result is limited by the precision to which the concentration of the formic acid solution is known at equilibrium.

11.71   Explain the action of smelling salts, which is ammonium carbonate [$(NH_4)_2CO_3$]. (*Hint*: The thin film of aqueous solution that lines the nasal passage is slightly basic.)

In inhaling the smelling salts, some of the powder dissolves in the basic solution lining the nasal passages, and as a strong electrolyte, dissociates into ammonium and carbonate ions. The weakly acidic ammonium ions react with the basic solution,

$$NH_4^+(aq) + OH^-(aq) \rightarrow NH_3(aq) + H_2O(l)$$

$$NH_3(aq) \rightarrow NH_3(g)$$

The pungent smell of the gaseous ammonia so produced prevents a person from fainting.

11.72   Acid–base reactions usually go to completion. Confirm this statement by calculating the equilibrium constant for each of the following cases: **(a)** a strong acid reacting with a strong base, **(b)** a strong acid reacting with a weak base (NH_3), **(c)** a weak acid (CH_3COOH) reacting with a strong base, and **(d)** a weak acid (CH_3COOH) reacting with a weak base (NH_3). (*Hint*: Strong acids exist as H^+ ions and strong bases exist as OH^- ions in solution. You need to look up K_a, K_b, and K_w values.)

(a) The reaction is $H^+ + OH^- \rightleftharpoons H_2O$ for which $K = \dfrac{1}{K_w} = 1.0 \times 10^{14}$.

(b) The reaction is $H^+ + NH_3 \rightleftharpoons NH_4^+$ for which $K = \dfrac{1}{K_a} = \dfrac{1}{5.6 \times 10^{-10}} = 1.8 \times 10^9$.

(c) The reaction,

$$CH_3COOH + OH^- \rightleftharpoons CH_3COO^- + H_2O$$

may be considered as the sum of the two equations, each with its own equilibrium constant

$$CH_3COOH \rightleftharpoons CH_3COO^- + H^+ \qquad K_a$$

$$H^+ + OH^- \rightleftharpoons H_2O \qquad 1/K_w$$

The overall equilibrium constant is

$$K = K_a \left(1/K_w\right) = \frac{K_a}{K_w} = \frac{1.8 \times 10^{-5}}{1.0 \times 10^{-14}} = 1.8 \times 10^9$$

(d) Again, the reaction,

$$CH_3COOH + NH_3 \rightleftharpoons CH_3COO^- + NH_4^+$$

may be considered as the sum of the two equations, each with its own equilibrium constant

$$CH_3COOH \rightleftharpoons CH_3COO^- + H^+ \qquad K_a$$

$$NH_3 + H^+ \rightleftharpoons NH_4^+ \qquad 1/K_a'$$

The overall equilibrium constant is

$$K = K_a \left(1/K_a'\right) = \frac{K_a}{K_a'} = \frac{1.8 \times 10^{-5}}{5.6 \times 10^{-10}} = 3.2 \times 10^4$$

In each case the overall equilibrium constant is large enough that the reaction proceeds to completion.

11.73 When lemon juice is squirted into tea, the color becomes lighter. In part, the color change is due to dilution, but the main reason for the change is an acid–base reaction. What is the reaction? (*Hint*: Tea contains "polyphenols," which are weak acids, and lemon juice contains citric acid.)

The weak acid polyphenols form an indicator. That is, the acid form and base form have different colors. The base form is dark colored, and upon reaction with the citric acid in lemon juice, the lighter colored acid form is produced.

11.74 One of the most common antibiotics is penicillin G (benzylpenicillinic acid), which has the following structure:

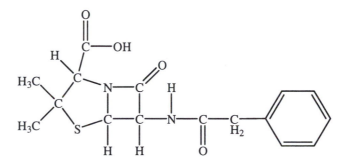

It is a weak monoprotic acid:

$$HP \rightleftharpoons H^+ + P^- \qquad K_a = 1.64 \times 10^{-3}$$

where HP denotes the parent acid and P^- the conjugate base. Penicillin G is produced by growing molds in fermentation tanks at 25°C and a pH range of 4.5 to 5.0. The crude form of this antibiotic is obtained by extracting the fermentation broth with an organic solvent in which the acid is soluble. **(a)** Identify the acidic hydrogen atom. **(b)** In one stage of purification, the organic extract of the crude penicillin G is treated with a buffer solution at pH = 6.50. What is the ratio of the conjugate base of penicillin G to the acid at this pH? Would you expect the conjugate base to be more soluble in water than the acid? **(c)** Penicillin G is not suitable for oral administration, but the sodium salt (NaP) is because it is soluble. Calculate the pH of a 0.12 M NaP solution formed when a tablet containing the salt is dissolved in a glass of water.

(a) The acidic hydrogen is from the carboxyl group (-COOH).

(b) The ratio of the conjugate base and the acid can be obtained from the Henderson–Hasselbalch equation:

$$pH = pK_a + \log \frac{[P^-]}{[HP]}$$

$$6.50 = -\log\left(1.64 \times 10^{-3}\right) + \log \frac{[P^-]}{[HP]}$$

$$\frac{[P^-]}{[HP]} = 10^{3.715} = 5.2 \times 10^3$$

Thus, nearly all of the penicillin G will be in the ionized form. The ionized form is more soluble in water because it bears a net charge; penicillin G is largely nonpolar and therefore much less soluble in water. (Both penicillin G and its salt are effective antibiotics.)

(c) NaP dissociates into Na^+ and P^-, and P^- undergoes hydrolysis:

	$P^-(aq)$	+	$H_2O(l)$	⇌	$HP(aq)$	+	OH^-	
Initial	0.12				0		0	M
At equilibrium	0.12 − x				x		x	M

The equilibrium expression associated with the hydrolysis reaction is

$$K_b = \frac{K_w}{K_a} = \frac{x^2}{0.12 - x}$$

$$\frac{1.00 \times 10^{-14}}{1.64 \times 10^{-3}} = 6.098 \times 10^{-12} = \frac{x^2}{0.12 - x}$$

Since K_b is very small, $0.12 - x$ can be approximated by 0.12. The expression above simplifies to

$$6.098 \times 10^{-12} = \frac{x^2}{0.12}$$

$$x = 8.55 \times 10^{-7}$$

Checking the assumption,

$$\frac{x}{0.12} \times 100\% = 7.1 \times 10^{-4}\% < 5\%$$

Thus, the assumption is valid.

Therefore,

$$[H^+] = \frac{K_w}{[OH^-]} = \frac{1.00 \times 10^{-14}}{8.55 \times 10^{-7}} = 1.17 \times 10^{-8}$$

$$pH = 7.93$$

Because HP is a relatively strong acid, P^- is a weak base. Consequently, only a small fraction of P^- undergoes hydrolysis and the solution is slightly basic.

CHAPTER 12

Chemical Kinetics

PROBLEMS AND SOLUTIONS

12.1 Write the rates for the following reactions in terms of the disappearance of reactants and appearance of products:

(a) $3O_2 \rightarrow 2O_3$

(b) $C_2H_6 \rightarrow C_2H_4 + H_2$

(c) $ClO^- + Br^- \rightarrow BrO^- + Cl^-$

(d) $(CH_3)_3CCl + H_2O \rightarrow (CH_3)_3COH + H^+ + Cl^-$

(e) $2AsH_3 \rightarrow 2As + 3H_2$

(a) Rate $= -\dfrac{1}{3}\dfrac{d\,[O_2]}{dt} = \dfrac{1}{2}\dfrac{d\,[O_3]}{dt}$

(b) Rate $= -\dfrac{d\,[C_2H_6]}{dt} = \dfrac{d\,[C_2H_4]}{dt} = \dfrac{d\,[H_2]}{dt}$

(c) Rate $= -\dfrac{d\,[ClO^-]}{dt} = -\dfrac{d\,[Br^-]}{dt} = \dfrac{d\,[BrO^-]}{dt} = \dfrac{d\,[Cl^-]}{dt}$

(d) Rate $= -\dfrac{d\,[(CH_3)_3CCl]}{dt} = -\dfrac{d\,[H_2O]}{dt} = \dfrac{d\,[(CH_3)_3COH]}{dt} = \dfrac{d\,[H^+]}{dt} = \dfrac{d\,[Cl^-]}{dt}$

(e) Rate $= -\dfrac{1}{2}\dfrac{d\,[AsH_3]}{dt} = \dfrac{1}{2}\dfrac{d\,[As]}{dt} = \dfrac{1}{3}\dfrac{d\,[H_2]}{dt}$

12.2 The rate law for the reaction

$$NH_4^+(aq) + NO_2^-(aq) \rightarrow N_2(g) + 2H_2O(l)$$

is given by rate $= k\,[NH_4^+]\,[NO_2^-]$. At 25°C, the rate constant is $3.0 \times 10^{-4}\ M^{-1}\,s^{-1}$. Calculate the rate of the reaction at this temperature if $[NH_4^+] = 0.26\ M$ and $[NO_2^-] = 0.080\ M$.

Rate $= k\,[NH_4^+]\,[NO_2^-] = \left(3.0 \times 10^{-4}\ M^{-1}\,s^{-1}\right)(0.26\ M)\,(0.080\ M) = 6.2 \times 10^{-6}\ M\,s^{-1}$

12.3   What are the units of the rate constant for a third-order reaction?

The units of reaction rate are $M\,s^{-1}$, while the units of all the concentration terms in a third-order reaction rate law are M^3. Therefore, the units of the rate constant are $M\,s^{-1}/M^3$, or $M^{-2}\,s^{-1}$.

12.4   The following reaction is found to be first order in A:

$$A \rightarrow B + C$$

If half of the starting quantity of A is used up after 56 s, calculate the fraction that will be used up after 6.0 min.

From the half-life (56 s), k can be determined:

$$k = \frac{\ln 2}{t_{1/2}} = \frac{\ln 2}{56\ \text{s}} = 1.24 \times 10^{-2}\ \text{s}^{-1}$$

The fraction of A that will remain after 6.0 min (3.6×10^2 s) is

$$\frac{[A]}{[A]_0} = e^{-kt} = e^{-\left(1.24\times10^{-2}\,\text{s}^{-1}\right)\left(3.6\times10^2\,\text{s}\right)} = 0.0115$$

Thus, the fraction that will be used up after 6.0 min is $1 - 0.0115 = 0.99$.

12.5   A certain first-order reaction is 34.5% complete in 49 min at 298 K. What is its rate constant?

$$\frac{[A]}{[A]_0} = e^{-kt}$$

$$k = -\frac{1}{t}\ln\frac{[A]}{[A]_0} = -\frac{1}{49\ \text{min}}\ln\left(1 - 0.345\right) = 8.6 \times 10^{-3}\ \text{min}^{-1}$$

12.6   (a) The half-life of the first-order decay of radioactive ^{14}C is about 5720 years. Calculate the rate constant for the reaction. (b) The natural abundance of ^{14}C isotope is 1.1×10^{-13} mol % in living matter. Radiochemical analysis of an object obtained in an archaeological excavation shows that the ^{14}C isotope content is 0.89×10^{-14} mol %. Calculate the age of the object. State any assumptions.

(a)

$$k = \frac{\ln 2}{t_{1/2}} = \frac{\ln 2}{5720\ \text{yr}} = 1.212 \times 10^{-4}\ \text{yr}^{-1} = 1.21 \times 10^{-4}\ \text{yr}^{-1}$$

(b)

$$\frac{[^{14}C]}{[^{14}C]_0} = e^{-kt}$$

$$t = -\frac{1}{k} \ln \frac{[^{14}C]}{[^{14}C]_0}$$

Due to constant exchange of material with the surroundings, the mol % of ^{14}C of all living matter is assumed to be the same. However, when the object ceases to live, it no longer exchanges material with the environment and the mol % of ^{14}C will decrease according to first-order decay kinetics. Therefore, the ratio between $[^{14}C]$ and $[^{14}C]_0$ depends on the time elapsed since the object's "death." Thus, t in the equation above gives the age of the object.

$$t = -\frac{1}{1.212 \times 10^{-4} \text{ yr}^{-1}} \ln \frac{0.89 \times 10^{-14}}{1.1 \times 10^{-13}} = 2.1 \times 10^4 \text{ yr}$$

A key assumption in radiocarbon dating is that the natural abundance of ^{14}C has remained constant throughout the ages. Since the production of terrestrial ^{14}C is due to bombardment of ^{14}N by cosmic rays, variations in cosmic ray flux have in fact led to variations in the natural abundance of ^{14}C.

12.7 The first-order rate constant for the gas-phase decomposition of dimethyl ether,

$$(CH_3)_2O \rightarrow CH_4 + H_2 + CO$$

is $3.2 \times 10^{-4} \text{ s}^{-1}$ at 450°C. The reaction is carried out in a constant-volume container. Initially, only dimethyl ether is present and the pressure is 0.350 atm. What is the pressure of the system after 8.0 min? Assume ideal-gas behavior.

At constant temperature and volume, the number of moles (and hence the concentration) of a gas is proportional to its pressure. Therefore, the integrated rate law becomes

$$P = P_0 e^{-kt}$$

where P and P_0 are the pressures of dimethyl ether at time t and 0 sec, respectively. After 8.0 min $(4.8 \times 10^2 \text{ s})$,

$$P = (0.350 \text{ atm}) \, e^{-\left(3.2 \times 10^{-4} \text{ s}^{-1}\right)\left(4.8 \times 10^2 \text{ s}\right)} = 0.3002 \text{ atm}$$

Thus, the pressure of dimethyl ether decreases by $(0.350 - 0.3002)$ atm $= 0.0498$ atm. For each atmosphere of dimethyl ether decomposed, an equal pressure of each product is generated according to the chemical equation. Thus, the total pressure of the system after 8.0 min is

$$P = P_{(CH_3)_2O} + P_{CH_4} + P_{H_2} + P_{CO}$$

$$= 0.3002 \text{ atm} + 0.0498 \text{ atm} + 0.0498 \text{ atm} + 0.0498 \text{ atm}$$

$$= 0.450 \text{ atm}$$

12.8   When the concentration of A in the reaction A $\rightarrow$ B was changed from 1.20 M to 0.60 M, the half-life increased from 2.0 min to 4.0 min at 25°C. Calculate the order of the reaction and the rate constant.

The half life is related to the initial concentration of A by

$$t_{1/2} \propto \frac{1}{[A]_0{}^{n-1}}$$

According to the data, the half-life doubled when $[A]_0$ was halved. This is only possible if the half-life is inversely proportional to $[A]_0$, or the reaction order, $n = 2$, indicating a second-order reaction.

The rate constant can be calculated using either $[A]_0 = 1.20$ M or 0.60 M and the corresponding half-life.

$$k = \frac{1}{[A]_0 t_{1/2}} = \frac{1}{(1.20\ M)\ (2.0\ \text{min})} = 0.42\ M^{-1}\ \text{min}^{-1}$$

12.9   The progress of a reaction in the aqueous phase was monitored by the absorbance of a reactant at various times:

Time/s	0	54	171	390	720	1010	1190
Absorbance	1.67	1.51	1.24	0.847	0.478	0.301	0.216

Determine the order of the reaction and the rate constant.

The absorbance of a reactant is proportional to its concentration. The following plot shows absorbance as a function of time for the reactant.

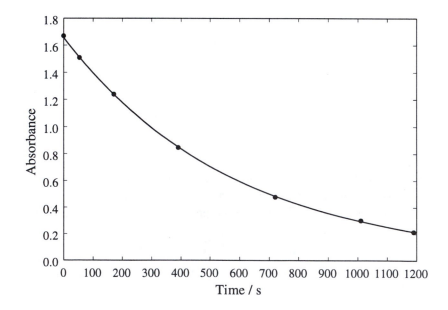

According to the plot, the absorbance decreases from 1.4 to 0.70 as t goes from 100 s to 505 s. That is, the half-life is 405 s when the absorbance is 1.4. The absorbance decreases from 0.70 to 0.35, over the period t equals 505 s to 910 s. That is, the half-life is 405 s when the absorbance is 0.70. Since the half-life is independent of reactant concentration, this is a first order reaction.

Let $[C]$ and $[C]_0$ be the concentrations of the reactant at time t and at the beginning of the reaction, respectively; and let A and A_0 be the absorbance of the reactant at time t and at the beginning of the reaction, respectively. Since this is a first-order reaction,

$$[C] = [C]_0 e^{-kt}$$

$$\ln \frac{[C]}{[C]_0} = \ln \frac{A}{A_0} = -kt$$

A plot of $\ln \left(A/A_0 \right)$ vs t will give a straight line with a slope of $-k$.

Time/s	0	54	171	390	720	1010	1190
$\ln \left(A/A_0 \right)$	0.0000	−0.1007	−0.2977	−0.6789	−1.251	−1.713	−2.045

The equation of the line is $y = -1.71 \times 10^{-3}x - 6.94 \times 10^{-3}$. Therefore, $k = 1.71 \times 10^{-3} \text{ s}^{-1}$.

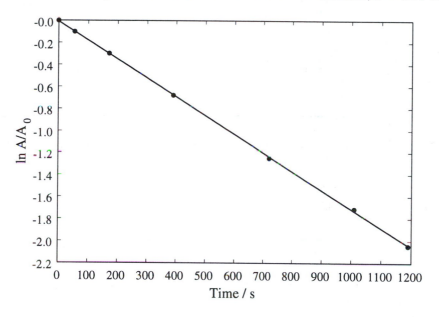

12.10   Cyclobutane decomposes to ethylene according to the equation

$$C_4H_8(g) \rightarrow 2C_2H_4(g)$$

Determine the order of the reaction and the rate constant based on the following pressures, which were recorded when the reaction was carried out at 430°C in a constant-volume vessel:

Time/s	$P_{C_4H_8}$/mmHg
0	400
2000	316
4000	248
6000	196
8000	155
10000	122

At constant temperature and volume, the pressure of cyclobutane is proportional to its concentration. A plot of $\ln P$ vs t shows a straight line with an equation of $y = -1.19 \times 10^{-4}x + 5.99$. Thus, the reaction is first order.

The slope of the line is $-k$. In other words, $k = 1.19 \times 10^{-4}\,\text{s}^{-1}$.

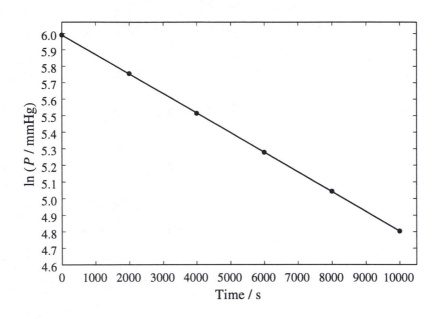

12.11 What is the half-life of a compound if 75% of a given sample of the compound decomposes in 60 min? Assume first-order kinetics.

The rate constant is first calculated.

$$\frac{[A]}{[A]_0} = e^{-kt}$$

$$k = -\frac{1}{t} \ln \frac{[A]}{[A]_0} = -\frac{1}{60 \text{ min}} \ln (1 - 0.75) = 2.31 \times 10^{-2} \text{ min}^{-1}$$

The half-life is

$$t_{1/2} = \frac{\ln 2}{k} = \frac{\ln 2}{2.31 \times 10^{-2} \text{ min}^{-1}} = 30 \text{ min}$$

12.12 The rate constant for the second-order reaction

$$2NO_2(g) \rightarrow 2NO(g) + O_2(g)$$

is 0.54 $M^{-1}\,\text{s}^{-1}$ at 300°C. How long (in seconds) would it take for the concentration of NO_2 to decrease from 0.62 M to 0.28 M?

$$\frac{1}{[A]} - \frac{1}{[A]_0} = kt$$

$$t = \frac{1}{k}\left(\frac{1}{[A]} - \frac{1}{[A]_0}\right) = \frac{1}{0.54\ M^{-1}\ s^{-1}}\left(\frac{1}{0.28\ M} - \frac{1}{0.62\ M}\right) = 3.6\ s$$

12.13   The decomposition of N_2O to N_2 and O_2 is a first-order reaction. At 730°C, the half-life of the reaction is 3.58×10^3 min. If the initial pressure of N_2O is 2.10 atm at 730°C, calculate the total gas pressure after one half-life. Assume that the volume remains constant.

The reaction is

$$N_2O(g) \longrightarrow N_2(g) + \tfrac{1}{2}O_2(g)$$

After one half-life, the pressure of N_2O becomes 1.05 atm. Since an equal amount, 1.05 atm, of N_2O reacts, 1.05 atm of N_2 and 1.05 atm/2 = 0.525 atm of O_2 are produced. Therefore, the total gas pressure is

$$P = P_{N_2O} + P_{N_2} + P_{O_2} = 1.05\ atm + 1.05\ atm + 0.525\ atm = 2.63\ atm$$

12.14   The integrated rate law for the zero-order reaction A → B is $[A] = [A]_0 - kt$. **(a)** Sketch the following plots: **(i)** rate versus [A] and **(ii)** [A] versus t. **(b)** Derive an expression for the half-life of the reaction. **(c)** Calculate the time in half-lives when the integrated rate law is no longer valid, that is, when [A] = 0.

(a) (i) The rate law is

$$Rate = k$$

Therefore, the rate of reaction is independent of the concentration of A.

(a) (ii) Since

$$[A] = [A]_0 - kt$$

a plot of [A] vs t is a straight line with a slope of $-k$.

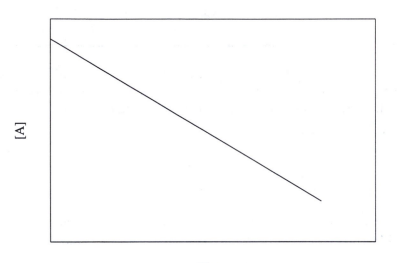

Time

(b) At $t = t_{1/2}$, $[A] = [A]_0/2$. Therefore,

$$\frac{[A]_0}{2} = [A]_0 - kt_{1/2}$$

$$t_{1/2} = \frac{1}{2k}[A]_0$$

(c) When $[A] = 0$,

$$[A] = 0 = [A]_0 - kt$$

According to part (b),

$$k = \frac{1}{2t_{1/2}}[A]_0$$

Therefore, the time it takes to consume all the reactant is

$$t = \frac{[A]_0}{k} = \frac{[A]_0}{\frac{1}{2t_{1/2}}[A]_0} = 2t_{1/2}$$

The integrated rate law is no longer valid after 2 half-lives.

12.15 In the nuclear industry, workers use a rule of thumb that the radioactivity from any sample will be relatively harmless after ten half-lives. Calculate the fraction of a radioactive sample that remains after this time period. (*Hint*: Radioactive decays obey first-order kinetics.)

When $t = 10t_{1/2}$,

$$\frac{[A]}{[A]_0} = e^{-kt} = e^{-k\left(10t_{1/2}\right)}$$

$$= e^{\left(\frac{\ln 2}{t_{1/2}}\right)\left(10t_{1/2}\right)} = e^{-10\ln 2}$$

$$= 9.8 \times 10^{-4}$$

Thus, only 9.8×10^{-2} % of radioactive sample is left after 10 half-lives.

12.16 Many reactions involving heterogeneous catalysis are zero order; that is, rate = k. An example is the decomposition of phosphine (PH_3) over tungsten (W):

$$4PH_3(g) \rightarrow P_4(g) + 6H_2(g)$$

The rate for this reaction is independent of $[PH_3]$ as long as phosphine's pressure is sufficiently high (≥ 1 atm). Explain.

With sufficient PH_3, all the catalytic sites on the tungsten surface are occupied. Further increases in the amount of phosphine cannot affect the reaction, and the rate is independent of $[PH_3]$.

12.17 If the first half-life of a zero-order reaction is 200 s, what will be the duration of the next half-life?

Since the first half-life is 200 s,

$$\frac{[A]_0}{2k} = 200 \text{ s}$$

The next half-life is

$$\frac{\frac{[A]_0}{2}}{2k} = \frac{[A]_0}{4k} = \frac{200 \text{ s}}{2} = 100 \text{ s}$$

12.18 Consider the following nuclear decay

$$^{64}Cu \rightarrow\, ^{64}Zn +\, _{-1}^{0}\beta \qquad t_{1/2} = 12.8 \text{ hr}$$

Starting with one mole of ^{64}Cu, calculate the number of grams of ^{64}Zn formed after 25.6 hours.

Two half-lives would have elapsed after 25.6 hours. After the first half-life, 1/2 mole of ^{64}Cu would remain. After the second half-life, 1/4 mol of ^{64}Cu would remain. Thus, a total of 3/4 mole of ^{64}Cu decays to form 3/4 mole of ^{64}Zn. The mass of ^{64}Zn produced is

$$m = \left(\frac{3}{4}\ \text{mol}\right)(63.93\ \text{g mol}^{-1}) = 47.95\ \text{g}$$

12.19   The reaction $S_2O_8^{2-} + 2I^- \rightarrow 2SO_4^{2-} + I_2$ proceeds slowly in aqueous solution, but it can be catalyzed by the Fe^{3+} ion. Given that Fe^{3+} can oxidize I^- and Fe^{2+} can reduce $S_2O_8^{2-}$, write a plausible two-step mechanism for this reaction. Explain why the uncatalyzed reaction is slow.

A plausible mechanism is

Fe^{3+} oxidizes I^-: $\qquad\qquad\qquad\qquad 2Fe^{3+} + 2I^- \longrightarrow 2Fe^{2+} + I_2$

Fe^{2+} reduces $S_2O_8^{2-}$: $\qquad\qquad\quad 2Fe^{2+} + S_2O_8^{2-} \longrightarrow 2Fe^{3+} + 2SO_4^{2-}$

Fe^{3+} undergoes a redox cycle: $Fe^{3+} \rightarrow Fe^{2+} \rightarrow Fe^{3+}$ in this mechanism.

The uncatalyzed reaction is slow because both I^- and $S_2O_8^{2-}$ are negatively charged which makes their mutual approach unfavorable.

12.20   Derive Equation 12.22 using the steady-state approximation for both the H and Br atoms.

Apply the steady state approximation to both the H and Br atoms:

$$\frac{d\,[\text{H}]}{dt} = k_2[\text{Br}][\text{H}_2] - k_3[\text{H}][\text{Br}_2] - k_4[\text{H}][\text{HBr}] = 0 \qquad (12.20.1)$$

$$\frac{d\,[\text{Br}]}{dt} = 2k_1[\text{Br}_2] - k_2[\text{Br}][\text{H}_2] + k_3[\text{H}][\text{Br}_2] + k_4[\text{H}][\text{HBr}] - k_5[\text{Br}]^2 = 0 \qquad (12.20.2)$$

Summing Eqs. 12.20.1 and 12.20.2,

$$2k_1[\text{Br}_2] - k_5[\text{Br}]^2 = 0$$

$$[\text{Br}] = \left(\frac{2k_1[\text{Br}_2]}{k_5}\right)^{1/2} \qquad (12.20.3)$$

Solving for [H] using Eq. 12.20.1,

$$[\text{H}] = \frac{k_2[\text{Br}][\text{H}_2]}{k_3[\text{Br}_2] + k_4[\text{HBr}]} \qquad (12.20.4)$$

Substitute Eq. 12.20.3 into Eq. 12.20.4 to obtain an expression for [H] in terms of the concentrations of the reactants and products, and of the rate constants.

$$[\text{H}] = \frac{k_2\left(\frac{2k_1}{k_5}\right)^{1/2}[\text{Br}_2]^{1/2}\,[\text{H}_2]}{k_3[\text{Br}_2] + k_4[\text{HBr}]} \qquad (12.20.5)$$

According to the reaction mechanism,

$$\frac{d\,[\text{HBr}]}{dt} = k_2[\text{Br}][\text{H}_2] + k_3[\text{H}][\text{Br}_2] - k_4[\text{H}][\text{HBr}]$$

$$= k_2[\text{Br}][\text{H}_2] + [\text{H}]\left(k_3[\text{Br}_2] - k_4[\text{HBr}]\right) \tag{12.20.6}$$

Substitute Eqs. 12.20.3 and 12.20.5 into Eq. 12.20.6,

$$\frac{d\,[\text{HBr}]}{dt} = k_2\left(\frac{2k_1}{k_5}\right)^{1/2}[\text{Br}_2]^{1/2}\,[\text{H}_2] + \frac{k_2\left(\frac{2k_1}{k_5}\right)^{1/2}[\text{Br}_2]^{1/2}\,[\text{H}_2]}{k_3[\text{Br}_2] + k_4[\text{HBr}]}\left(k_3[\text{Br}_2] - k_4[\text{HBr}]\right)$$

$$= k_2\left(\frac{2k_1}{k_5}\right)^{1/2}[\text{Br}_2]^{1/2}\,[\text{H}_2]\left\{1 + \frac{k_3[\text{Br}_2] - k_4[\text{HBr}]}{k_3[\text{Br}_2] + k_4[\text{HBr}]}\right\}$$

$$= k_2\left(\frac{2k_1}{k_5}\right)^{1/2}[\text{Br}_2]^{1/2}\,[\text{H}_2]\left\{\frac{2k_3[\text{Br}_2]}{k_3[\text{Br}_2] + k_4[\text{HBr}]}\right\}$$

$$= k_2\left(\frac{2k_1}{k_5}\right)^{1/2}[\text{Br}_2]^{1/2}\,[\text{H}_2]\left\{\frac{2}{1 + \frac{k_4}{k_3}\frac{[\text{HBr}]}{[\text{Br}_2]}}\right\} \tag{12.20.7}$$

Setting

$$\alpha = 2k_2\left(\frac{2k_1}{k_5}\right)^{1/2}$$

$$\beta = \frac{k_4}{k_3}$$

Eq. 12.20.7 becomes

$$\frac{d\,[\text{HBr}]}{dt} = \frac{\alpha[\text{H}_2][\text{Br}_2]^{1/2}}{1 + \beta[\text{HBr}]/[\text{Br}_2]}$$

12.21 An excited ozone molecule, O_3^*, in the atmosphere can undergo one of the following reactions:

$$\text{O}_3^* \xrightarrow{k_1} \text{O}_3 \qquad\qquad (1)\ \text{fluorescence}$$

$$\text{O}_3^* \xrightarrow{k_2} \text{O} + \text{O}_2 \qquad\qquad (2)\ \text{decomposition}$$

$$\text{O}_3^* + \text{M} \xrightarrow{k_3} \text{O}_3 + \text{M} \qquad\qquad (3)\ \text{deactivation}$$

where M is an inert molecule. Calculate the fraction of ozone molecules undergoing decomposition in terms of the rate constants.

The fraction of ozone molecules undergoing decomposition is

$$\frac{\text{Rate of decomposition}}{\text{Rate of fluorescence} + \text{Rate of decomposition} + \text{Rate of deactivation}}$$

$$= \frac{k_1[\text{O}_3^*]}{k_1[\text{O}_3^*] + k_2[\text{O}_3^*] + k_3[\text{O}_3^*][\text{M}]}$$

$$= \frac{k_2}{k_1 + k_2 + k_3[\text{M}]}$$

12.22 The following data were collected for the reaction between hydrogen and nitric oxide at 700°C:

$$2H_2(g) + 2NO(g) \rightarrow 2H_2O(g) + N_2(g)$$

Experiment	$[H_2]/M$	$[NO]/M$	Initial rate/$M \cdot s^{-1}$
1	0.010	0.025	2.4×10^{-6}
2	0.0050	0.025	1.2×10^{-6}
3	0.010	0.0125	0.60×10^{-6}

(a) What is the rate law for the reaction? **(b)** Calculate the rate constant for the reaction. **(c)** Suggest a plausible reaction mechanism that is consistent with the rate law. (*Hint:* Assume that the oxygen atom is the intermediate.) **(d)** More careful studies of the reaction show that the rate law over a wide range of concentrations of reactants should be

$$\text{rate} = \frac{k_1[NO]^2[H_2]}{1 + k_2[H_2]}$$

What happens to the rate law at very high and very low hydrogen concentrations?

(a) Comparing Experiment 1 and Experiment 2, the concentration of NO is constant and the concentration of H_2 has decreased by one-half. The initial rate has also decreased by one-half. Therefore, the initial rate is directly proportional to the concentration of H_2. That is, the reaction is first order in H_2.

Comparing Experiment 1 and Experiment 3, the concentration of H_2 is constant and the concentration of NO has decreased by one-half. The initial rate has decreased by one-fourth. Therefore, the initial rate is proportional to the squared concentration of NO. That is, the reaction is second order in NO.

Therefore, the rate law is

$$\text{Rate} = k[NO]^2[H_2]$$

(b) Using Experiment 1 to calculate the rate constant,

$$k = \frac{\text{Rate}}{[NO]^2[H_2]} = \frac{2.4 \times 10^{-6} \, M \, s^{-1}}{(0.025 \, M)^2 \, (0.010 \, M)} = 0.38 \, M^{-2} \, s^{-1}$$

(c) The rate law suggests that the slow step in the reaction mechanism will probably involve one H_2 molecule and two NO molecules. Additionally, the hint suggests that the O atom is an intermediate. A plausible mechanism is

$$H_2 + 2NO \longrightarrow N_2 + H_2O + O \qquad \text{slow step}$$
$$O + H_2 \longrightarrow H_2O \qquad \text{fast step}$$

(d) At very high hydrogen concentrations, $k_2 [H_2] \gg 1$. Therefore, the rate law becomes

$$\text{rate} = \frac{k_1[NO]^2[H_2]}{k_2[H_2]} = \frac{k_1}{k_2}[NO]^2$$

At very low hydrogen concentrations, $k_2[H_2] \ll 1$. Therefore, the rate law becomes

$$\text{rate} = k_1[NO]^2[H_2]$$

12.23   The rate law for the decomposition of ozone to molecular oxygen

$$2O_3(g) \rightarrow 3O_2(g)$$

is rate $= k\dfrac{[O_3]^2}{[O_2]}$. The mechanism proposed for this process is

$$O_3 \underset{k_{-1}}{\overset{k_1}{\rightleftharpoons}} O + O_2$$

$$O + O_3 \overset{k_2}{\longrightarrow} 2O_2$$

Derive the rate law from these elementary steps. Clearly state the assumptions you use in the derivation. Explain why the rate decreases with increasing O_2 concentration.

The first step involves forward and reverse reactions that are much faster than the second step. The rates of the reaction in the first step are given by

$$\text{Forward rate} = k_1[O_3]$$

$$\text{Reverse rate} = k_{-1}[O][O_2]$$

Assume that these two processes rapidly reach a state of dynamic equilibrium in which the rates of the forward and reverse reactions are equal:

$$k_1[O_3] = k_{-1}[O][O_2]$$

Solving for [O],

$$[O] = \frac{k_1[O_3]}{k_{-1}[O_2]}$$

The rate for the second step gives the rate of reaction,

$$\text{Rate} = k_2[O][O_3] = k_2\frac{k_1[O_3]}{k_{-1}[O_2]}[O_3] = \frac{k_1k_2}{k_{-1}}\frac{[O_3]^2}{[O_2]} = k\frac{[O_3]^2}{[O_2]}$$

The rate law shows that higher concentrations of O_2 will decrease the reaction rate. This is due to the reverse reaction in the first step of the mechanism. If more O_2 molecules are present, they will serve to scavenge free O atoms and thus slow the disappearance of O_3.

12.24   The gas-phase reaction between H_2 and I_2 to form HI involves a two-step mechanism:

$$I_2 \rightleftharpoons 2I$$

$$H_2 + 2I \rightarrow 2HI$$

The rate of formation of HI increases with the intensity of visible light. **(a)** Explain why this fact supports the two-step mechanism given. (*Hint:* The color of I_2 vapor is purple.) **(b)** Explain why the visible light has no effect on the formation of H atoms.

(a) In this two-step mechanism, the rate determining step is the second one where a hydrogen molecule collides with two iodine atoms. The absorption of visible light by the colored molecular iodine vapor weakens the I_2 bond and increases the number of I atoms present, which in turn increases the reaction rate.

(b) Hydrogen gas is colorless and does not absorb visible light. Ultraviolet light is required to photodissociate H_2 molecules.

12.25   In recent years, ozone in the stratosphere has been depleted at an alarmingly fast rate by chlorofluorocarbons (CFCs). A CFC molecule such as $CFCl_3$ is first decomposed by UV radiation:

$$CFCl_3 \longrightarrow CFCl_2 + Cl$$

The chlorine radical then reacts with ozone as follows:

$$Cl + O_3 \longrightarrow ClO + O_2$$

$$ClO + O \longrightarrow Cl + O_2$$

(a) Write the overall reaction for the last two steps. **(b)** What are the roles of Cl and ClO? **(c)** Why is the fluorine radical not important in this mechanism? **(d)** One suggestion for reducing the concentration of chlorine radicals is to add hydrocarbons such as ethane (C_2H_6) to the stratosphere. How will this approach work? **(e)** Draw potential energy versus reaction progress diagrams for the uncatalyzed and catalyzed (by Cl) destruction of ozone: $O_3 + O \rightarrow 2O_2$. Use the thermodynamic data in Appendix B to determine whether the reaction is exothermic or endothermic.

(a) The sum of the following reactions

$$Cl + O_3 \longrightarrow ClO + O_2$$

$$ClO + O \longrightarrow Cl + O_2$$

gives

$$O_3 + O \longrightarrow 2O_2$$

(b) Cl is a catalyst; ClO is an intermediate.

(c) The C–F bond is stronger than the C–Cl bond. Thus, Cl is formed more easily than F.

(d) Ethane will remove the Cl atom:

$$C_2H_6 + Cl \longrightarrow C_2H_5 + HCl$$

(e)

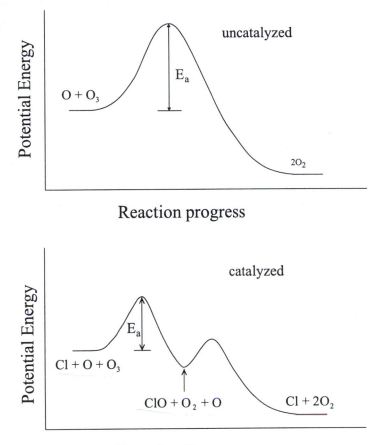

The reaction is exothermic.

12.26   Use Equation 12.23 to calculate the rate constant at 300 K for $E_a = 0$, 2, and 50 kJ mol^{-1}. Assume that $A = 10^{11}$ s^{-1} in each case.

For $E_a = 0$ kJ mol^{-1},

$$k = \left(10^{11} \text{ s}^{-1}\right) e^{-\left(0 \text{ J mol}^{-1}\right)/\left(8.314 \text{ J K}^{-1} \text{ mol}^{-1}\right)(300 \text{ K})} = 10^{11} \text{ s}^{-1}$$

For $E_a = 2$ kJ mol^{-1},

$$k = \left(10^{11} \text{ s}^{-1}\right) e^{-\left(2 \times 10^3 \text{ J mol}^{-1}\right)/\left(8.314 \text{ J K}^{-1} \text{ mol}^{-1}\right)(300 \text{ K})} = 4.5 \times 10^{10} \text{ s}^{-1}$$

For $E_a = 50$ kJ mol^{-1},

$$k = \left(10^{11} \text{ s}^{-1}\right) e^{-\left(50 \times 10^3 \text{ J mol}^{-1}\right)/\left(8.314 \text{ J K}^{-1} \text{ mol}^{-1}\right)(300 \text{ K})} = 2.0 \times 10^2 \text{ s}^{-1}$$

12.27   Many reactions double their rates with every 10° rise in temperature. Assume that such a reaction takes place at 305 K and 315 K. What must its activation energy be for this statement to hold?

If the rate doubles, the rate constant doubles, too.

$$\ln \frac{k_{315}}{k_{305}} = -\frac{E_a}{R}\left(\frac{1}{315\ \text{K}} - \frac{1}{305\ \text{K}}\right)$$

$$\ln 2 = -\frac{E_a}{8.314\ \text{J\,K}^{-1}\,\text{mol}^{-1}}\left(\frac{1}{315\ \text{K}} - \frac{1}{305\ \text{K}}\right)$$

$$E_a = 5.54 \times 10^4\ \text{J\,mol}^{-1}$$

12.28   Over a range of about $\pm 3°C$ from normal body temperature the metabolic rate, M_T, is given by $M_T = M_{37}\,(1.1)^{\Delta T}$, where M_{37} is the normal rate and ΔT is the change in T. Discuss this equation in terms of a possible molecular interpretation. [Source: "Eco-Chem," J. A. Campbell, *J. Chem. Educ.* **52**, 327 (1975).]

Converting to kelvin, and using the Arrhenius equation,

$$\ln \frac{M_T}{M_{37}} = -\frac{E_a}{R}\left(\frac{1}{T} - \frac{1}{310\ \text{K}}\right)$$

Since the temperature range is so small, $f(T) = \frac{1}{T} - \frac{1}{310\ \text{K}}$ may be expanded in a Taylor series about $T_0 = 310$ K. Keeping only the first non-zero term results in $f(T) \approx -\frac{\Delta T}{T_0^2}$, where $\Delta T = 310\ \text{K} - T$. Thus,

$$\ln \frac{M_T}{M_{37}} = \frac{E_a}{R}\frac{\Delta T}{T_0^2}$$

or

$$M_T = M_{37}e^{\frac{E_a}{RT_0^2}\Delta T} = M_{37}\left(e^{\frac{E_a}{RT_0^2}}\right)^{\Delta T} = M_{37}(\text{constant})^{\Delta T}$$

which is of the observed form, providing an implicit factor of $1\ \text{K}^{-1}$ is incorporated into the argument of the exponential function and the ΔT is interpreted as a unitless number. Specifically, it must be true that

$$e^{\frac{E_a}{RT_0^2}} = 1.1$$

$$\frac{E_a}{RT_0^2} = \ln 1.1 = 0.0953$$

$$E_a = \left(1\ \text{K}^{-1}\right)\left(8.314\ \text{J\,K}^{-1}\,\text{mol}^{-1}\right)(310\ \text{K})^2\,(0.0953) = 7.6 \times 10^4\ \text{J\,mol}^{-1}$$

This activation energy is consistent with a single rate determining step controlling the metabolic rate within this temperature range.

12.29 The rate of bacterial hydrolysis of fish muscle is twice as great at 2.2°C as at −1.1°C. Estimate a ΔE_a value for this reaction. Is there any relation to the problem of storing fish for food? [Source: "Eco-Chem," J. A. Campbell, *J. Chem. Educ.* **52,** 390 (1975).]

Using the Arrhenius equation, with temperatures converted to the kelvin scale,

$$\ln \frac{k_{275.4}}{k_{272.1}} = -\frac{E_a}{R} \left(\frac{1}{275.4 \text{ K}} - \frac{1}{272.1 \text{ K}} \right)$$

$$\ln 2 = -\frac{E_a}{8.314 \text{ J K}^{-1} \text{mol}^{-1}} \left(\frac{1}{275.4 \text{ K}} - \frac{1}{272.1 \text{ K}} \right)$$

$$E_a = 1.3 \times 10^5 \text{ J mol}^{-1}$$

(Note that there is a math error in the solution given in the reference that makes the answer there too large by a factor of 3.3^2.) Nevertheless, this is a relatively large E_a, and temperature will have a large effect on reaction rate. Thus, refrigeration is an essential, effective method of preserving fish and preventing spoilage. Note that the rate of bacterial hydrolysis at room temperature (298 K) is 74 times greater than the rate at 275.4 K.

12.30 The rate constants for the first-order decomposition of an organic compound in solution are measured at several temperatures:

k/s^{-1}	4.92×10^{-3}	0.0216	0.0950	0.326	1.15
$t/°C$	5.0	15	25	35	45

Determine graphically the pre-exponential factor and the energy of activation for the reaction.

Since

$$\ln k = \ln A - \frac{E_a}{RT}$$

A plot of $\ln k$ vs $1/T$ gives a slope of $-E_a/R$ and an intercept of $\ln A$. The following data are used for the plot:

10^3 K/T	3.595	3.470	3.353	3.245	3.143
$\ln(k/s^{-1})$	−5.314	−3.835	−2.354	−1.121	0.140

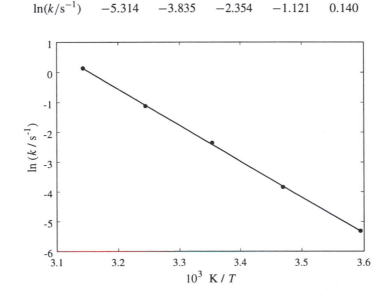

The equation for the line that best fits these points is $y = -1.207 \times 10^4 x + 38.06$. Therefore,

$$E_a = -\left(-1.207 \times 10^4 \text{ K}^{-1}\right)\left(8.314 \text{ J K}^{-1} \text{ mol}^{-1}\right) = 1.00 \times 10^5 \text{ J mol}^{-1}$$

$$A = e^{38.06} = 3.38 \times 10^{16} \text{ s}^{-1}$$

12.31   The energy of activation for the reaction $2HI \rightarrow H_2 + I_2$ is 180 kJ mol^{-1} at 556 K. Calculate the rate constant using Equation 12.23. The collision diameter for HI is 3.5×10^{-8} cm. Assume that the pressure is 1 atm.

The frequency factor in the equation is the number of binary collisions, $Z_{\text{HI-HI}}$, which is then multiplied by the activation energy factor, $e^{-E_a/RT}$. The expression for the number of binary collisions, Equation 12.26, contains the square of the number of HI molecules per cubic meter. In forming the rate constant, however, the concentration term cancels, since rate $= kN_{\text{HI}}^2$. Finally, although two HI molecules are consumed in each reactive collision, the rate of the reaction is given by rate $= \frac{1}{2}\frac{dN_{\text{HI}}}{dt}$ and the factor of 1/2 cancels the factor of 2. Putting everything together gives,

$$k = \frac{Z_{\text{HI-HI}}}{N_{\text{HI}}^2} e^{-E_a/RT}$$

$$= 2d^2 \sqrt{\frac{\pi k_B T}{m_{\text{HI}}}} e^{-E_a/RT}$$

$$= 2\left(3.5 \times 10^{-10} \text{ m}\right)^2 \sqrt{\frac{\pi \left(1.381 \times 10^{-23} \text{ J K}^{-1}\right)(556 \text{ K})}{(127.9 \text{ amu})\left(1.661 \times 10^{-27} \text{ kg amu}^{-1}\right)}} \, e^{-\frac{180 \times 10^3 \text{ J mol}^{-1}}{\left(8.314 \text{ J K}^{-1} \text{ mol}^{-1}\right)(556 \text{ K})}}$$

$$= 1.01 \times 10^{-33} \text{ m}^3 \text{ molecule}^{-1} \text{ s}^{-1}$$

Where the dimensionless "unit" molecule^{-1} has been added as a reminder that this rate constant has been calculated on a molecular basis. It can be converted to more "chemical" units,

$$k = 1.01 \times 10^{-33} \text{ m}^3 \text{ molecule}^{-1} \text{ s}^{-1} \left(\frac{1000 \text{ L}}{1 \text{ m}^3}\right)\left(6.022 \times 10^{23} \text{ molecules mol}^{-1}\right)$$

$$= 6.10 \times 10^{-7} \, M^{-1} \text{ s}^{-1}$$

12.32   The rate constant of a first-order reaction is 4.6×10^{-4} s^{-1} at 350°C. If the activation energy is 104 kJ mol^{-1}, calculate the temperature at which its rate constant is 8.80×10^{-4} s^{-1}.

$$\ln \frac{k_2}{k_1} = -\frac{E_a}{R}\left(\frac{1}{T_2} - \frac{1}{T_1}\right)$$

$$\ln \frac{8.80 \times 10^{-4}}{4.60 \times 10^{-4}} = -\frac{104 \times 10^3 \text{ J mol}^{-1}}{8.314 \text{ J K}^{-1} \text{ mol}^{-1}}\left(\frac{1}{T_2} - \frac{1}{623.2 \text{ K}}\right)$$

$$T_2 = 644.0 \text{ K} = 371°C$$

12.33   The rate at which tree crickets chirp is 2.0×10^2 per minute at 27°C but only 39.6 per minute at 5°C. From these data, calculate the "activation energy" for the chirping process. (*Hint*: The ratio of rates is equal to the ratio of rate constants.) Find the chirping rate at 15°C.

First find the "activation energy" for the chirping process.

$$\ln \frac{k_2}{k_1} = \ln \frac{\text{Rate}_2}{\text{Rate}_1} = -\frac{E_a}{R}\left(\frac{1}{T_2} - \frac{1}{T_1}\right)$$

$$\ln \frac{2.0 \times 10^2 \text{ min}^{-1}}{39.6 \text{ min}^{-1}} = -\frac{E_a}{8.314 \text{ J K}^{-1} \text{mol}^{-1}}\left(\frac{1}{300.2 \text{ K}} - \frac{1}{278.2 \text{ K}}\right)$$

$$E_a = 5.111 \times 10^4 \text{ J mol}^{-1} = 5.11 \times 10^4 \text{ J mol}^{-1}$$

Now use one of the known rates and E_a to find the chirping rate at 15°C (288.2 K).

$$\ln \frac{\text{Rate}_2}{\text{Rate}_1} = -\frac{E_a}{R}\left(\frac{1}{T_2} - \frac{1}{T_1}\right)$$

$$\ln \frac{\text{Rate}_2}{39.6 \text{ min}^{-1}} = -\frac{5.111 \times 10^4 \text{ J mol}^{-1}}{8.314 \text{ J K}^{-1} \text{mol}^{-1}}\left(\frac{1}{288.2 \text{ K}} - \frac{1}{278.2 \text{ K}}\right)$$

$$\text{Rate}_2 = 85.2 \text{ min}^{-1}$$

12.34   Consider the following parallel reactions

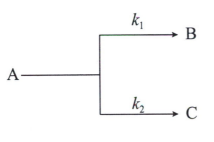

The activation energies are 45.3 kJ mol^{-1} for k_1 and 69.8 kJ mol^{-1} for k_2. If the rate constants are equal at 320 K, at what temperature will $k_1/k_2 = 2.00$?

The ratio of the rate constants is

$$\frac{k_1}{k_2} = \frac{A_1 e^{-E_{a1}/RT}}{A_2 e^{-E_{a2}/RT}}$$

$$= \frac{A_1}{A_2} e^{(E_{a2}-E_{a1})/RT} = \frac{A_1}{A_2} e^{\left(69.8\times10^3 \text{ J mol}^{-1} - 45.3\times10^3 \text{ J mol}^{-1}\right)/\left[\left(8.314 \text{ J K}^{-1} \text{mol}^{-1}\right)T\right]}$$

$$= \frac{A_1}{A_2} e^{2.947\times10^3 \text{ K}/T}$$

First use data at 320 K to calculate A_1/A_2:

$$\frac{k_1}{k_2} = 1.00 = \frac{A_1}{A_2}e^{2.947 \times 10^3 \text{ K}/320 \text{ K}}$$

$$\frac{A_1}{A_2} = 1.001 \times 10^{-4}$$

When $k_1/k_2 = 2.00$,

$$2.00 = \frac{A_1}{A_2}e^{2.947 \times 10^3 \text{ K}/T} = \left(1.001 \times 10^{-4}\right)e^{2.947 \times 10^3 \text{ K}/T}$$

$$\frac{1}{T} = \frac{1}{2.947 \times 10^3 \text{ K}} \ln \frac{2.00}{1.001 \times 10^{-4}} = 3.360 \times 10^{-3} \text{ K}^{-1}$$

$$T = 298 \text{ K}$$

12.35 The thermal isomerization of cyclopropane to propene in the gas phase has a rate constant of 5.95×10^{-4} s^{-1} at 500°C. Calculate the value of $\Delta G^{\circ\ddagger}$ for the reaction.

---

k is related to $\Delta G^{\circ\ddagger}$ by

$$k = \frac{k_B T}{h}e^{-\Delta G^{\circ\ddagger}/RT}M^{1-m}$$

Since this is a unimolecular process, $m = 1$.

$$\Delta G^{\circ\ddagger} = -RT \ln \frac{kh}{k_B T}$$

$$= -\left(8.314 \text{ J K}^{-1} \text{ mol}^{-1}\right)(773.2 \text{ K}) \ln \frac{\left(5.95 \times 10^{-4} \text{ s}^{-1}\right)\left(6.626 \times 10^{-34} \text{ J s}\right)}{\left(1.381 \times 10^{-23} \text{ J K}^{-1}\right)(773.2 \text{ K})}$$

$$= 2.43 \times 10^5 \text{ J mol}^{-1}$$

12.36 The rate of the electron-exchange reaction between naphthalene ($C_{10}H_8$) and its anion radical ($C_{10}H_8^-$) are diffusion-controlled:

$$C_{10}H_8^- + C_{10}H_8 \rightleftharpoons C_{10}H_8 + C_{10}H_8^-$$

The reaction is bimolecular and second order. The rate constants are

T/K	307	299	289	273
$k/10^9 \cdot M^{-1} \cdot \text{s}^{-1}$	2.71	2.40	1.96	1.43

Calculate the values of E_a, $\Delta H^{\circ\ddagger}$, $\Delta S^{\circ\ddagger}$ and $\Delta G^{\circ\ddagger}$ at 307 K for the reaction. [*Hint*: Rearrange Equation 12.36 and plot $\ln k/T$ versus $1/T$.]

Equation 12.36 gives

$$k = \frac{k_B T}{h} e^{\Delta S^{\circ\ddagger}/R} e^{-\Delta H^{\circ\ddagger}/RT}$$

or

$$\ln \frac{k}{T} = \ln \frac{k_B}{h} + \frac{\Delta S^{\circ\ddagger}}{R} - \frac{\Delta H^{\circ\ddagger}}{RT}$$

A plot of $\ln k/T$ vs $1/T$ gives a slope of $-\Delta H^{\circ\ddagger}/R$ and an intercept of $\ln k_B/h + \Delta S^{\circ\ddagger}/R$. The data used for the plot are

10^3 K/T	3.257	3.344	3.460	3.663
$\ln \frac{k}{T}$	15.9934	15.8983	15.7298	15.4715

The best fit line has a formula of $y = -1302.0x + 20.24$. Therefore,

$$\Delta H^{\circ\ddagger} = -(-1302.0 \text{ K}) \left(8.314 \text{ J K}^{-1} \text{ mol}^{-1}\right)$$

$$= 1.082 \times 10^4 \text{ J mol}^{-1}$$

$$= 1.08 \times 10^4 \text{ J mol}^{-1}$$

and

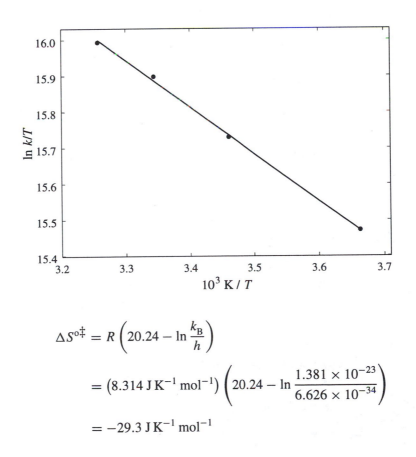

$$\Delta S^{\circ\ddagger} = R \left(20.24 - \ln \frac{k_B}{h}\right)$$

$$= \left(8.314 \text{ J K}^{-1} \text{ mol}^{-1}\right) \left(20.24 - \ln \frac{1.381 \times 10^{-23}}{6.626 \times 10^{-34}}\right)$$

$$= -29.3 \text{ J K}^{-1} \text{ mol}^{-1}$$

From Equation 12.38 and the discussion following it (see also the answer to Problem 12.37), the activation energy for this reaction, which occurs in solution (condensed phase) is

$$E_a = \Delta H^{o\ddagger} + RT$$

$$= 1.082 \times 10^4 \, \text{J mol}^{-1} + \left(8.314 \, \text{J K}^{-1} \, \text{mol}^{-1}\right)(307 \, \text{K})$$

$$= 1.34 \times 10^4 \, \text{J mol}^{-1}$$

From $\Delta H^{o\ddagger}$ and $\Delta S^{o\ddagger}$, $\Delta G^{o\ddagger}$ at 307 K is calculated.

$$\Delta G^{o\ddagger} = \Delta H^{o\ddagger} - T\Delta S^{o\ddagger}$$

$$= 1.082 \times 10^4 \, \text{J mol}^{-1} - (307 \, \text{K})\left(-29.3 \, \text{J K}^{-1} \, \text{mol}^{-1}\right)$$

$$= 1.98 \times 10^4 \, \text{J mol}^{-1}$$

12.37 (a) The pre-exponential factor and activation energy for the hydrolysis of t-butyl chloride are $2.1 \times 10^{16} \, \text{s}^{-1}$ and $102 \, \text{kJ mol}^{-1}$, respectively. Calculate the values of $\Delta S^{o\ddagger}$ and $\Delta H^{o\ddagger}$ at 286 K for the reaction. (b) The pre-exponential factor and activation energy for the gas-phase cycloaddition of maleic anhydride and cyclopentadiene are $5.9 \times 10^7 \, M^{-1} \, \text{s}^{-1}$ and $51 \, \text{kJ mol}^{-1}$, respectively. Calculate the values of $\Delta S^{o\ddagger}$ and $\Delta H^{o\ddagger}$ at 293 K for the reaction.

(a) This reaction occurs in the condensed phase. From Equation 12.38 and the discussion following it, and since $\Delta U^{o\ddagger} \approx \Delta H^{o\ddagger}$ in condensed phases,

$$E_a = \Delta U^{o\ddagger} + RT$$

$$\approx \Delta H^{o\ddagger} + RT$$

or

$$\Delta H^{o\ddagger} \approx E_a - RT$$

$$= 1.02 \times 10^5 \, \text{J mol}^{-1} - \left(8.314 \, \text{J K}^{-1} \, \text{mol}^{-1}\right)(286 \, \text{K})$$

$$= 9.96 \times 10^4 \, \text{J mol}^{-1}$$

From Equation 12.37,

$$A = \frac{k_B T}{h} e^{\Delta S^{o\ddagger}/R}$$

or

$$\Delta S^{o\ddagger} = R \ln \frac{hA}{k_B T}$$

$$= \left(8.314 \, \text{J K}^{-1} \, \text{mol}^{-1}\right) \ln \frac{\left(6.626 \times 10^{-34} \, \text{J s}\right)\left(2.1 \times 10^{16} \, \text{s}^{-1}\right)}{\left(1.381 \times 10^{-23} \, \text{J K}^{-1}\right)(286 \, \text{K})}$$

$$= 68 \, \text{J K}^{-1} \, \text{mol}^{-1}$$

(b) This is a bimolecular, gas-phase reaction. Using Equation 12.41 with $\Delta n^{\ddagger} = -1$,

$$E_a = \Delta H^{o\ddagger} + 2RT$$

or

$$\Delta H^{o\ddagger} = E_a - 2RT$$

$$= 5.1 \times 10^4 \text{ J mol}^{-1} - 2 \,(1 \text{ mol}) \left(8.314 \text{ J K}^{-1} \text{ mol}^{-1}\right) (293 \text{ K})$$

$$= 4.6 \times 10^4 \text{ J mol}^{-1}$$

From Equation 12.37,

$$A = \frac{k_B T}{h} e^{\Delta S^{o\ddagger}/R}$$

or

$$\Delta S^{o\ddagger} = R \ln \frac{hA}{k_B T}$$

$$= \left(8.314 \text{ J K}^{-1} \text{ mol}^{-1}\right) \ln \frac{(1 \, M) \left(6.626 \times 10^{-34} \text{ J s}\right) \left(5.9 \times 10^7 \, M^{-1} \text{ s}^{-1}\right)}{\left(1.381 \times 10^{-23} \text{ J K}^{-1}\right) (293 \text{ K})}$$

$$= -96 \text{ J K}^{-1} \text{ mol}^{-1}$$

As expected, $\Delta S^{o\ddagger}$ is negative for this bimolecular reaction.

12.38 A person may die after drinking D_2O instead of H_2O for a prolonged period (on the order of days). Explain. Since D_2O has practically the same properties as H_2O, how would you test the presence of large quantities of the former in a victim's body?

Because of the lower zero-point energy for bonds in which D is substituted for H, there is a higher activation energy required for reactions in which this bond breaks. Thus, the rate of H^+ ion exchange is faster than that for the D^+ ion. Additionally, the dissociation constants of deuterated acids are smaller than the corresponding acid with the normal H^+ ion. These differences will affect the delicate acid-base balance in the body as well as the kinetics of biological processes, and could lead to death. A mass spectrum of a body fluid sample should reveal the presence of a larger than natural abundance of the heavier isotope of hydrogen.

12.39 The rate-determining step of the bromination of acetone involves breaking a carbon–hydrogen bond. Estimate the ratio of the rate constants $k_{C\text{-}H}/k_{C\text{-}D}$ for the reaction at 300 K. The frequencies of vibration for the particular bonds are $\tilde{v}_{C\text{-}H} = 3000 \text{ cm}^{-1}$ and $\tilde{v}_{C\text{-}D} = 2100 \text{ cm}^{-1}$. The wavenumber $(\tilde{v})$ is given by v/c, where v is the frequency and c is the velocity of light.

This is analogous to the example given in the text and Figure 12.16b is relevant to this solution. The difference in rates is due to the smaller zero-point energy for the C-D bond that results in a larger activation energy.

$$E_{C\text{-}H} = E_{\text{stretch}} - E^o_{C\text{-}H}$$

$$E_{C\text{-}D} = E_{\text{stretch}} - E^o_{C\text{-}D}$$

The zero-point energies are given by

$$E^o_{C\text{-}H} = \frac{1}{2}h\nu_{C\text{-}H}$$

$$= \frac{1}{2}hc\tilde{\nu}_{C\text{-}H}$$

$$= \frac{1}{2}(6.626 \times 10^{-34}\,\text{J s})(3000\,\text{cm}^{-1})\left(\frac{100\,\text{cm}}{1\,\text{m}}\right)(2.9979 \times 10^8\,\text{m s}^{-1})(6.022 \times 10^{23}\,\text{mol}^{-1})$$

$$= 1.794 \times 10^4\,\text{J mol}^{-1}$$

and

$$E^o_{C\text{-}D} = \frac{1}{2}h\nu_{C\text{-}D}$$

$$= \frac{1}{2}hc\tilde{\nu}_{C\text{-}D}$$

$$= \frac{1}{2}(6.626 \times 10^{-34}\,\text{J s})(2100\,\text{cm}^{-1})\left(\frac{100\,\text{cm}}{1\,\text{m}}\right)(2.9979 \times 10^8\,\text{m s}^{-1})(6.022 \times 10^{23}\,\text{mol}^{-1})$$

$$= 1.256 \times 10^4\,\text{J mol}^{-1}$$

The rate constant ratio is given by

$$\frac{k_{C\text{-}H}}{k_{C\text{-}D}} = e^{(E^o_{C\text{-}H} - E^o_{C\text{-}D})/RT}$$

$$= e^{\left(1.794\times10^4\,\text{J mol}^{-1} - 1.256\times10^4\,\text{J mol}^{-1}\right)/\left(8.314\,\text{J K}^{-1}\,\text{mol}^{-1}\right)(300\,\text{K})}$$

$$= 8.6$$

12.40 Lubricating oils for watches or other mechanical objects are made of long-chain hydrocarbons. Over long-periods of time they undergo auto-oxidation to form solid polymers. The initial step in this process involves hydrogen abstraction. Suggest a chemical means for prolonging the life of these oils.

Deuterating the oils, that is, replacing the H atoms with D atoms, will slow down the rate of hydrogen abstraction.

12.41 A flask contains a mixture of compounds A and B. Both compounds decompose by first-order kinetics. The half-lives are 50.0 min for A and 18.0 min for B. If the concentrations of A and B are equal initially, how long will it take for the concentration of A to be four times that of B?

Let the initial concentrations of A and B be c_0. The concentrations of A and B at a later time, t, satisfy

$$\ln \frac{[A]}{c_0} = -k_A t$$

$$\ln \frac{[B]}{c_0} = -k_B t$$

Subtracting the second equation from the first,

$$\ln \left(\frac{[A]}{c_0} \frac{c_0}{[B]} \right) = \ln \frac{[A]}{[B]} = - \left(k_A - k_B \right) t$$

When [A] becomes four times [B], the above expression becomes

$$\ln 4 = - \left(\frac{\ln 2}{50.0 \text{ min}} - \frac{\ln 2}{18.0 \text{ min}} \right) t$$

$$t = 56.2 \text{ min}$$

12.42 The term *reversible* is used in both thermodynamics (see Chapter 4) and in this chapter. Does it convey the same meaning in these two instances?

The term has a different meaning in kinetics than it does in thermodynamics. In kinetics, a reversible reaction is one in which both the forward and the backwards reaction occur. In thermodynamics, a reversible process is one that is in equilibrium at every point along the path connecting the initial and final states.

12.43 The recombination of iodine atoms in an organic solvent, such as carbon tetrachloride, is a diffusion-controlled process:

$$I + I \rightarrow I_2$$

Given that the viscosity of CCl_4 is $9.69 \times 10^{-4} \text{ N s m}^{-2}$ at 20°C, calculate the rate of recombination at this temperature.

$$k_D = \frac{8}{3} \frac{RT}{\eta}$$

$$= \frac{8}{3} \frac{\left(8.314 \text{ J K}^{-1} \text{ mol}^{-1} \right) (293.2 \text{ K})}{9.69 \times 10^{-4} \text{ N s m}^{-2}}$$

$$= \left(6.71 \times 10^6 \text{ m}^3 \text{ mol}^{-1} \text{ s}^{-1} \right) \left(\frac{1000 \text{ L}}{1 \text{ m}^3} \right)$$

$$= 6.71 \times 10^9 \text{ } M^{-1} \text{ s}^{-1}$$

12.44   The equilibrium between dissolved CO_2 and carbonic acid can be represented by

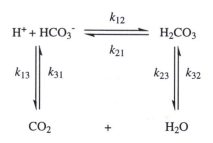

Show that

$$-\frac{d\,[CO_2]}{dt} = (k_{31} + k_{32})\,[CO_2] - \left(k_{13} + \frac{k_{23}}{K}\right)[H^+]\,[HCO_3^-]$$

where $K = [H^+]\,[HCO_3^-]/[H_2CO_3]$.

From the equilibrium,

$$\frac{d\,[CO_2]}{dt} = k_{13}[H^+][HCO_3^-] - k_{31}[CO_2] + k_{23}[H_2CO_3] - k_{32}[CO_2]$$

Since H_2O is present in a great quantity, the effectively constant concentration, $[H_2O]$, is incorporated into the constants k_{31} and k_{32}. Rearranging the expression gives

$$\frac{d\,[CO_2]}{dt} = -\,(k_{31} + k_{32})\,[CO_2] + k_{13}[H^+][HCO_3^-] + k_{23}[H_2CO_3] \tag{12.44.1}$$

Since H^+ and HCO_3^- are in equilibrium with H_2CO_3, let

$$K = \frac{[H^+][HCO_3^-]}{[H_2CO_3]}$$

$$[H_2CO_3] = \frac{[H^+][HCO_3^-]}{K} \tag{12.44.2}$$

Substitute Eq. 12.44.2 into Eq. 12.44.1,

$$\frac{d\,[CO_2]}{dt} = -\,(k_{31} + k_{32})\,[CO_2] + k_{13}[H^+][HCO_3^-] + k_{23}\frac{[H^+][HCO_3^-]}{K}$$

$$= -\,(k_{31} + k_{32})\,[CO_2] + \left(k_{13} + \frac{k_{23}}{K}\right)[H^+][HCO_3^-]$$

$$-\frac{d\,[CO_2]}{dt} = (k_{31} + k_{32})\,[CO_2] - \left(k_{13} + \frac{k_{23}}{K}\right)[H^+][HCO_3^-]$$

12.45   Polyethylene is used in many items, including water pipes, bottles, electrical insulation, toys, and mailing envelopes. It is a *polymer*, a molecule with a very high molar mass made by joining many ethylene molecules (the basic unit is called a *monomer*) together. The initiation step is

$$R_2 \xrightarrow{k_i} 2R\bullet \qquad \text{initiation}$$

The R• species (called a radical) reacts with an ethylene molecule (M) to generate another radical

$$R\bullet + M \longrightarrow M_1\bullet$$

Reaction of $M_1\bullet$ with another monomer leads to the growth or propagation of the polymer chain:

$$M_1\bullet + M \xrightarrow{k_p} M_2\bullet \qquad \text{propagation}$$

This step can be repeated with hundreds of monomer units. The propagation terminates when two radicals combine

$$M'\bullet + M''\bullet \xrightarrow{k_t} M'-M'' \qquad \text{termination}$$

The initiator in the polymerization of ethylene commonly is benzoyl peroxide $[(C_6H_5COO)_2]$:

$$[(C_6H_5COO)_2] \longrightarrow 2C_6H_5COO\bullet$$

This is a first-order reaction. The half-life of benzoyl peroxide at 100°C is 19.8 min. **(a)** Calculate the rate constant (in min^{-1}) of the reaction. **(b)** If the half-life of benzoyl peroxide is 7.30 h, or 438 min, at 70°C, what is the activation energy (in kJ/mol) for the decomposition of benzoyl peroxide? **(c)** Write the rate laws for the elementary steps in the above polymerization process and identify the reactant, product, and intermediates. **(d)** What condition would favor the growth of long high-molar-mass polyethylenes?

(a)

$$k = \frac{\ln 2}{t_{1/2}} = \frac{\ln 2}{19.8\ \text{min}} = 3.501 \times 10^{-2}\ \text{min}^{-1} = 3.50 \times 10^{-2}\ \text{min}^{-1}$$

(b) At 70°C,

$$k = \frac{\ln 2}{t_{1/2}} = \frac{\ln 2}{438\ \text{min}} = 1.583 \times 10^{-3}\ \text{min}^{-1}$$

The activation energy can be calculated using the rate constants at 100°C and 70°C.

$$\ln\frac{k_2}{k_1} = -\frac{E_a}{R}\left(\frac{1}{T_2} - \frac{1}{T_1}\right)$$

$$\ln\frac{3.501 \times 10^{-2}\ \text{min}^{-1}}{1.583 \times 10^{-3}\ \text{min}^{-1}} = -\frac{E_a}{8.314\ \text{J K}^{-1}\,\text{mol}^{-1}}\left(\frac{1}{373.2\ \text{K}} - \frac{1}{343.2\ \text{K}}\right)$$

$$E_a = 1.10 \times 10^5\ \text{J mol}^{-1} = 110\ \text{kJ mol}^{-1}$$

(c) Since all steps are elementary steps, we can deduce the rate laws simply from the equations representing the steps. The rate laws are

Initiation   Rate $= k_i[R_2]$
Propagation   Rate $= k_p[M_1\bullet][M]$
Termination   Rate $= k_t[M_1'\bullet][M_1''\bullet]$

The reactant molecules are ethylene monomers, the product is polyethylene, and the intermediates are the radicals $R\bullet$, $M_1\bullet$, $M'_1\bullet$, $M''_1\bullet$, etc.

(d) The growth of long polymers would be favored by a high rate of propagation and a slow rate of termination. Since the rate law of propagation depends on the concentration of monomer, an increase in the concentration of ethylene would increase the propagation (growth) rate. From the rate law for termination, a low concentration of the radical fragment $M'_1\bullet$ or $M''_1\bullet$ would lead to a slower rate of termination. This can be accomplished by using a low concentration of the initiator, R_2.

12.46   In a certain industrial process involving a heterogeneous catalyst, the volume of the catalyst (in the shape of a sphere) is 10.0 cm³. **(a)** Calculate the surface area of the catalyst. **(b)** If the sphere is broken down into eight spheres, each of which has a volume of 1.25 cm³, what is the total surface area of the spheres? **(c)** Which of the two geometric configurations is the more effective catalyst? (*Hint:* The surface area of a sphere is $4\pi r^2$, where r is the radius of the sphere.)

(a) One 10.0 cm³ sphere

First calculate the radius of the sphere.

$$V = 10.0 \text{ cm}^3 = \frac{4}{3}\pi r^3$$

$$r = 1.337 \text{ cm}$$

The surface area is

$$A = 4\pi r^2 = 4\pi \left(1.337 \text{ cm}\right)^2 = 22.5 \text{ cm}^2$$

(b) Eight 1.25 cm³ spheres

First calculate the radius of one sphere.

$$V = 1.25 \text{ cm}^3 = \frac{4}{3}\pi r^3$$

$$r = 0.6683 \text{ cm}$$

The total surface area of the spheres is

$$A = 8\left(4\pi r^2\right) = 32\pi \left(0.6683 \text{ cm}\right)^2 = 44.9 \text{ cm}^2$$

(c) Since a greater surface area promotes the catalyzed reaction more effectively, the eight smaller spheres are more effective than one large sphere.

12.47   Explain why grain dust in elevators can be explosive.

The answer here is related to that of Problem 12.46. A finely dispersed dust presents a very large surface area to the atmosphere and combustion can occur with extreme rapidity. Just about any organic material can serve as a fuel in this manner, and if it is in a fine enough form, dry enough, and in contact with sufficient air in a confined area, it will be explosive. Wheat and corn starch dusts are among the most explosive grain dusts.

12.48 At a certain elevated temperature, ammonia decomposes on the surface of tungsten metal as follows:

$$NH_3 \longrightarrow \tfrac{1}{2}N_2 + \tfrac{3}{2}H_2$$

The kinetic data are expressed as the variation of the half-life with the initial pressure of NH_3:

P/torr	264	130	59	16
$t_{1/2}$/s	456	228	102	60

(a) Determine the order of the reaction. **(b)** How does the order depend on the initial pressure? **(c)** How does the mechanism of the reaction vary with pressure?

(a) The half-life of a reaction and the initial concentration are related by

$$t_{1/2} = C \frac{1}{[A]_0^{n-1}}$$

where C is a constant. Taking the common logarithm of both sides of the equation,

$$\log t_{1/2} = \log C - (n-1) \log [A]_0$$

Since pressure is proportional to concentration at constant temperature, the above equation can also be written as

$$\log t_{1/2} = \log C' - (n-1) \log P$$

A plot of $\log t_{1/2}$ vs $\log P$ gives a slope of $n - 1$. The data used for the plot are

$\log(P/\text{torr})$	2.422	2.114	1.77	1.20
$\log(t_{1/2}/\text{s})$	2.659	2.358	2.009	1.78

There are clearly two types of behavior exhibited in the graph. At pressures above 50 torr the graph appears to be a straight line, and fitting to these three points results in a best fit line with an equation of $y = 1.00x + 0.24$. Thus, $1 = -(n-1)$, or $n = 0$, and the reaction is zero-order.

Although the data are limited, it is clear that there is a change in slope below 50 torr, indicating a change in reaction order. It does appear that the limiting slope as pressure approaches zero is itself zero. Thus, $0 = -(n-1)$, or $n = 1$, and the limiting behavior is that of a first-order reaction.

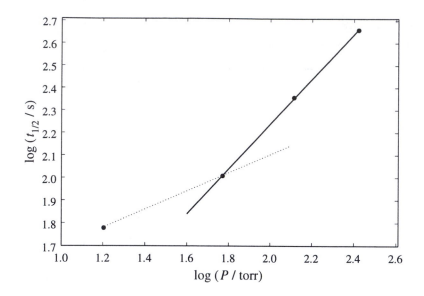

$$\log (P / \text{torr})$$

(b) As discovered in part (a), the reaction is first order at low pressures and zero order at pressures above 50 torr.

(c) The mechanism is actually the same at all pressures considered. At low pressures, the fraction of the tungsten surface covered is proportional to the pressure of NH_3, so the rate of decomposition will have a first order dependence on ammonia pressure. As discussed in Problem 12.16, however, at increased pressures, all the catalytic sites are occupied by, in this case, NH_3 molecules and the rate becomes independent of the ammonia pressure and zero order in NH_3.

12.49   The *activity* of a radioactive sample is the number of nuclear disintegrations per second, which is equal to the first-order rate constant times the number of radioactive nuclei present. The fundamental unit of radioactivity is the *curie* (Ci), where 1 Ci corresponds to exactly 3.70×10^{10} disintegrations per second. This decay rate is equivalent to that of 1 g of radium-226. Calculate the rate constant and half-life for the radium decay. Starting with 1.0 g of the radium sample, what is the activity after 500 years? The molar mass of Ra-226 is 226.03 g mol^{-1}.

For nuclear decay,

$$\text{rate} = \lambda N$$

where λ is the first order rate constant and N is the number of nuclei. In 1.0 g of ^{226}Ra, there are

$$N = (1.0 \text{ g}) \left(\frac{1 \text{ mol}}{226.06 \text{ g}} \right) (6.022 \times 10^{23} \text{ mol}^{-1}) = 2.66 \times 10^{21} \text{ nuclei}$$

Since this sample has an activity of 1 Ci,

$$3.70 \times 10^{10} \text{ s}^{-1} = \lambda (2.66 \times 10^{21})$$

$$\lambda = 1.39 \times 10^{-11} \text{ s}^{-1} = 1.4 \times 10^{-11} \text{ s}^{-1}$$

$$t_{1/2} = \frac{\ln 2}{\lambda} = \frac{\ln 2}{1.39 \times 10^{-11} \text{ s}^{-1}} = 5.0 \times 10^{10} \text{ s} = 1.6 \times 10^3 \text{ yr}$$

For first-order decay, $N = N_0 e^{-\lambda t}$, but since $N = \text{rate}/\lambda = R/\lambda$,

$$\frac{R}{\lambda} = \frac{R_0}{\lambda} e^{-\lambda t}$$

$$R = R_0 e^{-\lambda t}$$

$$= 3.70 \times 10^{10} e^{-\left(1.39 \times 10^{-11} \text{s}^{-1}\right)(500 \text{ yr})\left(3.15 \times 10^7 \text{ s yr}^{-1}\right)}$$

$$= 3.0 \times 10^{10} \text{ s}^{-1}$$

12.50   The reaction $X \rightarrow Y$ has a reaction enthalpy of -64 kJ mol^{-1} and an activation energy of 22 kJ mol^{-1}. What is the activation energy for the $Y \rightarrow X$ reaction?

Referring to the figure, the activation energy for the reverse reaction, $Y \rightarrow X$, is seen to be $22 \text{ kJ mol}^{-1} + 64 \text{ kJ mol}^{-1} = 86 \text{ kJ mol}^{-1}$.

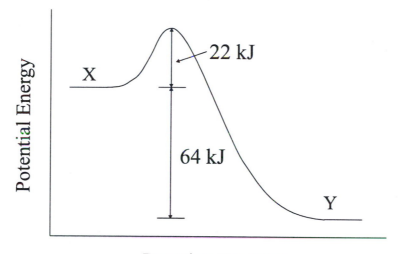

Reaction Progress

12.51   Consider the following parallel first-order reactions:

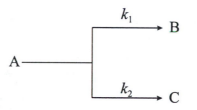

(a) Write the expression for $d\,[B]/dt$ at time t, given that $[A]_0$ is the concentration of A at $t = 0$.
(b) What is the ratio of $[B]/[C]$ upon the completion of the reactions?

(a) As long as the back reactions can be ignored, $\dfrac{d[B]}{dt} = k_1[A] = k_1 \left([A]_0 - [B] - [C]\right)$.

Likewise, $\dfrac{d[C]}{dt} = k_2[A] = k_2 \left([A]_0 - [B] - [C]\right)$

(b) Regardless of the importance of the back reactions,

$$-\frac{d[A]}{dt} = \frac{d[B]}{dt} + \frac{d[C]}{dt}$$

but if the back reactions can be ignored,

$$-\frac{d[A]}{dt} = \left(k_1 + k_2\right)[A]$$

and

$$[A] = [A]_0 e^{-(k_1+k_2)t}$$

From part **(a)**,

$$\frac{d[B]}{dt} = k_1[A]$$

$$= k_1[A]_0 e^{-(k_1+k_2)t}$$

$$d[B] = k_1[A]_0 e^{-(k_1+k_2)t} dt$$

$$\int_0^t d[B] = \int_0^t k_1[A]_0 e^{-(k_1+k_2)t} dt$$

$$[B]\Big|_0^t = -\frac{k_1[A]_0}{k_1 + k_2} e^{-(k_1+k_2)t}\Big|_0^t$$

$$[B] = \frac{k_1[A]_0}{k_1 + k_2}\left[1 - e^{-(k_1+k_2)t}\right]$$

since $[B]_0 = 0$. Similarly,

$$[C] = \frac{k_2[A]_0}{k_1 + k_2}\left[1 - e^{-(k_1+k_2)t}\right]$$

Thus, at any point during the reactions, as long as the back reactions are ignored,

$$\frac{[B]}{[C]} = \frac{k_1}{k_2}$$

At completion, when the reaction reaches equilibrium, the ratio of concentrations is determined by thermodynamic considerations, specifically the difference in Gibbs energy between B and C. At this point, the back reactions can no longer be ignored, as the principle of microscopic reversibility requires that they occur at the same rate as the respective forward reactions.

12.52   As a result of being exposed to the radiation released during the Chernobyl nuclear accident, a person had a level of iodine-131 in his body equal to 7.4 mCi ($1\ \text{mCi} = 1 \times 10^{-3}$ Ci). Calculate the number of atoms of I-131 to which this radioactivity corresponds. Why were people living close to the nuclear reactor site urged to take large amounts of potassium iodide after the accident?

For nuclear decay

$$\text{rate} = \lambda N = \frac{\ln 2}{t_{1/2}} N$$

where N is the number of radioactive atoms. The rate is given by the activity,

$$\text{rate} = (7.4 \text{ mCi}) \left(\frac{1 \text{ Ci}}{1000 \text{ mCi}} \right) \left(\frac{3.70 \times 10^{10} \text{ s}^{-1}}{1 \text{ Ci}} \right)$$

$$= 2.74 \times 10^{8} \text{ s}^{-1}$$

where the activity corresponding to 1 Ci is found in Problem 12.49. The half-life for I-131 is found in Table 12.1.

$$N = \left(\frac{t_{1/2}}{\ln 2} \right) \times \text{rate}$$

$$= \left(\frac{8.05 \text{ d}}{\ln 2} \right) \left(\frac{24 \text{ hr}}{1 \text{ d}} \right) \left(\frac{3600 \text{ s}}{1 \text{ hr}} \right) (2.74 \times 10^{8} \text{ s}^{-1})$$

$$= 2.7 \times 10^{14}$$

The human body concentrates iodine in the thyroid gland. Large doses of (non-radioactive) KI will displace the radioactive iodine from the thyroid and allow its excretion from the body.

12.53 A certain protein molecule P of molar mass M dimerizes when it is allowed to stand in solution at room temperature. A plausible mechanism is that the protein molecule is first denatured before it dimerizes:

$$\text{P} \xrightarrow{k} \text{P*} \text{ (denatured)} \qquad \text{slow}$$
$$2\text{P*} \longrightarrow \text{P}_2 \qquad \text{fast}$$

The progress of this reaction can be followed by making viscosity measurements of the average molar mass $\overline{M}$. Derive an expression for $\overline{M}$ in terms of the initial concentration $[\text{P}]_0$ and the concentration at time t, $[\text{P}]$, and M. Write a rate equation consistent with this scheme.

The average molar mass is given by

$$\overline{M} = \frac{[\text{P}]M + 2[\text{P}_2]M}{[\text{P}] + [\text{P}_2]}$$

The stoichiometry of the reaction requires $[\text{P}_2] = ([\text{P}]_0 - [\text{P}])/2$

$$\overline{M} = \frac{[\text{P}]M + [\text{P}]_0 M - [\text{P}]M}{[\text{P}] + \frac{1}{2}[\text{P}]_0 - \frac{1}{2}[\text{P}]}$$

$$= \frac{2M[\text{P}]_0}{[\text{P}]_0 + [\text{P}]}$$

In the proposed mechanism the denaturation step is rate determining. Thus,

$$-\frac{d[\text{P}]}{dt} = k[\text{P}]$$

This first order rate equation has solution

$$[P] = [P]_0 e^{-kt}$$

Using this result for [P] in the expression for average molar mass,

$$\overline{M} = \frac{2M[P]_0}{[P]_0 + [P]_0 e^{-kt}}$$

$$= \frac{2M}{1 + e^{-kt}}$$

or

$$\frac{2M - \overline{M}}{\overline{M}} = e^{-kt}$$

$$\ln\left(\frac{2M - \overline{M}}{\overline{M}}\right) = -kt$$

Thus a plot of $\ln\left(\frac{2M-\overline{M}}{\overline{M}}\right)$ versus t will give a straight line with slope $-k$.

12.54   The bromination of acetone is acid-catalyzed:

$$CH_3COCH_3 + Br_2 \xrightarrow{H^+} CH_3COCH_2Br + H^+ + Br^-$$

The rate of disappearance of bromine was measured for several different concentrations of acetone, bromine, and H^+ ions at a certain temperature:

	$[CH_3COCH_3]/M$	$[Br_2]/M$	$[H^+]/M$	Rate of disappearance of Br_2 $/M \cdot s^{-1}$
(1)	0.30	0.050	0.050	5.7×10^{-5}
(2)	0.30	0.10	0.050	5.7×10^{-5}
(3)	0.30	0.050	0.1	1.2×10^{-4}
(4)	0.40	0.050	0.2	3.1×10^{-4}
(5)	0.40	0.050	0.050	7.6×10^{-5}

(a) What is the rate law for the reaction? **(b)** Determine the rate constant. **(c)** The following mechanism has been proposed for the reaction:

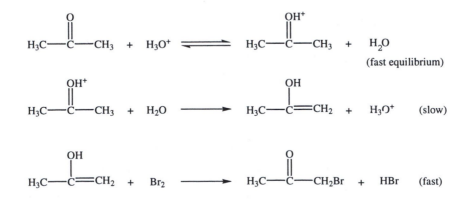

Show that the rate law deduced from the mechanism is consistent with that shown in (a).

(a) Comparing Experiment (1) and Experiment (5), the concentrations of Br_2 and H^+ are constant and the concentration of CH_3COCH_3 has increased 1.33 times. The rate has also increased 1.33 times. Therefore, the rate is directly proportional to the concentration of CH_3COCH_3. That is, the reaction is first order in CH_3COCH_3.

Comparing Experiment (1) and Experiment (2), the concentrations of CH_3COCH_3 and H^+ are constant and the concentration of Br_2 has increased 2.0 times but the rate has not changed. Therefore, the rate is independent of the concentration of Br_2. That is, the reaction is zeroth order in Br_2.

Comparing Experiment (1) and Experiment (3), the concentrations of CH_3COCH_3 and Br_2 are constant and the concentration of H^+ has increased 2.0 times. The rate has increased 2.1 times. Therefore, the rate is directly proportional to the concentration of H^+. That is, the reaction is first order in H^+.

Therefore, the rate law is

$$Rate = k[CH_3COCH_3][H^+]$$

(b) The rate constant can be calculated using data from any experiment. Using Exp. 1,

$$Rate = 5.7 \times 10^{-5} \, M\,s^{-1} = k \, (0.30 \, M) \, (0.050 \, M)$$

$$k = 3.8 \times 10^{-3} \, M^{-1} \, s^{-1}$$

(c) Since step 2 is the slow step,

$$Rate = k_2[CH_3COHCH_3^+][H_2O] \tag{12.54.1}$$

The intermediate, $CH_3COHCH_3^+$ can be written in terms of the reactants using the first fast equilibrium step.

$$Forward \; rate \; of \; step \; 1 = Reverse \; rate \; of \; step \; 1$$

$$k_1[CH_3COCH_3][H_3O^+] = k_{-1}[CH_3COHCH_3^+][H_2O]$$

$$[CH_3COHCH_3^+][H_2O] = \frac{k_1}{k_{-1}}[CH_3COCH_3][H_3O^+] \tag{12.54.2}$$

Substituting Eq. 12.54.2 into Eq. 12.54.1, the rate law becomes

$$Rate = \frac{k_2 k_1}{k_{-1}}[CH_3COCH_3][H_3O^+] = \frac{k_2 k_1}{k_{-1}}[CH_3COCH_3][H^+]$$

which has the same form as that shown in (a).

12.55   The rate law for the reaction $2NO_2(g) \rightarrow N_2O_4(g)$ is rate $= k \, [NO_2]^2$. Which of the following changes will alter the value of k? (a) The pressure of NO_2 is doubled. (b) The reaction is run in an organic solvent. (c) The volume of the container is doubled. (d) The temperature is decreased. (e) A catalyst is added to the container.

(a) Changing the concentration of a reactant has no effect on k.

(b) If a reaction is run in a solvent other than in the gas phase, then the reaction mechanism will probably change and will thus change k.

(c) Doubling the volume simply changes the concentration. There is no effect on k, as in (a).

(d) The value of k will change with temperature.

(e) A catalyst changes the reaction mechanism and therefore changes k.

12.56   For the cyclic reaction shown in on p. 465, show that $k_{-1}k_2k_3 = k_1k_{-2}k_{-3}$.

According to the reaction,

$$k_2[A] = k_{-2}[B]$$

$$k_{-1}[C] = k_1[A]$$

$$k_3[B] = k_{-3}[C]$$

Therefore,

$$k_2[A]k_{-1}[C]k_3[B] = k_{-2}[B]k_1[A]k_{-3}[C]$$

$$k_{-1}k_2k_3 = k_1k_{-2}k_{-3}$$

12.57   Oxygen for metabolism is taken up by hemoglobin (Hb) to form oxyhemoglobin (HbO$_2$) according to the simplified equation

$$Hb(aq) + O_2(aq) \xrightarrow{k} HbO_2(aq)$$

where the second-order rate constant is $2.1 \times 10^6 \ M^{-1} \, s^{-1}$ at 37°C. For an average adult, the concentrations of Hb and O$_2$ in the blood in the lungs are $8.0 \times 10^{-6} \ M$ and $1.5 \times 10^{-6} \ M$, respectively. (a) Calculate the rate of formation of HbO$_2$. (b) Calculate the rate of consumption of O$_2$. (c) The rate of formation of HbO$_2$ increases to $1.4 \times 10^{-4} \ M \, s^{-1}$ during exercise to meet the demand of an increased metabolic rate. Assuming the Hb concentration remains the same, what oxygen concentration is necessary to sustain this rate of HbO$_2$ formation?

(a)

$$\text{Rate of formation of HbO}_2 = \frac{d\,[\text{HbO}_2]}{dt} = k[\text{Hb}][\text{O}_2]$$

$$= \left(2.1 \times 10^6 \ M^{-1}\,s^{-1}\right)\left(8.0 \times 10^{-6} \ M\right)\left(1.5 \times 10^{-6} \ M\right)$$

$$= 2.5 \times 10^{-5} \ M\,s^{-1}$$

(b) The rate of consumption of O_2 is

$$-\frac{d\,[O_2]}{dt} = \frac{d\,[HbO_2]}{dt} = 2.5 \times 10^{-5}\,M\,s^{-1}$$

(c)

$$\text{Rate of formation of } HbO_2 = k[Hb][O_2]$$

$$[O_2] = \frac{\text{Rate of formation of } HbO_2}{k[Hb]}$$

$$= \frac{1.4 \times 10^{-4}\,M\,s^{-1}}{\left(2.1 \times 10^6\,M^{-1}\,s^{-1}\right)\left(8.0 \times 10^{-6}\,M\right)}$$

$$= 8.3 \times 10^{-6}\,M$$

Thus, 5.6 times as much oxygen is required to sustain the increase in metabolism.

12.58 Sucrose ($C_{12}H_{22}O_{11}$), commonly called table sugar, undergoes hydrolysis (reaction with water) to produce fructose ($C_6H_{12}O_6$) and glucose ($C_6H_{12}O_6$):

$$C_{12}H_{22}O_{11} \quad + \quad H_2O \quad \longrightarrow \quad \underset{\text{fructose}}{C_6H_{12}O_6} \quad + \quad \underset{\text{glucose}}{C_6H_{12}O_6}$$

This reaction has particular significance in the candy industry. First, fructose is sweeter than sucrose. Second, a mixture of fructose and glucose, called *invert* sugar, does not crystallize, so candy made with this combination is chewier and not brittle as crystalline sucrose is. Sucrose is dextrorotatory (+) whereas the mixture of glucose and fructose resulting from inversion in levorotatory (−). Thus, a decrease in the concentration of sucrose will be accompanied by a proportional decrease in the optical rotation. **(a)** From the following kinetic data, show that the reaction is first order and determine the rate constant.

time/min	0	7.20	18.0	27.0	∞
optical rotation (α)	+24.08°	+21.40°	+17.73°	+15.01°	−10.73°

(b) Explain why the rate law does not include [H_2O] even though water is a reactant.

(a) The total change in rotation (from $t = 0$ to $t = \infty$), given by ($\alpha_0 - \alpha_\infty$), will be proportional to the decrease in concentration of sucrose by that time. Therefore, the concentration of sucrose remaining at time t will be proportional to

$$\left(\alpha_0 - \alpha_\infty\right) - \left(\alpha_0 - \alpha_t\right) = \alpha_t - \alpha_\infty$$

If the reaction is first order, then

$$\ln\frac{[\text{sucrose}]}{[\text{sucrose}]_0} = \ln\frac{\alpha_t - \alpha_\infty}{\alpha_0 - \alpha_\infty} = -kt$$

A plot of $\left[\ln\left(\alpha_t - \alpha_\infty\right)/\left(\alpha_0 - \alpha_\infty\right)\right]$ vs t should give a straight line with a slope of $-k$. The plot uses the following data:

time/min	0	7.20	18.0	27.0
$\ln\left(\alpha_t - \alpha_\infty\right)/\left(\alpha_0 - \alpha_\infty\right)$	0	-8.0115×10^{-2}	-0.20141	-0.30186

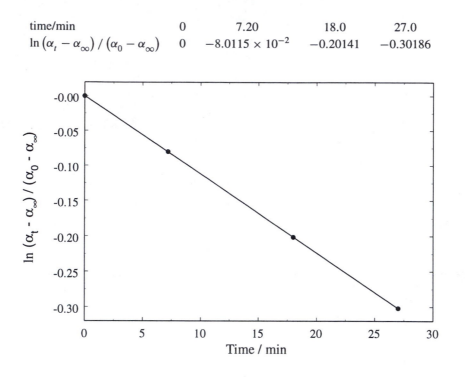

The plot indeed results in a line described by $y = -1.112 \times 10^{-2}x + 1.5 \times 10^{-4}$. Therefore, the rate constant is 1.11×10^{-2} min^{-1}.

(b) The rate law does not include [H$_2$O] because water is present in very high concentration (ca. 55.5 M) and this is a pseudo first-order reaction.

12.59 Thallium(I) is oxidized by cerium(IV) in solution as follows:

$$Tl^+ + 2Ce^{4+} \longrightarrow Tl^{3+} + 2Ce^{3+}$$

The elementary steps, in the presence of Mn(II), are

$$Ce^{4+} + Mn^{2+} \longrightarrow Ce^{3+} + Mn^{3+}$$

$$Ce^{4+} + Mn^{3+} \longrightarrow Ce^{3+} + Mn^{4+}$$

$$Tl^+ + Mn^{4+} \longrightarrow Tl^{3+} + Mn^{2+}$$

(a) Identify the catalyst, intermediates, and the rate-determining step if the rate law is rate = $k\,[Ce^{4+}]\,[Mn^{2+}]$. **(b)** Explain why the reaction is slow without the catalyst. **(c)** Classify the type of catalysis (homogeneous or heterogeneous).

(a) The catalyst is Mn^{2+}. It participates in the reaction but is regenerated at the end. The intermediates are Mn^{3+} and Mn^{4+}.

The first step is the rate-determining step because the rate depends on the concentrations of the reactants for that step.

(b) Without the catalyst, the reaction would be a termolecular one involving 3 cations (Tl^+ and two Ce^{4+}). The reaction would be slow.

(c) The catalyst is a homogeneous catalyst because it has the same phase (aqueous) as the reactants.

12.60 Under certain conditions the gas-phase decomposition of ozone is found to be second order in O_3 and inhibited by molecular oxygen. Apply the steady-state approximation to the following mechanism to show that the rate law is consistent with the experimental observations:

$$O_3 \underset{k_{-1}}{\overset{k_1}{\rightleftharpoons}} O_2 + O$$

$$O + O_3 \xrightarrow{k_2} 2O_2$$

State any assumption made in the derivation.

The rate of decomposition of O_3 is

$$-\frac{d[O_3]}{dt} = k_1[O_3] - k_{-1}[O_2][O] + k_2[O][O_3]$$

$$= k_1[O_3] + (k_2[O_3] - k_{-1}[O_2])[O] \tag{12.60.1}$$

Apply the steady state approximation to O:

$$\frac{d[O]}{dt} = k_1[O_3] - k_{-1}[O_2][O] - k_2[O][O_3] = 0$$

$$[O] = \frac{k_1[O_3]}{k_{-1}[O_2] + k_2[O_3]} \tag{12.60.2}$$

Substitute Eq. 12.60.2 into Eq. 12.60.1:

$$-\frac{d[O_3]}{dt} = k_1[O_3] + (k_2[O_3] - k_{-1}[O_2])\frac{k_1[O_3]}{k_{-1}[O_2] + k_2[O_3]}$$

$$= \frac{k_1 k_{-1}[O_3][O_2] + k_1 k_2[O_3]^2}{k_{-1}[O_2] + k_2[O_3]} + \frac{k_1 k_2[O_3]^2}{k_{-1}[O_2] + k_2[O_3]} - \frac{k_1 k_{-1}[O_3][O_2]}{k_{-1}[O_2] + k_2[O_3]}$$

$$= \frac{2k_1 k_2[O_3]^2}{k_{-1}[O_2] + k_2[O_3]} \tag{12.60.3}$$

If the rate of the second step is assumed to be much slower than the rate of the reverse reaction for the first step, then

$$k_2[O][O_3] \ll k_{-1}[O_2][O]$$

$$k_2[O_3] \ll k_{-1}[O_2]$$

Equation 12.60.3 becomes

$$-\frac{d[O_3]}{dt} = \frac{2k_1 k_2[O_3]^2}{k_{-1}[O_2]}$$

Since the rate for the reaction $2O_3 \rightarrow 3O_2$ is $-\frac{1}{2}\frac{d[O_3]}{dt}$, the rate law predicted by this mechanism is

$$\text{Rate} = -\frac{1}{2}\frac{d[O_3]}{dt} = \frac{k_1 k_2 [O_3]^2}{k_{-1}[O_2]}$$

which is consistent with experimental observations.

12.61   The rate constants for the reaction

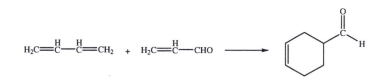

have been measured at several temperatures:

$10^3 k/M^{-1} \cdot s^{-1}$	0.138	1.63	7.2	36.8	81
$t/°C$	155.3	208.3	246.5	295.8	330.8

Calculate the values of the pre-exponential factor, E_a, $\Delta S^{o\ddagger}$, and $\Delta H^{o\ddagger}$ for the reaction. Use 516 K as the mean temperature for your calculation. [Data taken from G. B. Kistiakowsky and J. R. Lacher, *J. Am. Chem. Soc.* **58**, 123 (1936).]

Since

$$k = Ae^{-E_a/RT}$$

$$\ln k = \ln A - \frac{E_a}{RT}$$

A plot of $\ln k$ vs $1/T$ gives a straight line with a slope of $-E_a/R$ and an intercept of $\ln A$.

10^3 K/T	2.3337	2.0768	1.9242	1.7575	1.6556
$\ln(k/M^{-1} \cdot s^{-1})$	−8.8883	−6.4192	−4.934	−3.3023	−2.513

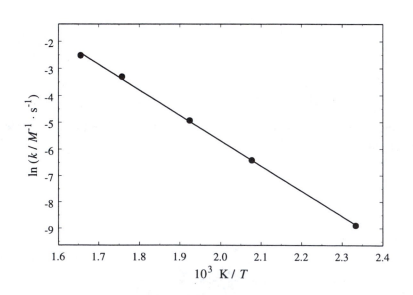

The best fit line has a formula of $y = -9499x + 13.31$. Therefore,

$$E_a = -\left(-9499\ \text{K}\right)\left(8.314\ \text{J K}^{-1}\ \text{mol}^{-1}\right) = 7.897 \times 10^4\ \text{J mol}^{-1} = 7.90 \times 10^4\ \text{J mol}^{-1}$$

$$A = e^{13.31} = 6.032 \times 10^5\ M^{-1}\ \text{s}^{-1} = 6.03 \times 10^5\ M^{-1}\ \text{s}^{-1}$$

For this gas-phase, bimolecular reaction, (see Equation 12.43)

$$A = e^2 \frac{k_\text{B} T}{h} e^{\Delta S^{\text{o}\ddagger}/R}$$

$$\Delta S^{\text{o}\ddagger} = R\left(\ln \frac{hA}{k_\text{B} T} - 2\right)$$

$$= \left(8.314\ \text{J K}^{-1}\ \text{mol}^{-1}\right)\left[\ln \frac{\left(1\ M\right)\left(6.626 \times 10^{-34}\ \text{J s}\right)\left(6.032 \times 10^5\ M^{-1}\ \text{s}^{-1}\right)}{\left(1.381 \times 10^{-23}\ \text{J K}^{-1}\right)\left(516\ \text{K}\right)} - 2\right]$$

$$= -155\ \text{J K}^{-1}\ \text{mol}^{-1}$$

and

$$\Delta H^{\text{o}\ddagger} = E_a - 2RT$$

$$= 7.897 \times 10^4\ \text{J mol}^{-1} - 2\left(8.314\ \text{J K}^{-1}\ \text{mol}^{-1}\right)\left(516\ \text{K}\right)$$

$$= 7.04 \times 10^4\ \text{J mol}^{-1}$$

CHAPTER 13

Enzyme Kinetics

PROBLEMS AND SOLUTIONS

13.1 Explain why a catalyst must affect the rate of a reaction in both directions.

A catalyst speeds up a reaction by lowering the Gibbs energy of the transition state. Therefore, it lowers the activation energy for both the forward and the reverse reaction and affects the rates of both.

13.2 Measurements of a certain enzyme-catalyzed reaction give $k_1 = 8 \times 10^6 \ M^{-1} \, s^{-1}$, $k_{-1} = 7 \times 10^4 \ s^{-1}$, and $k_2 = 3 \times 10^3 \ s^{-1}$. Does the enzyme-substrate binding follow the equilibrium or steady-state scheme?

The dissociation constant, K_S, and the Michaelis constant, K_M must be compared.

$$K_S = \frac{k_{-1}}{k_1}$$

$$= \frac{7 \times 10^4 \ s^{-1}}{8 \times 10^6 \ M^{-1} \, s^{-1}}$$

$$= 9 \times 10^{-3} \ M$$

and

$$K_M = \frac{k_{-1} + k_2}{k_1}$$

$$= \frac{7 \times 10^4 \ s^{-1} + 3 \times 10^3 \ s^{-1}}{8 \times 10^6 \ M^{-1} \, s^{-1}}$$

$$= 9 \times 10^{-3} \ M$$

Within the precision of the measurements, the two constants are equal. Thus, the binding follows the equilibrium scheme. That is, k_{-1} is sufficiently greater than k_2 so that the binding reaches equilibrium.

13.3   The hydrolysis of acetylcholine is catalyzed by the enzyme acetylcholinesterase, which has a turnover rate of 25,000 s^{-1}. Calculate how long it takes for the enzyme to cleave one acetylcholine molecule.

The time required for the enzyme to cleave one acetylcholine molecule (one turnover) is the reciprocal of the turnover rate.

$$t = \frac{1}{k_2} = \frac{1}{25000 \text{ s}^{-1}} = 4.0 \times 10^{-5} \text{ s} = 40 \ \mu\text{s}$$

13.4   Derive the following equation from Equation 13.10

$$\frac{v_0}{[\text{S}]} = \frac{V_{max}}{K_M} - \frac{v_0}{K_M}$$

and show how you would obtain values of K_M and V_{max} graphically from this equation.

Starting with Equation 13.10, multiply both sides by $K_M + [\text{S}]$, then divide by $K_M[\text{S}]$ and rearrange.

$$v_0 = \frac{V_{max}[\text{S}]}{K_M + [\text{S}]}$$

$$v_0 K_M + v_0[\text{S}] = V_{max}[\text{S}]$$

$$v_0 K_M = V_{max}[\text{S}] - v_0[\text{S}]$$

$$\frac{v_0}{[\text{S}]} = \frac{V_{max}}{K_M} - \frac{v_0}{K_M}$$

Thus, a plot of $v_0/[\text{S}]$ vs. v_0 will have a slope of $-1/K_M$ and a y-intercept of V_{max}/K_M. The same data, however, when plotted in a Eadie-Hofstee plot, v_0 vs. $v_0/[\text{S}]$, gives more straightforward results (see Problem 13.6).

13.5   An enzyme that has a K_M value of 3.9×10^{-5} M is studied at an initial substrate concentration of 0.035 M. After 1 min, it is found that 6.2 μM of product has been produced. Calculate the value of V_{max} and the amount of product formed after 4.5 min.

With 6.2 μM of product formed in 1 min, the initial rate, $v_0 = 6.2 \times 10^{-6}$ M min^{-1}. Then solve

$$v_0 = \frac{V_{max}[\text{S}]}{K_M + [\text{S}]}$$

$$6.2 \times 10^{-6} \ M \text{min}^{-1} = \frac{V_{max}(0.035 \ M)}{3.9 \times 10^{-5} \ M + 0.035 \ M}$$

to find $V_{max} = 6.21 \times 10^{-6}$ M min$^{-1} = 6.2 \times 10^{-6}$ M min^{-1}. The observed rate is V_{max}, indicating that the substrate concentration is so large that the enzyme is saturated with substrate.

Over 4.5 min, the substrate concentration stays essentially constant, and the rate does not change. Thus, after 4.5 min, $(4.5\ \text{min})(6.21 \times 10^{-6}\ M\ \text{min}^{-1}) = 2.8 \times 10^{-5}\ M$ of product has formed.

13.6 The hydrolysis of N-glutaryl-L-phenylalanine-p-nitroanilide (GPNA) to p-nitroaniline and N-glutaryl-L-phenylalanine is catalyzed by α-chymotrypsin. The following data are obtained:

$[S]/10^{-4} \cdot M$	2.5	5.0	10.0	15.0
$v_0/10^{-6} \cdot M \cdot \text{min}^{-1}$	2.2	3.8	5.9	7.1

where $[S] = [GPNA]$. Assuming Michaelis–Menten kinetics, calculate the values of V_{max}, K_M, and k_2 using the Lineweaver–Burk plot. Another way to treat the data is to plot v_0 versus $v_0/[S]$, which is the Eadie–Hofstee plot. Calculate the values of V_{max}, K_M, and k_2 from the Eadie–Hofstee treatment, given that $[E]_0 = 4.0 \times 10^{-6}\ M$. [*Source:* J. A. Hurlbut, T. N. Ball, H. C. Pound, and J. L. Graves, *J. Chem. Educ.* **50**, 149 (1973).]

For the Lineweaver–Burk plot, the following data are needed.

$(1/[S])/10^3 \cdot M^{-1}$	4.00	2.00	1.00	0.667
$(1/v_0)/10^5 \cdot M^{-1} \cdot \text{min}$	4.55	2.63	1.69	1.41

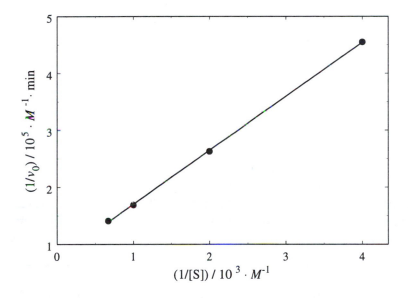

The best-fit line to the data has an equation of $y = 94.6x + 7.56 \times 10^4$. The intercept of a Lineweaver–Burk plot is $1/V_{max}$ giving

$$V_{max} = \frac{1}{7.56 \times 10^4\ M^{-1}\ \text{min}}$$

$$= 1.32 \times 10^{-5}\ M\ \text{min}^{-1}$$

$$= 1.3 \times 10^{-5}\ M\ \text{min}^{-1}$$

The slope is K_M/V_{max} so that

$$K_M = (94.6\ \text{min})(1.32 \times 10^{-5}\ M\ \text{min}^{-1})$$

$$= 1.2 \times 10^{-3}\ M$$

Finally,

$$k_2 = \frac{V_{max}}{[E]_0}$$

$$= \frac{1.32 \times 10^{-5} \, M \, min^{-1}}{4.0 \times 10^{-6} \, M}$$

$$= 3.3 \, min^{-1}$$

The Eadie–Hofstee plot uses the following data,

$(v_0/[S])/10^{-3} \cdot min^{-1}$	8.80	7.60	5.90	4.73
$v_0 /10^{-6} \cdot M \cdot min^{-1}$	2.2	3.8	5.9	7.1

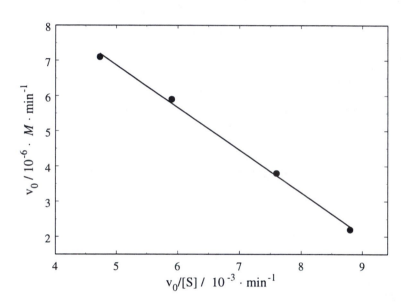

The best-fit line to the data has an equation of $y = -1.21 \times 10^{-3}x + 1.29 \times 10^{-5}$. In a Eadie–Hofstee plot the slope is $-K_M$ and the y-intercept is V_{max}. Thus, $V_{max} = 1.3 \times 10^{-5} \, M \, min^{-1}$ and $K_M = 1.2 \times 10^{-3} \, M$. $k_2 = 3.3 \, min^{-1}$ is found as above. These are the same values as found from the Lineweaver–Burk plot, which given good data is as expected. The two plots weight the data differently, so that the values determined may be different depending on the quality of the data.

13.7   The K_M of lysozyme is $6.0 \times 10^{-6} \, M$ with hexa-N-acetylglucosamine as a substrate. It is assayed at the following substrate concentrations: **(a)** $1.5 \times 10^{-7} \, M$, **(b)** $6.8 \times 10^{-5} \, M$, **(c)** $2.4 \times 10^{-4} \, M$, **(d)** $1.9 \times 10^{-3} \, M$, and **(e)** $0.061 \, M$. The initial rate measured at $0.061 \, M$ was $3.2 \, \mu M \, min^{-1}$. Calculate the initial rates at the other substrate concentrations.

V_{max} is calculated from the initial rate given for $0.061 \, M$ substrate.

$$v_0 = \frac{V_{max}[S]}{K_M + [S]}$$

$$3.2 \times 10^{-6} \, M \, min^{-1} = \frac{V_{max}(0.061 \, M)}{6.0 \times 10^{-6} \, M + 0.061 \, M}$$

$$V_{max} = 3.20 \times 10^{-6} \, M \, min^{-1}$$

Thus, the substrate concentration is sufficient to have reached V_{max}.

The initial rates at the other substrate concentrations are then found using

$$v_0 = \frac{V_{max}[S]}{K_M + [S]}$$

$$= \frac{(3.20 \times 10^{-6}\, M\,min^{-1})[S]}{6.0 \times 10^{-6}\, M + [S]}$$

For the five concentrations given,

(a) $v_0 = 7.8 \times 10^{-8}\, M\,min^{-1}$ (b) $v_0 = 2.9 \times 10^{-6}\, M\,min^{-1}$ (c) $v_0 = 3.1 \times 10^{-6}\, M\,min^{-1}$
(d) $v_0 = 3.2 \times 10^{-6}\, M\,min^{-1}$ (e) $v_0 = 3.2 \times 10^{-6}\, M\,min^{-1}$

13.8 The hydrolysis of urea,

$$(NH_2)_2CO + H_2O \rightarrow 2NH_3 + CO_2$$

has been studied by many researchers. At 100°C, the (pseudo) first-order rate constant is $4.2 \times 10^{-5}\, s^{-1}$. The reaction is catalyzed by the enzyme urease, which at 21°C has a rate constant of $3 \times 10^4\, s^{-1}$. If the enthalpies of activation for the uncatalyzed and catalyzed reactions are 134 kJ mol^{-1} and 43.9 kJ mol^{-1}, respectively, **(a)** calculate the temperature at which the nonenzymatic hydrolysis of urea would proceed at the same rate as the enzymatic hydrolysis at 21°C; **(b)** calculate the lowering of $\Delta G^{\ddagger}$ due to urease; and **(c)** comment on the sign of $\Delta S^{\ddagger}$. Assume that $\Delta H^{\ddagger} = E_a$ and that $\Delta H^{\ddagger}$ and $\Delta S^{\ddagger}$ are independent of temperature.

(a) The Arrhenius equation relates reaction rate and activation energy via $k = Ae^{-E_a/RT}$. Requiring that the rates of the catalyzed and uncatalyzed reactions be equal at their respective temperatures then means (assuming A to be the same for both the catalyzed and uncatalyzed reactions),

$$k_{cat} = k_{uncat}$$

$$Ae^{-E_a^{cat}/RT_1} = Ae^{-E_a^{uncat}/RT_2}$$

$$\frac{E_a^{cat}}{T_1} = \frac{E_a^{uncat}}{T_2}$$

taking $E_a \approx \Delta H^{\ddagger}$,

$$\frac{43.9 \times 10^3\, J\,mol^{-1}}{294.15\, K} = \frac{134 \times 10^3\, J\,mol^{-1}}{T_2}$$

$$T_2 = 898\, K$$

At this temperature the solvent would be vaporized and the urea thermally decomposed, so that it is in fact impossible to achieve the enzymatic rate without the catalyst.

(b) From Equation 12.36, $k = \frac{k_B T}{h}e^{-\Delta G^{\ddagger}/RT}$, or $\Delta G^{\ddagger} = -RT \ln \frac{kh}{k_B T}$.

For the uncatalyzed reaction at 373 K,

$$\Delta G^{\ddagger} = -(8.314\, J\,K^{-1}\,mol^{-1})(373.15\, K) \ln \frac{(4.2 \times 10^{-5}\, s^{-1})(6.626 \times 10^{-34}\, J\,s)}{(1.381 \times 10^{-23}\, J\,K^{-1})(373.15\, K)}$$

$$= 1.234 \times 10^5\, J\,mol^{-1}$$

For the catalyzed reaction at 294 K,

$$\Delta G^{\ddagger} = -\left(8.314\,\mathrm{J\,K^{-1}\,mol^{-1}}\right)(294.15\,\mathrm{K})\ln\frac{\left(3\times10^{4}\,\mathrm{s^{-1}}\right)\left(6.626\times10^{-34}\,\mathrm{J\,s}\right)}{\left(1.381\times10^{-23}\,\mathrm{J\,K^{-1}}\right)(294.15\,\mathrm{K})}$$

$$= 4.70\times10^{4}\,\mathrm{J\,mol^{-1}}$$

Thus, $\Delta G^{\ddagger}$ is lowered by $1.234\times10^{5}\,\mathrm{J\,mol^{-1}} - 4.70\times10^{4}\,\mathrm{J\,mol^{-1}} = 7.64\times10^{4}\,\mathrm{J\,mol^{-1}}$, although the comparison is being made at two different temperatures.

(c) Since $\Delta G^{\ddagger} = \Delta H^{\ddagger} - T\Delta S^{\ddagger}$, $\Delta S^{\ddagger} = \left(\Delta H^{\ddagger} - \Delta G^{\ddagger}\right)/T$.

For the uncatalyzed reaction,

$$\Delta S^{\ddagger} = \frac{134\times10^{3}\,\mathrm{J\,mol^{-1}} - 1.234\times10^{5}\,\mathrm{J\,mol^{-1}}}{373.15\,\mathrm{K}} = 28.4\,\mathrm{J\,K^{-1}\,mol^{-1}}$$

There is an increase in entropy upon approaching the transition state as would be expected in a case where a single molecule is breaking apart in two or more fragments in the transition state.

For the catalyzed reaction,

$$\Delta S^{\ddagger} = \frac{43.9\times10^{3}\,\mathrm{J\,mol^{-1}} - 4.70\times10^{4}\,\mathrm{J\,mol^{-1}}}{294.15\,\mathrm{K}} = -11\,\mathrm{J\,K^{-1}\,mol^{-1}}$$

Here there is a decrease in entropy upon approaching the transition state, since the rate determining step now involves the binding of two molecules, enzyme and substrate.

13.9   An enzyme is inactivated by the addition of a substance to a solution containing the enzyme. Suggest three ways to find out whether the substance is a reversible or an irreversible inhibitor.

---

(1) Binding of reversible inhibitors to an enzyme reaches equilibrium rapidly (usually within one second) while irreversible inhibition proceeds much more slowly. Therefore, a time-dependent study of enzyme activity and inhibition would distinguish between the two types of inhibition.

(2) A reversible inhibitor can be removed by dialysis while an irreversible one cannot be so removed.

(3) Dilution will affect reversible inhibition by altering the enzyme-inhibitor binding equilibrium while the effect of an irreversible inhibitor is not altered upon dilution.

13.10   Silver ions are known to react with the sulfhydryl groups of proteins and therefore can inhibit the action of certain enzymes. In one reaction, 0.0075 g of $AgNO_3$ is needed to completely inactivate a 5-mL enzyme solution. Estimate the molar mass of the enzyme. Explain why the molar mass obtained represents the minimum value. The concentration of the enzyme solution is such that 1 mL of the solution contains 75 mg of the enzyme.

The number of moles of $AgNO_3$ used to inactivate the enzyme is

$$\frac{7.5\times10^{-3}\,\mathrm{g}}{169.9\,\mathrm{g\,mol^{-1}}} = 4.41\times10^{-5}\,\mathrm{mol}$$

Assuming 1:1 binding between the silver ions and the protein, this is also the number of moles of enzyme present in the 5 mL solution which contains $(5 \text{ mL}) (75 \times 10^{-3} \text{ g mL}^{-1}) = 0.375 \text{ g}$ of enzyme. Thus, the molar mass of the enzyme is

$$\frac{0.375 \text{ g}}{4.41 \times 10^{-5} \text{ mol}} = 8.5 \times 10^3 \text{ g mol}^{-1}$$

This is the minimum value for the molar mass because of the assumption of 1:1 binding. If there were more than one Ag^+ ion binding site per enzyme, there would be fewer moles of enzyme present leading to a larger value for the molar mass.

13.11   The initial rates at various substrate concentrations for an enzyme-catalyzed reaction are as follows:

[S]/M	$v_0/10^{-6} \cdot M \cdot min^{-1}$
2.5×10^{-5}	38.0
4.00×10^{-5}	53.4
6.00×10^{-5}	68.6
8.00×10^{-5}	80.0
16.0×10^{-5}	106.8
20.0×10^{-5}	114.0

(a) Does this reaction follow Michaelis–Menten kinetics? **(b)** Calculate the value of V_{max} of the reaction. **(c)** Calculate the K_M value of the reaction. **(d)** Calculate the initial rates at [S] = 5.00×10^{-5} M and [S] = 3.00×10^{-1} M. **(e)** What is the total amount of product formed during the first 3 min at [S] = 7.2×10^{-5} M? **(f)** How would an increase in the enzyme concentration by a factor of 2 affect each of the following quantities: K_M, V_{max}, and v_0 (at [S] = 5.00×10^{-5} M)?

(a) A Lineweaver–Burk plot ($1/v_0$ versus $1/[S]$) using the following data results in a straight line, indicating that the reaction does indeed follow Michaelis–Menten kinetics.

$(1/[S])/10^3 \cdot M^{-1}$	$(1/v_0)/10^3 \cdot M^{-1} \cdot min$
40.0	26.32
25.00	18.73
16.67	14.58
12.50	12.50
6.250	9.3633
5.000	8.7719

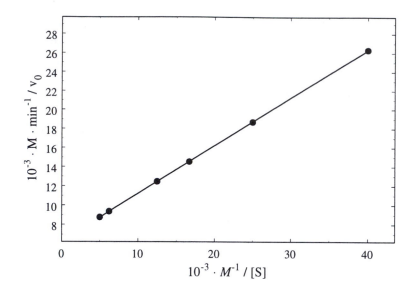

The equation of the best fit line to the data is $y = 5.013 \times 10^{-1}x + 6.23 \times 10^{3}$. Thus, since in a Lineweaver–Burk plot the slope is K_M/V_{max} and the intercept is $1/V_{max}$,

(b)

$$V_{max} = \frac{1}{6.23 \times 10^{3}\ M^{-1}\ \text{min}} = 1.605 \times 10^{-4}\ M\,\text{min}^{-1} = 1.61 \times 10^{-4}\ M\,\text{min}^{-1}$$

(c)

$$K_M = \frac{5.013 \times 10^{-1}\ \text{min}}{6.23 \times 10^{3}\ M^{-1}\ \text{min}} = 8.047 \times 10^{-5}\ M = 8.05 \times 10^{-5}\ M$$

(d) For Michaelis–Menten kinetics,

$$v_0 = \frac{V_{max}[S]}{K_M + [S]}$$

Thus at $[S] = 5.00 \times 10^{-5}\ M$,

$$v_0 = \frac{\left(1.605 \times 10^{-4}\ M\,\text{min}^{-1}\right)\left(5.00 \times 10^{-5}\ M\right)}{8.047 \times 10^{-5}\ M + 5.00 \times 10^{-5}\ M} = 6.15 \times 10^{-5}\ M\,\text{min}^{-1}$$

and at $[S] = 3.00 \times 10^{-1}\ M$,

$$v_0 = \frac{\left(1.605 \times 10^{-4}\ M\,\text{min}^{-1}\right)\left(3.00 \times 10^{-1}\ M\right)}{8.047 \times 10^{-5}\ M + 3.00 \times 10^{-1}\ M} = 1.60 \times 10^{-4}\ M\,\text{min}^{-1}$$

(e) At $[S] = 7.2 \times 10^{-5}\ M$,

$$v_0 = \frac{\left(1.605 \times 10^{-4}\ M\,\text{min}^{-1}\right)\left(7.2 \times 10^{-5}\ M\right)}{8.047 \times 10^{-5}\ M + 7.2 \times 10^{-5}\ M} = 7.6 \times 10^{-5}\ M\,\text{min}^{-1}$$

Although the reaction rate will slow down as substrate is converted into product, the combination of the rapid rate and small concentration of substrate indicates that three minutes will be sufficient time to convert all of the substrate to product.

(f) K_M is independent of [E], but since $V_{max} = k_2[E]_0$, doubling $[E]_0$ will double V_{max}. This in turn leads also to a doubling in v_0 [see equations in parts **(d)** and **(e)**].

13.12 An enzyme has a K_M value of 2.8×10^{-5} M and a V_{max} value of 53 μM min^{-1}. Calculate the value of v_0 if $[S] = 3.7 \times 10^{-4}$ M and $[I] = 4.8 \times 10^{-4}$ M for **(a)** a competitive inhibitor, **(b)** a noncompetitive inhibitor, and **(c)** an uncompetitive inhibitor. ($K_I = 1.7 \times 10^{-5}$ M for all three cases.)

(a) For a competitive inhibitor, from Equation 13.17,

$$v_0 = \frac{V_{max}[S]}{K_M \left(1 + \frac{[I]}{K_I}\right) + [S]}$$

$$= \frac{\left(53 \; \mu M \; \text{min}^{-1}\right)\left(3.7 \times 10^{-4} \; M\right)}{\left(2.8 \times 10^{-5} \; M\right)\left(1 + \frac{4.8 \times 10^{-4} \; M}{1.7 \times 10^{-5} \; M}\right) + 3.7 \times 10^{-4} \; M}$$

$$= 16.5 \; \mu M \; \text{min}^{-1}$$

$$= 16 \; \mu M \; \text{min}^{-1}$$

(b) For a noncompetitive inhibitor, Equation 13.19 gives,

$$v_0 = \frac{\frac{V_{max}}{\left(1 + \frac{[I]}{K_I}\right)}[S]}{K_M + [S]}$$

$$= \frac{\frac{53 \; \mu M \; \text{min}^{-1}}{\left(1 + \frac{4.8 \times 10^{-4} \; M}{1.7 \times 10^{-5} \; M}\right)}\left(3.7 \times 10^{-4} \; M\right)}{2.8 \times 10^{-5} \; M + 3.7 \times 10^{-4} \; M}$$

$$= 1.69 \; \mu M \; \text{min}^{-1}$$

$$= 1.7 \; \mu M \; \text{min}^{-1}$$

(c) For an uncompetitive inhibitor, Equation 13.22 is appropriate,

$$v_0 = \frac{\frac{V_{max}}{\left(1 + \frac{[I]}{K_I}\right)}[S]}{\frac{K_M}{\left(1 + \frac{[I]}{K_I}\right)} + [S]}$$

$$= \frac{\frac{53 \; \mu M \; \text{min}^{-1}}{\left(1 + \frac{4.8 \times 10^{-4} \; M}{1.7 \times 10^{-5} \; M}\right)}\left(3.7 \times 10^{-4} \; M\right)}{\frac{2.8 \times 10^{-5} \; M}{\left(1 + \frac{4.8 \times 10^{-4} \; M}{1.7 \times 10^{-5} \; M}\right)} + 3.7 \times 10^{-4} \; M}$$

$$= 1.81 \; \mu M \; \text{min}^{-1}$$

$$= 1.8 \; \mu M \; \text{min}^{-1}$$

13.13 The degree of inhibition i is given by $i\% = (1 - \alpha)\,100\%$, where $\alpha = (v_0)_{\text{inhibition}}/v_0$. Calculate the percent inhibition for each of the three cases in Problem 13.12.

First v_0 in the absence of inhibitor must be found.

$$v_0 = \frac{V_{\text{max}}[S]}{K_M + [S]}$$

$$= \frac{(53\ \mu M\,\text{min}^{-1})\,(3.7 \times 10^{-4}\ M)}{2.8 \times 10^{-5}\ M + 3.7 \times 10^{-4}\ M}$$

$$= 49.3\ \mu M\,\text{min}^{-1}$$

(a)

$$\alpha = \frac{16.5\ \mu M\,\text{min}^{-1}}{49.3\ \mu M\,\text{min}^{-1}} = 0.335$$

$$\text{percent inhibition} = (1 - 0.335)\,(100\%) = 67\%$$

(b)

$$\alpha = \frac{1.69\ \mu M\,\text{min}^{-1}}{49.3\ \mu M\,\text{min}^{-1}} = 3.43 \times 10^{-2}$$

$$\text{percent inhibition} = (1 - 3.43 \times 10^{-2})\,(100\%) = 96.7\%$$

(c)

$$\alpha = \frac{1.81\ \mu M\,\text{min}^{-1}}{49.3\ \mu M\,\text{min}^{-1}} = 3.67 \times 10^{-2}$$

$$\text{percent inhibition} = (1 - 3.67 \times 10^{-2})\,(100\%) = 96.3\%$$

13.14 An enzyme-catalyzed reaction ($K_M = 2.7 \times 10^{-3}\ M$) is inhibited by a competitive inhibitor I ($K_I = 3.1 \times 10^{-5}\ M$). Suppose that the substrate concentration is $3.6 \times 10^{-4}\ M$. How much of the inhibitor is needed for 65% inhibition? How much does the substrate concentration have to be increased to reduce the inhibition to 25%?

Expressions for the initial rate in the absence and presence of a competitive inhibitor are given by Equations 13.10 and 13.17, respectively. Dividing the former by the latter gives

$$\frac{v_0}{(v_0)_{\text{inhibition}}} = \frac{K_M\left(1 + \frac{[I]}{K_I}\right) + [S]}{K_M + [S]}$$

$$= 1 + \frac{K_M[I]}{(K_M + [S])\,K_I}$$

This can be solved for [I],

$$[I] = K_I \left(\frac{v_0}{(v_0)_{\text{inhibition}}} - 1 \right) \left(1 + \frac{[S]}{K_M} \right)$$

It can also be solved for [S],

$$[S] = K_M \left(\frac{[I]}{K_I \left(\frac{v_0}{(v_0)_{\text{inhibition}}} - 1 \right)} - 1 \right)$$

The expression for [I] is used in answering the first part of the question. For 65% inhibition, $(v_0)_{\text{inhibition}} = (1 - 0.65)v_0 = 0.35v_0$, and

$$[I] = \left(3.1 \times 10^{-5} \ M \right) \left(\frac{1}{0.35} - 1 \right) \left(1 + \frac{3.6 \times 10^{-4} \ M}{2.7 \times 10^{-3} \ M} \right) = 6.52 \times 10^{-5} \ M = 6.5 \times 10^{-5} \ M$$

To reduce the inhibition to 25%, where $(v_0)_{\text{inhibition}} = 0.75v_0$, at this concentration of inhibitor, use the expression for [S] to find the required substrate concentration.

$$[S] = \left(2.7 \times 10^{-3} \ M \right) \left[\frac{6.52 \times 10^{-5} \ M}{\left(3.1 \times 10^{-5} \ M \right) \left(\frac{1}{0.75} - 1 \right)} - 1 \right] = 1.4 \times 10^{-2} \ M$$

13.15   Calculate the concentration of a noncompetitive inhibitor ($K_I = 2.9 \times 10^{-4} \ M$) needed to yield 90% inhibition of an enzyme-catalyzed reaction.

Dividing the expression for the initial rate of an enzyme-catalyzed reaction in the absence of an inhibitor (Equation 13.10) by that for a noncompetitively inhibited reaction [Equation (13.19)] gives

$$\frac{v_0}{(v_0)_{\text{inhibition}}} = 1 + \frac{[I]}{K_I}$$

which can be solved for [I].

$$[I] = \left(\frac{v_0}{(v_0)_{\text{inhibition}}} - 1 \right) K_I$$

$$= \left(\frac{1}{1 - 0.90} - 1 \right) \left(2.9 \times 10^{-4} \ M \right)$$

$$= 2.6 \times 10^{-3} \ M$$

13.16   Derive Equation 13.22.

Mass balance on the enzyme concentration gives $[E]_0 = [E] + [ES] + [ESI]$, or solving for [E] and using $K_I = \frac{[ES][I]}{[ESI]}$,

$$[E] = [E]_0 - [ES] - \frac{[ES][I]}{K_I} = [E]_0 - [ES] \left(1 + \frac{[I]}{K_I} \right)$$

Next, the steady-state approximation is applied to ES, although since the equilibrium $ES + I \rightleftharpoons ESI$ is so rapidly established, it does not affect the rates of appearance and disappearance of ES.

$$\frac{d[ES]}{dt} = 0 = k_1[E][S] - k_{-1}[ES] - k_2[ES]$$

$$= k_1[E]_0[S] - k_1[ES]\left(1 + \frac{[I]}{K_I}\right)[S] - k_{-1}[ES] - k_2[ES]$$

$$= k_1[E]_0[S] - \left[k_1\left(1 + \frac{[I]}{K_I}\right)[S] + k_{-1} + k_2\right][ES]$$

where the result from the mass balance relation has been used. This last equation is solved for [ES], using $K_M = \frac{k_{-1}+k_2}{k_1}$

$$[ES] = \frac{k_1[E]_0[S]}{k_1\left(1 + \frac{[I]}{K_I}\right)[S] + k_{-1} + k_2}$$

$$= \frac{k_1[E]_0[S]}{k_1\left(1 + \frac{[I]}{K_I}\right)[S] + k_1 K_M}$$

$$= \frac{[E]_0[S]}{\left(1 + \frac{[I]}{K_I}\right)[S] + K_M}$$

Finally, the rate of reaction is found using this expression for [ES]

$$v_0 = k_2[ES]$$

$$= \frac{k_2[E]_0[S]}{\left(1 + \frac{[I]}{K_I}\right)[S] + K_M}$$

$$= \frac{V_{max}[S]}{\left(1 + \frac{[I]}{K_I}\right)[S] + K_M}$$

$$= \frac{\frac{V_{max}}{\left(1+\frac{[I]}{K_I}\right)}[S]}{[S] + \frac{K_M}{\left(1+\frac{[I]}{K_I}\right)}}$$

13.17   The metabolism of ethanol in our bodies is catalyzed by liver alcohol dehydrogenase (LADH) to acetaldehyde and finally to acetate. In contrast, methanol is converted to formaldehyde (also catalyzed by LADH), which can cause blindness or even death. An antidote for methanol is ethanol, which acts as a competitive inhibitor for LADH. The excess methanol can then be safely discharged from the body. How much absolute (100%) ethanol would a person have to consume after ingesting 50 mL of methanol (a lethal dosage) to reduce the activity of LADH to 3% of the original value? Assume that the total fluid volume in the person's body is 38 liters and that the densities of ethanol and methanol are 0.789 g mL^{-1} and 0.791 g mL^{-1}, respectively. The K_M value for methanol is 1.0×10^{-2} M and the K_I value for ethanol is 1.0×10^{-3} M. State any assumptions.

As found in the solution for Problem 13.14, the concentration of a competitive inhibitor required to reach a certain level of inhibition is given by

$$[I] = K_I \left(\frac{v_0}{(v_0)_{\text{inhibition}}} - 1 \right) \left(1 + \frac{[S]}{K_M} \right)$$

In this case $(v_0)_{\text{inhibition}} = 0.030 v_0$, and the substrate is methanol with concentration

$$[S] = \frac{(50 \text{ mL}) \left(0.791 \text{ g mL}^{-1}\right) \left(\frac{1 \text{ mol}}{32.04 \text{ g}}\right)}{38 \text{ L}} = 3.25 \times 10^{-2} \ M$$

The concentration of inhibitor (ethanol) required is then

$$[I] = \left(1.0 \times 10^{-3} \ M\right) \left(\frac{1}{0.030} - 1 \right) \left(1 + \frac{3.25 \times 10^{-2} \ M}{1.0 \times 10^{-2} \ M} \right) = 0.137 \ M = 0.14 \ M$$

The volume required, assuming that the alcohols are rapidly and uniformly mixed in the body, is

$$(0.137 \ M) \ (38 \text{ L}) \left(46.07 \text{ g mol}^{-1}\right) \left(\frac{1 \text{ mL}}{0.789 \text{ g}} \right) = 3.0 \times 10^2 \text{ mL} = 0.30 \text{ L}$$

Alternatively, the person would have to consume $2 \times 0.30 \text{ L} = 0.60 \text{ L}$ of 100 proof hard liquor.

13.18   **(a)** What is the physiological significance of cooperative O_2 binding by hemoglobin? Why is O_2 binding by myoglobin not cooperative? **(b)** Compare the concerted model with the sequential model for the binding of oxygen with hemoglobin.

(a) Cooperative O_2 binding enables hemoglobin to be a more efficient oxygen transporter than myoglobin. As discussed in detail in Section 13.6 of the text, nearly twice as much oxygen is delivered to the tissues than would be if O_2 binding to hemoglobin were not cooperative.

(b) The concerted model, with it's "all-or-none" limitation on the relaxed and tense forms of the four subunits in hemoglobin does not allow for the existence of mixed forms with some subunits in one form and the rest in the other. Although not relevant for the binding of oxygen with hemoglobin, the concerted model is unable to account for negative homotropic cooperativity. Nevertheless, it does allow the characterization of the allotropic behavior of hemoglobin (and enzymes) in terms of just three equilibrium constants.

The sequential model does allow for the exisitence of mixed forms of the subunits comprising the oligomer, since the binding of substrate (O_2 in the case of hemoglobin) affects only the conformation of the subunit bound to substrate. The actual mechanism of oxygen binding to hemoglobin is more complex than the limiting cases presented by the two models. The sequential model, however, does have the advantage of being able to account for negative homotropic cooperativity in enzymes displaying such behavior.

13.19   Fatality usually results when more than 50% of a human being's hemoglobin is complexed with carbon monoxide. Yet a person whose hemoglobin content is diminished by anemia to half its original content can often function normally. Explain.

Both O_2 and CO bind cooperatively with hemoglobin. When Hb binds CO at one site, its affinity for O_2 at other sites is increased, making it more difficult for O_2 to be released to the tissues. Thus, the ability of Hb to deliver oxygen is reduced by more than 50%. On the other hand, if half the Hb were absent due to anemia, the affinity of the remaining Hb for oxygen remains unchanged.

13.20 Competitive inhibitors, when present in small amounts, often act as an activators to allosteric enzymes. Why?

Competitive inhibitors bind at the same active site as normal substrate. Due to the competitive nature of the binding, the presence of the inhibitor enhances the affinity of the enzyme for normal substrate.

13.21 What is the advantage of having the heme group in a hydrophobic region in the myoglobin and hemoglobin molecule?

The hydrophobic region protects Fe^{2+} in the heme from oxidation to Fe^{3+}. Without such protection, oxidation via a mechanism involving a bridging O_2 readily occurs. Hemes containing Fe^{3+} do not bind O_2.

13.22 What is the effect of each of the following actions on oxygen affinity of adult hemoglobin (Hb A) *in vitro*? **(a)** Increase pH, **(b)** increase partial pressure of CO_2, **(c)** decrease [BPG], **(d)** dissociate the tetramer into monomers, **(e)** oxidize Fe(II) to Fe(III).

(a) Affinity increases, **(b)** affinity decreases, **(c)** affinity increases, **(d)** affinity increases, **(e)** affinity decreases.

13.23 Although it is possible to carry out X-ray diffraction studies of fully deoxygenated hemoglobin and fully oxygenated hemoglobin, it is much more difficult, if not impossible, to obtain crystals in which each hemoglobin molecule is bound to only one, two, or three oxygen molecules. Explain.

Although it is possible to obtain deoxygenated hemoglobin in completely anaerobic conditions and completely oxygenated hemoglobin via saturation with an excess of O_2, attempts to prepare the intermediate cases will lead to a statistical distribution of the different partially oxygenated species.

13.24 When deoxyhemoglobin crystals are exposed to oxygen, they shatter. On the other hand, deoxymyoglobin crystals are unaffected by oxygen. Explain.

When oxygen binds to deoxyhemoglobin, the protein undergoes a conformational change. Since a crystal does not possess much flexibility, the strain caused by the molecular motion breaks the crystal. Myoglobin does not exhibit cooperativity, hence there is no conformational change upon oxygen binding, and the crystal remains intact.

13.25 An enzyme contains a single dissociable group at its active site. For catalysis to occur, this group must be in the dissociated (that is, negative) form. The substrate bears a net positive charge. The reaction scheme can be represented by

$$EH \rightleftharpoons H^+ + E^-$$

$$E^- + S^+ \rightleftharpoons ES \rightarrow E + P$$

(a) What kind of inhibitor is H^+? **(b)** Write an expression for the initial rate of the reaction in the presence of the inhibitor.

(a) Since the dissociable group is at the active site, H^+ is a competitive inhibitor.

(b) The initial rate of the reaction is given by Equation 13.17,

$$v_0 = \frac{V_{max}[S]}{K_M\left(1 + \frac{[I]}{K_I}\right) + [S]}$$

13.26 The discovery in the 1980s that certain RNA molecules (the ribozymes) can act as enzymes was a surprise to many chemists. Why?

Prior to the discovery of ribozymes, all known enzymes were proteins with an immense array of varied and complex structures. RNA's on the other hand, all have relatively simple structures.

13.27 The activation energy for the decomposition of hydrogen peroxide

$$2H_2O_2(aq) \rightarrow 2H_2O(l) + O_2(g)$$

is 42 kJ mol^{-1}, whereas when the reaction is catalyzed by the enzyme catalase, it is 7.0 kJ mol^{-1}. Calculate the temperature that would cause the nonenzymatic catalysis to proceed as rapidly as the enzyme-catalyzed decomposition at 20°C. Assume the pre-exponential factor to be the same in both cases.

The Arrhenius equation relates reaction rate and activation energy via $k = Ae^{-E_a/RT}$. Requiring that the rates of the catalyzed and uncatalyzed reactions be equal at their respective temperatures then means (assuming A to be the same for both the catalyzed and uncatalyzed reactions),

$$k_{cat} = k_{uncat}$$

$$Ae^{-E_a^{cat}/RT_1} = Ae^{-E_a^{uncat}/RT_2}$$

$$\frac{E_a^{cat}}{T_1} = \frac{E_a^{uncat}}{T_2}$$

$$\frac{7.0 \times 10^3 \text{ J mol}^{-1}}{293 \text{ K}} = \frac{42 \times 10^3 \text{ J mol}^{-1}}{T_2}$$

$$T_2 = 1.8 \times 10^3 \text{ K}$$

At this temperature, the solution would be totally evaporated, making it, in fact, impossible to achieve such a rate in the uncatalyzed case.

13.28   Referring to the concerted model discussed on p. 541, show that $K_1 = cL_0$.

The dissociation of the tense state with one O_2 bound (T_1) to free O_2 and the tense state with no O_2 bound (T_0) is represented by

$$T_1O_2 \rightleftharpoons T_0 + O_2$$

and has dissociation constant

$$K_T = \frac{[T_0][O_2]}{[T_1O_2]}$$

Likewise, for the relaxed form

$$R_1O_2 \rightleftharpoons R_0 + O_2$$

with dissociation constant

$$K_R = \frac{[R_0][O_2]}{[R_1O_2]}$$

Solving each of these equations for the bound form and dividing leads to

$$K_1 = \frac{[T_1O_2]}{[R_1O_2]}$$

$$= \frac{\frac{[T_0][O_2]}{K_T}}{\frac{[R_0][O_2]}{K_R}}$$

$$= \frac{[T_0]}{[R_0]} \frac{K_R}{K_T}$$

$$= cL_0$$

A similar procedure leads to

$$K_2 = c^2 L_0$$

$$K_3 = c^3 L_0$$

$$K_4 = c^4 L_0$$

13.29   The following data were obtained for the variation of V_{max} with pH for a reaction catalyzed by α-amylase at 24°C. What can you conclude about the pK_a values of the ionizing groups at the active site?

pH	V_{max} (arbitary units)
3.0	200
3.5	501
4.0	1584
4.5	1778
5.0	3300
5.5	5248
6.0	5250
6.5	5251
7.0	2818
7.5	2510
8.0	1585
8.5	398
9.0	158

As suggested by Equations 13.34 through 13.36, a graph of log V_{max} vs. pH is prepared using the following data

pH	log V_{max}
3.0	2.301
3.5	2.700
4.0	3.1998
4.5	3.2499
5.0	3.5185
5.5	3.7200
6.0	3.7202
6.5	3.7202
7.0	3.4499
7.5	3.3997
8.0	3.2000
8.5	2.600
9.0	2.200

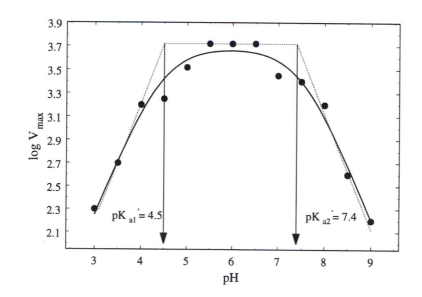

A line with slope = 1 is drawn along the low pH data and one with slope = -1 along the high pH data. At intermediate pH, V_{max} is independent of pH and a horizontal line is drawn here. The lines intersect at pH = 4.5 and pH = 7.4. Thus, pK'_{a1} = 4.5 and pK'_{a2} = 7.4. A more careful fit to the data using Equation 13.33 directly results in the solid curve and the values pK'_{a1} = 4.42 and pK'_{a2} = 7.53, but a closer look at the data shows systematic deviations from the fit, suggesting that there is more going on than the simple scheme leading to Equation 13.33. Since the data is for V_{max}, with the enzyme saturated with substrate, these pK_a's refer to dissociation of the enzyme-substrate complex.

13.30   **(a)** Comment on the following data obtained for an enzyme-catalyzed reaction (no calculations are needed):

$t/°C$	10	15	20	25	30	35	40	45
V_{max} (arbitrary units)	1.0	1.7	2.3	2.6	3.2	4.0	2.6	0.2

(b) Referring to Equation 13.8, under what conditions will an Arrhenius plot (that is, $\ln k$ versus $1/T$) yield a straight line?

(a) From 10°C to 35°C, the reaction rate increases with temperature as expected according to the Arrhenius equation. Above 35°C, the enzyme denatures, losing its catalytic ability, and the reaction rate slows.

(b) Simple Arrhenius behavior, resulting in a straight-line Arrhenius plot, occurs only for reactions whose rate is governed by a single rate constant. Thus, Equation 13.8, which contains the three rate constants, k_1, k_{-1}, and k_2, will not yield a straight line. At high substrate concentration, however, Equation 13.8 becomes $v_0 = k_2[E]_0[S]$, and a plot of $\ln k_{observed}$ vs. $1/T$ will give a straight line with slope related to the activation energy of the second (product forming) step.

13.31   Crocodiles can be submerged in water for a prolonged period of time (up to an hour), while drowning their preys. It is know that BPG does not bind to the crocodile deoxyhemoglobin but the bicarbonate ion does. Explain how this action enables crocodiles to utilize practically all of the oxygen bound to hemoglobin.

As a result of hydrolysis,

$$HCO_3^- + H_2O \rightleftharpoons H_2CO_3 + OH^-$$

the presence of bicarbonate ion increases the pH of a solution. Conversely, removing bicarbonate ion from the blood through binding with deoxyhemoglobin will decrease the pH. This increase in H^+ ion concentration, according to the Bohr Effect, will decrease the affinity of hemoglobin for O_2. Thus, there will be an unusually high concentration of free oxygen in the crocodile blood.

13.32   Give an explanation for the Lineweaver–Burk plot for a certain enzyme-catalyzed reaction show below.

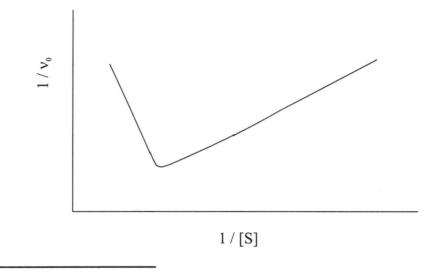

The plot shows that at high substrate concentration (low values of $1/[S]$), the initial rate of the reaction decreases ($1/v_0$ increases). Thus, the substrate must act as an inhibitor to the enzyme.

13.33 The following Arrhenius plot has been obtained for a certain enzyme. Account for the shape of the plot.

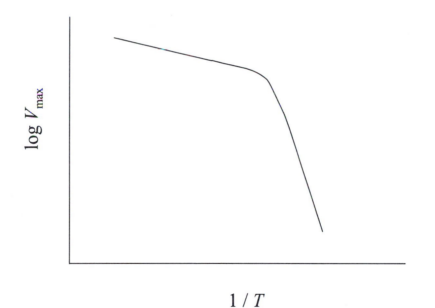

The Arrhenius plot shows two distinct linear regions with different slopes, indicating two different activation energies. At higher temperatures (low values of $1/T$) the slope is less steep than that at lower temperatures, indicating a lower activation energy at higher temperatures and a higher activation energy at lower temperatures. The enzyme probably exists in two interconvertible, active forms.

CHAPTER 14

Quantum Mechanics and Atomic Structure

PROBLEMS AND SOLUTIONS

14.1 Calculate the energy associated with a quantum (photon) of light with a wavelength of 500 nm.

$$E = h\nu = \frac{hc}{\lambda}$$

$$= \frac{\left(6.626 \times 10^{-34} \text{ J s}\right)\left(3.00 \times 10^8 \text{ m s}^{-1}\right)}{500 \times 10^{-9} \text{ m}}$$

$$= 3.98 \times 10^{-19} \text{ J}$$

14.2 The threshold frequency for dislodging an electron from a zinc metal surface is 8.54×10^{14} Hz. Calculate the minimum amount of energy required to remove an electron from the metal.

At the threshold frequency, the speed of electron, v, is 0. That is,

$$h\nu = \Phi + \frac{1}{2}m_e v^2 = \Phi$$

Therefore, the minimum amount of energy required to remove an electron from the metal is

$$\Phi = \left(6.626 \times 10^{-34} \text{ J s}\right)\left(8.54 \times 10^{14} \text{ Hz}\right) = 5.66 \times 10^{-19} \text{ J}$$

14.3 Calculate radii for the Bohr orbits with $n = 2$ and 3 for atomic hydrogen.

The radius of a hydrogen ($Z = 1$) orbit is

$$r_n = \frac{n^2 h^2 \epsilon_0}{\pi m_e e^2}$$

For $n = 2$,

$$r_2 = \frac{4\left(6.626 \times 10^{-34}\,\text{J s}\right)^2 \left(8.8542 \times 10^{-12}\,\text{C}^2\,\text{N}^{-1}\,\text{m}^{-2}\right)}{\pi\left(9.109 \times 10^{-31}\,\text{kg}\right)\left(1.602 \times 10^{-19}\,\text{C}\right)^2} = 2.117 \times 10^{-10}\,\text{m} = 2.117\,\text{Å}$$

For $n = 3$,

$$r_3 = \frac{9\left(6.626 \times 10^{-34}\,\text{J s}\right)^2 \left(8.8542 \times 10^{-12}\,\text{C}^2\,\text{N}^{-1}\,\text{m}^{-2}\right)}{\pi\left(9.109 \times 10^{-31}\,\text{kg}\right)\left(1.602 \times 10^{-19}\,\text{C}\right)^2} = 4.764 \times 10^{-10}\,\text{m} = 4.764\,\text{Å}$$

14.4 Calculate the frequency and wavelength associated with the transition from the $n = 5$ to the $n = 3$ level in atomic hydrogen.

The wavenumber of the emitted radiation is

$$\tilde{\nu} = \left(109737\,\text{cm}^{-1}\right)\left|\left(\frac{1}{n_i^2} - \frac{1}{n_f^2}\right)\right| = \left(109737\,\text{cm}^{-1}\right)\left|\left(\frac{1}{5^2} - \frac{1}{3^2}\right)\right| = 7803.520\,\text{cm}^{-1}$$

Therefore, the wavelength and the frequency of this radiation are

$$\lambda = \frac{1}{\tilde{\nu}} = \frac{1}{7803.520\,\text{cm}^{-1}} = 1.28147 \times 10^{-4}\,\text{cm} = 1.28147 \times 10^3\,\text{nm}$$

$$\nu = c\tilde{\nu} = \left(3.00 \times 10^{10}\,\text{cm s}^{-1}\right)\left(7803.520\,\text{cm}^{-1}\right) = 2.34 \times 10^{14}\,\text{Hz}$$

14.5 What are the wavelengths associated with **(a)** an electron moving at $1.50 \times 10^8\,\text{cm s}^{-1}$, and **(b)** a 60-g tennis ball moving at $1500\,\text{cm s}^{-1}$?

The wavelengths can be calculated using

$$\lambda = \frac{h}{m\upsilon}$$

(a)

$$\lambda = \frac{6.626 \times 10^{-34}\,\text{J s}}{\left(9.109 \times 10^{-31}\,\text{kg}\right)\left(1.50 \times 10^6\,\text{m s}^{-1}\right)} = 4.85 \times 10^{-10}\,\text{m} = 4.85\,\text{Å}$$

(b)

$$\lambda = \frac{6.626 \times 10^{-34}\,\text{J s}}{\left(60 \times 10^{-3}\,\text{kg}\right)\left(1500 \times 10^{-2}\,\text{m}\right)} = 7.36 \times 10^{-34}\,\text{m}$$

The wavelength of the tennis ball is too small to be of practical significance, whereas the wavelength of the electron is too large to be ignored.

14.6   A photoelectric experiment was performed by separately shining a laser at 450 nm (blue light) and a laser at 560 nm (yellow light) on a clean metal surface and measuring the number and kinetic energy of the ejected electrons. Which light would generate more electrons? Which light would eject electrons with greater kinetic energy? Assume that the same number of photons is delivered to the metal surface by each laser and that the frequencies of the laser lights exceed the threshold frequency.

Since each laser is delivering the same number of photons to the metal surface, the two experiments will each generate the same number of electrons. Since each blue light photon has greater energy than each yellow light photon, those electrons ejected by the blue light will have greater kinetic energy.

14.7   Explain how scientists are able to estimate the temperature on the surface of the sun. (*Hint:* Treat solar radiation like radiation from a blackbody.)

By measuring the emission spectrum of the sun and matching it to the measured, laboratory emission spectra of blackbodies of known temperatures, the surface temperature of the sun is estimated to be 6000 K.

14.8   In a photoelectric experiment, a student, uses a light source whose frequency is greater than that needed to eject electrons from a certain metal. After continuously shining the light on the same area of the metal for a long period of time, however, the student notices that the maximum kinetic energy of ejected electrons begins to decrease, even though the frequency of the light is held constant. How would you account for this behavior?

In an photoelectric experiment where the ejected electrons are not replaced (perhaps by making the photocathode part of an electric circuit), the metal surface will become positively charged due to the loss of the negatively charged electrons. Eventually, this positive charge is sufficient to cause a noticeable attraction between the surface and the ejected electrons which lowers their kinetic energy.

14.9   A proton is accelerated through a potential difference of 3.0×10^6 V, starting from rest. Calculate its final wavelength.

The wavelength of the proton can be calculated with the mass of an electron, m_e replaced by the mass of a proton, m_p in Equation 14.21:

$$\lambda = \frac{h}{\sqrt{2m_p eV}} = \frac{6.626 \times 10^{-34}\,\text{J s}}{\sqrt{2\left(1.673 \times 10^{-27}\,\text{kg}\right)\left(1.602 \times 10^{-19}\,\text{C}\right)\left(3.0 \times 10^6\,\text{V}\right)}} = 1.7 \times 10^{-14}\,\text{m}$$

14.10   Suppose that the uncertainty in determining the position of an electron circling an atom in an orbit is 0.4 Å. What is the uncertainty in its velocity?

The uncertainty in momentum is

$$\Delta p \geq \frac{h}{4\pi \Delta x} = \frac{6.626 \times 10^{-34} \text{ J s}}{4\pi \left(0.4 \times 10^{-10} \text{ m}\right)} = 1.3 \times 10^{-24} \text{ kg m s}^{-1}$$

Therefore, the uncertainty in the velocity of the electron is

$$\Delta v = \frac{\Delta p}{m} \geq \frac{1.3 \times 10^{-24} \text{ kg m s}^{-1}}{9.109 \times 10^{-31} \text{ kg}} = 1 \times 10^{6} \text{ m s}^{-1}$$

The uncertainty principle, when applied to a microscopic object such as an electron, results in a significant uncertainty in velocity.

14.11   A person weighing 77 kg jogs at 1.5 m s^{-1}. **(a)** Calculate the momentum and wavelength of this person. **(b)** What is the uncertainty in determining his position at any given instant if we can measure his momentum to ±0.05%? **(c)** Predict the changes that would take place in this problem if the Planck constant were 1 J s.

(a) The momentum of the person is

$$p = mv = (77 \text{ kg}) \left(1.5 \text{ m s}^{-1}\right) = 116 \text{ kg m s}^{-1} = 1.2 \times 10^{2} \text{ kg m s}^{-1}$$

The wavelength of the person is

$$\lambda = \frac{h}{p} = \frac{6.626 \times 10^{-34} \text{ J s}}{116 \text{ kg m s}^{-1}} = 5.7 \times 10^{-36} \text{ m}$$

(b) The uncertainty in momentum is

$$\Delta p = (0.05\%) \left(116 \text{ kg m s}^{-1}\right) = 0.0580 \text{ kg m s}^{-1}$$

The uncertainty in position is

$$\Delta x \geq \frac{h}{4\pi \Delta p} = \frac{6.626 \times 10^{-34} \text{ J s}}{4\pi \left(0.0580 \text{ kg m s}^{-1}\right)} = 9.1 \times 10^{-34} \text{ m}$$

This uncertainty in position is not physically significant. Thus, the uncertainty principle, when applied to a macroscopic object, produces a negligible effect.

(c) If the Planck constant were 1 J s, the wavelength of the person would be

$$\lambda = \frac{h}{p} = \frac{1 \text{ J s}}{116 \text{ kg m s}^{-1}} = 8.6 \times 10^{-3} \text{ m}$$

and the uncertainty in the position of the person would be

$$\Delta x \geq \frac{h}{4\pi \Delta p} = \frac{1 \text{ J s}}{4\pi \left(0.0580 \text{ kg m s}^{-1}\right)} = 1.4 \text{ m}$$

If this were the value of Planck's constant, there would be a noticeable effect on macroscopic objects due to the uncertainty principle. It is the extremely small value of Planck's constant that makes quantum effects significant only for microscopic objects.

14.12 The diffraction phenomenon can be observed whenever the wavelength is comparable in magnitude to the size of the slit opening. To be "diffracted," how fast must a person weighing 84 kg move through a door 1 m wide?

The person would need a wavelength comparable to 1 m to be diffracted. The momentum of the person would be

$$p = \frac{h}{\lambda} = \frac{6.626 \times 10^{-34}\,\text{J s}}{1\,\text{m}} = 6.626 \times 10^{-34}\,\text{kg m s}^{-1}$$

The velocity of the person is therefore

$$v = \frac{p}{m} = \frac{6.626 \times 10^{-34}\,\text{kg m s}^{-1}}{84\,\text{kg}} = 7.9 \times 10^{-36}\,\text{m s}^{-1}$$

At this rate, it would take 1.3×10^{35} s or 4.1×10^{27} years to move 1 m!

14.13 **(a)** Show that for the hydrogen atom, the first term on the right side of Equation 14.16 is 2.18×10^{-18} J. **(b)** Use Equation 14.18 to calculate the Rydberg constant in cm^{-1}. Use the constants on the inside of the back cover. (*Hint:* To obtain the full six significant figures used in the text for R_H, you would need at least seven significant figures in all your constants. In fact, R_H is known to eleven significant figures. See table of fundamental constants in inside back cover.)

(a)

$$\frac{m_e Z^2 e^4}{8 h^2 \epsilon_0^2} = \frac{(9.1093897 \times 10^{-31}\,\text{kg})\,(1)^2\,(1.60217733 \times 10^{-19}\,\text{C})^4}{8\,(6.6260755 \times 10^{-34}\,\text{J s})^2\,(8.854187817 \times 10^{-12}\,\text{C}^2\,\text{N}^{-1}\,\text{m}^{-2})^2}$$

$$= 2.17987406 \times 10^{-18}\,\text{J} = 2.18 \times 10^{-18}\,\text{J}$$

(b)

$$R_H = \frac{m_e Z^2 e^4}{8 c h^3 \epsilon_0^2} = \left(\frac{m_e Z^2 e^4}{8 h^2 \epsilon_0^2}\right)\left(\frac{1}{ch}\right)$$

$$= (2.17987406 \times 10^{-18}\,\text{J})\left[\frac{1}{(2.99792458 \times 10^8\,\text{m s}^{-1})\,(6.6260755 \times 10^{-34}\,\text{J s})}\right]$$

$$= 10973731\,\text{m}^{-1} = 109737.31\,\text{cm}^{-1}$$

14.14 Spectral lines of the Lyman and Balmer series do not overlap. Verify this statement by calculating the longest wavelength associated with the Lyman series and the shortest wavelength associated with the Balmer series (in nm).

The longest wavelength associated with the Lyman series corresponds to $n_i = 2$ and $n_f = 1$. First calculate the wavenumber of the emitted radiation.

$$\tilde{\nu} = (109737 \text{ cm}^{-1}) \left| \left(\frac{1}{n_i^2} - \frac{1}{n_f^2} \right) \right| = (109737 \text{ cm}^{-1}) \left| \left(\frac{1}{2^2} - \frac{1}{1^2} \right) \right| = 82302.75 \text{ cm}^{-1}$$

Therefore, the wavelength of the radiation is

$$\lambda = \frac{1}{\tilde{\nu}} = \frac{1}{82302.75 \text{ cm}^{-1}} = 1.21503 \times 10^{-5} \text{ cm} = 121.503 \text{ nm}$$

The shortest wavelength associated with the Balmer series corresponds to $n_i = \infty$ and $n_f = 2$. First calculate the wavenumber of the emitted radiation.

$$\tilde{\nu} = (109737 \text{ cm}^{-1}) \left| \left(\frac{1}{n_i^2} - \frac{1}{n_f^2} \right) \right| = (109737 \text{ cm}^{-1}) \left| \left(\frac{1}{\infty^2} - \frac{1}{2^2} \right) \right| = 27434.25 \text{ cm}^{-1}$$

Therefore, the wavelength of the radiation is

$$\lambda = \frac{1}{\tilde{\nu}} = \frac{1}{27434.25 \text{ cm}^{-1}} = 3.64508 \times 10^{-5} \text{ cm} = 364.508 \text{ nm}$$

Therefore, spectral lines of the two series do not overlap.

14.15 The He^+ ion contains only one electron and is therefore a hydrogenlike ion. Calculate the wavelengths, in increasing order, of the first four transitions in the Balmer series of the He^+ ion. Compare these wavelengths with the same transitions in a H atom. Comment on the differences. (The Rydberg constant for He^+ is 8.72×10^{-18} J.)

Convert the Rydberg constant for He^+ into cm^{-1}:

$$R_H = \frac{8.72 \times 10^{-18} \text{ J}}{hc} = \frac{8.72 \times 10^{-18} \text{ J}}{(6.626 \times 10^{-34} \text{ J s}) (3.00 \times 10^8 \text{ m s}^{-1})}$$

$$= 4.387 \times 10^7 \text{ m}^{-1} = 4.387 \times 10^5 \text{ cm}^{-1}$$

The wavenumbers and wavelengths for first four transitions ($n_i = 3, 4, 5, 6$) in the Balmer series can be calculated using

$$\tilde{\nu} = R_H \left| \left(\frac{1}{n_i^2} - \frac{1}{2^2} \right) \right|$$

$$\lambda = \frac{1}{\tilde{\nu}}$$

where $R_H = 109737 \text{ cm}^{-1}$ for H and $4.387 \times 10^5 \text{ cm}^{-1}$ for He^+. The wavelengths are

n_i	λ in nm for He^+	λ in nm for H
3	164	656
4	122	486
5	109	434
6	103	410

All the Balmer transitions for He$^+$ are in the ultraviolet region whereas the transitions for H are all in the visible region. The wavelength for a transition in He$^+$ is 1/4 that of the corresponding transition in H due to the factor of Z^2 in the R_H expression.

14.16   An electron in an excited state in a hydrogen atom can return to the ground state in two different ways: first, via a direct transition in which a photon of wavelength λ_1 is emitted and second, via an intermediate excited state reached by the emission of a photon of wavelength λ_2. This intermediate excited state then decays to the ground state by emitting another photon of wavelength λ_3. Derive an equation that relates λ_1 to λ_2 and λ_3.

The energy of the photon of wavelength λ_1, E_1, equals the sum of the energy of the photon of wavelength λ_2, E_2, and the energy of the photon of wavelength λ_3, E_3. That is,

$$E_1 = E_2 + E_3$$

$$\frac{hc}{\lambda_1} = \frac{hc}{\lambda_2} + \frac{hc}{\lambda_3}$$

$$\frac{1}{\lambda_1} = \frac{1}{\lambda_2} + \frac{1}{\lambda_3}$$

14.17   The retina of a human eye can detect light when radiant energy incident on it is at least 4.0×10^{-17} J. For light of 600-nm wavelength, how many photons does this correspond to?

The energy of a single 600 nm photon is

$$E = h\nu$$

$$= \frac{hc}{\lambda}$$

$$= \frac{\left(6.626 \times 10^{-34} \text{ J s}\right)\left(3.00 \times 10^8 \text{ m s}^{-1}\right)}{600 \times 10^{-9} \text{ m}}$$

$$= 3.313 \times 10^{-19} \text{ J}$$

The number of these photons required to provide 4.0×10^{-17} J so that the light can be detected by human eyes is

$$\frac{4.0 \times 10^{-17} \text{ J}}{3.313 \times 10^{-19} \text{ J}} = 1.2 \times 10^2$$

14.18   A 368-g sample of water absorbs infrared radiation at 1.06×10^4 nm from a carbon dioxide laser. Suppose all the absorbed radiation is converted to heat. Calculate the number of photons at this wavelength required to raise the temperature of the water by 5.00°C.

The energy required to raise the temperature of water by 5.00°C is

$$ms\,\Delta T = (368\text{ g})\left(4.184\text{ J g}^{-1}\,{}^\circ\text{C}^{-1}\right)(5.00^\circ\text{C}) = 7.699 \times 10^3\text{ J}$$

The energy of one photon is

$$h\nu = \frac{hc}{\lambda} = \frac{\left(6.626 \times 10^{-34}\text{ J s}\right)\left(3.00 \times 10^8\text{ m s}^{-1}\right)}{1.06 \times 10^{-5}\text{ m}} = 1.875 \times 10^{-20}\text{ J}$$

The number of photons required to raise the temperature of water by 5.00°C is

$$\frac{7.699 \times 10^3\text{ J}}{1.875 \times 10^{-20}\text{ J}} = 4.11 \times 10^{23}$$

14.19   Ozone (O_3) in the stratosphere absorbs the harmful radiation from the sun by undergoing decomposition: $O_3 \rightarrow O + O_2$. **(a)** Referring to Appendix B, calculate the $\Delta_r H^\circ$ for this process. **(b)** Calculate the maximum wavelength of photons (in nm) that possess this energy to bring about the decomposition of ozone photochemically.

(a)

$$\Delta_r H^\circ = \Delta_f \overline{H}^\circ\,[\text{O}(g)] + \Delta_f \overline{H}^\circ\,[\text{O}_2(g)] - \Delta_f \overline{H}^\circ\,[\text{O}_3(g)]$$

$$= 249.4\text{ kJ mol}^{-1} + 0\text{ kJ mol}^{-1} - 142.7\text{ kJ mol}^{-1}$$

$$= 106.7\text{ kJ mol}^{-1}$$

(b) Assuming that the decomposition requires a single photon and proceeds with 100% efficiency, the decomposition of 1 mole of O_3 requires 1 mole of photons. The energy of one photon is

$$E = h\nu = \frac{hc}{\lambda}$$

so that the energy of 1 mole of photons is

$$E = \frac{hc}{\lambda}\left(6.022 \times 10^{23}\right)$$

The wavelength required so that the photons possess the necessary energy is

$$\lambda = \frac{hc}{E}\left(6.022 \times 10^{23}\right) = \frac{\left(6.626 \times 10^{-34}\text{ J s}\right)\left(3.00 \times 10^8\text{ m s}^{-1}\right)}{106.7 \times 10^3\text{ J}}\left(6.022 \times 10^{23}\right)$$

$$= 1.12 \times 10^{-6}\text{ m} = 1.12 \times 10^3\text{ nm}$$

14.20   Scientists have found interstellar hydrogen atoms with quantum number n in the hundreds. Calculate the wavelength of light emitted when a hydrogen atom undergoes a transition from $n = 236$ to $n = 235$. In what region of the electromagnetic spectrum does this wavelength fall?

The wavenumber of the emitted radiation is

$$\tilde{\nu} = \left(109737 \text{ cm}^{-1}\right)\left|\left(\frac{1}{n_i^2} - \frac{1}{n_f^2}\right)\right| = \left(109737 \text{ cm}^{-1}\right)\left|\left(\frac{1}{236^2} - \frac{1}{235^2}\right)\right| = 1.680406 \times 10^{-2} \text{ cm}^{-1}$$

Therefore, the wavelength of this radiation is

$$\lambda = \frac{1}{\tilde{\nu}} = \frac{1}{1.680406 \times 10^{-2} \text{ cm}^{-1}} = 59.5094 \text{ cm}$$

This wavelength is in the radio frequency region.

14.21   A student records an emission spectrum of hydrogen and notices that one spectral line in the Balmer series cannot be accounted for by the Bohr theory. Assuming that the gas sample is pure, suggest a species that might be responsible for this line.

The line is most likely due to the emission of H_2.

14.22   In the mid-19th century, physicists studying the solar emission spectrum (a continuum) noticed a set of dark lines that did not match any of the emission lines (bright lines) on Earth. They concluded that the lines came from a yet unknown element. Later this element was identified as helium. **(a)** What is the origin of the dark lines? How were these lines correlated with the emission lines of helium? **(b)** Why was helium so difficult to detect in Earth's atmosphere? **(c)** Where is the most likely place to detect helium on Earth?

(a) The lines are absorption lines. For atomic systems, the lines occur at the same wavelengths as the emission lines (of helium in this case).

(b) Because of the light mass of helium, its thermal velocity is actually greater than the escape velocity of Earth's gravitational field. (The same is true for H_2, but unlike hydrogen, helium does not form compounds, whose greater mass traps the element on Earth.)

(c) α decay produces helium, so uranium mines are a likely source of helium. Helium is also found in natural gas deposits, where it has been trapped by layers of rock.

14.23   How many photons at 660 nm must be absorbed to melt 5.0×10^2 g of ice? On average, how many H_2O molecules does one photon convert from ice to water? (*Hint*: It takes 334 J to melt 1 g of ice at 0°C.)

The amount of energy that must be absorbed to melt 5.0×10^2 g of ice is

$$\left(5.0 \times 10^2 \text{ g}\right)\left(334 \text{ J g}^{-1}\right) = 1.67 \times 10^5 \text{ J}$$

The energy of 1 photon at 660 nm is

$$h\nu = \frac{hc}{\lambda} = \frac{\left(6.626 \times 10^{-34} \text{ J s}\right)\left(3.00 \times 10^8 \text{ m s}^{-1}\right)}{660 \times 10^{-9} \text{ m}} = 3.012 \times 10^{-19} \text{ J}$$

so that the number of photons required to melt the water is

$$\frac{1.67 \times 10^5 \text{ J}}{3.012 \times 10^{-19} \text{ J}} = 5.54 \times 10^{23} = 5.5 \times 10^{23}$$

The number of H_2O molecules converted in the 5.0×10^2 g sample from ice to water is

$$\frac{5.0 \times 10^2 \text{ g}}{18.0 \text{ g mol}^{-1}} \frac{6.022 \times 10^{23}}{1 \text{ mol}} = 1.67 \times 10^{25}$$

Since it took 5.54×10^{23} photons to melt the entire sample, the number of H_2O molecules converted from ice to water by 1 photon is

$$\frac{1.67 \times 10^{25}}{5.54 \times 10^{23}} = 30$$

14.24   Show that Equation 14.32 is dimensionally correct.

The units for the term $\dfrac{n^2 h^2}{8mL^2}$ are

$$\frac{\text{J}^2 \text{ s}^2}{\text{kg m}^2} = \frac{\text{J}^2}{\text{kg m}^2 \text{ s}^{-2}} = \frac{\text{J}^2}{\text{J}} = \text{J}$$

which are indeed energy units.

14.25   According to Equation 14.32, the energy is inversely proportional to the square of the length of the box. How would you account for this dependence in terms of the Heisenberg uncertainty principle?

As the length of the box is decreased, the particle is located with less uncertainty. According to the uncertainty principle,

$$\Delta x \, \Delta p \geq \frac{h}{4\pi}$$

a decrease in Δx, here due to a shortening of the box, requires an increase in Δp and hence p itself. Consequently, the kinetic energy of the particle, $p^2/2m$ must also increase.

14.26   What is the probability of locating a particle in a one-dimensional box between $L/4$ and $3L/4$, where L is the length of the box? Assume the particle to be in the lowest level.

The probability is

$$P = \int_{L/4}^{3L/4} \psi^2 \, dx$$

$$= \int \left(\frac{2}{L}\right) \sin^2 \frac{\pi}{L} x \, dx$$

$$= \frac{2}{L} \left(\frac{x}{2} - \frac{\sin \frac{2\pi x}{L}}{\frac{4\pi}{L}}\right)\Bigg|_{L/4}^{3L/4}$$

$$= \frac{2}{L} \left[\left(\frac{3L}{8} - \frac{\sin \frac{3\pi}{2}}{\frac{4\pi}{L}}\right) - \left(\frac{L}{8} - \frac{\sin \frac{\pi}{2}}{\frac{4\pi}{L}}\right)\right]$$

$$= \frac{2}{L} \left[\left(\frac{3L}{8} + \frac{L}{4\pi}\right) - \left(\frac{L}{8} - \frac{L}{4\pi}\right)\right]$$

$$= \frac{2}{L} \left(\frac{L}{4} + \frac{L}{2\pi}\right)$$

$$= \frac{1}{2} + \frac{1}{\pi}$$

$$= 0.82$$

14.27 Derive Equation 14.32 using de Broglie's relation. (*Hint*: First you must express the wavelength of the particle in the *n*th level in terms of the length of the box.)

According to Figure 14.18(a), the wavelength of the particle is given by

$$\lambda = \frac{2L}{n}$$

where $n = 1, 2, 3...$ The wavelength is also related to the momentum of the particle:

$$\lambda = \frac{h}{p} = \frac{h}{mv}$$

Equating the two expressions for λ and solving for v gives

$$\frac{2L}{n} = \frac{h}{mv}$$

$$v = \frac{nh}{2mL}$$

The particle has only kinetic energy so that

$$E = \frac{1}{2}mv^2 = \frac{1}{2}m\frac{n^2h^2}{4m^2L^2} = \frac{n^2h^2}{8mL^2}$$

14.28   An important property of the wave functions of the particle in a one-dimensional box is that they are orthogonal, that is,

$$\int_0^L \psi_n \psi_m \, dx = 0 \qquad m \neq n$$

Prove this statement using ψ_1 and ψ_2 and Equation 14.33.

Let $m = 1$ and $n = 2$.

$$\int_0^L \psi_2 \psi_1 \, dx = \frac{2}{L} \int_0^L \sin \frac{2\pi x}{L} \sin \frac{\pi x}{L} \, dx$$

A table of integrals gives, for $a^2 \neq b^2$,

$$\int \sin ax \, \sin bx \, dx = \frac{\sin (a - b) x}{2 (a - b)} - \frac{\sin (a + b) x}{2 (a + b)}$$

Setting $a = 2\pi/L$ and $b = \pi/L$,

$$\int_0^L \psi_2 \psi_1 \, dx = \frac{2}{L} \left[\frac{\sin \frac{\pi x}{L}}{\frac{2\pi}{L}} - \frac{\sin \frac{3\pi x}{L}}{\frac{6\pi}{L}} \right]_0^L$$

$$= \frac{2}{L} \left[\left(\frac{\sin \pi}{\frac{2\pi}{L}} - \frac{\sin 3\pi}{\frac{6\pi}{L}} \right) - \left(\frac{\sin 0}{\frac{2\pi}{L}} - \frac{\sin 0}{\frac{6\pi}{L}} \right) \right]$$

$$= \frac{2}{L} \left[(0 - 0) - (0 - 0) \right]$$

$$= 0$$

14.29   Use Equation 14.37 to calculate the wavelength of the electronic transition in polyenes for $N = 6, 8,$ and 10. Comment on the variation of λ with L, the length of the molecule.

Equation 14.37 gives the wavelength for the electronic transition in polyenes as

$$\lambda = \frac{8 m_e L^2 c}{h (N + 1)}$$

For $N = 6$, L is a sum of 3 double bonds, 2 single bonds and twice the radius of a C atom, that is,

$$L = 3 \, (1.35 \text{ Å}) + 2 \, (1.54 \text{ Å}) + 2 \, (0.77 \text{ Å}) = 8.67 \text{ Å}$$

The wavelength of the electronic transition for $N = 6$ is therefore

$$\lambda = \frac{8 \, (9.109 \times 10^{-31} \text{ kg}) \, (8.67 \times 10^{-10} \text{ m})^2 \, (3.00 \times 10^8 \text{ m s}^{-1})}{(6.626 \times 10^{-34} \text{ J s}) \, (7)} = 3.54 \times 10^{-7} \text{ m} = 354 \text{ nm}$$

For $N = 8$, L is a sum of 4 double bonds, 3 single bonds and twice the radius of a C atom, that is,

$$L = 4 \, (1.35 \text{ Å}) + 3 \, (1.54 \text{ Å}) + 2 \, (0.77 \text{ Å}) = 11.56 \text{ Å}$$

The wavelength of the electronic transition for $N = 8$ is therefore

$$\lambda = \frac{8 \left(9.109 \times 10^{-31} \text{ kg}\right) \left(11.56 \times 10^{-10} \text{ m}\right)^2 \left(3.00 \times 10^8 \text{ m s}^{-1}\right)}{\left(6.626 \times 10^{-34} \text{ J s}\right) (9)} = 4.90 \times 10^{-7} \text{ m} = 490 \text{ nm}$$

For $N = 10$, L is a sum of 5 double bonds, 4 single bonds and twice the radius of a C atom, that is,

$$L = 5 \left(1.35 \text{ Å}\right) + 4 \left(1.54 \text{ Å}\right) + 2 \left(0.77 \text{ Å}\right) = 14.45 \text{ Å}$$

The wavelength of the electronic transition for $N = 10$ is therefore

$$\lambda = \frac{8 \left(9.109 \times 10^{-31} \text{ kg}\right) \left(14.45 \times 10^{-10} \text{ m}\right)^2 \left(3.00 \times 10^8 \text{ m s}^{-1}\right)}{\left(6.626 \times 10^{-34} \text{ J s}\right) (11)} = 6.26 \times 10^{-7} \text{ m} = 626 \text{ nm}$$

As N increases and the molecule gets longer, λ increases and shifts the light from UV to visible.

14.30 Based on the particle-in-a-one-dimensional-box approximation for polyenes, suggest where along the box the $n = 1 \rightarrow n = 2$ electronic transition would most likely take place. Explain your choice.

For a transition to take place, both initial and final states must have non-zero probabilities of finding the electron at a given location. Consequently, the transition must take place at points where both ψ_1^2 and ψ_2^2 are non-zero. A likely place is shown in the figure

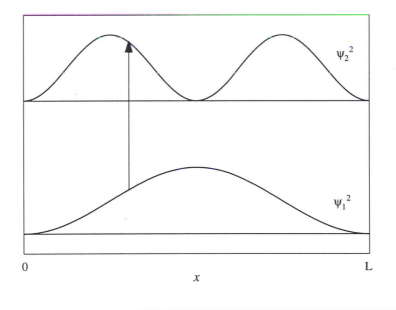

14.31 As stated in the chapter, the probability of locating a particle in a one-dimensional box is given by $\psi^2 \, dx$. Over a small distance, the probability can be calculated without integration. Consider an electron with $n = 1$ in a box of length 2.000 nm. Calculate the probability of locating the electron **(a)** between 0.500 nm and 0.502 nm and **(b)** between 0.999 nm and 1.001 nm. Comment on your results and on the validity of your approximation.

For $n = 1$,

$$\psi^2 \, dx = \frac{2}{L} \sin^2 \frac{\pi x}{L} \, dx$$

(a) x can be approximated by the average of 0.500 nm and 0.502 nm. That is, $x = 0.501$ nm. $dx = 0.502$ nm $-$ 0.500 nm $= 0.002$ nm. Therefore, the probability of locating the electron is

$$\frac{2}{2.000 \text{ nm}} \left[\sin^2 \frac{\pi \, (0.501 \text{ nm})}{2.000 \text{ nm}} \right] (0.002 \text{ nm}) = (0.002) \left[\sin^2 (0.2505\pi) \left(\frac{180°}{\pi} \right) \right]$$

$$= 1 \times 10^{-3}$$

(b) x can be approximated by the average of 0.999 nm and 1.001 nm. That is, $x = 1.000$ nm. $dx = 1.001$ nm $-$ 0.999 nm $= 0.002$ nm. Therefore, the probability of locating the electron is

$$\frac{2}{2.000 \text{ nm}} \left[\sin^2 \frac{\pi \, (1.000 \text{ nm})}{2.000 \text{ nm}} \right] (0.002 \text{ nm}) = (0.002) \left[\sin^2 (0.5000\pi) \left(\frac{180°}{\pi} \right) \right]$$

$$= 0.002$$

The result from part (b) is, in fact, for the most probable location to find the particle, the middle of the box. The approximation is valid for small dx, becoming better as dx gets smaller, and is equivalent to assuming that the value of the wave function is approximately constant over the interval.

14.32 Obtain an expression for the most probable radius at which an electron will be found when it occupies the $1s$ orbital.

The $1s$ radial distribution function is

$$f = 4\pi r^2 R(r)^2 = 4\pi r^2 \left(\frac{2}{\sqrt{a_0^3}} e^{-r/a_0} \right)^2 = \frac{16\pi}{a_0^3} r^2 e^{-2r/a_0}$$

The radial distribution function reaches a maximum with $df/dr = 0$ at the most probable radius, r_{mp}. Differentiating f with respect to r,

$$\frac{df}{dr} = \frac{16\pi}{a_0^3} \left[2r e^{-2r/a_0} + r^2 \left(-\frac{2}{a_0} \right) e^{-2r/a_0} \right]$$

$r = r_{mp}$ when $\dfrac{df}{dr} = 0$.

$$\frac{16\pi}{a_0^3} \left[2r_{mp} e^{-2r_{mp}/a_0} + r_{mp}^2 \left(-\frac{2}{a_0} \right) e^{-2r_{mp}/a_0} \right] = 0$$

$$2 - \frac{2r_{mp}}{a_0} = 0$$

$$r_{mp} = a_0$$

The most probable radius for the $1s$ orbital is the Bohr radius.

14.33 Use the $2s$ wavefunction given in Table 14.2 to calculate the value of r (other than $r = \infty$) at which this wavefunction becomes zero.

Only the radial portion of the $2s$ wave function depends on r, and ψ becomes zero when $R(r) = 0$.

$$\frac{1}{\sqrt{2a_0^3}}\left(1 - \frac{r}{2a_0}\right)e^{-r/2a_0} = 0$$

This requires (for $r \neq \infty$)

$$1 - \frac{r}{2a_0} = 0$$

$$r = 2a_0 = 2\,(0.529\ \text{Å}) = 1.058\ \text{Å}$$

14.34 Write the ground-state electron configurations of the following ions, which play important roles in biochemical processes in our bodies: **(a)** Na^+, **(b)** Mg^{2+}, **(c)** Cl^-, **(d)** K^+, **(e)** Ca^{2+}, **(f)** Fe^{2+}, **(g)** Cu^{2+}, **(h)** Zn^{2+}.

(a) Na^+: $1s^2 2s^2 2p^6 = [Ne]$ **(b)** Mg^{2+}: $1s^2 2s^2 2p^6 = [Ne]$ **(c)** Cl^-: $1s^2 2s^2 2p^6 3s^2 3p^6 = [Ar]$ **(d)** K^+: $1s^2 2s^2 2p^6 3s^2 3p^6 = [Ar]$ **(e)** Ca^{2+}: $1s^2 2s^2 2p^6 3s^2 3p^6 = [Ar]$ **(f)** Fe^{2+}: $1s^2 2s^2 2p^6 3s^2 3p^6 3d^6 = [Ar]3d^6$ **(g)** Cu^{2+}: $1s^2 2s^2 2p^6 3s^2 3p^6 3d^9 = [Ar]3d^9$ **(h)** Zn^{2+}: $1s^2 2s^2 2p^6 3s^2 3p^6 3d^{10} = [Ar]3d^{10}$

14.35 Explain, in terms of their electron configurations, why Fe^{2+} is more easily oxidized to Fe^{3+} than Mn^{2+} to Mn^{3+}.

The electron configurations of the species being considered are

Fe^{2+}: $[Ar]3d^6$

Fe^{3+}: $[Ar]3d^5$

Mn^{2+}: $[Ar]3d^5$

Mn^{3+}: $[Ar]3d^4$

A half-filled subshell has extra stability. On oxidizing Fe^{2+}, the product has a half-filled d subshell. In oxidizing Mn^{2+}, a half-filled d subshell is being lost, which requires more energy.

14.36 Ionization energy is the energy required to remove a ground state ($n = 1$) electron from an atom. It is usually expressed in units of $kJ\ mol^{-1}$. **(a)** Calculate the ionization energy for the hydrogen atom. **(b)** Repeat the calculation, assuming in this case that the electron is removed from the $n = 2$ state.

The ionization energy for the hydrogen atom ($Z = 1$) can be calculated using

$$\Delta E = \left(\frac{m_e Z^2 e^4}{8h^2 \epsilon_0^2}\right)\left(\frac{1}{n_i^2} - \frac{1}{n_f^2}\right)$$

$$= (2.1799 \times 10^{-18} \text{ J})\left(\frac{1}{n_i^2} - \frac{1}{n_f^2}\right) \text{ (See Problem 14.13 for the value of the constant)}$$

with $n_f = \infty$. Therefore, the ionization energy, IE, for one hydrogen atom is

$$\text{IE} = (2.1799 \times 10^{-18} \text{ J})\left(\frac{1}{n_i^2}\right)$$

(a) If $n_i = 1$, for the hydrogen atom,

$$\text{IE} = (2.1799 \times 10^{-18} \text{ J})\left(\frac{1}{1}\right) = 2.1799 \times 10^{-18} \text{ J}$$

For 1 mole of such hydrogen atoms,

$$\text{IE} = (2.1799 \times 10^{-18} \text{ J})\left(\frac{6.022 \times 10^{23}}{1 \text{ mol}}\right) = 1.313 \times 10^6 \text{ J mol}^{-1} = 1.313 \times 10^3 \text{ kJ mol}^{-1}$$

(b) If $n_i = 2$, for the hydrogen atom,

$$\text{IE} = (2.1799 \times 10^{-18} \text{ J})\left(\frac{1}{4}\right) = 5.44975 \times 10^{-19} \text{ J}$$

For 1 mole of such hydrogen atoms,

$$\text{IE} = (5.44975 \times 10^{-19} \text{ J})\left(\frac{6.022 \times 10^{23}}{1 \text{ mol}}\right) = 3.282 \times 10^5 \text{ J mol}^{-1} = 3.282 \times 10^2 \text{ kJ mol}^{-1}$$

14.37 The formula for calculating the energies of an electron in a hydrogenlike ion is given in Equation 14.13. This equation cannot be applied to many-electron atoms. One way to modify it for the more complex atoms is to replace Z with $(Z - \sigma)$, where Z is the atomic number and σ is a positive dimensionless quantity called the shielding constant. Consider the helium atom as an example. The physical significance of σ is that it represents the extent of shielding that the two $1s$ electrons exert on each other. Thus, the quantity $(Z - \sigma)$ is appropriately called the "effective nuclear charge." Calculate the value of σ if the first ionization energy of helium is 3.94×10^{-18} J per atom.

The ionization energy for the helium atom ($Z = 2$) can be calculated by using

$$\Delta E = \left[\frac{m_e (Z - \sigma)^2 e^4}{8h^2 \epsilon_0^2}\right]\left(\frac{1}{n_i^2} - \frac{1}{n_f^2}\right)$$

$$= (2.1799 \times 10^{-18} \text{ J})(2 - \sigma)^2\left(\frac{1}{n_i^2} - \frac{1}{n_f^2}\right) \text{ (See Problem 14.13 for the value of the constant)}$$

with $n_f = \infty$. The experimental value of the ionization energy for an electron in He with $n = 1$ is used to determine σ.

$$IE = \left(2.1799 \times 10^{-18} \text{ J}\right) (2 - \sigma)^2 \left(\frac{1}{1}\right) = 3.98 \times 10^{-18} \text{ J}$$

$$\sigma = 0.649$$

14.38 Plasma is a state of matter consisting of positive gaseous ions and electrons. In the plasma state, a mercury atom could be stripped of its 80 electrons and therefore would exist as Hg^{80+}. Calculate the energy required for the last ionization step, that is,

$$Hg^{79+}(g) \rightarrow Hg^{80+}(g) + e^-$$

Since Hg^{79+} is a hydrogenlike ion with $Z = 80$ and $n_i = 1$, the expression for ionization energy is similar to that shown in Problem 14.36:

$$IE = \left(\frac{m_e Z^2 e^4}{8h^2 \epsilon_0^2}\right) \left(\frac{1}{n_i^2}\right)$$

$$= \left(2.1799 \times 10^{-18} \text{ J}\right) (Z^2) \left(\frac{1}{n_i^2}\right)$$

$$= \left(2.1799 \times 10^{-18} \text{ J}\right) (80^2) \left(\frac{1}{1}\right)$$

$$= 1.3951 \times 10^{-14} \text{ J}$$

For 1 mole of Hg^{79+},

$$IE = \left(1.3951 \times 10^{-14} \text{ J}\right) \left(\frac{6.022 \times 10^{23}}{1 \text{ mol}}\right)$$

$$= 8.401 \times 10^9 \text{ J mol}^{-1} = 8.401 \times 10^6 \text{ kJ mol}^{-1}$$

14.39 A technique called photoelectron spectroscopy is used to measure the ionization energy of atoms. A sample is irradiated with UV light, which causes electrons to be ejected from the valence shell. The kinetic energies of the ejected electrons are measured. Knowing the energy of the UV photon and the kinetic energy of the ejected electron, we can write

$$h\nu = IE + \frac{1}{2}mv^2$$

where ν is the frequency of the UV light, and m and v are the mass and velocity of the electron, respectively. In one experiment, the kinetic energy of the ejected electron from potassium is found to be 5.34×10^{-19} J using a UV source of wavelength 162 nm. Calculate the ionization energy of potassium. How can you be sure that this ionization energy corresponds to the electron in the valence shell (that is, the most loosely held electron)?

The ionization energy for one potassium atom is

$$IE = h\nu - \frac{1}{2}mv^2$$

$$= \frac{hc}{\lambda} - KE$$

$$= \frac{(6.626 \times 10^{-34} \text{ J s}) (3.00 \times 10^8 \text{ m s}^{-1})}{162 \times 10^{-9} \text{ m}} - 5.34 \times 10^{-19} \text{ J}$$

$$= 6.930 \times 10^{-19} \text{ J}$$

The ionization energy for one mole of potassium atoms is

$$IE = (6.930 \times 10^{-19} \text{ J}) \left(\frac{6.022 \times 10^{23}}{1 \text{ mol}} \right)$$

$$= 4.173 \times 10^5 \text{ J mol}^{-1} = 4.173 \times 10^2 \text{ kJ mol}^{-1}$$

To ensure that the ejected electron is the valence electron, UV light of the longest wavelength (lowest energy) should be used that can still eject electrons.

14.40   The energy needed for the following process is $1.96 \times 10^4 \text{ kJ mol}^{-1}$:

$$Li(g) \rightarrow Li^{3+}(g) + 3e^-$$

If the first ionization of lithium is 520 kJ mol^{-1}, calculate the second ionization of lithium, that is, the energy required for the process

$$Li^+(g) \rightarrow Li^{2+}(g) + e^-$$

The overall process

$$Li(g) \rightarrow Li^{3+}(g) + 3e^-$$

is a sum of three processes:

$$Li(g) \rightarrow Li^+(g) + e^- \qquad \text{First ionization}$$
$$Li^+(g) \rightarrow Li^{2+}(g) + e^- \qquad \text{Second ionization}$$
$$Li^{2+}(g) \rightarrow Li^{3+}(g) + e^- \qquad \text{Third ionization}$$

Therefore, the energy for the overall process, I, is a sum of the first ionization energy, I_1, second ionization energy, I_2, and third ionization energy, I_3, of lithium. That is,

$$I = I_1 + I_2 + I_3$$

$$I_2 = I - I_1 - I_3$$

The third process represents the ionization of a hydrogenlike ion with $Z = 3$ and $n_i = 1$. The third ionization energy can be expressed using the equation shown in Problem 14.38.

$$I_3 = (2.1799 \times 10^{-18} \text{ J}) (Z^2) \left(\frac{1}{n_i^2}\right)$$

$$= (2.1799 \times 10^{-18} \text{ J}) (3^2) \left(\frac{1}{1}\right)$$

$$= 1.96191 \times 10^{-17} \text{ J}$$

For 1 mole of ions,

$$I_3 = (1.96191 \times 10^{-17} \text{ J}) \left(\frac{6.022 \times 10^{23}}{1 \text{ mol}}\right)$$

$$= 1.1815 \times 10^7 \text{ J mol}^{-1} = 1.1815 \times 10^4 \text{ kJ mol}^{-1}$$

Therefore,

$$I_2 = I - I_1 - I_2 = (1.96 \times 10^4 - 520 - 1.1815 \times 10^4) \text{ kJ mol}^{-1} = 7.3 \times 10^3 \text{ kJ mol}^{-1}$$

14.41 Experimentally, the electron affinity of an element can be determined by using a laser light to ionize the anion of the element in the gas phase:

$$X^-(g) + h\nu \rightarrow X(g) + e^-$$

Referring to Table 14.5, calculate the photon wavelength (in nanometers) corresponding to the electron affinity for chlorine. In what region of the electromagnetic spectrum does this wavelength fall?

The electron affinity for Cl is 349 kJ mol^{-1}. That is,

$$Cl(g) + e^- \rightarrow Cl^-(g) \qquad \Delta H = -349 \text{ kJ mol}^{-1}$$

The reverse reaction

$$Cl^-(g) + h\nu \rightarrow Cl(g) + e^-$$

occurs when the photon energy = 349 kJ mol^{-1}. The energy of one photon is

$$E = (349 \text{ kJ mol}^{-1}) \left(\frac{1 \text{ mol}}{6.022 \times 10^{23}}\right) = 5.795 \times 10^{-22} \text{ kJ} = 5.795 \times 10^{-19} \text{ J}$$

Since

$$E = h\nu = \frac{hc}{\lambda}$$

The wavelength corresponding to this energy is

$$\lambda = \frac{hc}{E}$$

$$= \frac{(6.626 \times 10^{-34} \text{ J s}) (3.00 \times 10^8 \text{ m s}^{-1})}{5.795 \times 10^{-19} \text{ J mol}^{-1}}$$

$$= 3.43 \times 10^{-7} \text{ m} = 343 \text{ nm}$$

This wavelength is in the UV region.

14.42   The standard enthalpy of atomization of an element is the energy required to convert 1 mole of an element in its most stable form at 25°C to 1 mole of monatomic gas. Given that the standard enthalpy of atomization for sodium is 108.4 kJ mol^{-1}, calculate the energy in kilojoules required to convert 1 mole of sodium metal at 25°C to 1 mole of gaseous Na$^+$ ions.

The equation

$$Na(s) \rightarrow Na^+(g) + e^- \qquad \Delta H_1$$

is a sum of the following two processes

$$Na(s) \rightarrow Na(g) \qquad\qquad \Delta H_2 = 108.4 \text{ kJ mol}^{-1}$$
$$Na(g) \rightarrow Na^+(g) + e^- \qquad \Delta H_3$$

Therefore,

$$\Delta H_1 = \Delta H_2 + \Delta H_3$$

ΔH_3 represents the ionization energy of 1 mole of Na(g), which is 495.9 kJ (Table 14.4). Thus, the energy required to convert 1 mole of sodium metal to 1 mole of gaseous Na$^+$ ions is

$$\Delta H_1 = 108.4 \text{ kJ mol}^{-1} + 495.9 \text{ kJ mol}^{-1} = 604.3 \text{ kJ mol}^{-1}$$

14.43   Explain why the electron affinity of nitrogen is approximately zero, while the elements on either side, carbon and oxygen, have substantial positive electron affinities.

The electron affinity depends on the Z_{eff} for the *empty* orbital into which the additional electron is placed. In general, Z_{eff} increases across a row in the periodic table, so that electrons are held more tightly and increasing electron affinity. Thus, carbon and oxygen have electron affinities on the order of 100 kJ mol^{-1}. In the case of nitrogen, however, the additional electron must go into an orbital that is already half-occupied, and breaks up the half-filled subshell. Consequently, there is little tendency for the atom to accept another electron.

14.44   Calculate the maximum wavelength of light (in nanometers) required to ionize a single sodium atom.

The ionization energy of sodium is 495.9 kJ mol^{-1} (Table 14.4). Thus, the minimum energy of photon required to ionize a sodium atom is

$$E = \left(495.9 \times 10^3 \text{ J mol}^{-1}\right) \left(\frac{1 \text{ mol}}{6.022 \times 10^{23}}\right) = 8.2348 \times 10^{-19} \text{ J}$$

The maximum wavelength of light corresponds to this energy and is calculated as follows.

$$E = 8.2348 \times 10^{-19} \text{ J} = h\nu = \frac{hc}{\lambda}$$

$$\lambda = \frac{\left(6.626 \times 10^{-34} \text{ J s}\right)\left(3.00 \times 10^8 \text{ m s}^{-1}\right)}{8.2348 \times 10^{-19} \text{ J}} = 2.41 \times 10^{-7} \text{ m} = 241 \text{ nm}$$

14.45 The first four ionization energies of an element are approximately 738 kJ mol^{-1}, 1450 kJ mol^{-1}, 7.7×10^3 kJ mol^{-1}, and 1.1×10^4 kJ mol^{-1}. To which periodic group does this element belong? Why?

The large jump between the second and third ionization energies indicates a change in the principal quantum number n. That is, if the first two electron removed have principal quantum number n, then the next two have principal quantum number $n - 1$. Thus, the element is in the second column of the periodic table, or Group 2A.

14.46 When two atoms collide, some of their kinetic energy may be converted into electronic energy in one or both atoms. If the average kinetic energy $\left(\frac{3}{2}k_B T\right)$ is about equal to the energy for some allowed electronic transition, an appreciable number of atoms can absorb enough energy through an inelastic collision to be raised to an excited electronic state. **(a)** Calculate the average kinetic energy per atom in a gas sample at 298 K. **(b)** Calculate the energy difference between the $n = 1$ and $n = 2$ levels in hydrogen. **(c)** At what temperature is it possible to excite a hydrogen atom from the $n = 1$ level to $n = 2$ level by collision?

(a) The average kinetic energy per atom is

$$\text{KE} = \frac{3}{2}k_B T = \frac{3}{2}\left(1.381 \times 10^{-23} \text{ J K}^{-1}\right)(298 \text{ K}) = 6.173 \times 10^{-21} \text{ J}$$

(b) The energy difference between the $n = 1$ and $n = 2$ levels in hydrogen ($Z = 1$) is

$$\Delta E = \left(\frac{m_e Z^2 e^4}{8h^2\epsilon_0^2}\right)\left(\frac{1}{n_i^2} - \frac{1}{n_f^2}\right)$$

$$= 2.180 \times 10^{-18} \text{ J}\left(\frac{1}{1} - \frac{1}{4}\right) \quad \text{(See Problem 14.13)}$$

$$= 1.635 \times 10^{-18} \text{ J}$$

(c) For a collision to excite a hydrogen atom from the $n = 1$ to $n = 2$ level,

$$KE = \Delta E$$

$$\frac{3}{2}k_B T = 1.635 \times 10^{-18} \text{ J}$$

$$T = \frac{(2)\left(1.635 \times 10^{-18} \text{ J}\right)}{(3)\left(1.381 \times 10^{-23} \text{ J K}^{-1}\right)} = 7.89 \times 10^4 \text{ K}$$

This is an extremely high temperature, and other means of excitation must be used for practical methods of exciting H atoms.

14.47   Photodissociation of water,

$$H_2O(g) + h\nu \rightarrow H_2(g) + \tfrac{1}{2}\,O_2(g)$$

has been suggested as a source of hydrogen. The $\Delta_r H^\circ$ value for the reaction, calculated from thermochemical data, is 285.8 kJ per mole of water decomposed. Calculate the maximum wavelength (in nm) that would provide the necessary energy. In principle, is it feasible to use sunlight as a source of energy for this process?

The minimum photon energy is 285.8 kJ mol^{-1}.

$$E = \left(285.8 \times 10^3 \text{ J mol}^{-1}\right)\left(\frac{1 \text{ mol}}{6.022 \times 10^{23}}\right) = 4.746 \times 10^{-19} \text{ J} = h\nu = \frac{hc}{\lambda}$$

$$\lambda = \frac{\left(6.626 \times 10^{-34} \text{ J s}\right)\left(3.00 \times 10^8 \text{ m s}^{-1}\right)}{4.746 \times 10^{-19} \text{ J}} = 4.19 \times 10^{-7} \text{ m} = 419 \text{ nm}$$

This wavelength is in the visible range of the electromagnetic spectrum. Since water is continuously being struck by visible radiation without decomposition, it seems unlikely that photodissociation of water by this method is possible.

14.48   Based on the discussion of decay and quantum mechanical tunneling, suggest a relation between the energy of emitted α particles and the half-life for the radioactive decay.

The greater the energy of the α particle, the more likely it will be able to tunnel through the potential barrier and leave the nucleus. Since this would correspond to a greater number of α particles ejected from a sample of the element in a given time, an inverse relationship between particle energy and half-life would be expected. That is, a shorter half-life should be found for those elements that emit more energetic α particles.

14.49   Only a fraction of the electrical energy supplied to a tungsten light bulb is converted to visible light. The rest of the energy shows up as infrared radiation (that is, heat). A 75-W light bulb converts 15.0% of the energy supplied to it into visible light. Assuming a wavelength of 550 nm, how many photons are emitted by the light bulb per second? (1 W = 1 J s^{-1}.)

The energy of visible light emitted by the light bulb per second is

$$E_{bulb} = (75 \text{ J s}^{-1}) (1 \text{ s}) (15\%) = 11.3 \text{ J}$$

The energy of a 550 nm photon is

$$E_{photon} = h\nu = \frac{hc}{\lambda} = \frac{(6.626 \times 10^{-34} \text{ J s}) (3.00 \times 10^8 \text{ m s}^{-1})}{550 \times 10^{-9} \text{ m}} = 3.614 \times 10^{-19} \text{ J}$$

The number of photons emitted by the light bulb per second is

$$\frac{11.3 \text{ J}}{3.614 \times 10^{-19} \text{ J}} = 3.1 \times 10^{19}$$

14.50 An electron in a hydrogen atom is excited from the ground state to the $n = 4$ state. State whether the following statements are true or false. **(a)** $n = 4$ is the first excited state. **(b)** It takes more energy to ionize (remove) the electron from the $n = 4$ state than from the ground state. **(c)** The electron is farther from the nucleus (on average) in the $n = 4$ state than in the ground state. **(d)** The wavelength of light emitted when the electron drops from $n = 4$ to $n = 1$ is longer than that from $n = 4$ to $n = 2$. **(e)** The wavelength the atom absorbs in going from $n = 1$ to $n = 4$ is the same as that emitted as it goes from $n = 4$ to $n = 1$.

(a) False. The first excited state is $n = 2$. **(b)** False. The electron in the $n = 4$ level is closer to the dissociation limit. **(c)** True. **(d)** False. Since the energy difference between the $n = 4$ and $n = 1$ levels is greater than that between the $n = 4$ and $n = 2$ levels, the wavelength of light emitted when the electron drops from the $n = 4$ level to $n = 1$ level is shorter than that emitted when the electron drops from the $n = 4$ level to $n = 2$ level. **(e)** True.

14.51 The ionization energy of a certain element is 412 kJ mol^{-1}. When the atoms of this element are in the first excited state, however, the ionization energy is only 126 kJ mol^{-1}. Based on this information, calculate the wavelength of light emitted in a transition from the first excited state to the ground state.

The ionization energy of 412 kJ mol^{-1} represents the energy difference between the ground state and the dissociation limit whereas the ionization energy of 126 kJ mol^{-1} represents the energy difference between the first excited state and the dissociation limit. Therefore, the energy difference between the ground state and the excited state is

$$\Delta E = (412 - 126) \text{ kJ mol}^{-1} = 286 \text{ kJ mol}^{-1}$$

The energy of light emitted in a transition from the first excited state to the ground state is therefore 286 kJ mol^{-1}. The wavelength emitted is calculated as follows.

$$E = (286 \times 10^3 \text{ J mol}^{-1}) \left(\frac{1 \text{ mol}}{6.022 \times 10^{23}} \right) = 4.749 \times 10^{-19} \text{ J} = h\nu = \frac{hc}{\lambda}$$

$$\lambda = \frac{(6.626 \times 10^{-34} \text{ J s}) (3.00 \times 10^8 \text{ m s}^{-1})}{4.749 \times 10^{-19} \text{ J}} = 4.19 \times 10^{-7} \text{ m} = 419 \text{ nm}$$

14.52 The UV light responsible for sun tanning falls in the 320- to 400-nm region. Calculate the total energy (in joules) absorbed by a person exposed to this radiation for 2.0 hours, given that there are 2.0×10^{16} photons hitting Earth's surface per square centimeter per second over an 80-nm (320-nm to 400-nm) range and that the exposed body area is 0.45 m². Assume that only half of the radiation is absorbed, and the other half is reflected by the body. (*Hint*: Use an average wavelength 0f 360 nm to calculate the energy of a photon.)

The energy of one 360 nm photon is

$$E = h\nu = \frac{hc}{\lambda} = \frac{\left(6.626 \times 10^{-34}\ \text{J s}\right)\left(3.00 \times 10^{8}\ \text{m s}^{-1}\right)}{360 \times 10^{-9}\ \text{m}} = 5.522 \times 10^{-19}\ \text{J}$$

The total energy absorbed by a person is

$$\left(\frac{1}{2}\right)\left(5.522 \times 10^{-19}\ \text{J photon}^{-1}\right)\left(2.0 \times 10^{16}\ \text{photons cm}^{-2}\,\text{s}^{-1}\right)(7200.0\ \text{s})\left(0.45 \times 10^{4}\ \text{cm}^{2}\right)$$

$$= 1.79 \times 10^{5}\ \text{J}$$

14.53 In 1996, physicists created an antiatom of hydrogen. In such an atom, which is the antimatter equivalent of an ordinary atom, the electrical charges of all the component particles are reversed. Thus, the nucleus of an antiatom is made of an antiproton, which has the same mass as a proton but bears a negative charge, while the electron is replaced by an antielectron (also called positron) with the same mass as an electron but bearing a positive charge. Would you expect the energy levels, emission spectra, and atomic orbitals of an antihydrogen atom to be different from those of a hydrogen atom? What would happen if an antiatom of hydrogen collided with a hydrogen atom?

The antiatom of hydrogen should show the same characteristics with regard to energy levels, emission spectra, and atomic orbitals as does ordinary hydrogen. Should an antiatom of hydrogen collide with an ordinary hydrogen atom, they would annihilate each other, and energy would be given off.

14.54 A student carried out a photoelectric experiment by shining visible light on a clean piece of cesium metal. She determined the kinetic energy of ejected electrons by applying a retarding voltage such that the current due to the electrons reads exactly zero. This condition was reached when $eV = (1/2)m_e v^2$, where e is electric charge and V is the retarding potential. Her results are shown below:

λ/nm	405	435.8	480	520	577.7	650
V/volt	1.475	1.268	1.027	0.886	0.667	0.381

Rearrange Equation 14.3 to read

$$\nu = \frac{\Phi}{h} + \frac{e}{h}V$$

Determine the values of h and Φ graphically.

A plot of v vs V gives a slope of $\dfrac{e}{h}$ and an intercept of $\dfrac{\Phi}{h}$. The data used in the plot are

$v/10^{14}\cdot\text{s}^{-1}$	7.407	6.884	6.250	5.769	5.193	4.615
V/volt	1.475	1.268	1.027	0.886	0.667	0.381

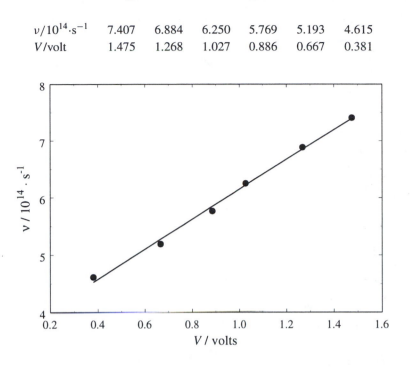

The best fit line is described by the equation $y = 2.618 \times 10^{14}x + 3.531 \times 10^{14}$. Therefore,

$$\frac{e}{h} = 2.618 \times 10^{14}\ \text{s}^{-1}\ \text{V}^{-1}$$

$$h = \frac{1.602 \times 10^{-19}\ \text{C}}{2.618 \times 10^{14}\ \text{s}^{-1}\ \text{V}^{-1}} = 6.119 \times 10^{-34}\ \text{J s} = 6.12 \times 10^{-34}\ \text{J s}$$

and

$$\frac{\Phi}{h} = 3.531 \times 10^{14}\ \text{s}^{-1}$$

$$\Phi = \left(3.531 \times 10^{14}\ \text{s}^{-1}\right)\left(6.119 \times 10^{-34}\ \text{J s}\right) = 2.16 \times 10^{-19}\ \text{J}$$

14.55 Use Equation 3.7 to calculate the de Broglie wavelength of a N_2 molecule at 300 K.

The rms speed of N_2 is

$$v_{\text{rms}} = \sqrt{\frac{3RT}{\mathcal{M}}} = \sqrt{\frac{3\left(8.314\ \text{J K}^{-1}\ \text{mol}^{-1}\right)(300\ \text{K})}{28.02 \times 10^{-3}\ \text{kg}}} = 516.8\ \text{m s}^{-1}$$

Using $v = v_{\text{rms}}$, the de Broglie wavelength of the molecule is

$$\lambda = \frac{h}{mv} = \frac{6.626 \times 10^{-34}\ \text{J s}}{(28.02\ \text{amu})\left(1.661 \times 10^{-27}\ \text{kg amu}^{-1}\right)\left(516.8\ \text{m s}^{-1}\right)} = 2.75 \times 10^{-11}\ \text{m}$$

14.56   Alveoli are tiny sacs of air in the lungs. Their average diameter is 5.0×10^{-5} m. Calculate the uncertainty in the velocity of an oxygen molecule (5.3×10^{-26} kg) trapped within a sac. (*Hint*: The maximum uncertainty in the position of the molecule is given by the diameter of the sac.)

Using $\Delta x = 5.0 \times 10^{-5}$ m, the uncertainty in the velocity of the oxygen molecule can be calculated.

$$\Delta x \, \Delta p \geq \frac{h}{4\pi}$$

$$\Delta x \, (m \, \Delta v) \geq \frac{h}{4\pi}$$

$$\Delta v \geq \frac{h}{4\pi \, m \, \Delta x}$$

$$\Delta v \geq \frac{6.626 \times 10^{-34} \text{ J s}}{4\pi \, (5.3 \times 10^{-26} \text{ kg}) \, (5.0 \times 10^{-5} \text{ m})} = 2.0 \times 10^{-5} \text{ m s}^{-1}$$

14.57   The sun is surrounded by a white circle of gaseous material called the corona, which becomes visible during a total eclipse of the sun. The temperature of the corona is in the millions of degrees Celsius, high enough to break up molecules and remove some or all of the electrons from atoms. One way astronomers have been able to estimate the temperature of the corona is by studying the emission lines of ions of certain elements. For example, the emission spectrum of Fe^{14+} ions has been recorded and analyzed. Knowing that it takes 3.5×10^4 kJ mol^{-1} to convert Fe^{13+} to Fe^{14+}, estimate the temperature of the sun's corona. (*Hint*: The average kinetic energy of 1 mole of a gas is $\frac{3}{2}RT$.)

The energy required to create the Fe^{14+} ion from Fe^{13+} must come from the thermal energy of the corona. That is, collisions with other species in the plasma provide the necessary energy. Thus, estimate the average kinetic energy in the plasma as being equal to the ionization energy of Fe^{13+}.

$$KE = IE$$

$$\frac{3}{2}RT = 3.5 \times 10^7 \text{ J mol}^{-1}$$

$$T = \frac{2 \, (3.5 \times 10^7 \text{ J mol}^{-1})}{3 \, (8.314 \text{ J K}^{-1} \text{ mol}^{-1})} = 2.8 \times 10^6 \text{ K}$$

14.58   Consider a particle in the one-dimensional box shown below.

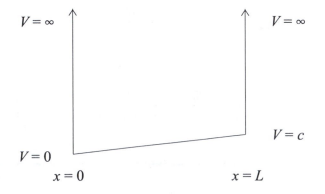

The potential energy of the particle is not zero everywhere in the box but varies according to its position, or x. **(a)** Write the Schrödinger wave equation for this system. **(b)** Obviously, ψ_n will not be symmetrical about the center of the box, regardless of the value of n. If the wave function for $n = 10$ in a one-dimensional box, where $V = 0$ throughout the box, looks like

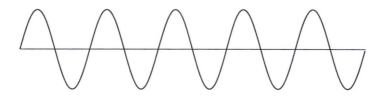

sketch the corresponding wavefunction for the above box. Give a physical interpretation for your sketch. (*Hint*: Neither the amplitude nor the wavelength of the wave will be constant along the box.)

(a) The potential energy is

$$V = \frac{cx}{L}$$

Therefore, the Schrödinger equation is

$$-\frac{h^2}{8\pi^2 m}\frac{d^2\psi}{dx^2} + \frac{cx}{L}\psi = E\psi$$

(b) The total energy of the particle (kinetic plus potential) is constant. Because the potential energy increases from left to right in the box, the kinetic energy must decrease along the same direction. Thus, the particle spends less time on the left, where it is moving faster, and the amplitude of the wavefunction must be smaller there. Additionally, the wavelength of the particle is inversely dependent on the momentum ($\lambda = h/mv$), so that the wavelength is shorter on the left as well. In fact this Schrödinger equation has analytical solutions, known as Airy functions, and the figure shows such a solution which demonstrates the effects discussed above. (The magnitude of the effects depend on the value of c.)

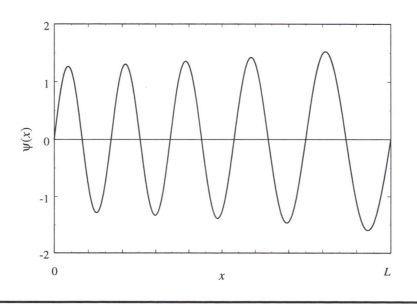

CHAPTER 15

The Chemical Bond

PROBLEMS AND SOLUTIONS

15.1 Which of the following molecules has the shortest nitrogen-to-nitrogen bond? Explain.

N_2H_4 N_2O N_2 N_2O_4

In considering the Lewis structures for all four molecules, shown below, only N_2 has a full N≡N triple bond. Therefore, it has the shortest nitrogen-to-nitrogen bond.

N_2H_4:

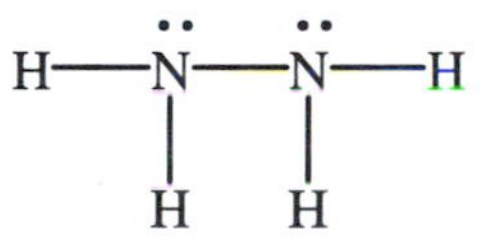

N_2O:

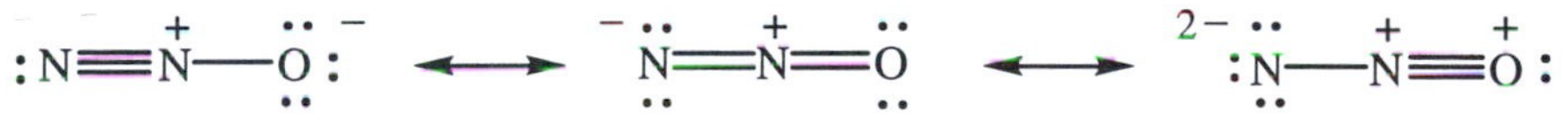

N_2:

:N≡N:

N_2O_4:

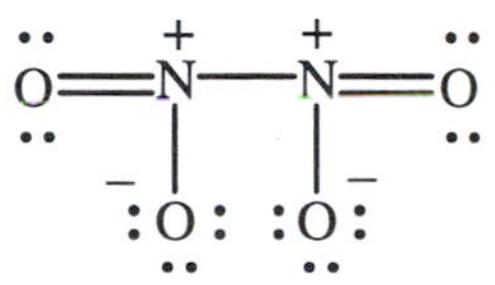

This is one of 4 possible, equivalent resonance structures for N_2O_4; all have a nitrogen-to-nitrogen single bond.

15.2   The chlorine nitrate molecule ($ClONO_2$) is believed to be involved in the destruction of ozone in the Antarctic stratosphere. Draw a plausible Lewis structure for this molecule.

Two equivalent resonance structures are possible for this molecule.

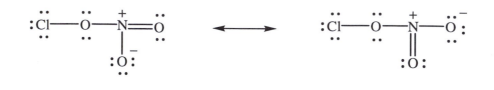

15.3   Draw resonance structures for N_2O. The atomic arrangement is NNO. Show formal charges. How would you distinguish this structure from the NON structure?

Three resonance structures for N_2O are shown in the answer to Problem 15.1 Those for NON are shown below. Note that the middle resonance structure is non-polar while the outer two, which are equivalent, although polar, contribute equal but opposite polarities resulting in a non-polar NON molecule. N_2O, on the other hand, is polar.

$$:N\equiv\overset{2+}{O}-\overset{2-}{\underset{\cdot\cdot}{N}}: \quad\longleftrightarrow\quad \overset{-}{N}=\overset{2+}{O}=\overset{-}{N} \quad\longleftrightarrow\quad \overset{2-}{:\underset{\cdot\cdot}{N}}-\overset{2+}{O}\equiv N:$$

15.4   Carbon monoxide has a rather small dipole moment ($\mu = 0.12$ D) even though the electronegativity difference between C and O is rather large ($X_C = 2.5$ and $X_O = 3.5$). How would you explain this fact in terms of resonance structures?

The resonance form that makes the major contribution to the electronic structure of CO is $^-:C \equiv O:^+$. The lone pair and negative formal charge on the carbon tends to counteract the imbalance in shared electron pairs due to the greater electronegativity of oxygen. This results in a small dipole moment. (In fact, experiment shows that the carbon end of the molecule is the more negative end.)

15.5   The resonance concept is sometimes described by analogy to a mule, which is a cross between a horse and a donkey. Compare this analogy with the description of a rhinoceros as a cross between a griffin and a unicorn. Which description is more appropriate? Why?

Both the horse and the donkey are real animals whereas the griffin and the unicorn, like the electron configurations described by individual resonance structures, are non-existent. Thus, the latter analogy is more appropriate.

15.6   Comment on the appropriateness of using the following resonance structure for O_2 intended to explain its paramagnetism.

This resonance structure indicates a single oxygen-to-oxygen bond in contradiction to the double bond that the molecule possesses. Furthermore, the oxygen atoms do not obey the octet rule.

15.7 Consider the Lewis structure for boron trifluoride (BF_3) shown below:

Does it satisfy the octet rule? If not, draw additional resonance structures that do satisfy the octet rule. Suggest an experimental measurement that would enable you to show the relative importance of the resonance structures.

Three resonance structures that satisfy the octet rule are shown below. These indicate partial double bond character for the boron-to-fluorine bond. This suggests that a measurement of the B-F bond distance in BF_3 followed by a comparison to the bond length for a B-F single bond would indicate the importance of these three resonance structures in describing the electronic structure of BF_3.

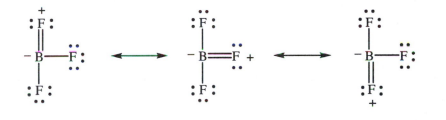

15.8 Disulfide bonds play an important role in the three-dimensional structure of protein molecules. Discuss the nature of the —S—S— bond.

The electron configuration of the sulfur atom is $[Ne]3s^2 3p^4$, suggesting that each S atom is sp^3 hybridized to allow for each atom to form 2 covalent bonds and to have 2 lone pairs on each S atom as shown below.

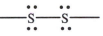

15.9 Consider the HCl molecule. Let ψ_{1s} be the hydrogen $1s$ wavefunction and ψ_{3p} be the chlorine $3p$ wavefunction. Write the VB wavefunction of HCl assuming that **(a)** the bond is purely covalent, **(b)** the bond is purely ionic, and the electron is transferred from H to Cl, and **(c)** the bond is polar.

(a) The purely covalent bond would have complete sharing of the electron pair:
$\psi = \psi_{1s}(1)\psi_{3p}(2) + \psi_{1s}(2)\psi_{3p}(1)$ where 1 and 2 denote electrons. (Normalization is ignored.)

(b) In the purely ionic bond, both electrons are on the more electronegative Cl atom:
$\psi = \psi_{3p}(1)\psi_{3p}(2)$.

(c) The polar bond has contributions from both the purely covalent form and the purely ionic form:
$\psi = \psi_{1s}(1)\psi_{3p}(2) + \psi_{1s}(2)\psi_{3p}(1) + \lambda[\psi_{3p}(1)\psi_{3p}(2)]$.

15.10   The unstable molecule carbene or methylene (CH_2) has been isolated and studied spectroscopically. Suggest two types of bonding that might be present in this molecule. How would you determine which type of bond is present in CH_2?

One possibility would be for the C atom not to be hybridized. In this case there would be no unpaired electrons and the molecule would be diamagnetic.

A second possibility would be for the C atom to be sp^3 hybridized, with two of the hybrid orbitals used for bonding to the H's and the remaining two each containing a single, unpaired electron. These two unpaired electrons would cause the molecule to be paramagnetic.

This suggests that a measurement of the magnetic properties of the molecule would distinguish between the two cases. In fact, CH_2 in its ground state is paramagnetic and (nearly) linear. This would imply bonding to the H's using sp hybrids with a single, unpaired electron in each of the two unhybridized p orbitals.

15.11   Describe the bonding in CO_2 and C_3H_4 (allene) in terms of hybridization. Draw diagrams to show the formation of σ bonds and π bonds in allene.

In CO_2, the central carbon atom is sp hybridized resulting in a linear molecule. The hybridization of the terminal oxygen atoms is less well defined, but each must have at least one unhybridized p orbital to participate in a π bond with the carbon.

For allene, the central C atom is sp hybridized, but the two terminal C atoms are each sp^2 hybridized. The planes containing the $-CH_2$ groups at the ends of the molecule make a dihedral angle of 90° with each other. This is a direct result of the requirement that there be 2 π bonds involving the central C atom, as indicated in the sketch below.

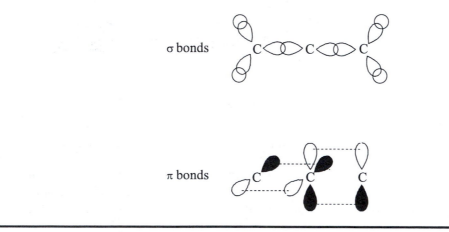

σ bonds

π bonds

15.12   Describe the bonding scheme in the following species in terms of molecular-orbital theory: H_2^+, H_2, He_2^+, and He_2. List the species in order of decreasing stability.

The molecular orbital electron configuration for the four species is shown below. H_2 has a bond order of 1, both H_2^+ and He_2^+ have bond order of 1/2, and He_2 has a bond order of 0. Thus, the stability decreases as follows, $H_2 > H_2^+$, $He_2^+ > He_2$, with H_2^+ slightly more stable than He_2^+.

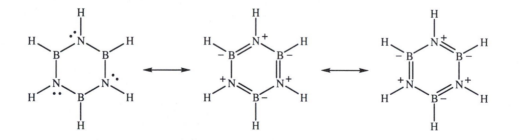

15.13 Which molecule would have the longer bond length: F_2 or F_2^+? Explain in terms of molecular-orbital theory.

To form F_2^+ from F_2, an electron is removed from a π^* orbital, which is antibonding. Therefore, F_2^+ has a higher bond order, is more stable, and has a shorter bond than F_2.

15.14 Which of the following species has the longest bond: CN^+, CN, CN^-?

The electron configuration of CN is $KK(\sigma_{2s})^2(\sigma_{2s}^*)^2(\pi_x)^2 (\pi_y)^2(\sigma_{2p})^1$. This results in a bond order of 2.5. In forming CN^+, an electron is removed from the bonding σ_{2p} orbital, which reduces the bond order to 2.0. When forming CN^-, the electron is added to the bonding σ_{2p} orbital, increasing the bond order to 3.0. Thus, CN^+ has the lowest bond order and the longest bond.

15.15 Borazine ($B_3N_3H_6$) is isoelectronic with benzene. Describe qualitatively the bonding in this molecule in terms of **(a)** resonance and **(b)** molecular-orbital theory.

(a) Three resonance structures for borazine are shown below.

(b) In molecular orbital theory, each N atom as well as each B atom is sp^2 hybridized. The unhybridized p orbitals overlap to form an extended orbital with electron delocalization similar to that in benzene. Since N and B have different electronegativities, the delocalization is not uniform, and there will be charge separation. The molecule is more reactive than benzene as a result.

15.16   Which of the following two molecules has a greater degree of π-electron delocalization: naphthalene or biphenyl?

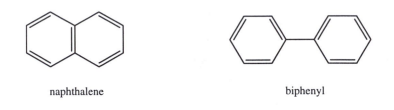

naphthalene   biphenyl

The rotation about the C–C single bond in biphenyl, shown below, partially destroys the electron delocalization between the two benzene rings, so that naphthalene has the greater degree of π-electron delocalization.

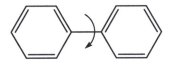

15.17   Use MO theory to describe the bonding in NO^+, NO, and NO^-. Compare their bond energies and bond lengths.

Since N and O are adjacent in the periodic table and have similar electronegativities, the molecular orbital energy diagram for NO and its ions is essentially the same as that for N_2. The three species have the following electron configurations.

NO^+: $KK(\sigma_{2s})^2(\sigma_{2s}^*)^2(\pi_x)^2(\pi_y)^2(\sigma_{2p})^2$ with a bond order of 3 (like N_2).

NO: $KK(\sigma_{2s})^2(\sigma_{2s}^*)^2(\pi_x)^2(\pi_y)^2(\sigma_{2p})^2(\pi_x^*)^1$ with a bond order of 2.5.

NO^-: $KK(\sigma_{2s})^2(\sigma_{2s}^*)^2(\pi_x)^2(\pi_y)^2(\sigma_{2p})^2(\pi_x^*)^1(\pi_y^*)^1$ with a bond order of 2.

The bond energies of the three species are expected to increase in the order $NO^- < NO < NO^+$. The bond lengths are expected to increase in the order $NO^+ < NO < NO^-$.

15.18   Compare the MO theory description for the H_2 molecule, where the wavefunction is given by

$$\psi = \left[\psi_A(1) + \psi_B(1)\right]\left[\psi_A(2) + \psi_B(2)\right]$$

with the VB theory treatment given by Equation 15.4. Under what condition do they become identical?

In Equation 15.4, the valence bond theory treatment gives the wavefunction

$$\psi_{VB} = \psi_A(1)\psi_B(2) + \psi_A(2)\psi_B(1) + \lambda\left[\psi_A(1)\psi_A(2) + \psi_B(1)\psi_B(2)\right]$$

The molecular orbital theory wavefunction as given above can be expanded to give

$$\psi_{MO} = \psi_A(1)\psi_B(2) + \psi_A(2)\psi_B(1) + \psi_A(1)\psi_A(2) + \psi_B(1)\psi_B(2)$$

The two wavefunctions are the same when $\lambda = 1$. The MO approach overemphasizes the ionic character, reckoning it equally as important as covalent character. In its first approximation, with $\lambda = 0$, the VB approach underestimates ionic character.

15.19   Acetylene (C_2H_2) has a tendency to lose two protons (H^+) and form the carbide ion (C_2^{2-}), which is present in several ionic compounds, such as CaC_2 and MgC_2. Describe the bonding scheme in the C_2^{2-} ion in terms of molecular-orbital theory. Compare the bond order in C_2^{2-} with that in C_2.

Table 15.3 has the electron configuration for C_2, and C_2^- is isoelectronic with N_2.

Species	Electron Configuration	Bond Order
C_2	$KK(\sigma_{2s})^2(\sigma_{2s}^*)^2(\pi_x)^2(\pi_y)^2$	2
C_2^-	$KK(\sigma_{2s})^2(\sigma_{2s}^*)^2(\pi_x)^2(\pi_y)^2(\sigma_{2p})^2$	3

15.20   Describe the bonding in the nitrate ion NO_3^- in terms of delocalized molecular orbitals.

The ion contains 24 valence electrons. Of these, six are involved in three sigma bonds, indicated below, between the nitrogen and the oxygen atoms. The hybridization of the nitrogen atom is sp^2. There are four valence electrons on each oxygen atom, for a total of 12, which are non-bonding. The remaining six electrons are in delocalized π molecular orbitals which result from the overlap of the p_z orbital of the nitrogen with the p_z orbitals on the three oxygen atoms as shown below. The molecular orbitals are similar to those of the carbonate ion.

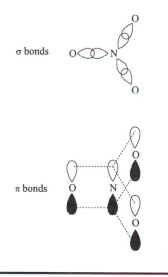

15.21   A single bond is usually a σ bond and a double bond is usually made up of a σ bond and a π bond. Can you identify the exceptions in the homonuclear diatomic molecules of the second period?

As can be noted from Table 15.3, the single bond in B_2 is a π bond, and for C_2, the double bond is made up of two π bonds.

15.22   Draw energy-level diagrams to show the low- and high-spin octahedral complexes of the transition-metal ions that have the electron configurations d^4, d^5, d^6, and d^7.

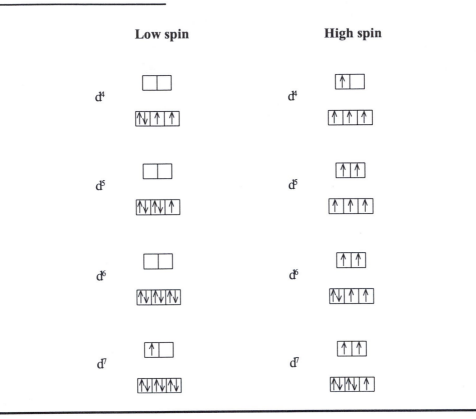

15.23   The Ni(CN)$_4^{2-}$ ion, which has square-planar geometry, is diamagnetic, whereas the NiCl$_4^{2-}$ ion, which has tetrahedral geometry, is paramagnetic. Show the crystal-field splitting diagrams for those two complexes.

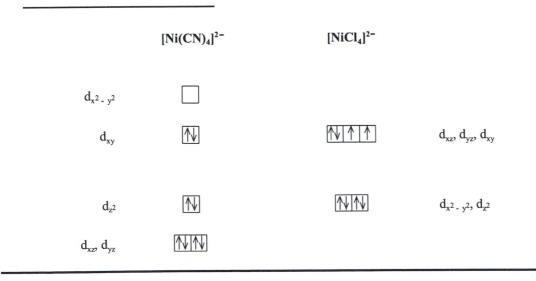

15.24   Predict the number of unpaired electrons in the following complex ions: **(a)** Cr(CN)$_6^{4-}$ and **(b)** Cr(H$_2$O)$_6^{2+}$.

(a) The electron configuration of Cr^{2+} is $[Ar]3d^4$. CN^- is a strong-field ligand, and the four electrons will occupy the three lower orbitals as shown. There will be two unpaired electrons.

$[Cr(CN)_6]^{4-}$

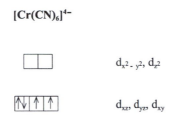

(b) Since H_2O is a weak-field ligand, the four $3d$ electrons of Cr^{2+} will be arranged to maximize the number of unpaired electrons as shown below. There will be four unpaired electrons.

$[Cr(H_2O)_6]^{2+}$

$\boxed{\uparrow\,|\,}$ $d_{x^2-y^2}, d_{z^2}$

$\boxed{\uparrow\,|\,\uparrow\,|\,\uparrow}$ d_{xz}, d_{yz}, d_{xy}

15.25   Transition metal complexes containing CN^- ligands are often yellow in color, whereas those containing H_2O ligands tend to be green or blue. Explain.

A substance that appears to be yellow is absorbing light from the blue-violet, high energy end of the visible spectrum. Substances that appear green or blue to the eye are absorbing light from the lower energy, red or orange part of the spectrum. CN^- is a strong field ligand and causes a larger crystal field splitting than water. Consequently, complexes containing CN^- ligands will absorb radiation of higher energy (shorter wavelength) when a d electron is excited to a higher energy d orbital than will complexes containing H_2O ligands.

15.26   The absorption maximum for the complex ion $Co(NH_3)_6^{3+}$ occurs at 470 nm. **(a)** Predict the color of the complex, and **(b)** calculate the crystal-field splitting in $kJ\,mol^{-1}$.

(a) The complex ion is absorbing 470 nm light, which is blue to blue-green. Referring to Table 15.6, this implies that the complex is between yellow and red, or orange.

(b) The energy of a 470 nm photon is found using Planck's relationship and converted to a molar basis.

$$\Delta E = \frac{hc}{\lambda}$$

$$= \frac{(6.626 \times 10^{-34}\,\text{J s})\,(3.00 \times 10^8\,\text{m s}^{-1})}{470 \times 10^{-9}\,\text{m}}$$

$$= 4.229 \times 10^{-19}\,\text{J}$$

This is the crystal-field splitting per molecule. Multiplying by Avogadro's constant gives the crystal-field splitting for a mole of molecules.

$$\left(4.229 \times 10^{-19} \text{ J}\right)\left(6.022 \times 10^{23} \text{ mol}^{-1}\right)\left(\frac{1 \text{ kJ}}{1000 \text{ J}}\right) = 255 \text{ kJ mol}^{-1}$$

15.27 The label of a certain brand of mayonnaise lists EDTA as a food preservative. How does EDTA prevent the spoilage of mayonnaise?

EDTA sequesters metal ions such as Ca^{2+} and Mg^{2+}, which are essential for bacterial growth and function.

15.28 Hydrated Mn^{2+} ions are practically colorless even though they possess five $3d$ electrons. Explain. (*Hint*: Electronic transitions in which there is a change in the number of unpaired electrons do not occur readily.)

In a high-spin complex with five d electrons, all the electrons are unpaired and each of the five $3d$ orbitals are singly occupied as shown in the diagram below. Any excitation of a $3d$ electron in such a complex would require that an orbital become doubly occupied and that, to satisfy the Pauli exclusion principle, an electron would need to change its spin. Such transitions that involve a change in spin state are forbidden and do not occur to any appreciable extent. Thus, hydrated Mn^{2+} ions do not absorb light in the visible region of the spectrum and, as a result, appear faintly colored (pink).

[Mn(H₂O)₆]²⁺

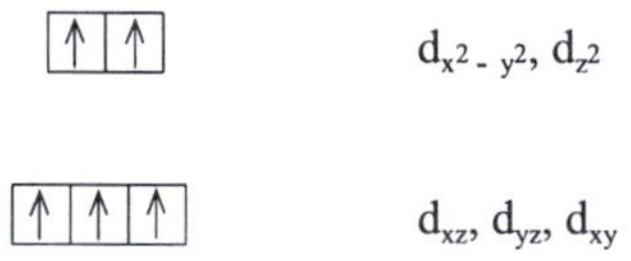

|↑|↑| $d_{x^2-y^2}, d_{z^2}$

|↑|↑|↑| d_{xz}, d_{yz}, d_{xy}

15.29 Oxyhemoglobin is bright red, whereas deoxyhemoglobin is purple. Explain the difference in color in terms of the electron configurations of iron in these two complexes.

Both oxyhemoglobin and deoxyhemoglobin contain the Fe^{2+} ion, which has a $3d^6$ electron configuration. Deoxyhemoglobin, however, is high spin while oxyhemoglobin is low spin, as indicated in the diagram below. This implies that the crystal-field splitting in deoxyhemoglobin is less than that of oxyhemoglobin and that deoxyhemoglobin will absorb light at longer wavelengths. Since deoxyhemoglobin is absorbing redder light, it will appear more purple, or bluish.

Deoxyhemoglobin
(paramagnetic)

Oxyhemoglobin
(diamagnetic)

↑	↑	

$d_{x^2-y^2}, d_{z^2}$

↑↓	↑	↑

d_{xz}, d_{yz}, d_{xy}

↑↓	↑↓	↑↓

15.30 Although both carbon and silicon are in Group 4A, very few Si=Si bonds are known. Account for the instability of silicon-to-silicon double bonds in general. (*Hint*: Compare the covalent radii of C and Si.)

The larger size of the Si atoms prevents effective sideways overlap of the $3p$ orbitals to form π bonds like those formed from the $2p$ orbitals in C atoms.

15.31 Compare the bonding in FeF_6^{3-} and $Fe(CN)_6^{3-}$ in terms of hybridization.

Valence bond theory assumes that each ligand donates a pair of electrons to the metal ion to form a coordinate covalent bond. This requires six vacant orbitals on the metal for hybridization. The Fe^{3+} ion has a $3d^5$ electron configuration. For the high spin FeF_6^{3-}, each of the five $3d$ orbitals are singly occupied, so that the six empty orbitals available for ligand electrons are the $4s$, $4p$ and two of the $4d$ orbitals; this is called sp^3d^2 hybridization. For the low spin $Fe(CN)_6^{3-}$, two of the $3d$ orbitals on the Fe remain empty so that these plus the $4s$ and $4p$ orbitals provide the necessary six empty orbitals; this is called d^2sp^3 hybridization.

High spin

Low spin

↑	↑

$d_{x^2-y^2}, d_{z^2}$

↑	↑	↑

d_{xz}, d_{yz}, d_{xy}

↑↓	↑↓	↑

15.32 Chemical analysis shows that hemoglobin is 0.34% Fe by mass. What is the minimum possible molar mass of hemoglobin? The actual molar mass of hemoglobin is about 65,000 g. How do you account for the discrepancy between your minimum value and the actual value?

The mass percent indicates that a 100.00 g sample of hemoglobin contains 0.34 g Fe, or in moles,

$$(0.34 \text{ g}) \left(\frac{1 \text{ mol}}{55.85 \text{ g}} \right) = 6.09 \times 10^{-3} \text{ mol}$$

There is a molar relationship between the moles of Fe and the moles of hemoglobin. The greatest possible number of moles of hemoglobin, implying the minimum possible molar mass, would be if there were 1 Fe atom per hemoglobin molecule. This would result in 100.00 g of hemoglobin likewise being 6.09×10^{-3} mol, or

$$\frac{100.00 \text{ g}}{6.09 \times 10^{-3} \text{ mol}} = 1.6 \times 10^4 \text{ g mol}^{-1}$$

This is only a fraction of the actual molar mass of 65,000 g mol^{-1}, since there are more than 1 Fe atom per hemoglobin molecule. Indeed the molar mass ratio provides the Fe to hemoglobin ratio.

$$6.5 \times 10^4 \text{ g hemoglobin mol}^{-1} \left(\frac{1 \text{ mol Fe}}{1.6 \times 10^4 \text{ g hemoglobin}} \right) = 4.1 \text{ mol Fe (mol hemoglobin)}^{-1}$$

$$\approx 4 \text{ mol Fe (mol hemoglobin)}^{-1}$$

15.33   Use the molecular-orbital energy-level diagram for O_2 to show that the following Lewis structure corresponds to an excited state.

The Lewis structure indicates that all electrons are paired. This would correspond to an electron configuration of $KK(\sigma_{2s})^2(\sigma_{2s}^*)^2 (\sigma_{2p})^2(\pi_x)^2(\pi_y)^2(\pi_x^*)^2$. Compared to the electron configuration of O_2 from Table 15.3, this configuration has paired the electrons in the π^* orbitals. Since it takes energy to pair electrons in the same orbital, the configuration implied by the Lewis structure is at a higher energy than that from Table 15.3, and is thus an excited state.

15.34   Co binds better to the heme group than Fe, and Co^{2+} has less of a tendency to be oxidized to Co^{3+} than Fe^{2+} does to Fe^{3+}. Why is Fe the metal in hemoglobin and myoglobin rather than Co?

Fe has a greater natural abundance in Earth's crust than does Co.

15.35   Suffocation victims usually look purple, but a person poisoned by carbon monoxide often has rosy cheeks. Explain.

The CO-hemoglobin complex has a bright red color (CO is a strong-field ligand, more so than O_2), but deoxyhemoglobin looks purple. (See Problem 15.29.)

15.36   The dipole moment of *cis*-dichloroethylene is 1.81 D at 25°C. On heating, its dipole moment begins to decrease. Give a reasonable explanation for this observation.

Upon heating *cis-trans* isomerization takes place, and the *trans* isomer has no dipole moment.

15.37 Although the hydroxyl radical (OH) is present only in a trace amount in the atmosphere, it plays an important role in atmospheric chemistry because it is a strong oxidant and can react with many pollutants. Assume that the radical is analogous to the HF molecule and that the molecular orbitals result from the overlap of an oxygen $2p_x$ orbital and a hydrogen $1s$ orbital. **(a)** Draw pictures of the σ and σ^* molecular orbitals in OH. **(b)** Which of the two molecular orbitals has more hydrogen $1s$ character? **(c)** Draw a molecular-orbital energy-level diagram, and write the electron configuration for the radical. Note that the electrons in the nonbonding orbitals on oxygen have π character and should be assigned as such. **(d)** Estimate the bond order of OH. Compare this value with that for OH^+.

(a)

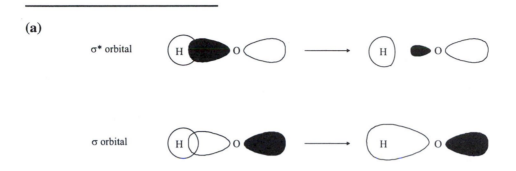

(b) Since the H $1s$ atomic orbital is higher in energy, the antibonding σ^* will have the greater hydrogen $1s$ character.

(c)

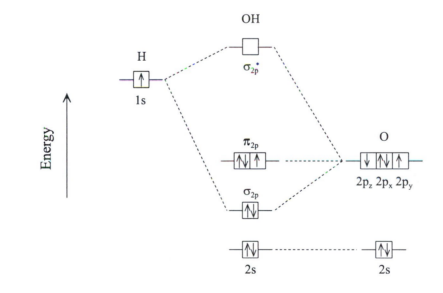

(d) There are two bonding electrons in OH and 5 nonbonding electrons. The bond order is 1. In forming OH^+, an electron is removed from a nonbonding orbital, and the bond order of OH^+ is also 1.

15.38 The H_3^+ ion is the simplest polyatomic molecule. It has equilateral geometry. **(a)** Draw three resonance structures to represent this species. **(b)** Use MO theory to describe the bonding molecular orbital for this ion. Write the wavefunction for the lowest-energy molecular orbital. Is it a σ or π delocalized molecular orbital? **(c)** Given that $\Delta_r H = -849$ kJ mol^{-1} for the reaction $2H + H^+ \rightarrow H_3^+$ and that $\Delta_r H = 436.4$ kJ mol^{-1} for $H_2 \rightarrow 2H$, calculate the value of $\Delta_r H$ for the reaction $H^+ + H_2 \rightarrow H_3^+$. Comment on the magnitude of $\Delta_r H$.

(a)

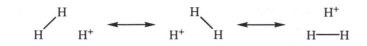

(b) In the MO theory, a molecular orbital is formed from the overlap of the three $1s$ orbitals, one from each H, or $\psi = \frac{1}{\sqrt{3}}\left(1s_a + 1s_b + 1s_c\right)$. Although neither the σ nor π label is strictly appropriate for this molecule, since the orbital is formed from the overlap of s orbitals it could be termed a σ MO. This is an example of a three-center, two electron bond.

(c) This is an application of Hess's Law.

$$2H + H^+ \longrightarrow H_3^+ \quad \Delta_r H = -849 \text{ kJ mol}^{-1}$$
$$H_2 \longrightarrow 2H \quad \Delta_r H = 436.4 \text{ kJ mol}^{-1}$$

The two reactions sum to give

$$H^+ + H_2 \longrightarrow H_3^+ \quad \Delta_r H = -413 \text{ kJ mol}^{-1}$$

The energy released in forming H_3^+ from H^+ and H_2 is almost as large as that released in the formation of H_2 from 2 H atoms. This result demonstrates the role of electron delocalization (σ electrons in this case) in producing stability in the H_3^+ ion.

15.39   A novel electron-transport protein is found to contain only zinc as the metal. Comment on this finding.

Since Zn cannot participate directly in oxidation-reduction reactions, its role is therefore a structural one. Most likely, it helps to maintain a conjugated π system for electron transport.

15.40   Does the molecule HBrC=C=CHBr have a dipole moment?

The two CBrH groups are in different (perpendicular) planes (see figure), and the two bond moments will not cancel. Thus, HBrC=C=CHBr is a polar molecule.

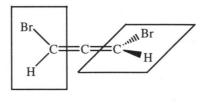

15.41   What is the state of hybridization of the central O atom in O_3? Describe the bonding in O_3 in terms of delocalized molecular orbitals.

The central O atom in O_3 is sp^2 hybridized. The unhybridized $2p_z$ orbital on the central O atom overlaps with the $2p_z$ orbitals on the end atoms to form delocalized π molecular orbitals for the molecule.

15.42   Oxalic acid, $H_2C_2O_4$, is sometimes used to remove rust stains from sinks and bathtubs. Explain the chemistry underlying this cleaning action.

The oxalate ion, from the oxalic acid, forms a water-soluble complex with the Fe^{3+} ion according to

$$Fe_2O_3(s) + 6H_2C_2O_4(aq) \longrightarrow 2Fe(C_2O_4)_3^{3-}(aq) + 3H_2O(l) + 6H^+(aq)$$

Note that bleach-based cleansers, which oxidize stains, would have little effect in removing a rust stain.

15.43   Use the particle-in-a-box model to explain the difference between the potential energy curves that result from Equations 15.2 and 15.3.

In going from Equation 15.2 to 15.3, the electrons are allowed to move between the two atoms. This has the effect of "lengthening the box" in which the electrons are confined. This leads to a lowering of energy, since the energy levels of the particle-in-a-box have an inverse dependence on length.

15.44   Draw qualitative diagrams for the crystal-field splitting in **(a)** a linear complex ion ML_2, **(b)** a trigonal-planar complex ion ML_3, and **(c)** a trigonal-bipyramidal complex ion ML_5.

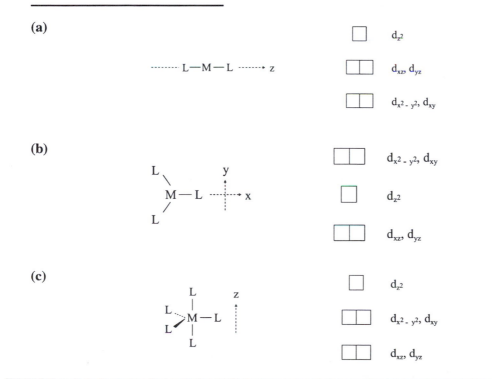

(a)

$$\cdots L-M-L \cdots \rightarrow z$$

d_{z^2}

d_{xz}, d_{yz}

$d_{x^2-y^2}, d_{xy}$

(b)

$d_{x^2-y^2}, d_{xy}$

d_{z^2}

d_{xz}, d_{yz}

(c)

d_{z^2}

$d_{x^2-y^2}, d_{xy}$

d_{xz}, d_{yz}

15.45 You are given two solutions containing $FeCl_2$ and $FeCl_3$ at the same concentration. One solution is light yellow and the other one is brown. Identify these solutions based only on color.

$FeCl_2$ contains the Fe^{2+} ion with a $3d^6$ electron configuration while the $FeCl_3$ solution has the Fe^{3+} ion with a $3d^5$ electron configuration. Since the $3d^5$ configuration leads to 5 singly occupied $3d$ orbitals, each with an unpaired electron, as shown below, d-d electronic transitions are spin-forbidden and the color is faint. (See Problem 15.28) Thus, the light yellow solution is $FeCl_3$ and the brown is $FeCl_2$.

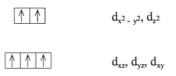

15.46 The geometries discussed in this chapter all lend themselves to fairly straightforward elucidation of bond angles. The exception is the tetrahedron, because its bond angles are hard to visualize. Consider the CCl_4 molecule, which has tetrahedral geometry and is nonpolar. By equating the bond moment of a particular C–Cl bond to the resultant bond moments of the other three C–Cl bonds in opposite directions, show that the bond angles are all equal to 109.5°.

Referring to the figure below, if the bond moment of the upward-pointing C–Cl bond is represented by ρ, then for the molecule to be nonpolar, the downward-pointing components of the three other C–Cl bonds, each of which is $\rho \cos\theta$, must sum to give the same value. That is,

$$3\rho \cos\theta = \rho$$

$$\cos\theta = \frac{1}{3}$$

$$\theta = 70.5°$$

θ is the supplement of the tetrahedral angle, which is thus $180° - 70.5° = 109.5°$.

CHAPTER 16

Intermolecular Forces

PROBLEMS AND SOLUTIONS

16.1 List all the intermolecular interactions that take place in each of the following kinds of molecules: Xe, SO_2, C_6H_5F, and LiF.

Xe:	Dispersion forces
SO_2:	Dipole–dipole and dispersion forces
C_6H_5F:	Dipole–dipole and dispersion forces
LiF:	Ionic and dispersion forces

16.2 Arrange the following species in order of decreasing melting points: Ne, KF, C_2H_6, MgO, H_2S.

$MgO > KF > H_2S > C_2H_6 > Ne$

16.3 The compounds Br_2 and ICl have the same number of electrons, yet Br_2 melts at $-7.2°C$, whereas ICl melts at $27.2°C$. Explain.

ICl is a polar molecule, but Br_2 is nonpolar. The dipole–dipole forces in ICl are responsible for the higher melting point.

16.4 If you lived in Alaska, which of the following natural gases would you keep in an outdoor storage tank in winter? Methane (CH_4), propane (C_3H_8), or butane (C_4H_{10}). Explain.

CH_4 has the weakest intermolecular forces and, as a result, the lowest boiling point, making it the best choice for a cold climate.

16.5 List the types of intermolecular forces that exist between molecules (or basic units) in each of the following species: **(a)** benzene (C_6H_6), **(b)** CH_3Cl, **(c)** PF_3, **(d)** NaCl, **(e)** CS_2.

(a) C_6H_6:   Dispersion forces
(b) CH_3Cl: Dipole–dipole and dispersion forces
(c) PF_3:   Dipole–dipole and dispersion forces
(d) NaCl: Ionic and dispersion forces
(e) CS_2:   Dispersion forces

16.6   The boiling points of the three different structural isomers of pentane (C_5H_{12}) are 9.5°C, 27.9°C, and 36.1°C. Draw their structures, and arrange them in order of decreasing boiling points. Justify your arrangement.

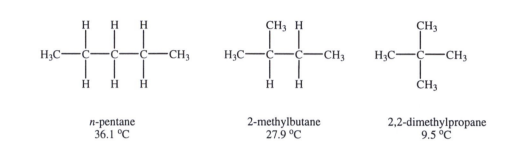

n-pentane   2-methylbutane   2,2-dimethylpropane
36.1 °C   27.9 °C   9.5 °C

The boiling points depend on the ease of packing the molecules together. The n-pentane packs together most easily, and it has the highest boiling point. The packing is least favorable for 2,2-dimethylpropane, which has the lowest boiling point.

16.7   Two water molecules are separated by 2.76 Å in air. Use Equation 16.9 to calculate the dipole–dipole interaction. The dipole moment of water is 1.82 D.

The dipole moment of H_2O is

$$\mu_{H_2O} = (1.82\,D)\left(\frac{3.336 \times 10^{-30}\,C\,m}{1\,D}\right) = 6.072 \times 10^{-30}\,C\,m$$

resulting in a dipole–dipole interaction of

$$V = -\frac{2\mu_{H_2O}\mu_{H_2O}}{4\pi\epsilon_0 r^3}$$

$$= -\frac{2\left(6.072 \times 10^{-30}\,C\,m\right)\left(6.072 \times 10^{-30}\,C\,m\right)}{4\pi\left(8.8542 \times 10^{-12}\,C^2\,N^{-1}\,m^{-2}\right)\left(2.76 \times 10^{-10}\,m\right)^3}$$

$$= -3.152 \times 10^{-20}\,J$$

$$= -3.15 \times 10^{-20}\,J$$

Expressing the energy on a per mole basis,

$$V = -\left(3.152 \times 10^{-20}\,J\right)\left(\frac{6.022 \times 10^{23}}{1\,mol}\right)\left(\frac{1\,kJ}{1000\,J}\right) = -19.0\,kJ\,mol^{-1}$$

16.8 Coulombic forces are usually referred to as long-range forces (they depend on $1/r^2$) whereas van der Waals forces are called short-range forces (they depend on $1/r^7$). **(a)** Assuming that the forces (F) depend only on distances, plot F as a function of r at $r = 1$ Å, 2 Å, 3 Å, 4 Å, and 5 Å. **(b)** Based on your results, explain the fact that although a 0.2 M nonelectrolyte solution usually behaves ideally, nonideal behavior is quite noticeable in a 0.02 M electrolyte solution.

(a) A plot with graphs of $1/r^2$ vs r and $1/r^7$ vs r is presented below. The forces will be proportional to these functions.

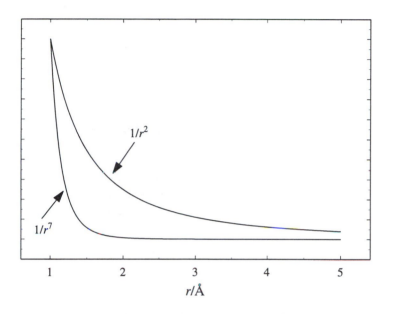

(b) In a nonelectrolyte solution, the attractive forces have a $1/r^7$ dependence, and as the graph shows, they fall off very rapidly with distance. In an electrolyte solution, the ionic (Coulombic) forces have a $1/r^2$ dependence that extends to large distances. These "long-range" forces are responsible for nonideal behavior, even at low concentrations.

16.9 Calculate the induced dipole moment of I_2 due to a Na^+ ion that is 5.0 Å away from the center of the I_2 molecule. The polarizability of I_2 is 12.5×10^{-30} m^3.

Using Equation 16.13 and the discussion following it in the text,

$$\mu_{induced} = \alpha' E$$

$$= \left(4\pi\epsilon_0\alpha\right)\left(\frac{q}{4\pi\epsilon_0 r^2}\right)$$

$$= \frac{\alpha q}{r^2}$$

$$= \frac{\left(12.5 \times 10^{-30}\ \text{m}^3\right)\left(1.602 \times 10^{-19}\ \text{C}\right)}{\left(5.0 \times 10^{-10}\ \text{m}\right)^2}$$

$$= \left(8.01 \times 10^{-30}\ \text{C m}\right)\left(\frac{1\ \text{D}}{3.336 \times 10^{-30}\ \text{C m}}\right)$$

$$= 2.4\ \text{D}$$

16.10   Differentiate Equation 16.21 with respect to r to obtain an expression for σ and ϵ. Express the equilibrium distance, r_e, in terms of σ and show that $V = -\epsilon$.

Starting with Equation 16.21,

$$V = 4\epsilon \left[\left(\frac{\sigma}{r} \right)^{12} - \left(\frac{\sigma}{r} \right)^{6} \right]$$

and differentiating gives

$$\frac{dV}{dr} = 4\epsilon \left[-\frac{12\sigma^{12}}{r^{13}} + \frac{6\sigma^{6}}{r^{7}} \right]$$

The minimum of the potential energy occurs when $r = r_e$ and $\dfrac{dV}{dr} = 0$.

$$4\epsilon \left[-\frac{12\sigma^{12}}{r_e^{13}} + \frac{6\sigma^{6}}{r_e^{7}} \right] = 0$$

$$-\frac{12\sigma^{12}}{r_e^{13}} + \frac{6\sigma^{6}}{r_e^{7}} = 0$$

$$-\frac{2\sigma^{6}}{r_e^{6}} + 1 = 0$$

$$r_e = 2^{1/6}\sigma$$

To calculate the potential energy at the equilibrium distance, substitute the expression for r_e into that for the potential energy.

$$V = 4\epsilon \left[\left(\frac{\sigma}{2^{1/6}\sigma} \right)^{12} - \left(\frac{\sigma}{2^{1/6}\sigma} \right)^{6} \right]$$

$$= 4\epsilon \left[\frac{1}{4} - \frac{1}{2} \right]$$

$$= -\epsilon$$

16.11   Calculate the bond enthalpy of LiF using the Born–Haber cycle. The bond length of LiF is 1.51 Å. See Tables 14.4 and 14.5 for other information. Use $n = 10$ in Equation 16.7.

The Born-Haber cycle is

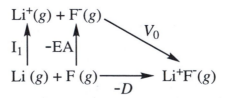

From Equation 16.7, the potential energy at $r = r_0$ is

$$V_0 = -\frac{q_{Na^+}q_{Cl^-}}{4\pi\epsilon_0 r_0}\left(1 - \frac{1}{n}\right)$$

$$= -\frac{(1.602 \times 10^{-19}\,\text{C})\,(1.602 \times 10^{-19}\,\text{C})}{4\pi\,(8.8542 \times 10^{-12}\,\text{C}^2\,\text{N}^{-1}\,\text{m}^{-2})\,(1.51 \times 10^{-10}\,\text{m})}\left(1 - \frac{1}{10}\right)\left(\frac{6.022 \times 10^{23}}{1\,\text{mol}}\right)$$

$$= -8.279 \times 10^5\,\text{J}\,\text{mol}^{-1} = -827.9\,\text{kJ}\,\text{mol}^{-1}$$

$I_1 = 520\,\text{kJ}\,\text{mol}^{-1}$ for Li (Table 14.4) and EA $= 328\,\text{kJ}\,\text{mol}^{-1}$ for F (Table 14.5). Therefore,

$$-D = I_1 - \text{EA} + V_0 = (520 - 328 - 827.9)\;\text{kJ}\,\text{mol}^{-1} = -636\,\text{kJ}\,\text{mol}^{-1}$$

The bond enthalpy of LiF is $636\,\text{kJ}\,\text{mol}^{-1}$

16.12 (a) From the data in Table 16.2, determine the van der Waals radius for argon. (b) Use this radius to determine the fraction of the volume occupied by 1 mole of argon at 25°C and 1 atm.

(a) Since σ gives the distance of closest approach of two argon atoms, the van der Waals radius for argon is

$$r = \frac{\sigma}{2} = \frac{3.40\,\text{Å}}{2} = 1.70\,\text{Å}$$

(b) The volume of 1 mole of Ar atoms is

$$\frac{4}{3}\pi r^3\left(\frac{6.022 \times 10^{23}}{1\,\text{mol}}\right) = \frac{4}{3}\pi\,(1.70 \times 10^{-10}\,\text{m})^3\left(\frac{6.022 \times 10^{23}}{1\,\text{mol}}\right)$$

$$= 1.239 \times 10^{-5}\,\text{m}^3\,\text{mol}^{-1}$$

$$= 1.239 \times 10^{-2}\,\text{L}\,\text{mol}^{-1}$$

The volume occupied by one mole of argon gas is

$$\frac{V}{n} = \frac{RT}{P} = \frac{(0.08206\,\text{L}\,\text{atm}\,\text{K}^{-1}\;\text{mol}^{-1})\,(298.2\,\text{K})}{1\,\text{atm}} = 24.47\,\text{L}\,\text{mol}^{-1}$$

The fraction of this volume occupied by the one mole of argon atoms is

$$\frac{1.239 \times 10^{-2}\,\text{L}\,\text{mol}^{-1}}{24.47\,\text{L}\,\text{mol}^{-1}} = 5.1 \times 10^{-4}$$

16.13 Diethyl ether ($C_2H_5OC_2H_5$) has a boiling point of 34.5°C, whereas 1-butanol (C_4H_9OH) boils at 117°C. These two compounds have the same type and number of atoms. Explain the difference in their boiling points.

Diethyl ether cannot form intermolecular hydrogen bonds, but 1-butanol can. The greater intermolecular forces in 1-butanol results in a much higher boiling point.

16.14   If water were a linear molecule, **(a)** would it still be polar and **(b)** would the water molecules still be able to form hydrogen bonds with one another?

(a) A "linear" water molecule would not be polar.

(b) Such a molecule could still form hydrogen bonds, although it would assume two-dimensional hydrogen bond structures.

16.15   Which of the following compounds is a stronger base: $(CH_3)_4NOH$ or $(CH_3)_3NHOH$? Explain.

A hydrogen bond can exist between $(CH_3)_3NH^+$ and OH^-, as shown in the picture below. Such a hydrogen bond does not exist between $(CH_3)_4N^+$ and OH^- so this species can dissociate more readily, and thus be a stronger base than $(CH_3)_3NHOH$.

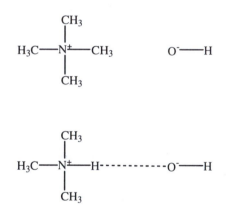

16.16   Explain why ammonia is soluble in water but nitrogen trichloride is not.

Ammonia, NH_3, can form hydrogen bonds with water, but NCl_3 cannot.

16.17   Acetic acid is miscible with water, but it also dissolves in nonpolar solvents such as benzene or carbon tetrachloride. Explain.

Acetic acid can form a dimer that is less polar than the monomer, as shown below. In nonpolar, non-hydrogen-bond-forming solvents, acetic acid exists as this dimeric species.

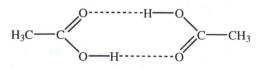

16.18 Which of the following molecules has a higher melting point? Explain your answer.

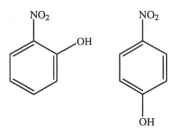

The *para* isomer can form intermolecular hydrogen bonds, while the *ortho* isomer can form only intramolecular hydrogen bonds as shown below. Thus, the *para* form with stronger intermolecular forces will have the higher melting point.

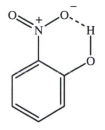

16.19 What type of chemical analysis might be used to test the A–T and C–G pairing in DNA?

If the DNA is denatured and broken up into individual bases for which concentrations are determined, then the following relations, known as Chargoff's rules, must hold.

$$\frac{[A]}{[T]} = \frac{[C]}{[G]}$$

and

$$[A] = [T]$$

$$[C] = [G]$$

or

$$[A] + [G] = [C] + [T]$$

16.20 Assume the energy of hydrogen bonds per base pair to be 10 kJ mol^{-1}. Given two complementary strands of DNA containing 100 base pairs each, calculate the ratio of two separate strands to hydrogen-bonded double helix in solution at 300 K.

For one pair of bases, the ratio of the two separate strands to hydrogen-bonded double helix is

$$\exp\left(-\frac{\Delta E}{RT}\right) = \exp\left[-\frac{10 \times 10^3 \text{ J mol}^{-1}}{(8.314 \text{ J K}^{-1} \text{ mol}^{-1})(300 \text{ K})}\right] = 1.8 \times 10^{-2}$$

For 100 base pairs, the ratio of the two separate strands to hydrogen-bonded double helix is

$$= \exp\left[-\frac{(100)\,(10 \times 10^3 \text{ J mol}^{-1})}{(8.314 \text{ J K}^{-1} \text{ mol}^{-1})\,(300 \text{ K})}\right] = 7.6 \times 10^{-175} \approx 0$$

16.21 The term "like dissolves like" has often been used to describe solubility. Explain what it means.

Molecules with similar electronic structures and functional groups will have similar intermolecular interactions.

16.22 List all the intra- and intermolecular forces that could exist between hemoglobin molecules in water.

All of the intermolecular interactions discussed in the chapter (dispersion, dipole–dipole, ionic) exist between hemoglobin molecules in water.

16.23 A small drop of oil in water usually assumes a spherical shape. Explain.

Oil molecules are hydrophobic, and a drop of oil of a given volume in water will assume a shape that minimizes the surface area exposed to the water. A spherical shape accomplishes this. For example, for a fixed volume of 1 cm^3, a cube would have sides 1 cm long and a total surface area of 6 cm^2, but a sphere would have a radius of 0.62 cm (use $V = \frac{4}{3}\pi r^3$) and a surface area ($4\pi r^2$) of just 4.8 cm^2.

16.24 Which of the following properties indicates very strong intermolecular forces in a liquid? **(a)** A very low surface tension, **(b)** a very low critical temperature, **(c)** a very low boiling point, **(d)** a very low vapor pressure.

Only **(d)** indicates very strong intermolecular forces in a liquid. The others indicate weak intermolecular forces.

16.25 Figure 16.10 shows that the average distance between base pairs measured parallel to the axis of a DNA molecule is 3.4 Å. The average molar mass of a pair of nucleotides is 650 g mol^{-1}. Estimate the length in cm of a DNA molecule of molar mass 5.0×10^9 g mol^{-1}. Roughly how many base pairs are contained in this molecule?

The number of base pairs is

$$\frac{5.0 \times 10^9 \text{ g mol}^{-1}}{650 \text{ g mol}^{-1}} = 7.69 \times 10^6 = 7.7 \times 10^6$$

The length of this DNA molecule is

$$(7.69 \times 10^6)\,(3.4 \times 10^{-8}\ \text{cm}) = 0.26\ \text{cm}$$

16.26   Using values listed in Table 16.1 and a handbook of chemistry, plot the polarizabilities of the noble gases versus their boiling points. On the same graph, also plot their molar masses versus boiling points. Comment on the trends.

The necessary data are in the table. The polarizability of Rn is not accurately known; this noble gas is not included.

Noble Gas	Molar Mass/g·mol^{-1}	$\alpha/10^{-30} \cdot$ m^3	b.p. / K
He	4.00	0.20	4.2
Ne	20.18	0.40	27.1
Ar	39.95	1.66	87.3
Kr	83.80	2.54	120.0
Xe	131.29	4.15	165.2

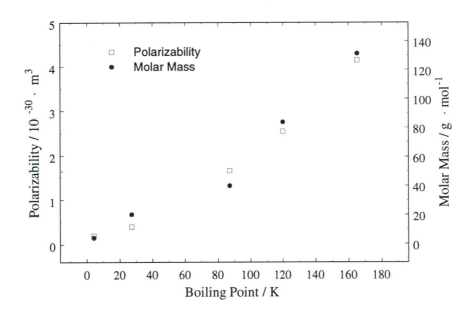

Both polarizability and molar mass seem to track the boiling point of the noble gas.

16.27 Given the following general properties of water and ammonia, comment on the problems that a biological system (as we know it) would have developing in an ammonia medium.

	H_2O	NH_3
Boiling point	373.15 K	239.65 K
Melting point	273.15 K	195.3 K
Molar heat capacity	75.3 J K^{-1} mol^{-1}	8.53 J K^{-1} mol^{-1}
Molar heat of vaporization	40.79 kJ mol^{-1}	23.3 kJ mol^{-1}
Molar heat of fusion	6.0 kJ mol^{-1}	5.9 kJ mol^{-1}
Dielectric constant	78.54	16.9
Viscosity	0.001 N s m^{-2}	0.0254 N s m^{-2} (at 240 K)
Surface tension	0.07275 N m^{-2} (293 K)	0.0412 N m^{-1} (at 244 K)
Dipole moment	1.82 D	1.46 D
Phase at 300 K	Liquid	Gas

The temperature range over which ammonia is a liquid is rather narrow (less than 45 K), although the lower melting point might be advantageous in a cooler environment than our own. Ammonia has a much lower dielectric constant than does water, which makes it much less suitable as a solvent for ionic compounds. The heat capacity of ammonia is just a fraction of that of water so that a biological system in an ammonia medium would have difficulty "thermostatting" the temperature of the system from the heat generated from metabolic processes. The molar heat of vaporization for ammonia is about half that of water, but that might not be much of a problem, since cooling also occurs radiatively through heat loss to the surroundings. The large viscosity of ammonia would make it difficult for molecules and ions to diffuse through an ammonia medium. On the other hand, the lower surface tension of ammonia might be advantageous towards the development of life, since it might not be necessary for organisms to develop surfactants to aid biological processes.

16.28 The HF_2^- ion exists as

$$\left[\text{F}\!-\!\!\text{H}\!\cdots\!\text{F}\right]^-$$

The fact that both HF bonds are the same length suggests that proton tunneling occurs. **(a)** Draw resonance structures for the ion. **(b)** Give a molecular orbital description (with an energy-level diagram) of hydrogen bonding in the ion.

(a)

(b) The 1s orbital on the H atom and a 2p orbital on each of the F atoms (the ones along the internuclear axis) combine to form 3 σ molecular orbitals: one bonding, one nonbonding, and one antibonding. There are four electrons to be accommodated in these molecular orbitals, and they are placed, paired, in the lowest two. Thus, there is a delocalized σ bond extending over the entire ion and a delocalized "lone pair" as a result of the nonbonding molecular orbital that has significant electron density at the fluorines, but a node at the hydrogen. (The other 12 valence electrons in the ion are in localized orbitals on the two fluorines.)

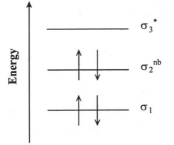

16.29 **(a)** Draw a potential-energy curve for two atoms, based on a hard-sphere model. **(b)** A potential intermediate between the hard-sphere and the Lennard-Jones potentials is the square-well potential, defined by $V = \infty$ for $r < \sigma$, $V = -\epsilon$ for $\sigma \leq r \leq a$, and $V = 0$ for $r > a$. Sketch this potential.

(a) Hard-sphere potential

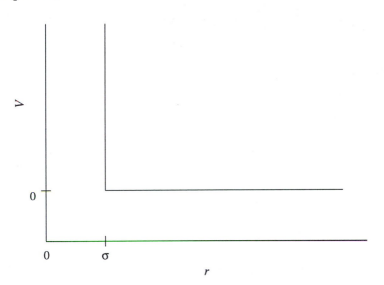

(b) Square-well potential

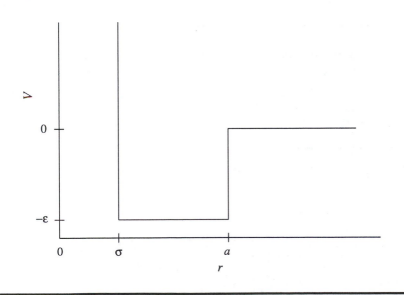

16.30   The potential energy of the helium dimer (He_2) is given by

$$V = \frac{B}{r^{13}} - \frac{C}{r^6}$$

where $B = 9.29 \times 10^4 \text{ kJ Å}^{13} \text{ (mol dimer)}^{-1}$ and $C = 97.7 \text{ kJ Å}^6 \text{ (mol dimer)}^{-1}$. **(a)** Calculate the equilibrium distance between the He atoms. **(b)** Calculate the binding energy of the dimer. **(c)** Would you expect the dimer to be stable at room temperature (300 K)?

(a) The equilibrium distance, r_e, can be calculated by setting $\dfrac{dV}{dr} = 0$.

$$\frac{dV}{dr} = -\frac{13B}{r^{14}} + \frac{6C}{r^7}$$

$$-\frac{13B}{r_e^{14}} + \frac{6C}{r_e^7} = 0$$

$$r_e = \left(\frac{13B}{6C} \right)^{1/7} = \left\{ \frac{13 \left[9.29 \times 10^4 \text{ kJ Å}^{13} \text{ (mol dimer)}^{-1} \right]}{6 \left[97.7 \text{ kJ Å}^6 \text{ (mol dimer)}^{-1} \right]} \right\}^{1/7} = 2.975 \text{ Å} = 2.98 \text{ Å}$$

(b) The binding energy, $V(r_e)$, is

$$V(r_e) = \frac{B}{r_e^{13}} - \frac{C}{r_e^6}$$

$$= \frac{9.29 \times 10^4 \text{ kJ Å}^{13} \text{ (mol dimer)}^{-1}}{\left(2.975 \text{ Å} \right)^{13}} - \frac{97.7 \text{ kJ Å}^6 \text{ (mol dimer)}^{-1}}{\left(2.975 \text{ Å} \right)^6}$$

$$= -7.60 \times 10^{-2} \text{ kJ (mol dimer)}^{-1}$$

(c) The thermal energy at 300 K is

$$RT = \left(8.314 \text{ J K}^{-1} \text{ mol}^{-1} \right) (300 \text{ K}) = 2.49 \times 10^3 \text{ J mol}^{-1} = 2.49 \text{ kJ mol}^{-1}$$

which is much larger than $7.60 \times 10^{-2} \text{ kJ (mol dimer)}^{-1}$. Thus the dimer would not be able to form at room temperature. This species has been observed and studied at low temperature.

16.31   The internuclear distance between two closest Ar atoms in solid argon is about 3.8 Å. The polarizability of argon is $1.66 \times 10^{-30} \text{ m}^3$, and the first ionization energy is 1521 kJ mol^{-1}. Estimate the boiling point of argon. [*Hint*: Calculate the potential energy due to dispersion interaction for solid argon, and equate this quantity to the average kinetic energy of 1 mole of argon gas, which is $(3/2)RT$.]

The potential energy due to the dispersion interaction for solid argon is

$$V = -\frac{3}{4} \frac{\alpha^2 I}{r^6}$$

$$= -\frac{3}{4} \left[\frac{\left(1.66 \times 10^{-30} \text{ m}^3 \right)^2 \left(1521 \text{ kJ mol}^{-1} \right)}{\left(3.8 \times 10^{-10} \text{ m} \right)^6} \right]$$

$$= -1.04 \text{ kJ mol}^{-1}$$

Argon boils when thermal energy, $\frac{3}{2}RT$, supplies $1.04\ \text{kJ mol}^{-1}$ to overcome the binding energy due to the dispersion interaction of argon atoms. That is, when

$$\frac{3}{2}RT = 1.04\ \text{kJ mol}^{-1}$$

$$T = \frac{2}{3}\left(\frac{1.04 \times 10^3\ \text{J mol}^{-1}}{8.314\ \text{J K}^{-1}\ \text{mol}^{-1}}\right) = 83\ \text{K}$$

The actual boiling point of argon is 88 K.

This analysis estimates the boiling point and not the melting point because in liquid argon most of the dispersion forces between atoms are still present. It is only in going to the gas phase that the interatomic forces become negligible.

CHAPTER 17

Spectroscopy

PROBLEMS AND SOLUTIONS

17.1 Convert 15,000 cm^{-1} to wavelength (nm) and frequency.

$$\tilde{\nu} = \frac{1}{\lambda} = \frac{\nu}{c}$$

$$\lambda = \frac{1}{\tilde{\nu}} = \frac{1}{15000 \text{ cm}^{-1}} \left(\frac{1 \text{ nm}}{1 \times 10^{-7} \text{ cm}} \right) = 670 \text{ nm}$$

$$\nu = c\tilde{\nu} = \left(3.00 \times 10^8 \text{ m s}^{-1} \right) \left(15000 \text{ cm}^{-1} \right) \left(\frac{100 \text{ cm}}{1 \text{ m}} \right) = 4.5 \times 10^{14} \text{ s}^{-1}$$

17.2 Convert 450 nm to wavenumber and frequency.

$$\lambda = \frac{c}{\nu} = \frac{1}{\tilde{\nu}}$$

$$\tilde{\nu} = \frac{1}{\lambda} = \frac{1}{450 \text{ nm}} \left(\frac{1 \text{ nm}}{1 \times 10^{-7} \text{ cm}} \right) = 2.2 \times 10^4 \text{ cm}^{-1}$$

$$\nu = \frac{c}{\lambda} = \frac{3.00 \times 10^8 \text{ m s}^{-1}}{450 \text{ nm}} \left(\frac{1 \text{ nm}}{1 \times 10^{-9} \text{ m}} \right) = 6.7 \times 10^{14} \text{ s}^{-1}$$

17.3 Convert the following percent transmittance to absorbance: **(a)** 100%, **(b)** 50%, and **(c)** 0%.

$$A = \log \frac{I_0}{I} = - \log T$$

(a) $T = 1$, and $A = - \log 1 = 0$. **(b)** $T = 0.5$, and $A = - \log 0.50 = 0.30$. **(c)** $T = 0$, and $A = - \log 0 = \infty$.

17.4   Convert the following absorbance to percent transmittance: **(a)** 0.0, **(b)** 0.12, and **(c)** 4.6.

$$T = \frac{I}{I_0} = 10^{\log \frac{I}{I_0}} = 10^{-A}$$

(a) $A = 0.0$, and $T = 10^{-0} = 1 = 100\%$. **(b)** $A = 0.12$, and $T = 10^{-0.12} = 0.76 = 76\%$.
(c) $A = 4.6$, and $T = 10^{-4.6} = 2.5 \times 10^{-5} = 0.0025\%$.

17.5   The absorption of radiation energy by a molecule results in the formation of an excited molecule. Given enough time, it would seem that all of the molecules in a sample would have been excited and no more absorption would occur. Yet in practice we find that the absorbance of a sample at any wavelength remains unchanged with time. Why?

Excited molecules lose their excess energy through a variety of relaxation mechanisms and return to the ground state. Under conditions of constant irradiation, a steady state is reached, and the absorbance of the sample remains unchanged in time. If the incident radiation is of sufficient intensity, so that the rate of excitation is faster than all relaxation processes, the sample approaches a state where half of the molecules have been excited, as indicated by Equation (17.46). At this point the populations of the two states involved in the transitions are equal, and the rate of stimulated absorption equals the rate of stimulated emission resulting in no net absorption.

17.6   The mean lifetime of an electronically excited molecule is 1.0×10^{-8} s. If the emission of the radiation occurs at 610 nm, what are the uncertainties in frequency ($\Delta\nu$) and wavelength ($\Delta\lambda$)?

The natural linewidth, or uncertainty in frequency, of a transition is related to the lifetime of the excited state through the uncertainty principle

$$\Delta\nu = \frac{1}{4\pi\,\Delta t}$$

$$= \frac{1}{4\pi\left(1.0 \times 10^{-8}\text{ s}\right)}$$

$$= 7.96 \times 10^6 \text{ s}^{-1}$$

$$= 8.0 \times 10^6 \text{ s}^{-1}$$

Since $\lambda = c/\nu$, then $|\Delta\lambda| = \dfrac{c}{\nu^2}|\Delta\nu| = \lambda\dfrac{|\Delta\nu|}{\nu}$.

With $\lambda = 610$ nm, $\nu = \left(3.00 \times 10^8 \text{ m s}^{-1}\right)/\left(610 \times 10^{-9}\text{ m}\right) = 4.92 \times 10^{14} \text{ s}^{-1}$, and

$$\Delta\lambda = \lambda\frac{\Delta\nu}{\nu}$$

$$= (610 \text{ nm})\left(\frac{7.96 \times 10^6 \text{ s}^{-1}}{4.92 \times 10^{14} \text{ s}^{-1}}\right)$$

$$= 9.9 \times 10^{-6} \text{ nm}$$

17.7 The familiar yellow D lines of sodium is actually a doublet at 589.0 nm and 589.6 nm. Calculate the difference in energy (in J) between these two lines.

$$\Delta E = \left(\frac{hc}{\lambda_1} - \frac{hc}{\lambda_2}\right)$$

$$= hc\left(\frac{1}{\lambda_1} - \frac{1}{\lambda_2}\right)$$

$$= (6.626 \times 10^{-34} \text{ J s})(3.00 \times 10^8 \text{ m s}^{-1})\left(\frac{1}{589.0 \times 10^{-9} \text{ m}} - \frac{1}{589.6 \times 10^{-9} \text{ m}}\right)$$

$$= 3.43 \times 10^{-22} \text{ J}$$

17.8 The resolution of visible and UV spectra can usually be improved by recording the spectra at low temperatures. Why does this procedure work?

The lower temperature reduces molecular speeds so that the effects of both Doppler and collisional broadening are reduced.

17.9 Assuming that the width of a spectral line is the result solely of lifetime broadening, estimate the lifetime of a state that gives rise to a line of width **(a)** 1.0 cm^{-1}, **(b)** 0.50 Hz.

The lifetime of an excited state is related to its natural linewidth via $\Delta \nu = \dfrac{1}{4\pi \Delta t}$.

(a) Since $\nu = c\tilde{\nu}$,

$$\Delta t = \frac{1}{4\pi \Delta \nu}$$

$$= \frac{1}{4\pi c \Delta \tilde{\nu}}$$

$$= \frac{1}{4\pi (3.00 \times 10^8 \text{ m s}^{-1})(1.0 \text{ cm}^{-1})\left(\frac{100 \text{ cm}}{1 \text{ m}}\right)}$$

$$= 2.7 \times 10^{-12} \text{ s}$$

(b)

$$\Delta t = \frac{1}{4\pi \Delta \nu}$$

$$= \frac{1}{4\pi (0.50 \text{ s}^{-1})}$$

$$= 0.16 \text{ s}$$

17.10   What is the molar absorptivity of a solute that absorbs 86% of a certain wavelength of light when the beam passes through a 1.0-cm cell containing a 0.16 M solution?

With 86% of light absorbed, the transmittance is $T = 1.00 - 0.86 = 0.14$, and the absorbance is $A = -\log T = -\log 0.14 = 0.854$. Then using the Beer–Lambert law, $A = \epsilon bc$,

$$\epsilon = \frac{A}{bc}$$

$$= \frac{0.854}{(1.0 \text{ cm})(0.16 \text{ } M)}$$

$$= 5.3 \text{ L mol}^{-1} \text{ cm}^{-1}$$

17.11   The molar absorptivity of a benzene solution of an organic compound is 1.3×10^2 L mol^{-1} cm^{-1} at 422 nm. Calculate the percentage reduction in light intensity when light of that wavelength passes through a 1.0-cm cell containing a solution of concentration 0.0033 M.

The Beer–Lambert law gives the absorbance,

$$A = \epsilon bc$$

$$= \left(1.3 \times 10^2 \text{ L mol}^{-1} \text{ cm}^{-1}\right)(1.0 \text{ cm})(0.0033 \text{ } M)$$

$$= 0.429$$

The transmittance is found via

$$A = -\log T$$

$$T = 10^{-A} = 10^{-0.429} = 0.372$$

This transmittance corresponds to a reduction in light intensity of $(1 - 0.372) \times 100\% = 63\%$.

17.12   A single NMR scan of a dilute sample exhibits a signal-to-noise (S/N) ratio of 1.8. If each scan takes 8.0 minutes, calculate the minimum time required to generate a spectrum with a S/N ratio of 20.

After n scans the signal intensity will increase by a factor of n, while the noise will increase by a factor of $\sqrt{n}$, so that, using $\left(\dfrac{S}{N}\right)_1$ for the signal-to-noise ratio after one scan, the ratio after n scans becomes

$$\left(\frac{S}{N}\right)_n = \frac{nS}{\sqrt{n}N} = \sqrt{n}\left(\frac{S}{N}\right)_1$$

or

$$\sqrt{n} = \frac{(S/N)_n}{(S/N)_1} = \frac{20}{1.8} = 11.1$$

$$n = 123$$

Acquiring 123 scans at 8.0 minutes per scan will require $123 \times 8 \text{ min} = 9.8 \times 10^2 \text{ min} = 16 \text{ hr}$.

17.13 Which of the following molecules are microwave active? C_2H_2, CH_3Cl, C_6H_6, CO_2, H_2O, HCN.

Those molecules with a permanent dipole moment are microwave active. These are CH_3Cl, H_2O, and HCN.

17.14 What is the degeneracy of the rotational energy level with $J = 7$ for a diatomic rigid rotor?

The degeneracy of a rotational energy level is given by $2J + 1$, so that for $J = 7$ the degeneracy is $2(7) + 1 = 15$.

17.15 The $J = 3 \rightarrow 4$ transition for a diatomic molecule occurs at 0.50 cm^{-1}. What is the wavenumber for the $J = 6 \rightarrow 7$ transition for this molecule? Assume the molecule is a rigid rotor.

Assuming a rigid rotor, the frequency of a rotational transition is given by $\nu = \Delta E_{rot}/h = 2BJ'$. The wavenumber of the $J = 3 \rightarrow 4$ transition with $J' = 4$ supplies the value of B.

$$B = \frac{\Delta E_{rot}/h}{2J'}$$

$$= \frac{hc\tilde{\nu}/h}{2J'}$$

$$= \frac{c\tilde{\nu}}{2J'}$$

$$= \frac{\left(3.00 \times 10^8 \text{ m s}^{-1}\right)\left(0.50 \text{ cm}^{-1}\right)\left(\frac{100 \text{ cm}}{1 \text{ m}}\right)}{2(4)}$$

$$= 1.88 \times 10^9 \text{ s}^{-1}$$

Then for the $J = 6 \rightarrow 7$ transition with $J' = 7$,

$$\tilde{\nu} = \frac{\nu}{c}$$

$$= \frac{\Delta E_{rot}/h}{c}$$

$$= \frac{2BJ'}{c}$$

$$= \frac{2\left(1.88 \times 10^9 \text{ s}^{-1}\right)(7)}{3.00 \times 10^8 \text{ m s}^{-1}}\left(\frac{1 \text{ m}}{100 \text{ cm}}\right)$$

$$= 0.88 \text{ cm}^{-1}$$

17.16 The equilibrium bond length in nitric oxide ($^{14}N^{16}O$) is 1.15 Å. Calculate **(a)** the moment of inertia of NO, and **(b)** the energy for the $J = 0 \rightarrow 1$ transition. How many times does the molecule rotate per second at the $J = 1$ level?

In finding the reduced mass of $^{14}N^{16}O$ it is important to use masses appropriate for the specific isotopes under consideration and not the average masses found in the periodic table.

$$\mu = \frac{m_N m_O}{m_N + m_O} = \frac{(14.00 \text{ amu}) (15.99 \text{ amu})}{14.00 \text{ amu} + 15.99 \text{ amu}} \left(1.661 \times 10^{-27} \text{ kg amu}^{-1}\right) = 1.2399 \times 10^{-26} \text{ kg}$$

(a) The moment of inertia is given by

$$I = \mu r^2$$

$$= \left(1.2399 \times 10^{-26} \text{ kg}\right) \left(1.15 \times 10^{-10} \text{ m}\right)^2$$

$$= 1.640 \times 10^{-46} \text{ kg m}^2$$

$$= 1.64 \times 10^{-46} \text{ kg m}^2$$

(b) The rotational constant for the molecule is

$$B = \frac{h}{8\pi^2 I} = \frac{6.626 \times 10^{-34} \text{ J s}}{8\pi^2 \left(1.640 \times 10^{-46} \text{ kg m}^2\right)} = 5.117 \times 10^{10} \text{ s}^{-1}$$

and the energy for the $J = 0 \rightarrow 1$ transition is

$$\Delta E_{0\rightarrow 1} = 2BhJ' = 2 \left(5.117 \times 10^{10} \text{ s}^{-1}\right) \left(6.626 \times 10^{-34} \text{ J s}\right) (1) = 6.781 \times 10^{-23} \text{ J}$$

$$= 6.78 \times 10^{-23} \text{ J}$$

The frequency of molecular rotation is equal to the frequency of the electromagnetic radiation that causes the transition, which is

$$v = \frac{\Delta E_{0\rightarrow 1}}{h} = \frac{6.781 \times 10^{-23} \text{ J}}{6.626 \times 10^{-34} \text{ J s}} = 1.02 \times 10^{11} \text{ s}^{-1}$$

17.17   Which of the following molecules are IR active: (a) N_2, (b) HBr, (c) CH_4, (d) Xe, (e) H_2O_2, (f) NO?

The molecules that are IR active are those that possess at least one normal mode of vibration whose motion changes the dipole moment of the molecule. These are (b), (c), (e), and (f).

17.18   Give the number of normal vibrational modes of (a) O_3, (b) C_2H_2, (c) CBr_4, (d) C_6H_6.

Molecules (a), (c), and (d) are non-linear and have $3N - 6$ normal modes, where N is the number of atoms in the molecule. Molecule (b) is linear and has $3N - 5$ normal modes.

(a) $3 \times 3 - 6 = 3$ (b) $3 \times 4 - 5 = 7$ (c) $3 \times 5 - 6 = 9$ (d) $3 \times 12 - 6 = 30$

17.19   Draw a vibrational mode of the BF_3 molecule that is IR inactive.

The symmetric stretching vibration of BF_3, where each B–F bond is stretched by the same amount, has no effect on the dipole moment of the trigonal planar molecule, and is IR inactive.

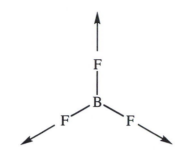

17.20 A 500-g object suspended from the end of a rubber band has a vibrational frequency of 4.2 Hz. Calculate the force constant of the rubber band.

The vibrational frequency is given by $\nu = \dfrac{1}{2\pi}\sqrt{\dfrac{k}{m}}$, so that

$$k = 4\pi^2 m \nu^2$$

$$= 4\pi^2 \left(500 \times 10^{-3}\ \text{kg}\right)\left(4.2\ \text{s}^{-1}\right)^2$$

$$= 3.5 \times 10^2\ \text{kg s}^{-2}$$

$$= 3.5 \times 10^2\ \text{N m}^{-1}$$

17.21 The fundamental frequency of vibration for carbon monoxide is 2143.3 cm^{-1}. Calculate the force constant of the carbon–oxygen bond.

Since $\nu = \dfrac{1}{2\pi}\sqrt{\dfrac{k}{\mu}}$, find

$$\nu = c\tilde{\nu} = \left(3.00 \times 10^8\ \text{m s}^{-1}\right)\left(2143.3\ \text{cm}^{-1}\right)\left(\frac{100\ \text{cm}}{1\ \text{m}}\right) = 6.430 \times 10^{13}\ \text{s}^{-1}$$

and, using the correct isotopic masses

$$\mu = \frac{m_C m_O}{m_C + m_O} = \frac{(12.00\ \text{amu})\,(15.99\ \text{amu})}{12.00\ \text{amu} + 15.99\ \text{amu}}\left(1.661 \times 10^{-27}\ \text{kg amu}^{-1}\right) = 1.1387 \times 10^{-26}\ \text{kg}$$

and then

$$k = 4\pi^2 \mu \nu^2$$

$$= 4\pi^2 \left(1.1387 \times 10^{-26}\ \text{kg}\right)\left(6.430 \times 10^{13}\ \text{s}^{-1}\right)^2$$

$$= 1.86 \times 10^3\ \text{kg s}^{-2}$$

$$= 1.86 \times 10^3\ \text{N m}^{-1}$$

17.22   If molecules did not possess zero-point energy, would they be able to undergo the $v = 0 \rightarrow 1$ transition?

Yes, since the oscillating electric field of the IR radiation would be able to induce motion of the centers of positive and negative charge in much the same way that the rotational motion of a polar molecule is excited. Of course, it is impossible to construct physically meaningful wavefunctions for an oscillating molecule with no zero-pont energy so that it might be argued that since such a molecule could never exist, neither could the transition take place.

17.23   Under what conditions can one observe a hot band in the IR spectrum?

There must be sufficient population in the $v = 1$ state so that the $v = 1 \rightarrow 2$ transition can be observed. Additionally, the anharmonicity must be great enough so that the $v = 1 \rightarrow 2$ transition is resolved from the $v = 0 \rightarrow 1$ fundamental transition.

17.24   Show all the fundamental vibration modes of **(a)** carbon disulfide (CS_2), and **(b)** carbonyl sulfide (OCS), and indicate which ones are IR active.

(a) The fundamental vibration modes of CS_2 are identical in form to those of CO_2 (see Figure 17.15). The asymmetric stretch and the (doubly degenerate) bending modes are IR active.

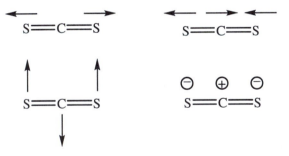

(b) OCS is also a linear molecule with 4 fundamental vibration modes. All four are IR active, leading to 3 IR peaks, since the bend is doubly degenerate.

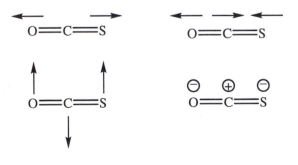

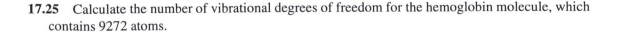

17.25   Calculate the number of vibrational degrees of freedom for the hemoglobin molecule, which contains 9272 atoms.

Hemoglobin is not a linear molecule, and the number of vibrational degrees of freedom is $3N - 6 = 3(9272) - 6 = 27810$.

17.26   Which of the following molecules has the highest fundamental frequency of vibration? H_2, D_2, HD.

Since the fundamental frequency of vibration is given by $v = \dfrac{1}{2\pi}\sqrt{\dfrac{k}{\mu}}$, and since the force constant for the bond is isotopically invariant, the molecule with the lowest reduced mass will have the highest fundamental frequency of vibration. The reduced masses are

$$\mu_{H_2} = \frac{(1.008 \text{ amu})(1.008 \text{ amu})}{1.008 \text{ amu} + 1.008 \text{ amu}} = 0.5040 \text{ amu}$$

$$\mu_{HD} = \frac{(1.008 \text{ amu})(2.014 \text{ amu})}{1.008 \text{ amu} + 2.014 \text{ amu}} = 0.6718 \text{ amu}$$

$$\mu_{D_2} = \frac{(2.014 \text{ amu})(2.014 \text{ amu})}{2.014 \text{ amu} + 2.014 \text{ amu}} = 1.007 \text{ amu}$$

Thus, H_2 has the highest fundamental vibration frequency.

17.27   The fundamental frequency of vibration for $D^{35}Cl$ is given by $\bar{v} = 2081.0 \text{ cm}^{-1}$. Calculate the force constant, k, and compare this value with the force constant obtained for $H^{35}Cl$ in Example 17.2. Comment on your result.

Since $v = \dfrac{1}{2\pi}\sqrt{\dfrac{k}{\mu}}$, find

$$v = c\tilde{v} = \left(3.00 \times 10^8 \text{ m s}^{-1}\right)\left(2081.0 \text{ cm}^{-1}\right)\left(\frac{100 \text{ cm}}{1 \text{ m}}\right) = 6.243 \times 10^{13} \text{ s}^{-1}$$

and (being sure to use the isotopic mass for ^{35}Cl, which differs from the average mass found in the periodic table)

$$\mu = \frac{m_D m_{Cl}}{m_D + m_{Cl}} = \frac{(2.014 \text{ amu})(34.97 \text{ amu})}{2.014 \text{ amu} + 34.97 \text{ amu}}\left(1.661 \times 10^{-27} \text{ kg amu}^{-1}\right) = 3.1631 \times 10^{-27} \text{ kg}$$

and then

$$k = 4\pi^2 \mu v^2$$

$$= 4\pi^2 \left(3.1631 \times 10^{-27} \text{ kg}\right)\left(6.243 \times 10^{13} \text{ s}^{-1}\right)^2$$

$$= 4.87 \times 10^2 \text{ kg s}^{-2}$$

$$= 4.87 \times 10^2 \text{ N m}^{-1}$$

Within the precision of the data used, this is the same as that found in Example 17.2, which is expected, since isotopic substitution does not affect the electron configuration and hence neither the force constant.

17.28   Anthracene is colorless, but tetracene is light orange. Explain.

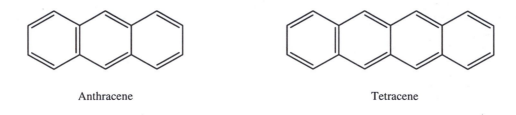

Anthracene   Tetracene

In tetracene, the electrons are delocalized over a greater space than those in anthracene. Recalling the particle-in-a-box model, the greater the length of the "box", the smaller the spacing between energy levels. The greater "length" of the box in tetracene causes the absorption wavelength to shift from the UV region (in anthracene) into the visible region.

17.29   Use the particle-in-a-one-dimensional-box model to calculate the longest-wavelength peak in the electronic absorption spectrum of hexatriene. (*Hint*: See Equation 14.37.)

According to Equation 14.37, $\lambda = \dfrac{8mL^2c}{h(N+1)}$. In hexatriene, $N = 6$, and the length of the box is given by the sum of the lengths of 2 C–C bonds plus 3 C=C bonds with the addition of $1/2 \times$ C–C at each end of the molecule, or

$$L = 2\left(1.54 \times 10^{-10}\ \text{m}\right) + 3\left(1.35 \times 10^{-10}\ \text{m}\right) + 2\left(0.77 \times 10^{-10}\ \text{m}\right) = 8.67 \times 10^{-10}\ \text{m}$$

and

$$\lambda = \frac{8\left(9.109 \times 10^{-31}\ \text{kg}\right)\left(8.67 \times 10^{-10}\ \text{m}\right)^2\left(3.00 \times 10^8\ \text{m s}^{-1}\right)}{\left(6.626 \times 10^{-34}\ \text{J s}\right)(7)} = 3.54 \times 10^{-7}\ \text{m} = 354\ \text{nm}$$

This is the transition from the HOMO to the LUMO ($\pi \rightarrow \pi^*$) and corresponds to the longest wavelength.

17.30   Many aromatic hydrocarbons are colorless, but their anion and cation radicals are often strongly colored. Give a qualitative explanation for this phenomenon. (*Hint*: Consider only the π molecular orbitals.)

In aromatic hydrocarbons, the HOMO is the highest energy π orbital and the LUMO is the lowest energy π^* orbital. The energy separation between these orbitals is so large that the lowest energy transition, the $\pi \rightarrow \pi^*$, lies in the UV, rendering the molecule colorless. In the cation and anion radicals, the lowest energy transition is a $\pi \rightarrow \pi$ or $\pi^* \rightarrow \pi^*$ transition, respectively. The smaller energy spacings among the π and π^* orbitals, compared to the HOMO-LUMO gap, result in the absorption of a photon in the visible region of the spectrum.

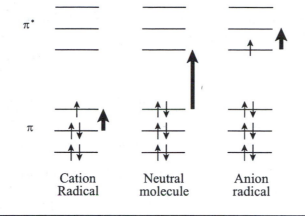

π^*

π

| Cation Radical | Neutral molecule | Anion radical |

17.31 Referring to Figure 17.24, explain why the value of T_m increases as the mole percent (C + G) increases.

There are 3 hydrogen bonds between each C–G pair compared to only 2 hydrogen bonds between each A–T pair.

17.32 The NMR signal of a compound is found to be 240 Hz downfield from the TMS peak using a spectrometer operating at 60 MHz. Calculate its chemical shift in ppm relative to TMS.

According to Equation 17.42

$$\delta = \frac{\nu - \nu_{ref}}{\nu_{ref}} \times 10^6 = \frac{240 \text{ Hz}}{60 \times 10^6 \text{ Hz}} \times 10^6 = 4.0 \text{ ppm}$$

17.33 Both NMR and ESR spectroscopy differ from other branches of spectroscopy discussed in this chapter in one important respect. Explain.

Both NMR and ESR spectroscopy utilize the interaction of the oscillating magnetic field of the electromagnetic radiation with a molecular property. The others forms of spectroscopy discussed involve the oscillating electric field of the electromagnetic radiation. Additionally, in both NMR and ESR spectroscopy, the energy levels are degenerate in the absence of an external magnetic field, and the separation of the energy levels depends on the strength of the external field. This is not a feature of the other spectroscopies discussed in the chapter, although a similar effect of an external electric or magnetic field may be observed on the degenerate rotational levels of molecules.

17.34 What is the field strength (in tesla) needed to generate a ^{1}H frequency of 600 MHz?

From Equation 17.39 and Table 17.4

$$B_0 = \frac{2\pi \nu}{\gamma} = \frac{2\pi \left(600 \times 10^6 \text{ s}^{-1}\right)}{26.75 \times 10^7 \text{ T}^{-1} \text{ s}^{-1}} = 14.1 \text{ T}$$

17.35   Suppose the NMR spectrum of acetaldehyde (see Figure 17.46) is recorded at 200 MHz and 400 MHz. State whether each of the following quantities remains unchanged or is different from 200 MHz to 400 MHz: **(a)** sensitivity of detection, **(b)** $\left|\delta_{CH_3} - \delta_H\right|$, **(c)** $\left|\nu_{CH_3} - \nu_H\right|$, **(d)** J.

(a) increased (due to more favorable Boltzmann distribution), **(b)** unchanged, **(c)** increased, **(d)** unchanged.

17.36   For an applied field of 9.4 T (used in a 400-MHz spectrometer), calculate the difference in frequencies for two protons whose δ values differ by 2.5.

$$\Delta\nu = \frac{\delta \times \nu_{ref}}{10^6} = \frac{2.5\left(400 \times 10^6 \text{ Hz}\right)}{10^6} = 1.0 \times 10^3 \text{ Hz}$$

17.37   For each of the following molecules, state how many proton NMR peaks occur and whether each peak is a singlet, doublet, triplet, etc. **(a)** CH_3OCH_3, **(b)** $C_2H_5OC_2H_5$, **(c)** C_2H_6, **(d)** CH_3F, **(e)** $CH_3COOC_2H_5$.

(a) One singlet, **(b)** two peaks, one triplet (1:2:1) and one quartet (1:3:3:1), **(c)** one singlet, **(d)** one doublet (1:1) due to the coupling of the protons to the fluorine nucleus which has $I = 1/2$, **(e)** three peaks, one singlet, one triplet (1:2:1), one quartet (1:3:3:1).

17.38   Sketch the NMR spectrum of isobutyl alcohol [$(CH_3)_2CHCH_2OH$], given the following chemical shift data: $-CH_3$: 0.89 ppm, $-C-H$: 1.67 ppm, $-CH_2$: 3.27 ppm, $-O-H$: 4.50 ppm.

The $-O-H$ proton is not split because of solvent exchange. The $-C-H$ proton at 1.67 ppm is split by the 6 protons on the 2 equivalent $-CH_3$ groups into 7 lines, which are each then further split into 3 lines by the two protons on the $-CH_2-$ for a total of 21 lines. In practice, however, free rotation about the C–C bonds causes the two different spin–spin couplings to have very nearly the same value. The 21 lines overlap so that except under the highest resolution they appear as a nonet.

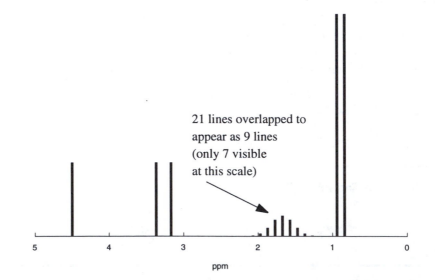

21 lines overlapped to appear as 9 lines (only 7 visible at this scale)

ppm

Under sufficiently high resolution, the multiplet at 1.67 ppm would appear as sketched below. (The differences in spin–spin coupling have been exaggerated for this sketch.)

High resolution view of
multiplet at 1.67 ppm

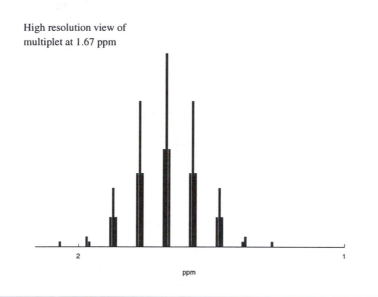

17.39   The toluene proton NMR spectrum, consisting of two peaks due to the methyl and aromatic protons, has been recorded at 60 MHz and 1.41 T. **(a)** What would be the magnetic field at 300 MHz? **(b)** At 60 MHz, the resonance frequencies are: methyl, 140 Hz; aromatic, 430 Hz. What would the frequencies be if recorded by a 300-MHz spectrometer? **(c)** Calculate the chemical shifts (δ) of the two signals, using both the 60 MHz and 300 MHz data.

(a) Equation (17.39) shows that the NMR frequency is linearly proportional to the magnetic field strength. Thus a 300 MHz proton NMR spectrum requires $B_0 = \dfrac{300 \text{ MHz}}{60 \text{ MHz}} \times 1.41 \text{ T} = 7.05 \text{ T}$.

(b) The frequencies of both the reference proton and the proton of interest are linearly proportional to the magnetic field strength. Thus,

$$\Delta \nu_{300} = \frac{300 \text{ MHz}}{60 \text{ MHz}} \nu_{60} - \frac{300 \text{ MHz}}{60 \text{ MHz}} \nu_{60}^{\text{ref}}$$

$$= \frac{300 \text{ MHz}}{60 \text{ MHz}} \left(\nu_{60} - \nu_{60}^{\text{ref}} \right)$$

$$= \frac{300 \text{ MHz}}{60 \text{ MHz}} \Delta \nu_{60}$$

For the methyl protons the 300 MHz frequency is $5 \times 140 \text{ Hz} = 700 \text{ Hz}$ relative to reference.

For the aromatic protons the 300 MHz frequency is $5 \times 430 \text{ Hz} = 2150 \text{ Hz}$ relative to reference.

(c) In each case $\delta = \dfrac{\nu - \nu_{\text{ref}}}{\nu_{\text{ref}}} \times 10^6$.

At 60 MHz:

$$\delta_{\text{methyl}} = \frac{140 \text{ Hz}}{60 \times 10^6 \text{ Hz}} \times 10^6 = 2.33 \text{ ppm}$$

$$\delta_{\text{aromatic}} = \frac{430 \text{ Hz}}{60 \times 10^6 \text{ Hz}} \times 10^6 = 7.17 \text{ ppm}$$

At 300 MHz:

$$\delta_{methyl} = \frac{700 \text{ Hz}}{300 \times 10^6 \text{ Hz}} \times 10^6 = 2.33 \text{ ppm}$$

$$\delta_{aromatic} = \frac{2150 \text{ Hz}}{300 \times 10^6 \text{ Hz}} \times 10^6 = 7.17 \text{ ppm}$$

The chemical shift is independent of frequency.

17.40   The methyl radical has a planar geometry. How many lines would you observe in the ESR spectrum of $\cdot CH_3$? Of $\cdot CD_3$?

Normal methyl radical, $\cdot CH_3$, has three equivalent H atoms, each with $I = 1/2$. Thus, there are $2nI + 1 = 2(3)(1/2) + 1 = 4$ lines in its ESR spectrum with intensity ratio 1:3:3:1.

The deuterated methyl radical, $\cdot CD_3$, has three equivalent D atoms, each with $I = 1$, giving $2(3)(1) + 1 = 7$ lines in its ESR spectrum. The intensity pattern does not follow the binomial distribution, which applies only to nuclei with $I = 1/2$.

17.41   Account for the number of lines observed in the ESR spectra of benzene and naphthalene anion radicals shown in Figure 17.36. How would you use isotopic substitution to assign the two hyperfine splitting constants in naphthalene?

In the benzene anion radical the splitting arises from 6 equivalent protons with $I = 1/2$, leading to $2nI + 1 = 2(6)(1/2) + 1 = 7$ lines.

In the naphthalene radical anion, there are two sets of 4 equivalent protons (those α to the common side of the two rings and those β to it). This gives rise to
$$\left(2n_1 I + 1\right)\left(2n_2 I + 1\right) = [2(4)(1/2) + 1][2(4)(1/2) + 1] = 25 \text{ lines.}$$

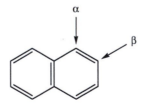

The isotopically substituted naphthalene shown below would have an ESR spectrum in which only the α hyperfine splitting would be affected, allowing the assignment of the observed hyperfine splitting constants in the normal isotopic species to the α and β protons.

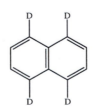

17.42   One way to study membrane structure (see Section 8.7) is to use a spin label, which is a
nitroxide radical that has the following structure

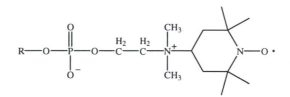

where R represents the hydrophobic tail part of the phosphatidic acid derivative (see Section 8.7).
The ESR spectrum of this spin label, like that of di-*tert*-butyl nitroxide, shows three lines of equal
intensities. The ESR signals disappear rapidly when the nitroxide comes in contact with a reducing
agent such as ascorbate. In one experiment, these spin-label molecules were incorporated in the
membrane lipid bilayer structure at about 5% concentration. The amplitude of the nitroxide ESR
signals decreased to 35% of the initial value within a few minutes of the addition of ascorbate. The
amplitude of the residual spectrum decayed exponentially with a half-life of about 7 hr. Explain
these observations.

The initial, rapid decrease of the ESR signal is due to the reduction of the spin label in the exterior
of the lipid bilayer. The ascorbate ion does not penetrate to the interior of the membrane, and the
spin label in the inner layer remains unaffected. The subsequent slow decay is due to the reduction
of spin-label bearing phospholipids that have flipped over to the outer region of the bilayer.

17.43   List some important differences between fluorescence and phosphorescence.

Fluorescence has much shorter lifetimes than phosphorescence. Fluorescence is decay from an
excited singlet state. Phosphorescence is decay from an excited triplet state.

17.44   The lowest triplet state in naphthalene ($C_{10}H_8$) is about 11,000 cm^{-1} below the lowest excited
singlet electronic level at 77 K. Calculate the ratio of the populations in these two states. (*Hint:*
The Boltzmann equation is given by $N_2/N_1 = (g_2/g_1)\exp(-\Delta E/k_B T)$, where g_1 and g_2 are the
degeneracies for levels 1 and 2.)

Assuming thermal equilibrium, the population ratio is given by the Boltzmann equation with
$\Delta E = hc\tilde{\nu} = \left(6.626 \times 10^{-34} \text{ J s}\right)\left(3.00 \times 10^8 \text{ m s}^{-1}\right)\left(11000 \text{ cm}^{-1}\right)\frac{100 \text{ cm}}{1 \text{ m}} = 2.187 \times 10^{-19}$ J.
The degeneracy of the triplet state is 3, while that for the singlet state is 1.

$$\frac{N_{\text{singlet}}}{N_{\text{triplet}}} = \frac{1}{3}\exp\left[-\frac{2.187 \times 10^{-19} \text{ J}}{\left(1.381 \times 10^{-23} \text{ J K}^{-1}\right)(77 \text{ K})}\right]$$

$$= \frac{e^{-206}}{3}$$

$$= 1.6 \times 10^{-90}$$

which is practically zero.

17.45   The luminescent first-order decay of a certain organic molecule yields the following data:

t/s	0	1	2	3	4	5	10
I	100	43.5	18.9	8.2	3.6	1.6	0.02

where I is the relative intensity. Calculate the mean lifetime, τ, for the process. Is the decay fluorescent or phosphorescent?

Since $I = I_0 e^{-t/\tau}$, where τ is the mean lifetime of the state, then $\ln\left(I/I_0\right) = -t/\tau$, and a graph of $\ln\left(I/I_0\right)$ vs. t will result in a straight line with slope equal to $-1/\tau$. The necessary data are

t/s	0	1	2	3	4	5	10
$\ln\left(I/I_0\right)$	0	-0.83	-1.67	-2.50	-3.32	-4.14	-8.52

The slope of the best fit line to the data is -0.851 s^{-1}, which implies that $\tau = -1/\left(-0.851 \text{ s}^{-1}\right) = 1.2 \text{ s}$. This is a long lifetime corresponding to phosphorescence.

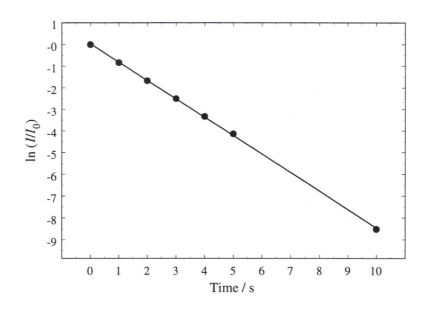

17.46   Give a qualitative explanation as to why POP absorbs light at a shorter wavelength than POPOP (see p. 745).

As was the case in Problem 17.28, greater delocalization of electrons leads to the $\pi \rightarrow \pi^*$ transition causing absorption at longer wavelength. There is greater delocalization in POPOP than in POP, so that POP is expected to absorb at the shorter wavelength.

17.47   The fluorescence of a protein is due to tryptophan, tyrosine, and phenylalanine (assuming that the protein does not contain a prosthetic group that is fluorescent). Iodide ions are known to quench the fluorescence of tryptophan. If a protein is known to contain only one tryptophan residue and iodide fails to quench its fluorescence, what can you conclude about the location of the tryptophan residue?

The tryptophan is most likely located in the interior of the protein in a region inaccessible to the iodide ion.

17.48   Name three characteristic properties of laser.

The three characteristic properties of laser light are high intensity, coherence, and monochromaticity.

17.49   Explain why we cannot produce laser light with a two-level system.

Any practical pumping process requires that there be more molecules in the lower level than the upper level to effect a net movement of population to the upper level. Laser action, however, requires greater population in the upper level than the lower level. Since these two conditions can obviously never be simultaneously true, laser action cannot be practically attained in a two-level system. (Note that in a true two-level system, all molecules must be in either one level or the other.)

17.50   For a three-level laser system, the wavelength for an absorption from level A to level C is found to be 466 nm, and the wavelength for a transition between levels B and C is 752 nm. What is the wavelength for a transition between levels A and B?

The wavelengths indicate that level B lies between levels A and C. It is then easy to see that

$$E_C - E_A = \left(E_C - E_B\right) + \left(E_B - E_A\right)$$

$$\frac{hc}{\lambda_{CA}} = \frac{hc}{\lambda_{CB}} + \frac{hc}{\lambda_{BA}}$$

$$\frac{1}{\lambda_{CA}} = \frac{1}{\lambda_{CB}} + \frac{1}{\lambda_{BA}}$$

$$\frac{1}{466 \text{ nm}} = \frac{1}{752 \text{ nm}} + \frac{1}{\lambda_{BA}}$$

$$\lambda_{AB} = 1225 \text{ nm}$$

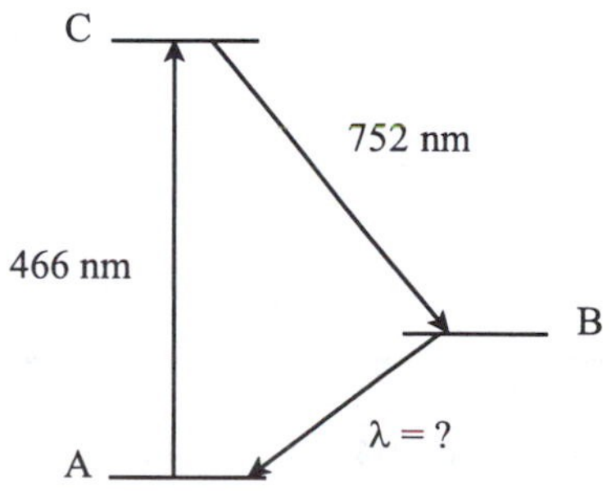

17.51   What is the difference between a one-photon and a multiphoton process? Why does the use of lasers make it favorable to observe, say, a two-photon process?

In a one-photon process, an atom or molecule absorbs a single photon of the same energy as that between the ground and the excited state. In a multiphoton process, the atom or molecule absorbs several photons whose energies sum to that between the ground and the excited state. A multiphoton process requires that these several photons pass essentially simultaneously through the region of space occupied by the atom or molecule. This requires a very intense beam of light, such as that provided by a high-intensity laser.

17.52 How many unpaired electrons are in a molecule in a quartet state?

A multiplicity of 4 requires that $2S + 1 = 4$, or $S = 3/2$, which implies three unpaired electrons.

17.53 The typical energy differences for transitions in the microwave, IR, and electronic spectroscopies are 5×10^{-22} J, 0.5×10^{-19} J, and 1×10^{-18} J, respectively. Calculate the ratio of the number of molecules in two adjacent energy levels (for example, the ground level and the first excited level) at 300 K in each case.

In each case the population ratio is given by the Boltzmann equation (see Problem 17.44).

$$\frac{N_1}{N_0} = \frac{g_1}{g_0} \exp\left(-\frac{\Delta E}{kT}\right)$$

The degeneracy ratio is important for microwave spectroscopy where $g_J = 2J + 1$, and may be important for IR and electronic spectroscopies as well, but for this problem the vibrational states are assumed singly degenerate, and the electronic states are assumed to be singlets.

Microwave:

$$\frac{N_1}{N_0} = \frac{3}{1} \exp\left[-\frac{5 \times 10^{-22}\,\text{J}}{(1.381 \times 10^{-23}\,\text{J K}^{-1})\,(300\,\text{K})}\right] = 2.7$$

IR:

$$\frac{N_1}{N_0} = \frac{1}{1} \exp\left[-\frac{0.5 \times 10^{-19}\,\text{J}}{(1.381 \times 10^{-23}\,\text{J K}^{-1})\,(300\,\text{K})}\right] = 5.7 \times 10^{-6}$$

Electronic:

$$\frac{N_1}{N_0} = \frac{1}{1} \exp\left[-\frac{1 \times 10^{-18}\,\text{J}}{(1.381 \times 10^{-23}\,\text{J K}^{-1})\,(300\,\text{K})}\right] = e^{-241} = 1.5 \times 10^{-105} \approx 0$$

17.54 The molar absorptivity of a solute at 664 nm is 895 $\text{L mol}^{-1}\,\text{cm}^{-1}$. When light at that wavelength is passed through a 2.0-cm cell containing a solution of the solute, 74.6 percent of the light is absorbed. Calculate the concentration of the solution.

With 74.6% of the light absorbed, the transmittance is $T = 1.000 - 0.746 = 0.254$, and the absorbance is $A = -\log T = -\log 0.254 = 0.5952$. Using the Beer–Lambert law, $A = \epsilon b c$,

$$c = \frac{A}{\epsilon b}$$

$$= \frac{0.5952}{(895\ \mathrm{L\,mol^{-1}\,cm^{-1}})\,(2.0\ \mathrm{cm})}$$

$$= 3.3 \times 10^{-4}\ M$$

17.55 The frequency of molecular collision in the liquid phase is about $1 \times 10^{13}\ \mathrm{s^{-1}}$. Ignoring all other mechanisms contributing to line width, calculate the width (in Hz) of vibrational transitions if **(a)** every collision is effective in deactivating the molecule vibrationally, and **(b)** that one collision in 40 is effective.

(a) If every collision is effective in deactivating the molecule, then the lifetime of the excited vibrational state is $\Delta t = 1/1 \times 10^{13}\ \mathrm{s^{-1}} = 1 \times 10^{-13}\ \mathrm{s}$, and the natural linewidth is

$$\Delta \nu = \frac{1}{4\pi\,\Delta t} = \frac{1}{4\pi\,(1 \times 10^{-13}\ \mathrm{s})} = 8 \times 10^{11}\ \mathrm{s^{-1}}$$

(b) If one collision in 40 is effective in deactivating the molecule, then the lifetime of the excited vibrational state is $\Delta t = 40/1 \times 10^{13}\ \mathrm{s^{-1}} = 4 \times 10^{-12}\ \mathrm{s}$, and the natural linewidth is

$$\Delta \nu = \frac{1}{4\pi\,\Delta t} = \frac{1}{4\pi\,(4 \times 10^{-12}\ \mathrm{s})} = 2 \times 10^{10}\ \mathrm{s^{-1}}$$

17.56 Consider the 2–propenenitrile molecule whose IR spectrum is shown in Figure 17.19. Which of the following types of energy has the largest number of energy levels appreciably occupied at 300 K? Electronic, C–H stretching vibration, C=C stretching vibration, HCH bending motion, or rotational.

Rotational, because the type of energy with the smallest spacing between quantized energy levels will have the largest number of energy levels populated at 300 K.

17.57 Analysis of lines broadened by the Doppler effect shows that the width at half-height, $\Delta\lambda$, is given by

$$\Delta\lambda = 2\left(\frac{\lambda}{c}\right)\left(\frac{2k_{\mathrm{B}}T}{m}\right)^{1/2}$$

where c is the speed of light, T is the temperature (in kelvin), and m is the mass of the species involved in the transition. The corona of the sun emits a spectral line at about 677 nm due to the presence of an ionized ^{57}Fe atom (molar mass: $0.0569\ \mathrm{kg\,mol^{-1}}$). If the line has a width of 0.053 nm, what is the temperature of the corona?

Solving for the temperature,

$$T = \frac{m}{8k_B} \left(\frac{c\Delta\lambda}{\lambda} \right)^2$$

$$= \frac{(0.0569 \text{ kg mol}^{-1})}{8 \left(1.381 \times 10^{-23} \text{ J K}^{-1}\right) \left(6.022 \times 10^{23} \text{ mol}^{-1}\right)} \left[\frac{(3.00 \times 10^8 \text{ m s}^{-1}) (0.053 \text{ nm})}{677 \text{ nm}} \right]^2$$

$$= 4.7 \times 10^5 \text{ K}$$

17.58   Derive an expression for the value of J corresponding to the most populous rotation energy level of a rigid diatomic rotor at temperature T. Evaluate the expression for HCl ($B = 10.59 \text{ cm}^{-1}$) at 25°C. (*Hint*: Remember to include the degeneracy.)

From the Boltzmann distribution law, the number of molecules in a given rotational level with quantum number J is

$$N_J = N g_J e^{-E_J/k_B T} = N (2J + 1) e^{-hcBJ(J+1)/k_B T}$$

where N is the total number of molecules and $E_{rot} = hcBJ(J + 1)$ when B is given in wavenumber units.

Differentiating with respect to J gives

$$\frac{dN_J}{dJ} = N \left[2 - \frac{hcB}{k_B T} (2J + 1)^2 \right] e^{-hcBJ(J+1)/k_B T}$$

Finding a maximum with respect to J requires $\frac{dN_J}{dJ} = 0$, or

$$N \left[2 - \frac{hcB}{k_B T} (2J + 1)^2 \right] e^{-hcBJ(J+1)/k_B T} = 0$$

$$2 - \frac{hcB}{k_B T} (2J + 1)^2 = 0$$

$$2J + 1 = \sqrt{\frac{2k_B T}{hcB}}$$

$$J = \sqrt{\frac{k_B T}{2hcB}} - \frac{1}{2}$$

For HCl at 298 K,

$$J = \sqrt{\frac{\left(1.381 \times 10^{-23} \text{ J K}^{-1}\right) (298 \text{ K})}{2 \left(6.626 \times 10^{-34} \text{ J s}\right) \left(3.00 \times 10^8 \text{ m s}^{-1}\right) \left(10.59 \text{ cm}^{-1}\right) \left(\frac{100 \text{ cm}}{1 \text{ m}}\right)}} - \frac{1}{2} = 3$$

17.59   Analyze the ^{31}P NMR spectrum of ATP shown in Figure 17.33.

Since the sugar contains no phosphorous, the structure of ATP can be represented as Sugar-O-P_α-O-P_β-O-P_γ-O. There are three different P nuclei, hence three peaks in the ^{31}P NMR spectrum. The α nucleus is split by β into a 1:1 doublet, and the γ nucleus is likewise split by β into a 1:1 doublet. In turn, the β nucleus is split by both α and γ into a "doublet of doublets" (each 1:1 doublet due to one nucleus is itself split into a 1:1 doublet due to the other). Since the coupling constants are about equal, however, the two center lines coincide to give what appears as a 1:2:1 triplet for the β nucleus.

17.60   An aqueous solution contains two species, A and B. The absorbance at 300 nm is 0.372 and at 250 nm is 0.478. The molar absorptivities of A and B are:

$$A:\ \epsilon_{300} = 3.22 \times 10^4\ L\,mol^{-1}\,cm^{-1}$$

$$\epsilon_{250} = 4.05 \times 10^4\ L\,mol^{-1}\,cm^{-1}$$

$$B:\ \epsilon_{300} = 2.86 \times 10^4\ L\,mol^{-1}\,cm^{-1}$$

$$\epsilon_{250} = 3.76 \times 10^4\ L\,mol^{-1}\,cm^{-1}$$

If the pathlength of the cell is 1.00 cm, calculate the concentrations of A and B in $mol\,L^{-1}$.

At each wavelength, the total absorbance is the sum of the absorbances due to each species, which can be written using the Beer–Lambert law, $A = \epsilon bc$. This gives a system of two equations in the two unknowns, c_A and c_B, the concentrations of A and B, respectively.

$$A_{300} = \epsilon_{300}^A bc_A + \epsilon_{300}^B bc_B$$
$$A_{250} = \epsilon_{250}^A bc_A + \epsilon_{250}^B bc_B$$

or, since $b = 1.0$ cm,

$$0.372 = 3.22 \times 10^4 c_A + 2.86 \times 10^4 c_B$$

$$0.478 = 4.05 \times 10^4 c_A + 3.76 \times 10^4 c_B$$

This system can be solved to give

$$c_A = 6.04 \times 10^{-6}\ M$$

$$c_B = 6.21 \times 10^{-6}\ M$$

Please note that this solution is very susceptible to rounding errors.

17.61   A molecule XY_2 is known to be linear, but it is not clear whether it is Y-X-Y or X-Y-Y. How would you use IR spectroscopy to determine its structure?

A linear molecule of the form Y-X-Y would have an IR spectrum similar to that of CO_2, with two peaks. There would be one peak for the asymmetric stretching vibration and one for the doubly degenerate bending vibrational mode. A linear, X-Y-Y molecule would have an IR spectrum similar to OCS (see Problem 17.24) with three peaks, of which one would be the doubly degenerate bending vibrational mode, and the other two corresponding to the two stretching vibrational modes.

17.62   The NMR spectrum of N,N'-dimethylformamide shows two methyl peaks at 25°C. When heated to 130°C, there is only one peak due to the methyl protons. Explain.

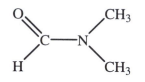

At the higher temperature, the rotation about the C-N bond becomes rapid enough to average out the chemical environment and thus the chemical shifts of the two methyl groups. Consequently, only one NMR peak for the methyl groups is seen. At 25°C, the rotation is slower than the NMR time scale, and the peaks are separately observed.

17.63   This problem deals with the amplitude of molecular vibration of a diatomic molecule in its ground vibrational state. **(a)** When the molecule is stretched by an extent x from the equilibrium position, the increase in the potential energy is given by the integral

$$\int_0^x kx\,dx$$

where k is the force constant. Evaluate this integral. **(b)** To calculate the amplitude of vibration, we equate the potential energy with the vibrational energy in the ground state. Use x_{max} to represent the maximum displacement. **(c)** Given that the force constant for $H^{35}Cl$ is 4.84×10^2 N m^{-1}, calculate the amplitude of vibration in the $v = 0$ state. **(d)** What is the percent of the amplitude compared to the bond length (1.27 Å)? **(e)** Repeat the calculations in **(c)** and **(d)** for carbon monoxide, given that the force constant is 1.85×10^3 N m^{-1} and the bond length is 1.13 Å. (^{35}Cl: 34.97 amu)

(a)

$$\int_0^x kx\,dx = \frac{kx^2}{2}$$

(b) Since $E_{vib} = \left(v + \frac{1}{2}\right)h\nu$ and $\nu = \frac{1}{2\pi}\sqrt{\frac{k}{\mu}}$, for the ground state with $v = 0$,

$$E_{vib} = \frac{1}{2}h\left(\frac{1}{2\pi}\sqrt{\frac{k}{\mu}}\right) = \frac{1}{2}kx_{max}^2$$

Solving for x_{max},

$$x_{max} = \left(\frac{h^2}{4\pi^2 k\mu}\right)^{1/4}$$

(c) For $H^{35}Cl$ the reduced mass is

$$\mu = \left[\frac{(1.008\text{ amu})(34.97\text{ amu})}{1.008\text{ amu} + 34.97\text{ amu}}\right](1.661 \times 10^{-27}\text{ kg amu}^{-1}) = 1.6274 \times 10^{-27}\text{ kg}$$

and

$$x_{max} = \left(\frac{h^2}{4\pi^2 k\mu}\right)^{1/4}$$

$$= \left[\frac{(6.626 \times 10^{-34} \text{ J s})^2}{4\pi^2 (4.84 \times 10^2 \text{ N m}^{-1}) (1.6274 \times 10^{-27} \text{ kg})}\right]^{1/4}$$

$$= 1.090 \times 10^{-11} \text{ m}$$

$$= 0.109 \text{ Å}$$

(d) This amplitude is $\dfrac{0.1090 \text{ Å}}{1.27 \text{ Å}} \times 100\% = 8.58\%$ of the bond length.

(e) For CO, the reduced mass is

$$\mu = \left[\frac{(12.00 \text{ amu}) (15.99 \text{ amu})}{12.00 \text{ amu} + 15.99 \text{ amu}}\right] (1.661 \times 10^{-27} \text{ kg amu}^{-1}) = 1.1387 \times 10^{-26} \text{ kg}$$

and

$$x_{max} = \left(\frac{h^2}{4\pi^2 k\mu}\right)^{1/4}$$

$$= \left[\frac{(6.626 \times 10^{-34} \text{ J s})^2}{4\pi^2 (1.85 \times 10^3 \text{ N m}^{-1}) (1.1387 \times 10^{-26} \text{ kg})}\right]^{1/4}$$

$$= 4.793 \times 10^{-12} \text{ m}$$

$$= 0.0480 \text{ Å}$$

This amplitude is $\dfrac{0.04793 \text{ Å}}{1.13 \text{ Å}} \times 100\% = 4.24\%$ of the bond length. A triple bond does not stretch as much as a single bond.

17.64 The IR spectrum of the carbon monoxide–hemoglobin complex shows a peak at about 1950 cm^{-1}, which is due to the carbonyl stretching frequency. (a) Compare this value with the fundamental frequency of free CO, which is 2143.3 cm^{-1}. Comment on the difference. (b) Convert this frequency to kJ mol^{-1}. (c) What conclusion can you draw from the fact that there is only one band present?

(a) The value of the carbonyl stretching frequency in the carbon monoxide–hemoglobin complex is lower than that of free CO. This is due to the carbon–oxygen π bond in the CO–Hb complex that donates electrons to the π^* orbitals in CO (see Figure 15.34).

(b) For each molecule,

$$E = hc\tilde{\nu}$$

$$= (6.626 \times 10^{-34} \text{ J s}) (3.00 \times 10^8 \text{ m s}^{-1}) (1950 \text{ cm}^{-1}) \left(\frac{100 \text{ cm}}{1 \text{ m}}\right)$$

$$= 3.876 \times 10^{-20} \text{ J}$$

So that on a molar basis,

$$E = 3.876 \times 10^{-20} \text{ J } (6.022 \times 10^{23} \text{ mol}^{-1}) \left(\frac{1 \text{ kJ}}{1000 \text{ J}} \right) = 23.3 \text{ kJ mol}^{-1}$$

(c) Since only one band is present, it may be concluded that there is only one type of CO binding site in hemoglobin. Indeed, although there are 4 heme groups per hemoglobin molecule, they are all equivalent.

17.65   True or false? **(a)** To be IR active, a polyatomic molecule must possess a permanent dipole moment. **(b)** The moment of inertia of a diatomic molecule measured from its microwave spectrum provides information about the force constant of the bond. **(c)** The fluorescence spectrum of a molecule occurs at a shorter wavelength than the absorption spectrum of the molecule. **(d)** A 600-MHz NMR spectrometer is more sensitive than a 400-MHz spectrometer. **(e)** Phosphorescence is a spin-forbidden process. **(f)** To observe hyperfine splittings in an ESR spectrum, the nucleus involved must have $I \neq 0$. **(g)** Whenever a molecule goes from one energy level to another, it must emit or absorb a photon whose energy is equal to the energy difference between the two levels.

(a) False, **(b)** false, **(c)** false, **(d)** true, **(e)** true, **(f)** true, **(g)** false.

CHAPTER 18

Molecular Symmetry and Optical Activity

PROBLEMS AND SOLUTIONS

18.1 List all the symmetry elements of the following molecules: CCl_4, CH_3Cl, CH_2Cl_2, $CHClBrI$.

In the lists below, only physically distinct symmetry elements are counted. In the mathematical theory of groups, not discussed in the text, it is necessary to consider that some symmetry elements have more than one operation associated with them. For example, one may rotate about a C_3 axis by either $120°$ *or* $240°$.

CCl_4: E, $4C_3$, $3C_2$, 6σ, $3S_4$ (coincident with C_2)

CH_3Cl: E, C_3, 3σ

CH_2Cl_2: E, C_2, 2σ

$CHClBrI$: E only

18.2 How many C_n axes does benzene possess?

Benzene contains one C_6 axis and 6 C_2 axes (through 3 pairs of opposing vertices and through the midpoints of 3 pairs of opposing sides in the hexagonal molecule). Notice that a rotation of any multiple of $60°$ about the C_6 axis leaves the molecule in an indistinguishable configuration. (See note to Problem 18.1.)

18.3 Optically active molecules can be classified into two types, asymmetric and dissymmetric. The former does not contain any symmetry elements (except C_1 and E), while the latter contains a single rotation axis C_n ($n > 1$). Classify the following molecules according to these two definitions:

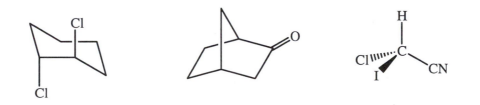

From left to right: dissymmetric (a C_2 axis), asymmetric, asymmetric.

18.4   Is 1,3-dichloroallene ($C_3H_2Cl_2$) chiral?

This molecule contains only the identity and a C_2 rotation axis. It contains no improper rotation axes and is therefore chiral.

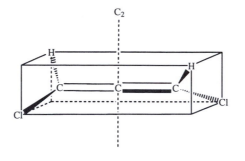

18.5   The optical rotation of a sucrose solution (concentration: 9.6 g in 100-mL soln) is +0.34° when measured in a 10-cm cell with the sodium D-line at room temperature. Calculate the specific rotation and molar rotation for sucrose.

The concentration of the solution is 9.6 g/100 cm³ = 0.096 g cm⁻³, so that the specific rotation is given by,

$$[\alpha]_D^{25} = \frac{\alpha}{lc}$$

$$= \frac{+0.34 \text{ deg}}{(1.0 \text{ dm})\left(0.096 \text{ g cm}^{-3}\right)}$$

$$= 3.5 \text{ deg dm}^{-1} \text{ cm}^3 \text{ g}^{-1}$$

The molar rotation is found using the molar mass of sucrose, namely 342.3 g mol⁻¹.

$$[\Phi]_D^{25} = \frac{[\alpha]_D^{25} M}{100}$$

$$= \frac{\left(3.5 \text{ deg dm}^{-1} \text{ cm}^3 \text{ g}^{-1}\right)\left(342.3 \text{ g mol}^{-1}\right)}{100}$$

$$= 12 \text{ deg dm}^{-1} \text{ cm}^3 \text{ mol}^{-1}$$

18.6   The rotation of a certain solution containing an optically active compound is 2.41° when measured in a 1.0-dm cell. Calculate the quantity $(n_L - n_R)$. Assume the sodium D-line was used for the measurement.

From Equation 18.2, and using 589.3 nm as the wavelength of the sodium D-line,

$$\alpha = \frac{180°}{\lambda}\left(n_L - n_R\right)l$$

$$2.41° = \frac{180°}{589.3 \times 10^{-7} \text{ dm}}\left(n_L - n_R\right)(1.0 \text{ dm})$$

$$\left(n_L - n_R\right) = 7.89 \times 10^{-7}$$

18.7 Two optical isomers, A and B, having specific rotations of $+27.6°$ and $-19.5°$, respectively, are in equilibrium in solution. If the specific rotation of the mixture is $16.2°$, calculate the equilibrium constant for the process $A \rightleftharpoons B$.

The specific rotation of the mixture is given by (with $c_{total} = c_A + c_B$ the total concentration of A and B)

$$[\alpha_{mix}]_\lambda^T = \frac{\alpha_{mix}}{lc_{total}}$$

$$= \frac{lc_A[\alpha_A]_\lambda^T + lc_B[\alpha_B]_\lambda^T}{l(c_A + c_B)}$$

$$= \frac{c_A}{c_A + c_B}[\alpha_A]_\lambda^T + \frac{c_B}{c_A + c_B}[\alpha_B]_\lambda^T$$

Thus, the contribution of each optical isomer toward the total specific rotation is directly proportional to the mole fraction of that isomer present. Representing $\dfrac{c_B}{c_A + c_B} = x$, then $\dfrac{c_A}{c_A + c_B} = 1 - x$, and

$$16.2° = (1 - x)(27.6°) + x(-19.5°)$$

$$x = 0.2420$$

The equilibrium constant is

$$K = \frac{c_B}{c_A}$$

$$= \frac{\frac{c_B}{c_A + c_B}}{\frac{c_A}{c_A + c_B}}$$

$$= \frac{x}{1 - x}$$

$$= \frac{0.2420}{1 - 0.2420}$$

$$= 0.319$$

18.8 Two substances, A and B, have identical absorption spectra and identical CD curves except that one CD curve is positive and the other is negative. What is the structural relationship between A and B?

The substances are mirror images of each other, or enantiomers.

18.9 In an optical-rotation measurement, the angle measured for a solution of concentration c in a cell of pathlength l is $-12.7°$. How can you be certain that this is the correct rotation and not

$(-12.7° + 360°)$ or 347.3°, considering that clockwise rotation by 347.3° is equivalent to counterclockwise rotation by 12.7°?

The best way to remove the ambiguity is to measure the optical rotation for a different concentration of the species. For example, using a concentration of $c/10$ will result in a rotation $\alpha = -1.27°$, which is clearly distinguishable from +34.73°. Another possibility would be to use a different pathlength, although most polarimeters use cells of fixed pathlength.

18.10 In what wavelength range would you expect to find the CD spectrum of tryptophan? (*Hint*: See Figure 17.22.)

The CD spectrum of tryptophan will be around 280 nm, which is the maximum of the molecule's absorption band.

18.11 The CD of a protein solution changes appreciably upon the addition of a certain achiral compound. What might have happened?

The compound has induced a conformation change of the protein molecule, most likely through binding of the protein to the molecule.

18.12 The optical rotation of a sample of α-D-glucose is +112.2° and that of β-D-glucose is +18.7°. A mixture of these two sugars has an optical rotation of 56.8°. Calculate the composition of the mixture.

Referring to the solution to Problem 18.7, let x be the mole fraction of β-D-glucose.

$$56.8° = (1 - x)(112.2°) + x(18.7°)$$

$$x = 0.5925$$

Thus, the mixture is 40.7% α-D-glucose and 59.3% β-D-glucose.

18.13 Winemakers often use a pocket polarimeter to check the maturity of grapes in their vineyards. Explain how it works.

Grape juice contains sugar (fructose) and other chiral compounds whose concentrations change as the grape matures. The observed rotation is the weighted rotation of all chiral compounds present (see Problem 18.7), so that this is a crude method, but nevertheless useful to the vinaculturist.

18.14 Sucrose ($C_{12}H_{22}O_{11}$) is known as cane sugar. In the confectionery industry, sucrose is hydrolyzed to glucose and fructose by dilute acids or the enzyme invertase as follows:

$$C_{12}H_{22}O_{11} \quad \rightarrow \quad C_6H_{12}O_6 \quad + \quad C_6H_{12}O_6$$
$$+66.48° \qquad\qquad +112.2° \qquad -132°$$

The specific rotations are all measured at 25°C with the sodium D-line. Both glucose and fructose have the same molecular formula; fructose has the negative specific rotation. One reason for the breakdown of sucrose is that fructose is the sweetest sugar known. **(a)** Why is the sugar manufactured from this process called "invert sugar?" **(b)** What additional advantage does a mixture of fructose and glucose have over pure sucrose in making candy?

(a) The original sucrose has a positive optical rotation (+66.48°), but the equimolar mixture of glucose and fructose comprising invert sugar will have a specific rotation of −9.9°(see Problem 18.7). The sign of the optical rotation has been "inverted" from (+) to (−).

(b) Sucrose forms crystals which are brittle. A mixture of glucose and fructose does not form crystals and results in a more chewy candy.

18.15   The optical rotation of the *d*-form of α-piene ($C_{10}H_{16}$), measured at 20°C with the sodium D-line in a cell of pathlength of 1.0 cm, is +4.4°. Given that the density of the liquid is 0.859 g mL^{-1}, calculate the specific rotation of α-piene. What does the positive sign of the rotation mean?

For a pure liquid,

$$[\alpha]_D^{20} = \frac{\alpha}{ld}$$

$$= \frac{+4.4 \text{ deg}}{(0.10 \text{ dm}) \left(0.859 \text{ g cm}^{-3}\right)}$$

$$= 51.2 \text{ deg dm}^{-1} \text{ cm}^3 \text{ g}^{-1}$$

$$= 51 \text{ deg dm}^{-1} \text{ cm}^3 \text{ g}^{-1}$$

The positive sign means that the plane of polarization for the light is rotated to the right when looking into the beam.

18.16   Using the results of Problem 18.15, calculate the molar rotation of α-piene.

The molar rotation is found using the molar mass of α-piene, 136.23 g mol^{-1}.

$$[\Phi]_D^{20} = \frac{[\alpha]_D^{20} M}{100}$$

$$= \frac{\left(51.2 \text{ deg dm}^{-1} \text{ cm}^3 \text{ g}^{-1}\right) \left(136.23 \text{ g mol}^{-1}\right)}{100}$$

$$= 70 \text{ deg dm}^{-1} \text{ cm}^3 \text{ mol}^{-1}$$

18.17   The rotation of a solution of L-ribulose measured at 25°C with the sodium D-line is −3.8°. The pathlength of the cell is 10 cm and $[\alpha]_D^{25}$ is −16.6 deg dm^{-1} cm^3 g^{-1}. What is the concentration of the solution in g mL^{-1}?

$$[\alpha]_D^{25} = \frac{\alpha}{lc}$$

$$-16.6 \text{ deg dm}^{-1} \text{ cm}^3 \text{ g}^{-1} = \frac{-3.8 \text{ deg}}{(1.0 \text{ dm}) \, c}$$

$$c = 0.23 \text{ g cm}^{-3} = 0.23 \text{ g mL}^{-1}$$

18.18 The specific rotation of (+)hexahelicene (see Figure 18.3) is given by $[\alpha]_D^{25} = 3750 \text{ deg dm}^{-1} \text{ cm}^3 \text{ g}^{-1}$. Explain why it is so much greater than that of most other compounds.

The entire molecule acts as a chromophore for the $\pi \to \pi^*$ transition.

CHAPTER 19

Photochemistry and Photobiology

PROBLEMS AND SOLUTIONS

19.1 In a photochemical reaction, 428.3 kJ mol^{-1} of energy input is required to break a chemical bond. What wavelength must be employed in the irradiation?

$$\lambda = \frac{c}{\nu} = \frac{ch}{E}$$

$$= \frac{(3.00 \times 10^8 \text{ m s}^{-1}) (6.626 \times 10^{-34} \text{ J s})}{(428.3 \times 10^3 \text{ J mol}^{-1}) \left(\frac{1 \text{ mol}}{6.022 \times 10^{23}}\right)} = 2.79 \times 10^{-7} \text{ m} = 279 \text{ nm}$$

19.2 Convert 450 nm to kJ einstein^{-1}.

$$E = h\nu = \frac{hc}{\lambda}$$

$$= \frac{(6.626 \times 10^{-34} \text{ J s}) (3.00 \times 10^8 \text{ m s}^{-1})}{450 \times 10^{-9} \text{ m}} \left(\frac{6.022 \times 10^{23}}{1 \text{ mol}}\right) \left(\frac{1 \text{ kJ}}{1000 \text{ J}}\right)$$

$$= 266 \text{ kJ einstein}^{-1}$$

19.3 Design an experiment that would allow you to measure the rate of absorption of light by a solution.

Use the ferrioxalate actinometer to measure the decomposition of iron oxalate with and without the solution under study. Since the quantum yield of the iron oxalate is known, the rate of light absorption by the solution can be determined from the difference in quantum yields.

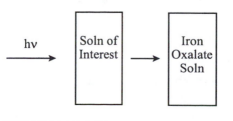

19.4   An organic molecule absorbs light at 549.6 nm. If 0.031 mole of the molecule is excited by 1.43 einsteins of light, what is the quantum efficiency for this process? Also, calculate the total energy taken up in the process.

The quantum efficiency is

$$\Phi = \frac{0.031 \text{ mol}}{1.43 \text{ einsteins}} = 0.022$$

The energy of each photon is

$$E = h\nu = \frac{hc}{\lambda} = \frac{(6.626 \times 10^{-34} \text{ J s})(3.00 \times 10^8 \text{ m s}^{-1})}{549.6 \times 10^{-9} \text{ m}} = 3.617 \times 10^{-19} \text{ J}$$

The energy taken up in the process is

$$(3.617 \times 10^{-19} \text{ J})(1.43 \text{ einsteins})\left(\frac{6.022 \times 10^{23}}{1 \text{ einstein}}\right) = 3.11 \times 10^5 \text{ J}$$

19.5   In the photochemical decomposition of a certain compound, light intensity of 5.4×10^{-6} einstein s^{-1} was employed. Assuming the most favorable conditions, estimate the time needed to decompose 1 mole of the compound.

The rate of a photochemical reaction is

$$\text{Rate} = IFf\Phi_p$$

The most favorable conditions are: $F = 1$ (all incident light is absorbed), $f = 1$ (all absorbed light produces the reactive state), and $\Phi_p = 1$ (quantum yield of product formation is 1). Under these conditions, the rate is

$$\text{Rate} = 5.4 \times 10^{-6} \text{ mol s}^{-1}$$

Thus, to decompose 1 mole of the compound, the time needed is

$$\frac{1 \text{ mol}}{5.4 \times 10^{-6} \text{ mol s}^{-1}} = 1.85 \times 10^5 \text{ s} = 2.1 \text{ days}$$

19.6   The first-order rate constants for the fluorescence and phosphorescence of naphthalene ($C_{10}H_8$) are 4.5×10^7 s^{-1} and 0.50 s^{-1}, respectively. Calculate how long it takes for 1.0% of fluorescence and phosphorescence to occur following termination of excitation.

For a first-order reaction, the number of photoexcited molecules at time t, N_t, is related to that at time 0, N_0, by

$$N_t = N_0 e^{-kt}$$

Therefore,

$$t = -\frac{1}{k} \ln \frac{N_t}{N_0}$$

The time for 1.0 % of fluorescence to occur is when 1.0% of the photoexcited molecules have decayed, leaving $N_t/N_0 = 0.990$,

$$t = -\frac{1}{4.5 \times 10^7 \text{ s}^{-1}} \ln 0.99 = 2.23 \times 10^{-10} \text{ s}$$

Similarly, the time for 1.0 % of phosphorescence to occur is

$$t = -\frac{1}{0.50 \text{ s}^{-1}} \ln 0.99 = 2.01 \times 10^{-2} \text{ s}$$

19.7   A barometer that has a cross-sectional area of 1 cm^2 at sea level measures a pressure of 76.0 cm of mercury. The pressure exerted by this column of mercury is equal to the pressure exerted by all the air molecules on 1 cm^2 of Earth's surface. Given that the density of mercury is 13.6 g mL^{-1} and the average radius of Earth is 6371 km, calculate the total mass of Earth's atmosphere in kilograms. Is your result an upper or lower estimate of the mass? Explain. (*Hint*: The surface area of a sphere of radius r is $4\pi r^2$.)

Total mass of Earth's atmosphere = (Surface area of Earth) (Mass of atmosphere per 1 cm^2 area)

$$= \left[4\pi \left(6371 \times 10^5 \text{ cm} \right)^2 \right] \left[(76.0 \text{ cm}) \left(13.6 \text{ g mL}^{-1} \right) \right]$$

$$= 5.27 \times 10^{21} \text{ g} = 5.27 \times 10^{18} \text{ kg}$$

This is an upper estimate, since it does not allow for the atmosphere displaced by land masses which rise above sea level.

19.8   Name the major source of heat that originates from Earth.

The major source of heat that is of terrestrial origin is radioactive decay.

19.9   The highly reactive OH radical (a species with an unpaired electron) is believed to be involved in some atmospheric processes. Table 4.4 lists the bond energy for the oxygen-to-hydrogen bond in OH as 460 kJ. What is the longest wavelength (in nm) of radiation that can bring about the following reaction:

$$OH(g) \rightarrow O(g) + H(g)$$

The longest wavelength corresponds to 460 kJ mol^{-1}.

$$\lambda = \frac{c}{\nu} = \frac{ch}{E}$$

$$= \frac{\left(3.00 \times 10^8 \text{ m s}^{-1}\right)\left(6.626 \times 10^{-34} \text{ J s}\right)}{\left(460 \times 10^3 \text{ J mol}^{-1}\right)\left(\frac{1 \text{ mol}}{6.022 \times 10^{23}}\right)}$$

$$= 2.60 \times 10^{-7} \text{ m} = 260 \text{ nm}$$

19.10 The hydroxyl radical in the atmosphere is most effectively removed by hydrocarbons such as methane according to the second order reaction

$$\text{HO·} + \text{CH}_4 \rightarrow \text{H}_2\text{O} + \text{CH}_3\text{·}$$

Given that the second-order rate constant is $4.6 \times 10^6 \text{ L mol}^{-1} \text{ s}^{-1}$, calculate the lifetime of the radical at 25°C if the concentration of CH_4 is 1.7×10^3 ppb by volume. (*Hint*: The lifetime of the radical is given by $1/k[\text{CH}_4]$.)

The volume ratio between CH_4 and the atmosphere is the same as their mole ratio. That is,

$$\frac{V_{\text{CH}_4}}{V_{\text{tot}}} = \frac{n_{\text{CH}_4}}{n_{\text{tot}}} = \frac{1.7 \times 10^3}{1 \times 10^9} = 1.7 \times 10^{-6}$$

The molar concentration of CH_4 is

$$[\text{CH}_4] = \frac{n_{\text{CH}_4}}{V_{\text{tot}}} = \frac{P_{\text{CH}_4}}{RT} = \frac{P_{\text{tot}}\left(\frac{n_{\text{CH}_4}}{n_{\text{tot}}}\right)}{RT}$$

$$= \frac{(1 \text{ atm})\left(1.7 \times 10^{-6}\right)}{\left(0.08206 \text{ L atm K}^{-1} \text{ mol}^{-1}\right)(298 \text{ K})}$$

$$= 6.95 \times 10^{-8} \text{ mol L}^{-1}$$

The lifetime of the radical is

$$\frac{1}{k[\text{CH}_4]} = \frac{1}{\left(4.6 \times 10^6 \text{ L mol}^{-1} \text{ s}^{-1}\right)\left(6.95 \times 10^{-8} \text{ mol L}^{-1}\right)} = 3.1 \text{ s}$$

19.11 Describe three human activities that generate carbon dioxide. List two major mechanisms for the uptake of carbon dioxide.

Three human activities that generate CO_2 are (1) combustion, (2) respiration, and (3) metallurgical processes (such as the heating of $CaCO_3$ or the roasting of ores).

Photosynthesis and the absorption of CO_2 by the oceans are the two major mechanisms for uptake of carbon dioxide.

19.12   Deforestation contributes to the greenhouse effect in two ways. What are they?

Combustion of the forest products directly releases CO_2 to the atmosphere, and the reduction in photosynthetic activity decreases the rate of removal of CO_2 from the atmosphere.

19.13   How does an increase in world population enhance the greenhouse effect?

With greater human population, the activities listed in the answer to Problem 19.11 will increase.

19.14   Is ozone a greenhouse gas? Sketch three ways an ozone molecule can vibrate.

The vibrational modes of O_3, as shown below, are similar to those of water and absorb strongly in the infrared. Thus, ozone is a greenhouse gas.

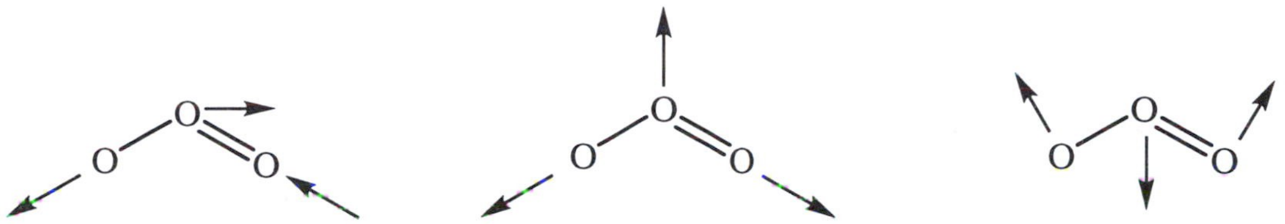

19.15   Suggest a gas other than carbon dioxide that scientists can study to substantiate the fact that CO_2 concentration is steadily increasing in the atmosphere.

CO_2 is formed at the expense of O_2, so that the amount of O_2 should be decreasing as CO_2 increases.

19.16   Which of the following settings is the most suitable for photochemical smog formation? **(a)** Gobi desert at noon in June, **(b)** New York city at 1 p.m. in July, **(c)** Boston at noon in January. Explain your choice.

Photochemical smog formation requires both heavy traffic and intense sunlight. These two conditions are met in **(b)**, an urban setting during a summer midday. (Note that daylight savings time is in effect in New York in July, so the sun is directly overhead at 1:00 p.m., not noon.)

19.17   On a smoggy day in a certain city, the ozone concentration was 0.42 ppm by volume. Calculate the partial pressure of ozone (in atm) and the number of ozone molecules per liter of air if the temperature and pressure were 20.0°C and 748 mmHg, respectively.

The volume ratio between O_3 and the atmosphere is the same as their mole ratio. That is,

$$\frac{V_{O_3}}{V_{tot}} = \frac{n_{O_3}}{n_{tot}} = \frac{0.42}{1 \times 10^6} = 4.2 \times 10^{-7}$$

The partial pressure of ozone is

$$P_{O_3} = P_{tot} \left(\frac{n_{O_3}}{n_{tot}} \right)$$

$$= (748 \text{ mmHg}) \left(4.2 \times 10^{-7} \right)$$

$$= 3.14 \times 10^{-4} \text{ mmHg}$$

$$= 4.13 \times 10^{-7} \text{ atm} = 4.1 \times 10^{-7} \text{ atm}$$

The number of moles of O_3 in 1 liter of air is

$$n_{O_3} = \frac{P_{O_3} V}{RT}$$

$$= \frac{\left(4.13 \times 10^{-7} \text{ atm} \right) (1 \text{ L})}{\left(0.08206 \text{ L atm K}^{-1} \text{ mol}^{-1} \right) (293.2 \text{ K})}$$

$$= 1.72 \times 10^{-8} \text{ mol}$$

The number of O_3 molecules in 1 liter of air is

$$\left(1.72 \times 10^{-8} \text{ mol} \right) \left(6.022 \times 10^{23} \text{ mol}^{-1} \right) = 1.0 \times 10^{16}$$

19.18   The gas-phase decomposition of peroxyacetyl nitrate (PAN) obeys first-order kinetics:

$$CH_3COOONO_2 \rightarrow CH_3COOO + NO_2$$

with a rate constant of $4.9 \times 10^{-4} \text{ s}^{-1}$. Calculate the rate of decomposition in $M \text{ s}^{-1}$ if the concentration of PAN is 0.55 ppm by volume. Assume STP conditions.

The volume ratio between PAN and the atmosphere is the same as their mole ratio. That is,

$$\frac{V_{PAN}}{V_{tot}} = \frac{n_{PAN}}{n_{tot}} = \frac{0.55}{1 \times 10^6} = 5.5 \times 10^{-7}$$

The molar concentration of PAN is

$$[PAN] = \frac{n_{PAN}}{V_{tot}} = \frac{P_{PAN}}{RT} = \frac{P_{tot} \left(\frac{n_{PAN}}{n_{tot}} \right)}{RT}$$

$$= \frac{(1 \text{ atm}) \left(5.5 \times 10^{-7} \right)}{\left(0.08206 \text{ L atm K}^{-1} \text{ mol}^{-1} \right) (273.2 \text{ K})}$$

$$= 2.45 \times 10^{-8} \text{ mol L}^{-1}$$

Thus, the rate of decomposition of PAN is

$$\text{Rate} = k[\text{PAN}] = \left(4.9 \times 10^{-4} \text{ s}^{-1}\right)\left(2.45 \times 10^{-8} \text{ mol L}^{-1}\right) = 1.2 \times 10^{-11} \, M\,\text{s}^{-1}$$

19.19   Assume that the formation of nitrogen dioxide,

$$2\text{NO}(g) + \text{O}_2(g) \rightarrow 2\text{NO}_2(g)$$

is an elementary reaction. **(a)** Write the rate law for this reaction. **(b)** A sample of air at a certain temperature is contaminated with 2.0 ppm of NO by volume. Under these conditions, can the rate law be simplified? If so, write the simplified rate law. **(c)** Under the conditions described in **(b)**, the half-life of the reaction has been estimated to be 6.4×10^3 min. What would the half-life be if the initial concentration of NO were 10 ppm?

(a) Rate = k $[\text{NO}]^2[\text{O}_2]$

(b) Since $[\text{O}_2]$ is very large compared to [NO], the reaction is a pseudo second-order reaction and the rate law can be simplified to

$$\text{Rate} = k'[\text{NO}]^2$$

where $k' = k[\text{O}_2]$.

(c) For a second-order reaction, the half life is inversely proportional to the initial concentration of the reactant:

$$t_{1/2} = \frac{1}{k[\text{A}]_0}$$

Let $(t_{1/2})_1$ be the half-life when 2.0 ppm of NO is present, and $(t_{1/2})_2$ be the half-life when 10 ppm of NO is present. The ratio of the half-lives is

$$\frac{(t_{1/2})_2}{(t_{1/2})_1} = \frac{k([\text{A}]_0)_1}{k([\text{A}]_0)_2}$$

$$(t_{1/2})_2 = \frac{([\text{A}]_0)_1}{([\text{A}]_0)_2}(t_{1/2})_1 = \left(\frac{2.0 \text{ ppm}}{10 \text{ ppm}}\right)\left(6.4 \times 10^3 \text{ min}\right) = 1.3 \times 10^3 \text{ min}$$

19.20   The safety limits of ozone and carbon monoxide are 120 ppb by volume and 9 ppm by volume, respectively. Why does ozone have a lower limit?

Ozone is much more damaging to lung tissues, and hence more toxic than carbon monoxide. Although carbon monoxide is quite dangerous, the human body can tolerate a fairly large amount of CO and still survive. For example, a person can still function even when 10% or so of the body's hemoglobin is complexed with CO.

19.21   Ozone in the troposphere is formed by the following steps:

$$NO_2 \rightarrow NO + O \qquad (1)$$
$$O + O_2 \rightarrow O_3 \qquad (2)$$

The first step is initiated by the absorption of visible light (NO_2 is a brown gas.) Calculate the longest wavelength required for step 1 at 25°C. (*Hint*: You need to first calculate the value of $\Delta_r H$ and hence the value of $\Delta_r U$ for step 1. Next, determine the wavelength for decomposing NO_2 from $\Delta_r U$.)

$\Delta_r H$ for the reaction described in step 1 is

$$\Delta_r H = \Delta_f \overline{H}°[NO(g)] + \Delta_f \overline{H}°[O(g)] - \Delta_f \overline{H}°[NO_2(g)]$$

$$= 90.4 \text{ kJ mol}^{-1} + 249.4 \text{ kJ mol}^{-1} - 33.9 \text{ kJ mol}^{-1}$$

$$= 305.9 \text{ kJ mol}^{-1}$$

Thus, $\Delta_r U$ for the dissociation of 1 mole of NO_2 is

$$\Delta_r U = \Delta_r H - \Delta(PV) \qquad \text{(From rearranging Equation (4.11)}$$

$$= \Delta_r H - \Delta(nRT) = \Delta_r H - RT\Delta n$$

$$= 305.9 \times 10^3 \text{ J mol}^{-1} - \left(8.314 \text{ J K}^{-1} \text{ mol}^{-1}\right)(298.2 \text{ K})$$

$$= 303.4 \times 10^3 \text{ J mol}^{-1}$$

The longest wavelength required for step 1 must have an energy equal to $\Delta_r U$.

$$\lambda = \frac{c}{\nu} = \frac{ch}{E}$$

$$= \frac{\left(3.00 \times 10^8 \text{ m s}^{-1}\right)\left(6.626 \times 10^{-34} \text{ J s}\right)}{\left(303.4 \times 10^3 \text{ J mol}^{-1}\right)\left(\frac{1 \text{ mol}}{6.022 \times 10^{23}}\right)}$$

$$= 3.95 \times 10^{-7} \text{ m} = 395 \text{ nm}$$

19.22   Given that the quantity of ozone in the stratosphere is equivalent to a 3.0-mm-thick layer of ozone on Earth at 1 atm and 25°C, calculate the number of O_3 molecules in the stratosphere and their mass in kilograms. See Problem 19.7 for other information.

The volume of ozone can be calculated from the Earth's average radius, r, and the thickness of the ozone layer, h.

$$V = 4\pi r^2 h = 4\pi \left(6371 \times 10^3 \text{ m}\right)^2 \left(3.0 \times 10^{-3} \text{ m}\right)\left(\frac{1000 \text{ L}}{1 \text{ m}^3}\right) = 1.53 \times 10^{15} \text{ L}$$

The number of moles of ozone is

$$n = \frac{PV}{RT} = \frac{(1 \text{ atm})\left(1.53 \times 10^{15} \text{ L}\right)}{\left(0.08206 \text{ L atm K}^{-1} \text{ mol}^{-1}\right)(298 \text{ K})} = 6.26 \times 10^{13} \text{ mol}$$

The number of O_3 molecules and the mass are

$$\text{Number of } O_3 \text{ molecules} = (6.26 \times 10^{13} \text{ mol}) \left(\frac{6.022 \times 10^{23} \text{ molecules}}{1 \text{ mol}} \right) = 3.8 \times 10^{37} \text{ molecules}$$

$$\text{Mass} = (6.26 \times 10^{13} \text{ mol}) \left(\frac{48.00 \text{ g}}{1 \text{ mol}} \right) \left(\frac{1 \text{ kg}}{1000 \text{ g}} \right) = 3.00 \times 10^{12} \text{ kg} = 3.0 \times 10^{12} \text{ kg}$$

19.23 Referring to the answer in Problem 19.22 and assuming that the level of ozone in the stratosphere has already fallen 6.0%, calculate the number of kilograms of ozone that must be manufactured on a daily basis so that we can restore the ozone to the original level in 100 years. If ozone is made according to the process $3O_2(g) \rightarrow 2O_3(g)$, how many kilojoules of energy input must be supplied to drive the reaction?

The quantity of ozone lost is

$$(3.00 \times 10^{12} \text{ kg}) (6.0\%) = 1.80 \times 10^{11} \text{ kg}$$

Assuming no further deterioration, the amount of O_3 that must be manufactured per day is

$$\left(\frac{1.80 \times 10^{11} \text{ kg}}{100 \text{ yr}} \right) \left(\frac{1 \text{ yr}}{365 \text{ days}} \right) = 4.9 \times 10^6 \text{ kg day}^{-1}$$

For the process $3O_2(g) \rightarrow 2O_3(g)$,

$$\Delta_r H = 2\Delta_f \overline{H}^{\circ}[O_3(g)] - 3\Delta_f \overline{H}^{\circ}[O_2(g)]$$

$$= 2 (142.7 \text{ kJ mol}^{-1}) - 3 (0 \text{ kJ mol}^{-1})$$

$$= 285.4 \text{ kJ mol}^{-1}$$

In the last result, kJ mol^{-1} means kJ (mol reaction)$^{-1}$ for the balanced equation $3O_2(g) \rightarrow 2O_3(g)$. Thus, it gives the numerical value for the amount of energy input required to generate 2 moles of O_3. To generate 1.80×10^{11} kg of ozone, the amount of energy needed is

$$(1.80 \times 10^{11} \text{ kg}) \left(\frac{1 \text{ mol}}{48.00 \times 10^{-3} \text{ kg}} \right) \left(\frac{285.4 \text{ kJ}}{2 \text{ mol}} \right) = 5.4 \times 10^{14} \text{ kJ}$$

19.24 Why are CFCs not decomposed by UV radiation in the troposphere?

The short-wavelength UV radiation with sufficient photon energy to decompose CFCs is absorbed by species in the upper atmosphere, and only UV radiation of longer wavelengths and lower photon energy reaches the troposphere.

19.25   The average bond enthalpies of the C–Cl and C–F bonds are 340 kJ mol^{-1} and 485 kJ mol^{-1}, respectively. Based on this information, explain why the C–Cl bond in a CFC molecule is preferentially broken by solar radiation at 250 nm.

The energy of 1 mole of photons at 250 nm is

$$E = h\nu = \frac{hc}{\lambda} = \frac{\left(6.626 \times 10^{-34} \text{ J s}\right)\left(3.00 \times 10^{8} \text{ m s}^{-1}\right)}{250 \times 10^{-9} \text{ m}}$$

$$= \left(7.951 \times 10^{-19} \text{ J}\right)\left(\frac{6.022 \times 10^{23}}{1 \text{ mol}}\right)$$

$$= 4.79 \times 10^{5} \text{ J mol}^{-1} = 479 \text{ kJ mol}^{-1}$$

This energy is high enough to break C–Cl bonds, but not enough to break C–F bonds.

19.26   Like CFCs, certain bromine-containing compounds, such as CF_3Br, can participate in the destruction of ozone by a similar mechanism starting with the Br atom:

$$CF_3Br \rightarrow CF_3 + Br$$

Given that the average C–Br bond enthalpy is 276 kJ mol^{-1}, estimate the longest wavelength required to break this bond. Will the decomposition of CF_3Br occur in the troposphere or in both the troposphere and stratosphere?

The energy of the radiation with the longest wavelength must be the same as the average C–Br bond enthalpy.

$$\lambda = \frac{c}{\nu} = \frac{ch}{E}$$

$$= \frac{\left(3.00 \times 10^{8} \text{ m s}^{-1}\right)\left(6.626 \times 10^{-34} \text{ J s}\right)}{\left(276 \times 10^{3} \text{ J mol}^{-1}\right)\left(\frac{1 \text{ mol}}{6.022 \times 10^{23}}\right)}$$

$$= 4.34 \times 10^{-7} \text{ m} = 434 \text{ nm}$$

This is light in the visible region of the spectrum, and the compound will be decomposed in both the troposphere and the stratosphere.

19.27   Draw Lewis structures for chlorine nitrate ($ClONO_2$) and chlorine monoxide (ClO).

Chlorine nitrate has two resonance structures.

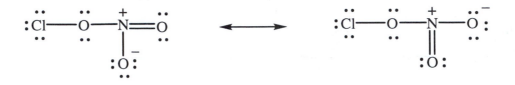

Chlorine monoxide has an odd number of electrons and an incomplete octet.

19.28   Why are CFCs more effective greenhouse gases than methane and carbon dioxide?

The greater polarity of the C–F and C–Cl bonds compared to the C–H and C=O bonds results in greater dipole derivatives for CFCs. This makes them stronger absorbers of IR radiation (see Section 17.3).

19.29   One suggestion for slowing down the destruction of ozone in the stratosphere is to spray the region with hydrocarbons such as ethane and propane. How does this method work? What is the drawback of this procedure if used on a large scale for an extended period of time?

The hydrocarbons would react with and scavenge Cl atoms as follows

$$C_2H_6 + Cl \longrightarrow C_2H_5Cl + H$$

$$C_3H_8 + Cl \longrightarrow C_3H_7Cl + H$$

The hydrocarbons are greenhouse gases, however, and their prolonged presence in the atmosphere would raise the temperature.

19.30   Calculate the standard enthalpy of formation ($\Delta_f \overline{H}^\circ$) of ClO from the following bond dissociation enthalpies: Cl_2: 242.7 kJ mol^{-1}; O_2: 498.8 kJ mol^{-1}; ClO: 206 kJ mol^{-1}.

First we esitmate the standard enthalpy of reaction for $Cl_2(g) + O_2(g) \rightarrow 2ClO(g)$ using bond dissociation enthalpies.

$$\Delta_r H^\circ = \sum BE(\text{reactants}) - \sum BE(\text{products})$$

$$= 242.7 \text{ kJ mol}^{-1} + 498.8 \text{ kJ mol}^{-1} - 2\left(206 \text{ kJ mol}^{-1}\right)$$

$$= 329.5 \text{ kJ mol}^{-1}$$

The standard enthalpy of reaction is also related to the standard enthalpies of formation of the reactants and products.

$$\Delta_r H^\circ = 2\Delta_f \overline{H}^\circ[ClO(g)] - \Delta_f \overline{H}^\circ[Cl_2(g)] - \Delta_f \overline{H}^\circ[O_2(g)]$$

$$329.5 \text{ kJ mol}^{-1} = 2\Delta_f \overline{H}^\circ[ClO(g)] - 0 \text{ kJ mol}^{-1} - 0 \text{ kJ mol}^{-1}$$

$$\Delta_f \overline{H}^\circ[ClO(g)] = \frac{329.5 \text{ kJ mol}^{-1}}{2} = 165 \text{ kJ mol}^{-1}$$

19.31   What is the biological significance of the fact that λ_{max} occurs at around 500 nm in the solar emission spectrum?

Both vision and photosynthesis have evolved to use light in this region of the spectrum. Presumably whatever region of the spectrum were to be used by vision would have been called "visible."

19.32   In the sea, light intensity decreases with depth. For example, at a depth of 20 m below the surface, light intensity is one-half of that at the sea level. In practice, total darkness sets in when 99% of the light is absorbed by water. Explain why green algae are found near the surface, but red algae are located as deep as 100 m.

The depth at which total darkness sets in can be found by applying the Beer–Lambert law to the absorption of light by sea water, since at 20 m, $I_0/I = 2$, and for total darkness, $I_0/I = 100$. Assuming the concentration of absorbing species to be uniform,

$$\log 2 = \epsilon \, (20 \text{ m}) \, c$$

$$\log 100 = \epsilon b c$$

and,

$$b = (20 \text{ m}) \frac{\log 100}{\log 2} = 133 \text{ m}$$

Thus, algae can find light at depths of 100 m, but since blue light penetrates the ocean depths to a greater extent than red light, only those algae that can absorb blue light for life-sustaining processes will be found at the greater depths. Such algae, since they absorb blue light will appear red. Green algae, on the other hand, absorb red light and cannot survive at the greater depths where red light does not penetrate.

19.33   Transition metals such as Fe, Cu, Co, and Mn are necessary for respiration and photosynthesis whereas nontransition metals such as Zn, Ca, and Na are not. Explain.

The transition metals have variable oxidation states and a rich oxidation-reduction chemistry that is lacking in the nontransition metals. This makes the transition metals much more suitable for the oxidation-reduction processes that need to occur in respiration and photosynthesis.

19.34   In photosynthesis, the term *quantum requirement* refers to the number of photons required to reduce one CO_2 molecule to (CH_2O):

$$H_2O + CO_2 \rightarrow (CH_2O) + O_2$$

The efficiency of this process depends on the wavelength of light employed. Assuming a quantum requirement of 8, calculate the efficiency under standard-state conditions for the synthesis of 1 mole of glucose if the wavelength of light employed is **(a)** 400 nm and **(b)** 700 nm.

The synthesis of 1 mole of glucose is described by

$$6H_2O + 6CO_2 \rightarrow C_6H_{12}O_6 + 6O_2$$

and the standard Gibbs energy of reaction is $\Delta_r G^\circ = 2879$ kJ for 1 mol of glucose. (Recall that $\Delta_r G$ corresponds to the nonexpansion work in the reversible limit that must be done at constant T and P in the reaction.) The quantum requirement is $6 \times 8 = 48$ since there are $6(CH_2O)$ units in glucose

(a) The energy for 1 mole of photons at 400 nm is

$$E = h\nu = \frac{hc}{\lambda}$$

$$= \frac{\left(6.626 \times 10^{-34} \, \text{J s}\right)\left(3.00 \times 10^8 \, \text{m s}^{-1}\right)}{400 \times 10^{-9} \, \text{m}} \left(\frac{6.022 \times 10^{23}}{1 \, \text{mol}}\right)$$

$$= 2.993 \times 10^5 \, \text{J mol}^{-1} = 299.3 \, \text{kJ mol}^{-1}$$

Therefore, the efficiency is

$$\frac{2879 \, \text{kJ mol}^{-1}}{48 \left(299.3 \, \text{kJ mol}^{-1}\right)} = 0.200 = 20.0\%$$

(b) The energy for 1 mole of photons at 700 nm is

$$E = h\nu = \frac{hc}{\lambda}$$

$$= \frac{\left(6.626 \times 10^{-34} \, \text{J s}\right)\left(3.00 \times 10^8 \, \text{m s}^{-1}\right)}{700 \times 10^{-9} \, \text{m}} \left(\frac{6.022 \times 10^{23}}{1 \, \text{mol}}\right)$$

$$= 1.710 \times 10^5 \, \text{J mol}^{-1} = 171.0 \, \text{kJ mol}^{-1}$$

Therefore, the efficiency is

$$\frac{2879 \, \text{kJ mol}^{-1}}{48 \left(171.0 \, \text{kJ mol}^{-1}\right)} = 0.351 = 35.1\%$$

19.35 At low light intensities, the rate of photosynthesis increases linearly with intensity. At high intensities, however, the rate is constant (saturation rate). Suggest an interpretation at the molecular level. The saturation rate varies with temperature. Explain.

At low light intensities, not all of the photocenters are absorbing photons. As the intensity, and number of photons increases, more and more photocenters are participating in photosynthesis until all of the photocenters are involved. At this point further increases in the light intensity will have no effect. The temperature can have either a positive or negative effect on the saturation rate, since it can affect the number of photocenters in an active configuration, or the progress of further reaction of molecules which have absorbed a photon.

19.36   Calculate the number of moles of ATP that can be synthesized at 80% efficiency by a photosynthetic organism upon the absorption of 2.1 einsteins of photons at 650 nm. (*Hint:* $\Delta_r G^{o\prime}$ for the synthesis of ATP from ADP and P_i is 31.4 kJ mol^{-1}.)

The energy for 1 mole of photons at 650 nm is

$$E = h\nu = \frac{hc}{\lambda}$$

$$= \frac{(6.626 \times 10^{-34}\,\mathrm{J\,s})\,(3.00 \times 10^{8}\,\mathrm{m\,s^{-1}})}{650 \times 10^{-9}\,\mathrm{m}} \left(\frac{6.022 \times 10^{23}}{1\,\mathrm{mol}} \right)$$

$$= 1.842 \times 10^{5}\,\mathrm{J\,mol^{-1}} = 184.2\,\mathrm{kJ\,mol^{-1}}$$

The photon energy utilized to synthesize ATP is

$$\left(184.2\,\mathrm{kJ\,mol^{-1}}\right)(2.1\ \text{einsteins})(80\%) = 309.5\,\mathrm{kJ}$$

The number of moles of ATP that can be synthesized is

$$\frac{309.5\,\mathrm{kJ}}{31.4\,\mathrm{kJ\,mol^{-1}}} = 9.9\,\mathrm{mol}$$

19.37   What is the advantage of the broad absorption spectrum of rhodopsin (see Figure 19.18)?

The broad absorption spectrum allows vision over a wide range of wavelengths, making the "visible" region of the spectrum as wide as possible.

19.38   In solution, 11-*cis*-retinal absorbs maximally at 380 nm. In rhodopsin, the maximum absorption occurs at 500 nm. Explain.

When bonded to opsin to form rhodopsin, the electrons of 11-*cis*-retinal are delocalized over a larger region. According to the particle-in-a-box model, greater delocalization leads to smaller energy spacings. Thus, the electronic transition occurs at a longer wavelength.

19.39   Why does one have to irradiate a sample for hours or even days to achieve acceptable yields in some photochemical reaction even though the lifetimes of excited electronic states are of the order of micro- or nanoseconds? Assume that the rate of light absorption is 2.0×10^{19} photons s^{-1}.

Even if every photon led directly to product, which would correspond to $F = f = \Phi_p = 1$ in Equation 19.4, the rate of product formation would still be limited by the rate of light absorption. At 2.0×10^{19} photons s^{-1} = 3.3×10^{-5} einstein s^{-1}, the rate of product formation is only 3.3×10^{-5} mol s^{-1}. At this rate, it would take 1 mol/3.3×10^{-5} mol s^{-1} = 3.0×10^{4} s = 8.4 hr to produce a mole of product.

19.40 The transparency of a certain type of sunglasses to light depends on the intensity of light in the environment. The lenses are clear in dimly lit rooms but darken when the wearer goes outdoors. The material responsible for this change is the very tiny AgCl crystals incorporated in the glass. Suggest a photochemical mechanism that would account for this change.

When irradiated with the appropriate wavelength (which must be in the visible region of the spectrum), AgCl decomposes according to

$$AgCl \longrightarrow Ag + Cl$$

The Ag and Cl atoms formed are trapped in the glass matrix. The small Ag particles diminish the amount of light transmitted. In the absence of intense light, the reactive atoms recombine to form transparent AgCl.

19.41 Suppose that an excited singlet, S_1, can be deactivated by three different mechanisms whose rate constants are k_1, k_2, and k_3. The rate of decay is given by $-d[S_1]/dt = (k_1 + k_2 + k_3)[S_1]$. **(a)** If τ is the mean lifetime, that is, the time required for $[S_1]$ to decrease to $1/e$ or 0.368 of the original value, show that $(k_1 + k_2 + k_3)\tau = 1$. **(b)** The overall rate constant, k, is given by

$$\frac{1}{\tau} = k = k_1 + k_2 + k_3 = \frac{1}{\tau_1} + \frac{1}{\tau_2} + \frac{1}{\tau_3}$$

Show that the quantum yield Φ_i is given by

$$\Phi_i = \frac{k_i}{\sum k_i} = \frac{\tau}{\tau_i}$$

where i denotes the ith decay mechanism. **(c)** If $\tau_1 = 10^{-7}$ s, $\tau_2 = 5 \times 10^{-8}$ s, and $\tau_3 = 10^{-8}$ s, calculate the lifetime of the singlet state and the quantum yield for the path that has τ_2.

(a) The integrated rate law is

$$[S_1]_t = [S_1]_0 e^{-(k_1+k_2+k_3)t}$$

When $t = \tau$, $[S_1]_t = 0.386[S_1]_0$. That is,

$$0.386[S_1]_0 = [S_1]_0 e^{-(k_1+k_2+k_3)\tau}$$

$$(k_1 + k_2 + k_3)\tau = -\ln 0.368 = 1$$

(b)

$$\Phi_i = \frac{\text{rate of the } i\text{th decay mechanism}}{\text{total rate of removal of the singlet state}}$$

$$= \frac{k_i[S_1]}{k_1[S_1] + k_2[S_1] + k_3[S_1]} = \frac{k_i}{k_1 + k_2 + k_3}$$

$$= \frac{\frac{1}{\tau_i}}{\frac{1}{\tau}} = \frac{\tau}{\tau_i}$$

(c) The lifetime of the singlet state is related to τ_1, τ_2, and τ_3.

$$\frac{1}{\tau} = \frac{1}{\tau_1} + \frac{1}{\tau_2} + \frac{1}{\tau_3} = \frac{1}{10^{-7} \text{ s}} + \frac{1}{5 \times 10^{-8} \text{ s}} + \frac{1}{10^{-8} \text{ s}} = 1.30 \times 10^8 \text{ s}^{-1}$$

$$\tau = 7.69 \times 10^{-9} \text{ s} = 7.7 \times 10^{-9} \text{ s}$$

The quantum yield for the path that has τ_2 is

$$\Phi_2 = \frac{\tau}{\tau_2} = \frac{7.69 \times 10^{-9} \text{ s}}{5 \times 10^{-8}} = 0.15$$

19.42   Consider the photochemical isomerization A $\rightleftharpoons$ B. At 650 nm, the quantum yields for the forward and reverse reactions are 0.73 and 0.44, respectively. If the molar absorptivities of A and B are 1.3×10^3 L mol^{-1} cm^{-1} and 0.47×10^3 L mol^{-1} cm^{-1}, respectively, what is the ratio [B]/[A] in the photostationary state?

The rate of formation of B from A is given by $I_A \Phi_A$, where I_A is the intensity of light *absorbed* by A. Likewise, the rate of formation of A from B is $I_B \Phi_B$. At the photostationary state, the two rates are equal, or

$$I_A \Phi_A = I_B \Phi_B$$

The intensity of light abosorbed by a sample is given by $I_0 - I$, where I is the transmitted light intensity. Since according to the Beer–Lambert law, $I = I_0 10^{-A} = I_0 10^{-\epsilon bc}$, then intensity of light absorbed is

$$I_0 - I = I_0 \left(1 - 10^{-\epsilon bc} \right)$$

Thus, at the photostationary state,

$$I_0 \left(1 - 10^{-\epsilon_A bc_A} \right) \Phi_A = I_0 \left(1 - 10^{-\epsilon_B bc_B} \right) \Phi_B$$

This is difficult to solve in the general case, but in the limit of small absorption of light ($\epsilon bc << 1$), the approximation $10^{-\epsilon bc} \approx 1 - 2.303 \epsilon bc$ may be made, leading to

$$I_0 \left(2.303 \epsilon_A bc_A \right) \Phi_A = I_0 \left(2.303 \epsilon_B bc_B \right) \Phi_B$$

$$\frac{c_B}{c_A} = \frac{[B]}{[A]} = \frac{\epsilon_A \Phi_A}{\epsilon_B \Phi_B} = \frac{\left(1.3 \times 10^3 \text{ L mol}^{-1} \text{ cm}^{-1} \right) (0.73)}{\left(0.47 \times 10^3 \text{ L mol}^{-1} \text{ cm}^{-1} \right) (0.44)} = 4.6$$

19.43   The molar heat capacity of a diatomic molecule is 29.1 J K^{-1} mol^{-1}. Assuming the atmosphere contains only nitrogen gas and there is no heat loss, calculate the total heat intake (in kilojoules) if the atmosphere warms up by 3°C during the next 50 years. Given that there are 1.8×10^{20} moles of diatomic molecules present, how many kilograms of ice (at the North and South Poles) will this quantity of heat melt at 0°C? (The molar heat of fusion of ice is 6.01 kJ mol^{-1}.)

The total heat intake is

$$\left(29.1 \text{ J K}^{-1} \text{ mol}^{-1}\right) (3 \text{ K}) \left(1.8 \times 10^{20} \text{ mol}\right) \left(\frac{1 \text{ kJ}}{1000 \text{ J}}\right) = 1.57 \times 10^{19} \text{ kJ} = 1.6 \times 10^{19} \text{ kJ}$$

The amount of ice that will be melted is

$$n = \frac{1.57 \times 10^{19} \text{ kJ}}{6.01 \text{ kJ mol}^{-1}} = 2.61 \times 10^{18} \text{ mol}$$

$$m = \left(2.61 \times 10^{18} \text{ mol}\right) \left(\frac{18.02 \text{ g}}{1 \text{ mol}}\right) \left(\frac{1 \text{ kg}}{1000 \text{ g}}\right) = 4.7 \times 10^{16} \text{ kg}$$

19.44   Carbon monoxide has a much higher affinity for hemoglobin than oxygen does. **(a)** Write the equilibrium constant expression (K_c) for the following process:

$$CO(g) + HbO_2(aq) \rightleftharpoons O_2(g) + HbCO(aq)$$

where HbO_2 and $HbCO$ are oxygenated hemoglobin and carboxylhemoglobin, respectively. **(b)** The composition of a breath of air inhaled by a person smoking a cigarette is 1.9×10^{-6} mol L^{-1} CO and 8.6×10^{-3} mol L^{-1} O_2. Calculate the ratio of [HbCO] to [HbO$_2$], given that K_c is 212 at 37°C.

(a)

$$K_c = \frac{[O_2][HbCO]}{[CO][HbO_2]}$$

(b)

$$\frac{[HbCO]}{[HbO_2]} = K_c \frac{[CO]}{[O_2]} = (212) \left(\frac{1.9 \times 10^{-6}}{8.6 \times 10^{-3}}\right) = 4.7 \times 10^{-2}$$

19.45   In 1991, it was discovered that nitrous oxide (N_2O) is produced in the synthesis of nylon. This compound, which is released into the atmosphere, contributes *both* to the depletion of ozone in the stratosphere and to the greenhouse effect. **(a)** Write equations representing the reactions between N_2O and oxygen atoms in the stratosphere to produce nitric oxide, which is then oxidized by ozone to form nitrogen dioxide. **(b)** Is N_2O a more effective greenhouse gas than carbon dioxide? Explain. **(c)** One of the intermediates in nylon manufacture is adipic acid [HOOC(CH$_2$)$_4$COOH]. About 2.2×10^9 kg of adipic acid are consumed every year. Estimates are that for every mole of adipic acid produced, 1 mole of N_2O is generated. What is the maximum number of moles of O_3 that can be destroyed as a result of this process per year?

(a) The individual reactions and the overall reaction are

$$N_2O + O \rightleftharpoons 2NO$$

$$2NO + 2O_3 \rightleftharpoons 2O_2 + 2NO_2$$

Overall: $$N_2O + O + 2O_3 \rightleftharpoons 2O_2 + 2NO_2$$

(b) As a polar molecule, the dipole derivatives associated with the vibrational motions of N_2O are larger than those for the nonpolar CO_2. This makes the IR absorptions of N_2O stronger than those of CO_2 (see Section 17.3), making N_2O a more effective greenhouse gas.

(c) The number of moles of N_2O generated is the same as the number of moles of adipic acid consumed.

$$n_{N_2O} = \frac{\text{mass of adipic acid consumed}}{\text{molar mass of adipic acid}} = \frac{2.2 \times 10^9 \text{ kg}}{146.1 \times 10^{-3} \text{ kg mol}^{-1}} = 1.51 \times 10^{10} \text{ mol}$$

For each mole of N_2O generated, 2 mol of O_3 can be destroyed. Thus, the maximum amount of O_3 destroyed is

$$n_{O_3} = 2n_{N_2O} = 2\left(1.51 \times 10^{10} \text{ mol}\right) = 3.01 \times 10^{10} \text{ mol}$$

19.46 The hydroxyl radical is formed by the following reactions:

$$O_3 \xrightarrow{\lambda < 320 \text{ nm}} O^* + O_2$$

$$O^* + H_2O \longrightarrow 2OH$$

where O^* denotes an electronically excited atom. (a) Explain why the concentration of OH is so small even though the concentrations of O_3 and H_2O are quite large in the troposphere. (b) What property makes OH a strong oxidizing agent? (c) The reaction between OH and NO_2 contributes to acid rain. Write an equation for this process. (d) The hydroxyl radical can oxidize SO_2 to H_2SO_4. The first step is the formation of a neutral HSO_3 species, followed by its reaction with O_2 and H_2O to form H_2SO_4 and the hydroperoxyl radical (HO_2). Write equations for these processes.

(a) The high reactivity of OH leads to a small concentration.

(b) The OH radical has a strong tendency to abstract an H atom (with its electron) from another compound due to the large O–H bond enthalpy.

(c) $NO_2 + OH \longrightarrow HNO_3$ (d) $OH + SO_2 \longrightarrow HSO_3$, $HSO_3 + O_2 + H_2O \longrightarrow H_2SO_4 + HO_2$

19.47 Given that the collision diameter of ozone is about 4.2 Å, calculate the mean free path of ozone at sea level (1 atm and 25°C) and in the stratosphere (3×10^{-3} atm and -23°C).

The mean free path can be calculated using Equation (3.19):

$$\lambda = \frac{RT}{\sqrt{2}\pi d^2 P N_A}$$

At sea level,

$$\lambda = \frac{\left(8.314 \text{ J K}^{-1} \text{ mol}^{-1}\right)(298 \text{ K})}{\sqrt{2}\pi \left(4.2 \times 10^{-10} \text{ m}\right)^2 \left[(1 \text{ atm})\left(\frac{1.01325 \times 10^5 \text{ Pa}}{1 \text{ atm}}\right)\right]\left(6.022 \times 10^{23} \text{ mol}^{-1}\right)}$$

$$= 5.2 \times 10^{-8} \text{ m}$$

In the stratosphere,

$$\lambda = \frac{\left(8.314 \, \text{J K}^{-1} \, \text{mol}^{-1}\right) (250 \, \text{K})}{\sqrt{2}\pi \left(4.2 \times 10^{-10} \, \text{m}\right)^2 \left[\left(3 \times 10^{-3} \, \text{atm}\right) \left(\frac{1.01325 \times 10^5 \, \text{Pa}}{1 \, \text{atm}}\right)\right] \left(6.022 \times 10^{23} \, \text{mol}^{-1}\right)}$$

$$= 1.4 \times 10^{-5} \, \text{m}$$

19.48 Comment on the comparison that the hydroxyl radical behaves in some ways like the white blood cells in our bodies.

The hydroxyl radical has sometimes been described as the "detergent" radical because of its ability, due to its high reactivity, to cleanse the atmosphere by removing undesirable species. In this way, it can be compared to the white blood cells that perform the same action in the bloodstream (although the mechanism of cleansing is very different in the two cases).

19.49 Account for the oscillation in atmospheric concentration CO_2 shown in Figure 19.7.

This is a seasonal oscillation caused by increased photosynthesis in the summer that reduces atmospheric CO_2. During the winter, the CO_2 level increases due to reduced photosynthesis.

19.50 Explain why phosphorescence of ethylene has never been observed.

In the triplet state of ethylene, the p orbitals on the two carbon atoms are oriented perpendicular to each other, and the planes formed by the two $-CH_2$'s are likewise perpendicular. There is little overlap between the wavefunction for this state and that for the planar, singlet ground state, making the probability of the $T_1 \rightarrow S_0$ transition very low.

19.51 A light source of power 2×10^{-16} W is sufficient to be detected by the human eye. Assuming the wavelength of the light is at 550 nm, calculate the number of photons that must be absorbed by rhodopsin per second. (*Hint*: Vision persists for only 1/30 of a second.)

The energy of 1 photon at 550 nm is

$$E = h\nu = \frac{hc}{\lambda} = \frac{\left(6.626 \times 10^{-34} \, \text{J s}\right) \left(3.00 \times 10^8 \, \text{m s}^{-1}\right)}{550 \times 10^{-9} \, \text{m}} = 3.614 \times 10^{-19} \text{J}$$

The light source produces $2 \times 10^{-16} \, \text{J s}^{-1}$. Therefore, The number of photons that must be absorbed by rhodopsin each second is

$$\frac{2 \times 10^{-16} \, \text{J}}{3.614 \times 10^{-19} \text{J}} = 5.53 \times 10^2 = 5.5 \times 10^2$$

Since vision persists for only 1/30 of a second, the number of photons detected is

$$\left(5.53 \times 10^2\right) \left(\frac{1}{30}\right) = 18$$

CHAPTER 20

The Solid State

PROBLEMS AND SOLUTIONS

20.1 Construct a table that lists the h, k, l, and $h^2 + k^2 + l^2$ values for the simple cubic, fcc, and bcc lattices. How would you use this table to deduce the nature of a crystal lattice from a series of experimentally determined hkl values?

Referring to Figure 20.7,

scc hkl	scc $h^2+k^2+l^2$	fcc hkl	fcc $h^2+k^2+l^2$	bcc hkl	bcc $h^2+k^2+l^2$
100	1				
110	2			110	2
111	3	111	3		
200	4	200	4	200	4
210	5				
211	6			211	6
220	8	220	8	220	8
221	9				
300	9				
310	10			310	10
311	11	311	11		
222	12	222	12	222	12
320	13				
321	14			321	14
		400	16	400	16

The value of $h^2 + k^2 + l^2$ determines the angle of X-ray scattering from a plane in a diffraction pattern,

$$\sin^2 \theta_{hkl} = \frac{\lambda^2}{4a^2} \left(h^2 + k^2 + l^2 \right)$$

Therefore for scattering from two different planes in a cubic crystal,

$$\frac{\sin^2 \theta_1}{\sin^2 \theta_2} = \frac{\left(h^2 + k^2 + l^2\right)_1}{\left(h^2 + k^2 + l^2\right)_2}$$

Looking at the first two lines in a diffraction pattern, which for simple cubic lattices are due to the 100 and 110 planes, it is apparent that for this lattice

$$\frac{\sin^2 \theta_{100}}{\sin^2 \theta_{110}} = \frac{1}{2} = 0.5$$

For the bcc lattice, the first two reflections are due to different planes, but the ratio is the same,

$$\frac{\sin^2 \theta_{110}}{\sin^2 \theta_{200}} = \frac{2}{4} = 0.5$$

These two cases are different, however, from the fcc lattice for which the first two lines in the diffraction pattern are due to the 111 and 200 planes,

$$\frac{\sin^2 \theta_{111}}{\sin^2 \theta_{200}} = \frac{3}{4} = 0.75$$

Thus, comparing $\sin^2 \theta$ for the first two lines in the pattern will distinguish between fcc on the one hand and either simple cubic or bcc on the other.

These latter two may be distinguished through comparison of the sixth and seventh lines in the pattern.

For simple cubic,

$$\frac{\sin^2 \theta_{211}}{\sin^2 \theta_{220}} = \frac{6}{8} = 0.75$$

For bcc,

$$\frac{\sin^2 \theta_{222}}{\sin^2 \theta_{321}} = \frac{12}{14} = 0.86$$

Note that these latter two lattice types give rise to very similar diffraction patterns with the first six lines given by $h^2 + k^2 + l^2$ in the ratio 1 : 2 : 3 : 4 : 5 : 6. Not until the seventh line do they differ.

20.2   When X rays with a wavelength of 0.85 Å are diffracted by a metallic crystal, the angle of first-order diffraction ($n = 1$) is measured to be 14.8°. What is the distance between the layers of atoms responsible for the diffraction?

$$2d \sin \theta = n\lambda$$

$$d = \frac{n\lambda}{2 \sin \theta} = \frac{(1)\left(0.85 \text{ Å}\right)}{2 \sin 14.8°} = 1.7 \text{ Å}$$

20.3 When X rays of wavelength 0.090 nm are diffracted by a metallic crystal, the angle of first-order diffraction ($n = 1$) is measured to be 15.2°. What is the distance (in pm) between the layers of atoms responsible for the diffraction?

$$2d \sin \theta = n\lambda$$

$$d = \frac{n\lambda}{2 \sin \theta} = \frac{(1)\,(0.090 \text{ nm})}{2 \sin 15.2°} = 0.17 \text{ nm} = 170 \text{ pm}$$

20.4 The distance between layers in a NaCl crystal is 282 pm. X rays are diffracted from these layers at an angle of 23.0°. Assuming that $n = 1$, calculate the wavelength of the X rays in nm.

$$2d \sin \theta = n\lambda$$

$$\lambda = \frac{2d \sin \theta}{n} = \frac{(2)\,(282 \text{ pm}) \sin 23.0°}{1} = 220 \text{ pm} = 0.220 \text{ nm}$$

20.5 Calculate the number of spheres in the simple cubic, body-centered cubic, and face-centered cubic cells. Also, calculate the packing efficiency of each type of cell.

Simple cubic

There are 8 spheres at the corners of the cell [Figure 20.18(a)]. Since each sphere at a corner is shared by 8 unit cells, the number of spheres is $8 \times 1/8 = 1$.

The length of the cube, a, is equivalent to $2r$. Thus,

$$PE = \frac{(4/3)\,\pi r^3}{(2r)^3} = 0.524 = 52.4\%$$

Body-centered cubic

There are 8 spheres at the corners of the cell and 1 sphere inside the cell [Figure 20.18(b)]. Since each sphere at a corner is shared by 8 unit cells, the total number of spheres in the cell is $8 \times 1/8 + 1 = 2$.

The length of the cube, a, is equivalent to $4r/\sqrt{3}$. Thus,

$$PE = \frac{(2)\,(4/3)\,\pi r^3}{\left(4r/\sqrt{3}\right)^3} = 0.680 = 68.0\%$$

Face-centered cubic

There are 8 spheres at the corners and 6 spheres on the faces of the cell [Figure 20.18(c)]. Since each sphere at a corner is shared by 8 unit cells and each sphere on a face is shared by 2 unit cells, the total number of spheres in the cell is $8 \times 1/8 + 6 \times 1/2 = 4$.

The length of the cube, a, is equivalent to $\sqrt{8}r$. Thus,

$$PE = \frac{(4)\,(4/3)\,\pi r^3}{\left(\sqrt{8}r\right)^3} = 0.740 = 74.0\%$$

20.6   Aluminum has a face-centered cubic lattice. The cell dimension is 4.05 Å. Calculate the closest interatomic distance and the density of the metal.

The face of the unit cell is shown below. The atomic radius is denoted by r. Thus, the diagonal of the face has a length of $4r$.

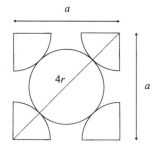

According to the figure and Pythagoras' theorem,

$$(4r)^2 = a^2 + a^2$$

$$r = \frac{\sqrt{2}a}{4} = \frac{\sqrt{2}\,(4.05\ \text{Å})}{4} = 1.432\ \text{Å}$$

The interatomic distance is $2r = 2.86$ Å.

The density of the metal is

$$d = \frac{\text{Mass of metal in one unit cell}}{\text{Volume of a unit cell}}$$

$$= \frac{4\,(\text{mass of 1 Al atom})}{a^3}$$

$$= \frac{(4)\,(26.98\ \text{amu})\left(1.661 \times 10^{-24}\ \text{g amu}^{-1}\right)}{\left(4.05 \times 10^{-8}\ \text{cm}\right)^3}$$

$$= 2.70\ \text{g cm}^{-3}$$

20.7   Silver crystallizes in a face-centered cubic lattice; the edge length of the unit cell is 4.08 Å, and the density of the metal is 10.5 g cm^{-3}. From these data, calculate the Avogadro constant.

The mass of Ag atoms in one unit cell is

$$m = dV = (10.5 \text{ g cm}^{-3})(4.08 \times 10^{-8} \text{ cm})^3 = 7.131 \times 10^{-22} \text{ g}$$

Since there are 4 Ag atoms in a fcc unit cell, the mass of one Ag atom is

$$\frac{7.131 \times 10^{-22} \text{ g}}{4} = 1.783 \times 10^{-22} \text{ g}$$

The Avogadro constant is

$$\frac{\text{Molar mass of Ag}}{\text{Mass of 1 Ag atom}} = \frac{107.9 \text{ g mol}^{-1}}{1.783 \times 10^{-22} \text{ g}} = 6.05 \times 10^{23} \text{ mol}^{-1}$$

20.8 Explain why diamond is harder than graphite. Why is graphite an electrical conductor but diamond is not?

In the diamond lattice, each carbon atom is covalently bonded to four other carbon atoms in a three-dimensional array. The bonds are strong and uniform, leading to a very hard substance. In graphite, the carbon atoms in each layer are covalently bonded to each other, but the individual layers are held together only by weak dispersion forces. Thus, graphite is easily cleaved between layers and is not hard.

In diamond, the bonding electrons are localized in covalent σ bonds so that the material is not an electrical conductor. Graphite contains delocalized π bonds in the planes of the layers, and the electrons in these delocalized orbitals are free to move throughout the layer, rendering the material electrically conductive.

20.9 Barium crystallizes in the body-centered arrangement. Assuming a hard-sphere model, calculate the "radius" of a barium atom if the unit cell edge length is 5.015 Å.

According to Figure 20.18(b),

$$a = \frac{4r}{\sqrt{3}}$$

$$r = \frac{\sqrt{3}a}{4} = \frac{\sqrt{3}(5.015 \text{ Å})}{4} = 2.172 \text{ Å}$$

20.10 Metallic iron crystallizes in a cubic lattice. The unit cell edge length is 287 pm. The density of iron is 7.87 g cm^{-3}. How many iron atoms are there within a unit cell?

The mass of iron atoms in a unit cell is

$$m = dV = (7.87 \text{ g cm}^{-3})(287 \times 10^{-10} \text{ cm})^3 = 1.860 \times 10^{-22} \text{ g}$$

The number of iron atoms in a unit cell is

$$\left(\frac{1.860 \times 10^{-22} \text{ g}}{55.85 \text{ g mol}^{-1}}\right)\left(\frac{6.022 \times 10^{23} \text{ atoms}}{1 \text{ mol}}\right) = 2 \text{ atoms}$$

Thus, iron crystallizes in a body-centered cubic lattice.

20.11 Crystalline silicon has a cubic structure. The unit cell edge length is 543 pm. The density of the solid is 2.33 g cm^{-3}. Calculate the number of Si atoms in one unit cell.

The mass of silicon atoms in a unit cell is

$$m = dV = \left(2.33 \text{ g cm}^{-3}\right)\left(543 \times 10^{-10} \text{ cm}\right)^3 = 3.730 \times 10^{-22} \text{ g}$$

The number of silicon atoms in one unit cell is

$$\left(\frac{3.730 \times 10^{-22} \text{ g}}{28.09 \text{ g mol}^{-1}}\right)\left(\frac{6.022 \times 10^{23} \text{ atoms}}{1 \text{ mol}}\right) = 8 \text{ atoms}$$

20.12 Barium metal crystallizes in a body-centered cubic lattice (the Ba atoms are at the lattice points only). The unit cell edge length is 502 pm, and the density of the metal is 3.50 g cm^{-3}. Using this information, calculate the Avogadro constant.

The mass of Ba atoms in one unit cell is

$$m = dV = \left(3.50 \text{ g cm}^{-3}\right)\left(502 \times 10^{-10} \text{ cm}\right)^3 = 4.428 \times 10^{-22} \text{ g}$$

Since there are 2 Ba atoms in a bcc unit cell, the mass of one Ba atom is

$$\frac{4.428 \times 10^{-22} \text{ g}}{2} = 2.214 \times 10^{-22} \text{ g}$$

The Avogadro constant is

$$\frac{\text{Molar mass of Ba}}{\text{Mass of 1 Ba atom}} = \frac{137.3 \text{ g mol}^{-1}}{2.214 \times 10^{-22} \text{ g}} = 6.20 \times 10^{23} \text{ mol}^{-1}$$

20.13 Vanadium crystallizes in a body-centered cubic lattice (the V atoms occupy only the lattice points). How many V atoms are present in a unit cell?

In a body-centered cubic cell, there is one sphere at the cube center and one at each of the eight corners. Each corner sphere is shared among eight adjacent unit cells, giving

$$1 \text{ center sphere} + 8 \text{ corner spheres} \times 1/8 = 2 \text{ spheres per cell}$$

There are two vanadium atoms per unit cell.

20.14 Europium crystallizes in a body-centered cubic lattice (the Eu atoms occupy only the lattice points). The density of Eu is 5.26 g cm^{-3}. Calculate the unit cell edge length in pm.

There are 2 Eu atoms in one unit cell. Thus, the mass of Eu in a unit cell is

$$(2 \text{ atoms}) \left(152.0 \text{ g mol}^{-1}\right) \left(\frac{1 \text{ mol}}{6.022 \times 10^{23} \text{ atoms}}\right) = 5.048 \times 10^{-22} \text{ g}$$

The volume of the unit cell is

$$V = \frac{m}{d} = \frac{5.048 \times 10^{-22} \text{ g}}{5.26 \text{ g cm}^{-3}} = 9.597 \times 10^{-23} \text{ cm}^3$$

Thus, the unit cell edge length is

$$a = V^{1/3} = \left(9.597 \times 10^{-23} \text{ cm}^3\right)^{1/3} = 4.58 \times 10^{-8} \text{ cm} = 458 \text{ pm}$$

20.15 Metallic iron can exist in the β form (bcc, cell dimension = 2.90 Å) and the γ form (fcc, cell dimension = 3.68 Å). The β form can be converted into the γ form by applying high pressures. Calculate the ratio of the densities of the β form to the γ form.

There are 2 atoms per unit cell in the β form and 4 atoms per unit cell in the γ form. The density of each form is

$$d_\beta = \frac{2 \text{ (mass of 1 Fe atom)}}{\left(2.90 \text{ Å}\right)^3}$$

$$d_\gamma = \frac{4 \text{ (mass of 1 Fe atom)}}{\left(3.68 \text{ Å}\right)^3}$$

Thus,

$$\frac{d_\beta}{d_\gamma} = \frac{1}{2} \left(\frac{3.68 \text{ Å}}{2.90 \text{ Å}}\right)^3 = 1.02$$

20.16 A face-centered cubic cell contains 8 X atoms at the corners of the cell and 6 Y atoms at the faces. What is the empirical formula of the solid?

Each corner atom is shared by 8 unit cells and each atom at a face is shared by 2 unit cells. Thus, in a unit cell, there are $8 \times 1/8 = 1$ X atom and $6 \times 1/2 = 3$ Y atoms. The empirical formula is XY_3.

20.17 Gold (Au) crystallizes in a cubic close-packed structure (the face-centered cube) and has a density of 19.3 g cm^{-3}. Calculate the atomic radius of gold.

There are 4 Au atoms in one unit cell. Thus, the mass of Au in a unit cell is

$$(4 \text{ atoms}) \left(197.0 \text{ g mol}^{-1}\right) \left(\frac{1 \text{ mol}}{6.022 \times 10^{23} \text{ atoms}}\right) = 1.309 \times 10^{-21} \text{ g}$$

The volume of the unit cell is

$$V = \frac{m}{d} = \frac{1.309 \times 10^{-21} \text{ g}}{19.3 \text{ g cm}^{-2}} = 6.782 \times 10^{-23} \text{ cm}^3$$

The unit cell edge length is

$$a = V^{1/3} = \left(6.782 \times 10^{-23} \text{ cm}^3\right)^{1/3} = 4.078 \times 10^{-8} \text{ cm}$$

The unit cell edge length is related to the atomic radius:

$$a = \sqrt{8}r$$

$$r = \frac{a}{\sqrt{8}} = \frac{4.078 \times 10^{-8} \text{ cm}}{\sqrt{8}} = 1.44 \times 10^{-8} \text{ cm} = 1.44 \text{ Å}$$

20.18   Argon crystallizes in the face-centered cubic arrangement. Given that the atomic radius of argon is 191 pm, calculate the density of solid argon.

There are 4 Ar atoms in one unit cell. Thus, the mass of Ar in a unit cell is

$$(4 \text{ atoms}) \left(39.95 \text{ g mol}^{-1}\right) \left(\frac{1 \text{ mol}}{6.022 \times 10^{23} \text{ atoms}}\right) = 2.654 \times 10^{-22} \text{ g}$$

The volume of the unit cell is

$$V = a^3 = \left(\sqrt{8}r\right)^3 = \left[\sqrt{8}\left(191 \times 10^{-10} \text{ cm}\right)\right]^3 = 1.577 \times 10^{-22} \text{ cm}^3$$

Thus, the density of solid Ar is

$$d = \frac{m}{V} = \frac{2.654 \times 10^{-22} \text{ g}}{1.577 \times 10^{-22} \text{ cm}^3} = 1.68 \text{ g cm}^{-3}$$

20.19   Given that the density of solid CsCl is 3.97 g cm^{-3}, calculate the distance between adjacent Cs$^+$ and Cl$^-$ ions.

Cs$^+$ occupies a cubic hole (Table 20.5, Figure 20.23). There are one Cs$^+$ and one Cl$^-$ in each unit cell [similar to Figure 20.18(b)] and the mass is

$$\left(168.35 \text{ g mol}^{-1}\right) \left(\frac{1 \text{ mol}}{6.022 \times 10^{23}}\right) = 2.7956 \times 10^{-22} \text{ g}$$

The volume of the unit cell is

$$V = \frac{m}{d} = \frac{2.7956 \times 10^{-22} \text{ g}}{3.97 \text{ g cm}^{-3}} = 7.042 \times 10^{-23} \text{ cm}^3$$

The length of each unit cell edge is

$$a = V^{1/3} = \left(7.042 \times 10^{-23} \text{ cm}^3\right)^{1/3} = 4.130 \times 10^{-8} \text{ cm}$$

The diagonal of the unit cell has a length of

$$d = r_{Cl^-} + 2r_{Cs^+} + r_{Cl^-} = 2(\text{distance between adjacent Cs}^+ \text{ and Cl}^- \text{ ions})$$

and can be calculated using Pythagoras' theorem,

$$d = \sqrt{a^2 + a^2 + a^2} = \sqrt{3}a = \sqrt{3}\left(4.130 \times 10^{-8} \text{ cm}\right) = 7.153 \times 10^{-8} \text{ cm}$$

Therefore, the distance between adjacent Cs^+ and Cl^- ions is

$$\frac{d}{2} = \frac{7.153 \times 10^{-8} \text{ cm}}{2} = 3.58 \times 10^{-8} \text{ cm} = 3.58 \text{ Å}$$

20.20   Use the Born–Haber cycle (see Section 16.2) to calculate the lattice energy of LiF. [The heat of sublimation of Li is 155.2 kJ mol^{-1} and $\Delta_f \overline{H}^\circ(\text{LiF}) = -594.1$ kJ mol^{-1}. Bond enthalpy for F_2 is 150.6 kJ mol^{-1}. Other data may be found in Tables 14.4 and 14.5]

The Born–Haber cycle is

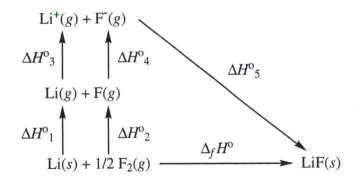

The lattice energy is $-\Delta H_5^\circ$. Since

$$\Delta_f \overline{H}^\circ = \Delta H_1^\circ + \Delta H_2^\circ + \Delta H_3^\circ + \Delta H_4^\circ + \Delta H_5^\circ$$

The lattice energy is

$$-\Delta H_5^\circ = \Delta H_1^\circ + \Delta H_2^\circ + \Delta H_3^\circ + \Delta H_4^\circ - \Delta_f \overline{H}^\circ$$

The individual enthalpy changes are

$\Delta H_1^\circ = 155.2 \text{ kJ mol}^{-1}$ [enthalpy of sublimation of Li(s)]

$\Delta H_2^\circ = \dfrac{(150.6 \text{ kJ mol}^{-1})}{2} = 75.3 \text{ kJ mol}^{-1}$ [bond enthalpy of $F_2(g)$, given]

$\Delta H_3^\circ = 520 \text{ kJ mol}^{-1}$ [first ionization energy of Li(g), Table 14.4]

$\Delta H_4^\circ = -328 \text{ kJ mol}^{-1}$ [electron affinity of F(g), Table 14.5]

$\Delta_f \overline{H}^\circ = -594.1 \text{ kJ mol}^{-1}$ [given]

Thus, the lattice energy is

$$(155.2 + 75.3 + 520 - 328 + 594.1) \text{ kJ mol}^{-1} = 1017 \text{ kJ mol}^{-1}$$

20.21 Calculate the lattice energy of calcium chloride, given that the heat of sublimation of Ca is 121 kJ mol^{-1} and $\Delta_f \overline{H}^\circ(\text{CaCl}_2) = -795 \text{ kJ mol}^{-1}$. (See Tables 14.4 and 14.5 for other data.)

The Born-Haber cycle is

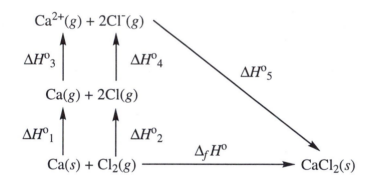

The lattice energy is $-\Delta H_5^\circ$. Since

$$\Delta_f \overline{H}^\circ = \Delta H_1^\circ + \Delta H_2^\circ + \Delta H_3^\circ + \Delta H_4^\circ + \Delta H_5^\circ$$

The lattice energy is

$$-\Delta H_5^\circ = \Delta H_1^\circ + \Delta H_2^\circ + \Delta H_3^\circ + \Delta H_4^\circ - \Delta_f \overline{H}^\circ$$

The various enthalpy changes are

$\Delta H_1^\circ = 121 \text{ kJ mol}^{-1}$ [enthalpy of sublimation of Ca(s)]

$\Delta H_2^\circ = 242.7 \text{ kJ mol}^{-1}$ [bond enthalpy of $Cl_2(g)$, Table 4.4]

$\Delta H_3^\circ = (589.5 + 1145) \text{ kJ mol}^{-1} = 1734.5 \text{ kJ mol}^{-1}$ [1st and 2nd ionization energies of Ca(g), Table 14.4]

$\Delta H_4^\circ = 2\left(-349 \text{ kJ mol}^{-1}\right) = -698 \text{ kJ mol}^{-1}$ [electron affinity of Cl(g), Table 14.5]

$\Delta_f \overline{H}^\circ = -795 \text{ kJ mol}^{-1}$ [given]

Thus, the lattice energy is

$$(121 + 242.7 + 1734.5 - 698 + 795) \text{ kJ mol}^{-1} = 2195 \text{ kJ mol}^{-1}$$

20.22   From the following data, explain why magnesium chloride in the solid state is $MgCl_2$ and not $MgCl$, whereas sodium chloride is $NaCl$ and not $NaCl_2$.

	Mg	Na
first ionization energy	$738\ kJ\ mol^{-1}$	$496\ kJ\ mol^{-1}$
second ionization energy	$1450\ kJ\ mol^{-1}$	$4560\ kJ\ mol^{-1}$

The lattice energy of $MgCl_2$ is $2527\ kJ\ mol^{-1}$.

The lattice energy for $MgCl_2$ is $2527\ kJ\ mol^{-1}$, which is more than enough to compensate for the energy needed, $(738\ kJ\ mol^{-1} + 1450\ kJ\ mol^{-1}) = 2188\ kJ\ mol^{-1}$, to remove the first two electrons from the Mg atom to form the Mg^{2+} cation.

For NaCl, even though a lattice with the Na^{2+} ion, such as would occur in $NaCl_2$, would have a lattice energy greater than that of NaCl, it is unable to compensate for the large second ionization energy of Na. The lattice energy of a hypothetical $NaCl_2$ crystal could be estimated as being similar to that for $MgCl_2$, namely $2527\ kJ\ mol^{-1}$, but the sum of the first two ionization energies for Na is $(496\ kJ\ mol^{-1} + 4560\ kJ\ mol^{-1}) = 5056\ kJ\ mol^{-1}$.

20.23   Calculate the temperature at which the wavelength of a neutron is 1.00 Å.

The temperature and wavelength of a neutron are related by Equation 20.8.

$$\lambda = \frac{h}{\sqrt{3mk_B T}}$$

Thus,

$$T = \frac{h^2}{3mk_B \lambda^2}$$

$$= \frac{\left(6.626 \times 10^{-34}\ J\,s\right)^2}{3\left(1.675 \times 10^{-27}\ kg\right)\left(1.381 \times 10^{-23}\ J\,K^{-1}\right)\left(1.00 \times 10^{-10}\ m\right)^2}$$

$$= 633\ K$$

20.24   Without referring to a handbook of chemistry, decide which of the following has a greater density: diamond or graphite.

Referring to Figure 20.27, it is apparent that with the relatively large spacing between layers in graphite, there is more "empty space" in this structure than in diamond. Thus, diamond has the greater density.

20.25   Predict the influence of temperature on X-ray diffraction patterns of crystals.

As temperature is increased, the vibrational motion of the atoms in the sample increases. This tends to blur the reflections and leads to a decrease in the intensity of the scattered X-rays.

20.26   Compare the temperature dependence of electrical conduction in an aqueous solution and in a metal.

In a solution, the electrical conductance depends on the motion of the ions so that increasing the temperature increases the conductance. In a metal, however, electrical conductance is a result of electron delocalization, which is disrupted by lattice vibrations. Thus, in a metal, increasing temperature results in a decrease in conductance.

20.27   Which of the following are molecular solids and which are covalent solids? Se_8, HBr, Si, CO_2, C, P_4O_6, B, SiH_4.

Molecular Solids: Se_8, HBr, CO_2, P_4O_6, SiH_4

Covalent Solids: Si, C, B

20.28   Classify the solid state of the following substances as ionic crystals, covalent crystals, molecular crystals, or metallic crystals: **(a)** SiO_2, **(b)** SiC, **(c)** S_8, **(d)** KBr, **(e)** Mg, **(f)** LiCl, **(g)** Cr.

(a) Covalent crystal, **(b)** covalent crystal, **(c)** molecular crystal, **(d)** ionic crystal, **(e)** metallic crystal, **(f)** ionic crystal, **(g)** metallic crystal.

20.29   Explain why most metals have a flickering appearance.

Metals have closely spaced energy levels and (referring to Figure 20.21) a very small energy gap between filled and empty levels. Consequently, many electronic transitions can take place with absorption and subsequent emission continually occurring. Some of these transitions fall in the visible region of the spectrum and give rise to the flickering appearance.

CHAPTER 21

The Liquid State

PROBLEMS AND SOLUTIONS

21.1 The viscosity of a gas increases with increasing temperatures (see Equation 3.22), yet the viscosity of a liquid decreases with increasing temperature. Explain.

Equation 3.22 was derived for an ideal gas with no intermolecular forces between molecules. The viscosity in this case arises from momentum transport, which increases with higher temperatures causing an increase in viscosity. For liquids, the viscosity is due to intermolecular forces that keep molecules from breaking away from each other. At higher temperatures, the increase in thermal motion results in molecules being able to more easily overcome these attractive forces, and the viscosity decreases.

21.2 At 293 K, the time of flow for water through an Ostwald viscometer is 342.5 s; for the same volume of an organic solvent, the time of flow is 271.4 s. Calculate the viscosity of the organic liquid relative to that of water. The density of the organic solvent is 0.984 g cm^{-3}.

Rearrange Equation (21.7),

$$\frac{\eta_{\text{organic}}}{\eta_{\text{water}}} = \frac{(\rho t)_{\text{organic}}}{(\rho t)_{\text{water}}}$$

to give

$$\eta_{\text{organic}} = \frac{(\rho t)_{\text{organic}}}{(\rho t)_{\text{water}}} \eta_{\text{water}}$$

$$= \frac{\left(0.984 \text{ g cm}^{-3}\right)(271.4 \text{ s})}{\left(1.00 \text{ g cm}^{-3}\right)(342.5 \text{ s})} \left(0.00101 \text{ N s m}^{-2}\right)$$

$$= 7.88 \times 10^{-4} \text{ N s m}^{-2}$$

21.3 For blood flowing in a capillary of radius 2.0×10^{-4} cm, estimate the maximum velocity for laminar flow at 37°C. (The density of whole blood is about 1.2 g cm^{-3}.)

The maximum velocity for laminar flow gives a Reynolds number of 2000.

$$2000 = \frac{2Rv\rho}{\eta}$$

$$v = \frac{2000\eta}{2R\rho} = \frac{2000\,(0.004\,\text{N s m}^{-2})}{2\,(2.0 \times 10^{-6}\,\text{m})\,(1.2 \times 10^{3}\,\text{kg m}^{-3})} = 2 \times 10^{3}\,\text{m s}^{-1}$$

21.4 An arteriole has a diameter of 2.4×10^{-5} m and a blood flow rate of 2.6×10^{-3} m s^{-1}. Calculate the pressure drop, ΔP, from one end to the other if the length of the arteriole is 5.0×10^{-3} m.

$$\Delta P = \frac{8\eta L Q}{\pi R^4} = \frac{8\eta L\,(\pi R^2 v)}{\pi R^4} = \frac{8\eta L v}{R^2}$$

$$= \frac{8\,(0.004\,\text{N s m}^{-2})\,(5.0 \times 10^{-3}\,\text{m})\,(2.6 \times 10^{-3}\,\text{m s}^{-1})}{\left(\frac{2.4 \times 10^{-5}\,\text{m}}{2}\right)^2}$$

$$= 3 \times 10^{3}\,\text{N m}^{-2}$$

21.5 The viscosity of a liquid usually decreases with increasing temperature. An empirical equation is $\log \eta = A/T + B$. Determine the constants, A and B, for water from the following data:

T/K	273	293	310	373
η/P	0.01787	0.0101	0.00719	0.00283

A plot of $\log \eta$ vs $1/T$ yields a line with a slope of A and an intercept of B. The data used for the plot are

10^3 K/T	3.663	3.413	3.226	2.681
$\log(\eta/\text{P})$	-1.74788	-1.9957	-2.1433	-2.5482

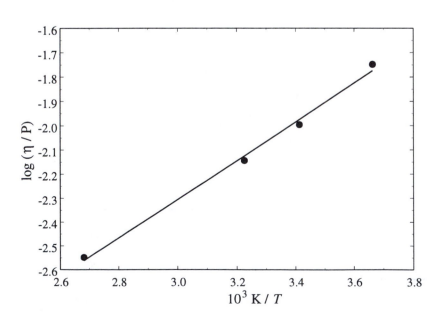

The line is described by $y = 802x - 4.72$. Thus, $A = 802$ K and $B = -4.72$.

21.6 Show that the Reynolds number (see Equation 21.4) is dimensionless.

The Reynolds number is given by

$$\frac{2Rv\rho}{\eta}$$

The units are

$$\frac{(\text{m})\left(\text{m s}^{-1}\right)\left(\text{kg m}^{-3}\right)}{\text{N s m}^{-2}} = \frac{(\text{m})\left(\text{m s}^{-1}\right)\left(\text{kg m}^{-3}\right)}{\text{kg m s}^{-2}\,\text{s m}^{-2}} = \text{dimensionless}$$

21.7 From the definition of Reynolds number (Equation 21.4), calculate the maximum value of v for the laminar flow of water at 293 K along a tube with a radius of 0.60 cm.

The maximum velocity for laminar flow gives a Reynolds number of 2000.

$$2000 = \frac{2Rv\rho}{\eta}$$

$$v = \frac{2000\eta}{2R\rho} = \frac{2000\left(0.00101\ \text{N s m}^{-2}\right)}{2\left(0.60 \times 10^{-2}\ \text{m}\right)\left(1.00 \times 10^{3}\ \text{kg m}^{-3}\right)} = 0.17\ \text{m s}^{-1}$$

21.8 The rate of flow of a liquid through a cylindrical tube that has an inner radius of 0.12 cm and a length of 26 cm is 364 cm^3 in 88 s. The pressure drop between the ends of the tube is 57 torr. Calculate the liquid's viscosity. Is the flow laminar? The density of the liquid is 0.98 g cm^{-3}.

The viscosity can be calculated by rearranging

$$\Delta P = \frac{8\eta L Q}{\pi R^4}$$

to give

$$\eta = \frac{\Delta P \pi R^4}{8LQ}$$

$$= \frac{(57\ \text{torr})\left(\frac{1\ \text{atm}}{760\ \text{torr}}\right)\left(\frac{10^5\ \text{Pa}}{1\ \text{atm}}\right)\pi\left(0.12 \times 10^{-2}\ \text{m}\right)^4}{8\left(26 \times 10^{-2}\ \text{m}\right)\left(\frac{364 \times 10^{-6}\ \text{m}^3}{88\ \text{s}}\right)}$$

$$= 5.68 \times 10^{-3}\ \text{N s m}^{-2}$$

$$= 5.7 \times 10^{-3}\ \text{N s m}^{-2}$$

Before calculating the Reynolds number, the velocity of the liquid needs to be determined. In time t, a volume $\pi R^2 L$ flows through the tube. The velocity is

$$v = \frac{L}{t} = \frac{\frac{V}{\pi R^2}}{t} = \frac{V}{\pi R^2 t} = \frac{Qt}{\pi R^2 t}$$

$$= \frac{Q}{\pi R^2}$$

$$= \frac{\frac{364 \times 10^{-6} \, \text{m}^3}{88 \, \text{s}}}{\pi \left(0.12 \times 10^{-2} \, \text{m}\right)^2} = 0.914 \, \text{m s}^{-1}$$

The Reynolds number is

$$\frac{2Rv\rho}{\eta} = \frac{(2)\left(0.12 \times 10^{-2} \, \text{m}\right)\left(0.914 \, \text{m s}^{-1}\right)\left(0.98 \times 10^3 \, \text{kg m}^3\right)}{5.68 \times 10^{-3} \, \text{N s m}^{-2}} = 378$$

Since the Reynolds number is less than 2000, the flow is laminar.

21.9 Water has an unusually large surface tension. Explain.

The large surface tension of water is due to the strong, extensive hydrogen bonding in the liquid.

21.10 Give a molecular interpretation for the decrease in surface tension of a liquid with increasing temperature.

The decrease in surface tension with temperature is analogous to the decrease in viscosity with temperature (see Problem 21.1). At higher temperatures, molecules have a greater average kinetic energy and are able to overcome the attractive intermolecular forces responsible for surface tension. Consequently, it is easier to stretch the surface of the liquid, since the molecules are not held together as tightly.

21.11 A glass capillary of diameter 0.10 cm is dipped into **(a)** water (contact angle 10°) at 293 K and **(b)** mercury (contact angle 170°) at 298 K. Calculate the level of the liquid in the capillary in each case.

The level of liquid in the capillary can be obtained by rearranging Equation 21.12 to give

$$h = \frac{2\gamma \cos\theta}{rg\rho}$$

(a)

$$h = \frac{2\left(0.07275 \, \text{N m}^{-1}\right)(\cos 10°)}{\left(\frac{0.10 \times 10^{-2} \, \text{m}}{2}\right)\left(9.81 \, \text{m s}^{-2}\right)\left(1.0 \times 10^3 \, \text{kg m}^{-3}\right)} = 2.9 \times 10^{-2} \, \text{m} = 2.9 \, \text{cm}$$

(b)

$$h = \frac{2\left(0.476\ \text{N m}^{-1}\right)(\cos 170°)}{\left(\frac{0.10 \times 10^{-2}\ \text{m}}{2}\right)\left(9.81\ \text{m s}^{-2}\right)\left(13.6 \times 10^3\ \text{kg m}^{-3}\right)} = -1.4 \times 10^{-2}\ \text{m} = -1.4\ \text{cm}$$

The minus sign denotes depression.

21.12 Both ethanol and mercury are used in thermometers. Explain the difference between the meniscus of the liquids in these two types of thermometers.

In ethanol, adhesion is stronger than cohesion giving rise to an upward-curving meniscus. The reverse holds true for mercury. (See Figure 21.7.)

21.13 The surface tension of liquid naphthalene at 127°C is 0.0288 N m^{-1}, and its density at this temperature is 0.96 g cm^{-3}. What is the radius of the largest capillary that will permit the liquid to rise 3.0 cm? Assume that the angle of contact is zero.

Rearrange Equation (21.12) to give

$$r = \frac{2\gamma \cos\theta}{hg\rho}$$

$$= \frac{(2)\left(0.0288\ \text{N m}^{-1}\right)(\cos 0°)}{\left(3.0 \times 10^{-2}\ \text{m}\right)\left(9.81\ \text{m s}^{-2}\right)\left(0.96 \times 10^3\ \text{kg m}^{-3}\right)}$$

$$= 2.0 \times 10^{-4}\ \text{m} = 2.0 \times 10^{-2}\ \text{cm}$$

21.14 The surface tension of quinoline is twice that of acetone at 20°C. If the capillary rise is 2.5 cm for quinoline, what is the rise for acetone in the same capillary? Assume that the angles of contact are zero. The densities of quinoline and acetone at 20°C are 1.09 g cm^{-3} and 0.79 g cm^{-3}, respectively.

Since the contact angles are assumed to be 0,

$$h_{\text{acetone}} = \frac{2\gamma_{\text{acetone}}}{rg\rho_{\text{acetone}}}$$

$$h_{\text{quinoline}} = \frac{2\gamma_{\text{quinoline}}}{rg\rho_{\text{quinoline}}}$$

Thus,

$$\frac{h_{acetone}}{h_{quinoline}} = \frac{\gamma_{acetone}\rho_{quinoline}}{\rho_{acetone}\gamma_{quinoline}}$$

$$h_{acetone} = \left(\frac{\gamma_{acetone}}{\gamma_{quinoline}}\right)\left(\frac{\rho_{quinoline}}{\rho_{acetone}}\right)h_{quinoline}$$

$$= \left(\frac{1}{2}\right)\left(\frac{1.09 \text{ g cm}^{-3}}{0.79 \text{ g cm}^{-3}}\right)(2.5 \text{ cm})$$

$$= 1.7 \text{ cm}$$

21.15 A capillary tube that has an inner diameter of 0.40 mm is inserted vertically into a pool of mercury at 20°C (see Figure 21.7 b). Calculate the depression in mercury given that the contact angle is 146°. The density of mercury is 13.6 g cm^{-3}.

The level of liquid in the capillary can be obtained by rearranging Equation 21.12 to give

$$h = \frac{2\gamma \cos\theta}{rg\rho}$$

$$= \frac{2\left(0.476 \text{ N m}^{-1}\right)(\cos 146°)}{\left(\frac{0.40\times 10^{-3} \text{ m}}{2}\right)\left(9.81 \text{ m s}^{-2}\right)\left(13.6\times 10^3 \text{ kg m}^{-3}\right)}$$

$$= -3.0\times 10^{-2} \text{ m} = -3.0 \text{ cm}$$

The minus sign denotes depression.

21.16 Two capillary tubes with inside diameters of 1.4 mm and 1.0 mm, respectively, are inserted into a liquid of density 0.95 g cm^{-3}. Calculate the surface tension of the liquid if the difference between the capillary rises in the tubes is 1.2 cm. Assume the contact angle is zero.

Let h_1 be the liquid rise for the capillary with diameter $r_1(= 1.4$ mm) and h_2 be the liquid rise for the capillary with diameter $r_2(= 1.0$ mm). Then, with contact angles of 0,

$$h_1 = \frac{2\gamma}{r_1 g\rho}$$

$$h_2 = \frac{2\gamma}{r_2 g\rho}$$

$$h_2 - h_1 = \frac{2\gamma}{g\rho}\left(\frac{1}{r_2} - \frac{1}{r_1}\right)$$

Therefore, the surface tension is

$$\gamma = \frac{g\rho\left(h_2 - h_1\right)}{2\left(\frac{1}{r_2} - \frac{1}{r_1}\right)}$$

$$= \frac{\left(9.81 \text{ m s}^{-2}\right)\left(0.95 \times 10^3 \text{ kg m}^{-3}\right)\left(1.2 \times 10^{-2} \text{ m}\right)}{(2)\left(\frac{1}{1.0 \times 10^{-3} \text{ m}} - \frac{1}{1.4 \times 10^{-3} \text{ m}}\right)}$$

$$= 0.20 \text{ N m}^{-1}$$

21.17 The diffusion coefficient of glucose is $5.7 \times 10^{-10} \text{ m}^2 \text{ s}^{-1}$. Calculate the time required for a glucose molecule to diffuse through **(a)** 10,000 Å and **(b)** 0.10 m.

The root-mean-square distance is related to time via

$$\sqrt{\overline{x^2}} = \sqrt{2Dt}$$

Thus,

$$t = \frac{\overline{x^2}}{2D}$$

(a)

$$t = \frac{\left(10,000 \times 10^{-10} \text{ m}\right)^2}{(2)\left(5.7 \times 10^{-10} \text{ m}^2 \text{ s}^{-1}\right)} = 8.8 \times 10^{-4} \text{ s}$$

(b)

$$t = \frac{(0.10 \text{ m})^2}{(2)\left(5.7 \times 10^{-10} \text{ m}^2 \text{ s}^{-1}\right)} = 8.8 \times 10^6 \text{ s} = 100 \text{ days}$$

21.18 The diffusion coefficient of sucrose in water at 298 K is $0.46 \times 10^{-5} \text{ cm}^2 \text{ s}^{-1}$, and the viscosity of water at the same temperature is $0.0010 \text{ N s m}^{-2}$. From these data, estimate the effective radius of a sucrose molecule.

Rearrange Equation 21.25 to give

$$r = \frac{k_B T}{6\pi\eta D}$$

$$= \frac{\left(1.381 \times 10^{-23} \text{ J K}^{-1}\right)(298 \text{ K})}{6\pi\left(0.0010 \text{ N s m}^{-2}\right)\left(0.46 \times 10^{-9} \text{ m}^2 \text{ s}^{-1}\right)}$$

$$= 4.7 \times 10^{-10} \text{ m} = 4.7 \text{ Å}$$

21.19 From the diffusion coefficients listed in Table 21.3, estimate the radius and molecular volume of myoglobin and hemoglobin. What conclusion can you draw from the results?

The effective radius of a molecule can be calculated by rearranging Equation 21.25. Once the radius is determined, the volume can be calculated assuming the molecule is a sphere.

For myoglobin,

$$r = \frac{k_B T}{6\pi \eta D}$$

$$= \frac{\left(1.381 \times 10^{-23} \, \text{J K}^{-1}\right) (298 \, \text{K})}{6\pi \left(0.00101 \, \text{N s m}^{-2}\right) \left(0.113 \times 10^{-9} \, \text{m}^2 \, \text{s}^{-1}\right)}$$

$$= 1.913 \times 10^{-9} \, \text{m} = 19.1 \, \text{Å}$$

$$V = \frac{4}{3}\pi r^3$$

$$= \frac{4}{3}\pi \left(1.913 \times 10^{-9} \, \text{m}\right)^3$$

$$= 2.932 \times 10^{-26} \, \text{m}^3 = 2.93 \times 10^4 \, \text{Å}^3$$

For hemoglobin,

$$r = \frac{k_B T}{6\pi \eta D}$$

$$= \frac{\left(1.381 \times 10^{-23} \, \text{J K}^{-1}\right) (298 \, \text{K})}{6\pi \left(0.00101 \, \text{N s m}^{-2}\right) \left(0.069 \times 10^{-9} \, \text{m}^2 \, \text{s}^{-1}\right)}$$

$$= 3.13 \times 10^{-9} \, \text{m} = 31 \, \text{Å}$$

$$V = \frac{4}{3}\pi r^3$$

$$= \frac{4}{3}\pi \left(3.13 \times 10^{-9} \, \text{m}\right)^3$$

$$= 1.28 \times 10^{-25} \, \text{m}^3 = 1.3 \times 10^5 \, \text{Å}^3$$

The volume of hemoglobin is $\dfrac{1.28 \times 10^{-25} \, \text{m}^3}{2.932 \times 10^{-26} \, \text{m}^3} = 4.4$ times greater than that of myoglobin. Since hemoglobin is a tetramer of myoglobin, this result makes sense.

21.20 Diffusion coefficients have been measured for many solid systems. If the diffusion coefficient of bismuth in lead is $1.1 \times 10^{-16} \, \text{cm}^2 \, \text{s}^{-1}$ at 20°C, calculate how long it will take (in years) for a bismuth atom to travel 1.0 cm.

Rearrange Equation 21.22 to give

$$t = \frac{\overline{x^2}}{2D}$$

$$t = \frac{\left(1.0 \times 10^{-2}\ m\right)^2}{(2)\left(1.1 \times 10^{-20}\ m^2\ s^{-1}\right)}$$

$$= \left(4.55 \times 10^{15}\ s\right)\left(\frac{1\ hr}{3600\ s}\right)\left(\frac{1\ day}{24\ hr}\right)\left(\frac{1\ year}{365\ days}\right)$$

$$= 1.4 \times 10^{8}\ years$$

21.21   What is the diffusion coefficient of a membrane-bound protein of molar mass 80,000 daltons at 37°C if the viscosity of the membrane is 1 poise ($0.10\ N\ s\ m^{-2}$)? What is the average distance traveled by this protein in 1.0 s? Assume that this protein is an unhydrated, rigid sphere that has a density of $1.4\ g\ cm^{-3}$.

Use the volume of 1 protein molecule,

$$V = \left(80,000\ g\ mol^{-1}\right)\left(\frac{1\ mol}{6.022 \times 10^{23}}\right)\frac{1}{\left(1.4\ g\ cm^{-3}\right)} = 9.49 \times 10^{-20}\ cm^3$$

to find the effective radius of the protein

$$V = \frac{4}{3}\pi r^3 = 9.49 \times 10^{-20}\ cm^3$$

$$r = \left[\frac{(3)\left(9.49 \times 10^{-20}\ cm^3\right)}{4\pi}\right]^{1/3} = 2.83 \times 10^{-7}\ cm$$

The diffusion coefficient can be calculated using Equation 21.25.

$$D = \frac{k_B T}{6\pi \eta r}$$

$$= \frac{\left(1.381 \times 10^{-23}\ J\,K^{-1}\right)(310\ K)}{6\pi \left(0.10\ N\,s\,m^{-2}\right)\left(2.83 \times 10^{-9}\ m\right)}$$

$$= 8.03 \times 10^{-13}\ m^2\ s^{-1} = 8.0 \times 10^{-13}\ m^2\ s^{-1}$$

The average distance traveled by this protein in 1.0 s is

$$\sqrt{\overline{x^2}} = \sqrt{2Dt}$$

$$= \sqrt{(2)\left(8.03 \times 10^{-13}\ m^2\ s^{-1}\right)(1.0\ s)}$$

$$= 1.3 \times 10^{-6}\ m = 1.3 \times 10^{4}\ \text{Å}$$

21.22   Two soap bubbles of radii r_1 and r_2 ($r_2 > r_1$) are connected by a piece of tubing with a stopcock. Predict how the size of the bubbles will change when the stopcock is opened.

The smaller bubble has the larger internal pressure (see Appendix 21.1). Thus, the smaller bubble will get smaller and the larger bubble gets larger.

21.23   Swimming coaches sometimes suggest that a drop of alcohol (ethanol) placed in an ear plugged with water "draws out the water." Comment from a molecular point of view. [*Source:* "Eco-Chem," J. A. Campbell, *J. Chem. Educ.* **52,** 655 (1975).]

The large surface tension of water suggests strong cohesion in the liquid (see Problem 21.9) while there is simultaneously little adhesion between the water and the wax-lined ear canal. Consequently, there is a large energy barrier opposing the increase of the water/earwax interface surface area, and the water will not flow. Adding alcohol (ethanol) both lowers the cohesion of the water (acting like a surfactant) and wets the earwax surface creating a surface to which the water can better adhere. Thus, the water is better able to flow out of the ear.

21.24   The carbon monoxide–hemoglobin complex has a diffusion coefficient of $0.062 \times 10^{-9} \ \text{m}^2 \ \text{s}^{-1}$ in water at 298 K. In the more viscous cytoplasm, the diffusion coefficient is only 0.013×10^{-9} $\text{m}^2 \ \text{s}^{-1}$. How long would it take for such a complex to travel the 3.0-μm length of a bacterial cell?

Rearrange Equation 21.22 to calculate travel time.

$$t = \frac{\overline{x^2}}{2D} = \frac{\left(3.0 \times 10^{-6} \ \text{m}\right)^2}{(2)\left(0.013 \times 10^{-9} \text{m}^2 \ \text{s}^{-1}\right)} = 0.35 \ \text{s}$$

21.25   Ozone (O_3) is a strong oxidizing agent that can oxidize all the common metals except gold and platinum. A convenient test for ozone is based on its action on mercury. When exposed to ozone, mercury becomes dull looking and sticks to glass tubing (instead of flowing freely through it). Write a balanced equation for the reaction. What property of mercury is altered by its interaction with ozone?

Ozone oxidizes mercury as follows.

$$3Hg(l) + O_3(g) \longrightarrow 3HgO(s)$$

The HgO covers the surface of the liquid mercury, preventing free flow and altering (lowering) its surface tension.

21.26   A hypodermic syringe is filled with a solution of viscosity $1.6 \times 10^{-3} \ \text{N m}^{-2}$ s. The plunger area of the syringe is $7.5 \times 10^{-5} \ \text{m}^2$, and the length of the needle is 0.026 m. The internal radius of the needle is 4.0×10^{-4} m. The gauge pressure in a vein is 1850 Pa (14 mmHg). Calculate the force in newtons that must be applied to the plunger so that $1.2 \times 10^{-6} \ \text{m}^3$ of the solution can be injected in 4.0 s.

First calculate ΔP by rearranging Equation 21.8.

$$\Delta P = \frac{8\eta L Q}{\pi R^4}$$

$$= \frac{(8)\left(1.6 \times 10^{-3}\,\mathrm{N\,m^2\,s}\right)(0.026\,\mathrm{m})\left(\frac{1.2 \times 10^{-6}\,\mathrm{m^3}}{4.0\,\mathrm{s}}\right)}{\pi\left(4.0 \times 10^{-4}\,\mathrm{m}\right)^4}$$

$$= 1.24 \times 10^3\,\mathrm{N\,m^{-2}}$$

The pressure that must be applied to the plunger must be ΔP above the gauge pressure in a vein; that is

$$1.24 \times 10^3\,\mathrm{N\,m^{-2}} + 1850\,\mathrm{N\,m^{-2}} = 3.09 \times 10^3\,\mathrm{N\,m^{-2}}$$

which corresponds to a force of

$$\left(3.09 \times 10^3\,\mathrm{N\,m^{-2}}\right)\left(7.5 \times 10^{-5}\,\mathrm{m^2}\right) = 0.23\,\mathrm{N}$$

21.27 A film of an organic liquid filled a rectangular wire loop similar to that shown in Figure 21.6. **(a)** Given that the wire loop is 9.0 cm wide and that a force of 7.2×10^{-3} N is needed to move the piston, calculate the surface tension of the liquid. **(b)** What is the work done in stretching the film to a distance of 0.14 cm?

(a)

$$\gamma = \frac{F}{2l} = \frac{7.2 \times 10^{-3}\,\mathrm{N}}{(2)\left(9.0 \times 10^{-2}\,\mathrm{m}\right)} = 0.040\,\mathrm{N\,m^{-1}}$$

(b)

$$\text{Work done} = F\Delta x = \left(7.2 \times 10^{-3}\,\mathrm{N}\right)\left(0.14 \times 10^{-2}\,\mathrm{m}\right) = 1.0 \times 10^{-5}\,\mathrm{J}$$

21.28 How much work is required to break up 1 mole of water at 20°C into spherical droplets that have a radius of 4.16×10^{-3} m? [*Hint*: The volume of a sphere is $(4/3)\pi r^3$, and the surface area of a sphere is $4\pi r^2$, where r is the radius of the sphere.] The density of water is $1.0\,\mathrm{g\,cm^{-3}}$.

Work must be done to increase the surface area in forming the droplets. A spherical molar volume of water has a mass of 18.02 g, and a volume of

$$V = \frac{m}{\rho} = \frac{18.02\,\mathrm{g}}{1.0\,\mathrm{g\,cm^{-3}}}\left(\frac{1\,\mathrm{m}}{100\,\mathrm{cm}}\right)^3 = 1.802 \times 10^{-5}\,\mathrm{m^3}$$

Since the volume of a sphere is given by $V = \frac{4}{3}\pi r^3$, the radius of the spherical molar volume of water is

$$r = \left(\frac{3V}{4\pi}\right)^{1/3} = \left[\frac{3\left(1.802 \times 10^{-5}\text{ m}^3\right)}{4\pi}\right]^{1/3} = 1.6264 \times 10^{-2}\text{ m}$$

and the surface area of this sphere is

$$A_i = 4\pi r^2 = 4\pi\left(1.6264 \times 10^{-2}\text{ m}\right)^2 = 3.3240 \times 10^{-3}\text{ m}^2$$

The volume of each small droplet is

$$V_{\text{drop}} = \frac{4}{3}\pi r^3 = \frac{4}{3}\pi\left(4.16 \times 10^{-3}\text{ m}\right)^3 = 3.016 \times 10^{-7}\text{ m}^3$$

The number of these droplets required to make a mole of water is

$$\frac{1.802 \times 10^{-5}\text{ m}^3}{3.016 \times 10^{-7}\text{ m}^3} = 59.75 \approx 60$$

The total surface area of 60 of these drops is

$$A_f = 60\left(4\pi r^2\right) = 60\left(4\pi\right)\left(4.16 \times 10^{-3}\text{ m}\right)^2 = 1.30 \times 10^{-2}\text{ m}^2$$

The increase in surface is opposed by the surface tension, and the work that must be done to overcome this resistance is

$$w = \gamma \Delta A = \left(0.07275\text{ N m}^{-1}\right)\left(1.30 \times 10^{-2}\text{ m}^2 - 3.3240 \times 10^{-3}\text{ m}^2\right) = 7.0 \times 10^{-4}\text{ J}$$

21.29 A sphere of volume V falling through a fluid experiences a downward gravitational force mg, where m is the mass of the sphere and g is the acceleration due to gravity. Simultaneously retarding the fall are the frictional force (see Equation 21.24) and an upward buoyant force given by $m_f g$ where m_f is the mass of the fluid of volume V. Calculate the terminal speed of fall of a steel ball of radius 1.2 mm and density 7.8 g cm^{-3} in water at 20°C. Based on your calculation, design an experiment that would allow you to measure the viscosity of a liquid.

The downward force on the ball is given by $F_d = mg = \rho_s Vg$, and the upward force is the sum of the frictional force and the buoyant force, $F_u = fv + m_f g = 6\pi \eta rv + \rho_f Vg$, where $V = \frac{4}{3}\pi r^3$ is the volume of the ball (and of the fluid it displaces), and ρ_s and ρ_f are the densities of the ball and the fluid, respectively. At the terminal velocity, these forces balance.

$$F_d = F_u$$

$$\rho_s \frac{4}{3}\pi r^3 g = 6\pi \eta rv + \rho_f \frac{4}{3}\pi r^3 g$$

Solving for the terminal velocity,

$$v = \frac{\frac{4}{3}\pi r^3 \left(\rho_s - \rho_f\right) g}{6\pi \eta r}$$

$$= \frac{2}{9} \frac{r^2 \left(\rho_s - \rho_f\right) g}{\eta}$$

$$= \frac{2}{9} \frac{\left(1.2 \times 10^{-3}\ \text{m}\right)^2 \left(7.8 \times 10^3\ \text{kg m}^{-3} - 1.0 \times 10^3\ \text{kg m}^{-3}\right) \left(9.81\ \text{m s}^{-2}\right)}{0.00101\ \text{N s m}^{-2}}$$

$$= 21\ \text{m s}^{-1}$$

This suggests that measuring the terminal velocity of a steel ball, of known radius and density, as it drops through the liquid under study will provide a value for the viscosity of the liquid, if its density is also known.

$$\eta = \frac{2}{9} \frac{r^2 \left(\rho_s - \rho_f\right) g}{v}$$

The temperature of the liquid must be kept constant.

21.30   The diffusion coefficient of oxygen in air is $0.20\ \text{cm}^2\ \text{s}^{-1}$; the diffusion coefficient of the same gas in water is about 10^4 times smaller. **(a)** Explain the huge difference in the magnitude of these diffusion coefficients. **(b)** Most animal cells are bathed in fluids, so that a hemoglobinlike molecule and a circulatory system are necessary for the purpose of transporting O_2 to their cells and carrying CO_2 away. (The diffusion coefficients of CO_2 in air and in water are comparable in magnitude to those of oxygen.) Because plants do not have a circulatory system, explain how the O_2 and CO_2 gases are transported efficiently in these systems. **(c)** Insects do possess a circulating system but lack a hemoglobinlike molecule. Considering the diffusion coefficients of CO_2 and O_2 in water, do you think it likely that ants, bees, and cockroaches can grow to human size, as they sometimes do in horror movies?

(a) Since diffusion involves molecular motion, it will depend greatly upon the medium through which the molecules in question must move.

(b) Plants have conspicuous intercellular air spaces by which they can take advantage of the large diffusion coefficients of O_2 and CO_2 in the gas phase.

(c) Without a means of efficient transport for O_2 and CO_2, insects are limited to small sizes so that their metabolic processes can obtain enough O_2 via diffusion through their circulating system. Because of this limitation, they can not grow to horror movie size.

CHAPTER 22

Macromolecules

PROBLEMS AND SOLUTIONS

22.1 A polydisperse solution has the following distribution:

Number of Molecules	Molar Mass / $g \cdot mol^{-1}$
10	25,000
7	17,000
24	31,000
16	49,000

Calculate the values of both $\overline{\mathcal{M}}_n$ and $\overline{\mathcal{M}}_w$ and the polydispersity of solution. (Polydispersity is defined as $\overline{\mathcal{M}}_w / \overline{\mathcal{M}}_n$.)

The molar masses determined using the number-average and the weighted-average are

$$\overline{\mathcal{M}}_n = \frac{(10)\,(25{,}000) + (7)\,(17{,}000) + (24)\,(31{,}000) + (16)\,(49{,}000)}{10 + 7 + 24 + 16}\ g\,mol^{-1}$$

$$= 33281\ g\,mol^{-1} = 3.3 \times 10^4\ g\,mol^{-1}$$

$$\overline{\mathcal{M}}_w = \frac{(10)\,(25{,}000)^2 + (7)\,(17{,}000)^2 + (24)\,(31{,}000)^2 + (16)\,(49{,}000)^2}{(10)\,(25{,}000) + (7)\,(17{,}000) + (24)\,(31{,}000) + (16)\,(49{,}000)}\ g\,mol^{-1}$$

$$= 36770\ g\,mol^{-1} = 3.7 \times 10^4\ g\,mol^{-1}$$

The polydispersity is

$$\frac{36770\ g\,mol^{-1}}{33281\ g\,mol^{-1}} = 1.1$$

22.2 Ceruloplasmin is a protein present in the blood plasma. It contains 0.33% copper by weight. **(a)** Calculate its minimum molar mass. **(b)** The actual molar mass of ceruloplasmin is 150,000 $g\,mol^{-1}$. How many copper atoms does each protein molecule contain?

(a) The minimum molar mass corresponds to a protein with 1 mole of Cu per mole of protein.

$$\frac{63.55 \text{ g mol}^{-1}}{\mathcal{M}_{min}} = 0.33\%$$

$$\mathcal{M}_{min} = \frac{63.55 \text{ g mol}^{-1}}{0.0033} = 19,258 \text{ g mol}^{-1} = 1.9 \times 10^4 \text{ g mol}^{-1}$$

(b) The number of copper atoms per protein molecule is

$$\frac{150,000 \text{ g mol}^{-1}}{19,258 \text{ g mol}^{-1}} = 7.8 \approx 8$$

22.3   Depending on experimental conditions, the measurement of the molar mass of hemoglobin in an aqueous solution may show that the solution is monodisperse or polydisperse. Explain.

Hemoglobin is a tetramer, and it may dissociate in solution to give rise to a polydisperse solution. If the tetramer remains together, the solution will be monodisperse.

22.4   An ultracentrifuge is spinning at 60,000 rpm. (a) Calculate the value of ω in radians s^{-1}. (b) Calculate the centrifugal acceleration, a, given by $\omega^2 r$, at a point 7.4 cm from the center of rotation. (c) How many "g's" is this acceleration equivalent to?

(a)

$$\omega = (60,000 \text{ rpm}) \left(\frac{2\pi \text{ radians}}{1 \text{ revolution}}\right) \left(\frac{1 \text{ min}}{60 \text{ s}}\right) = 6283 \text{ radians s}^{-1}$$

(b)

$$a = \omega^2 r = \left(6283 \text{ radians s}^{-1}\right)^2 (7.4 \text{ cm}) \left(\frac{1 \text{ m}}{100 \text{ cm}}\right) = 2.92 \times 10^6 \text{ m s}^{-2} = 2.9 \times 10^6 \text{ m s}^{-2}$$

Note that a radian has no units since it is simply the ratio of arc length to radius length.

(c) The ratio between a and g is

$$\frac{2.92 \times 10^6 \text{ m s}^{-2}}{9.81 \text{ m s}^{-2}} = 3.0 \times 10^5$$

Thus, the acceleration is equivalent to 3.0×10^5 "g's."

22.5   How does the sedimentation coefficient, s, depend on the mass of a protein (assumed to be spherical)? Compare the rates of sedimentation of two proteins with molar masses of 70,000 g and 35,000 g, respectively.

According to Equation 22.6, $s \propto m/f$, or the sedimentation coefficient is proportional to the ratio of the protein mass to the frictional coefficient (these are the only two quantities in the equation that depend on the size of the protein). Assuming a spherical shape and uniform density, the mass of the protein is proportional to its volume, and thus to r^3, where r is the radius of the sphere. The frictional coefficient, on the other hand, is proportional to r. Consequently, $s \propto r^3/r = r^2$, or $s \propto m^{2/3}$.

The relative sedimentation rates of two proteins are given by

$$\frac{\text{rate}_1}{\text{rate}_2} = \left(\frac{m_1}{m_2}\right)^{2/3} = \left(\frac{70000 \text{ g mol}^{-1}}{35000 \text{ g mol}^{-1}}\right)^{2/3} = 1.6$$

22.6 A protein with $\bar{v} = 0.74$ mL g^{-1} is sedimented in water at 20°C. If $s_{20,w} = 3.0 \times 10^{-13}$ s and $D = 1.5 \times 10^{-6}$ cm^2 s^{-1}, what is the molar mass of the protein? The density of the solution is 0.998 g mL^{-1}.

$$M = \frac{sRT}{D(1 - \bar{v}\rho)}$$

$$= \frac{\left(3.0 \times 10^{-13} \text{ s}\right)\left(8.314 \text{ J K}^{-1} \text{ mol}^{-1}\right)(293 \text{ K})}{\left(1.5 \times 10^{-6} \text{ cm}^2 \text{ s}^{-1}\right)\left(\frac{1 \text{ m}}{100 \text{ cm}}\right)^2 \left[1 - \left(0.74 \text{ mL g}^{-1}\right)\left(0.998 \text{ g mL}^{-1}\right)\right]}$$

$$= 19 \text{ kg mol}^{-1}$$

22.7 In a sedimentation equilibrium experiment carried out at 293 K, the following data were obtained for a certain protein molecule: $\omega = 19,000$ rpm, $s = 2.15 \times 10^{-13}$ s, $\bar{v} = 0.71$ mL g^{-1}, and $\rho = 1.1$ g mL^{-1}. The relative concentrations at distances r_1 and r_2 from the center of rotation are $c_1 = 4.72$ ($r_1 = 5.95$ cm) and $c_2 = 12.98$ ($r_2 = 6.23$ cm). What is the molar mass of the protein?

The molar mass can be calculated using Equation 22.14.

$$M = \frac{2RT \ln \frac{c_2}{c_1}}{(1 - \bar{v}\rho)\omega^2 \left(r_2^2 - r_1^2\right)}$$

$$= \frac{2\left(8.314 \text{ J K}^{-1} \text{ mol}^{-1}\right)(293 \text{ K}) \ln \frac{12.98}{4.72}}{\left[1 - \left(0.71 \text{ mL g}^{-1}\right)\left(1.1 \text{ g mL}^{-1}\right)\right]\left[(19,000 \text{ rpm})\left(\frac{2\pi}{1 \text{ revolution}}\right)\left(\frac{1 \text{ min}}{60 \text{ s}}\right)\right]^2}$$

$$\times \frac{1}{\left[\left(6.23 \times 10^{-2} \text{ m}\right)^2 - \left(5.95 \times 10^{-2} \text{ m}\right)^2\right]}$$

$$= 17 \text{ kg mol}^{-1}$$

22.8 What are the units for the various viscosities defined in Equations 22.15 to 22.18?

η_{rel} and η_{sp} are dimensionless, η_{red} and $[\eta]$ are in $mL\,g^{-1}$.

22.9 Will dissolving 1×10^{-3} g of glucose result in a greater relative viscisity in water or in a 10% glycerol solution?

Relative viscosity is defined as $\eta_{rel} = \eta/\eta_0$, where η_0 is the viscosity of the solvent. Since water has a smaller viscosity than glycerol, the solution in water will have the larger *relative* viscosity.

22.10 The intrinsic viscosity of ribonuclease is 3.4 at 20°C and 6 at 50°C. What can you say about the change in structure?

According to Table 22.2, the intrinsic viscosity of globular ribonuclease is $3.4\,mL\,g^{-1}$. Thus, at 20°C, the protein is in the globular form, the observed increase in viscosity to $6\,mL\,g^{-1}$ upon heating to 50°C is indicative of a change to a random coil conformation.

22.11 Show that the units for zeE in Equation 22.20 are those for force.

The units for zeE are

$$C\,V\,m^{-1} = J\,m^{-1} = kg\,m^2\,s^{-2}\,m^{-1} = kg\,m\,s^{-2} = N$$

22.12 At pH 6.5, the electrophoretic mobility of carboxyhemoglobin is $2.23 \times 10^{-5}\,cm^2\,s^{-1}\,V^{-1}$, and that of sickle-cell carboxyhemoglobin is $2.63 \times 10^{-5}\,cm^2\,s^{-1}\,V^{-1}$. Calculate how long it will take to separate these two proteins by 1.0 cm if the potential gradient is $5.0\,V\,cm^{-1}$.

The ionic velocity for carboxyhemoglobin is

$$v_{normal} = u_{normal}E = \left(2.23 \times 10^{-5}cm^2\,s^{-1}\,V^{-1}\right)\left(5.0\,V\,cm^{-1}\right) = 1.12 \times 10^{-4}\,cm\,s^{-1}$$

The ionic velocity for sickle-cell carboxyhemoglobin is

$$v_{sickle} = u_{sickle}E = \left(2.63 \times 10^{-5}\,cm^2\,s^{-1}\,V^{-1}\right)\left(5.0\,V\,cm^{-1}\right) = 1.32 \times 10^{-4}\,cm\,s^{-1}$$

Assuming both proteins start at the same place, the time it will take to separate these two proteins by 1.0 cm is

$$\frac{1.0\,cm}{1.32 \times 10^{-4}\,cm\,s^{-1} - 1.12 \times 10^{-4}\,cm\,s^{-1}} = 5.0 \times 10^4\,s = 14\,hr$$

22.13 In an electrophoretic study of an aqueous protein solution, two species were found with molar masses of 60,000 and 30,000, respectively. The solution contains 1.85% protein by weight. If the fraction of the larger protein is found to be 70%, calculate the values of $\overline{\mathcal{M}}_w$ and $\overline{\mathcal{M}}_n$.

There is 1.85 g of protein in 100 g of solution. In this sample, the mass of the large protein is $70\% \times 1.85$ g $= 1.295$ g, and the mass of the small protein is $1.85 - 1.295 = 0.555$ g. The numbers of moles of the large protein, n_l, and of the small protein, n_s are

$$n_l = \frac{1.295 \text{ g}}{60{,}000 \text{ g mol}^{-1}} = 2.158 \times 10^{-5} \text{ mol}$$

$$n_s = \frac{0.555 \text{ g}}{30{,}000 \text{ g mol}^{-1}} = 1.850 \times 10^{-5} \text{ mol}$$

Since the number of molecules is proportional to number of moles, the number-average molar mass and the weighted-average molar mass can be formulated using numbers of moles of various species.

$$\overline{\mathcal{M}}_n = \frac{(2.158 \times 10^{-5})\,(60{,}000) + (1.850 \times 10^{-5})\,(30{,}000)}{2.158 \times 10^{-5} + 1.850 \times 10^{-5}} \text{ g mol}^{-1} = 4.62 \times 10^4 \text{ g mol}^{-1}$$

$$\overline{\mathcal{M}}_w = \frac{(2.158 \times 10^{-5})\,(60{,}000)^2 + (1.850 \times 10^{-5})\,(30{,}000)^2}{(2.158 \times 10^{-5})\,(60{,}000) + (1.850 \times 10^{-5})\,(30{,}000)} \text{ g mol}^{-1} = 5.10 \times 10^4 \text{ g mol}^{-1}$$

22.14 The relative electrophoretic mobilities of several protein–SDS complexes in a polyacrylamide gel are as follows:

Protein	Molar Mass / g mol^{-1}	Relative Mobility
Myoglobin	17,200	0.95
Trypsin	23,300	0.82
Aldolase	40,000	0.59
Fumarase	49,000	0.50
Carbonic anhydrase	29,000	0.73

Plot log (molar mass) versus relative mobility. The relativity mobility of creatine kinase is 0.60. What is its molar mass? Compare your result with the molar mass of 80,000 obtained by ultracentrifugation. What conclusions can you draw?

The following data are used for the plot.

log(molar mass)	Relative mobility
4.236	0.95
4.367	0.82
4.602	0.59
4.690	0.50
4.462	0.73

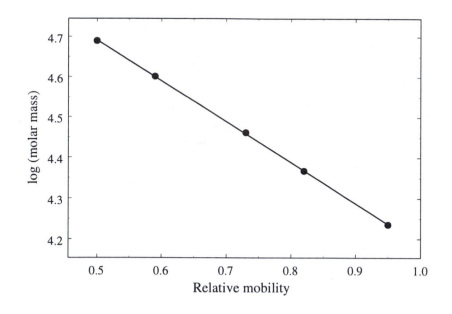

The plot can be described by a straight line: $y = -1.012x + 5.198$.

For a relative mobility (x) of 0.60, the log(molar mass) is

$$y = -(1.012)(0.60) + 5.198 = 4.591$$

Thus, the molar mass is $10^{4.591} = 3.90 \times 10^4$ g mol^{-1}

The molar mass by ultracentrifugation is twice that determined by the electrophoretic method. It appears that the protein dissociates into 2 subunits in the latter method.

22.15   The extent to which a protein molecule behaves like a spherical molecule can be tested by the frictional ratio, f/f_0, where f_0 is the frictional coefficient in Stokes law (Equation 21.24), and f is the frictional coefficient obtained from the diffusion coefficient (Equation 21.23). For spherical molecules, $f/f_0 = 1$; deviations from unity can be used as a measure of the nonspherical shape of the molecule. Consider hemoglobin and human fibrinogen (molar mass 339,700, $s = 7.63 \times 10^{-13}$ s, $D = 1.98 \times 10^{-7}$ cm^2 s^{-1}, and $\bar{v} = 0.725$ mL g^{-1}). What conclusions can you draw about the shape of the molecules? (The radius of an assumed spherical molecule, r, can be obtained from the equation $\mathcal{M} = 4\pi N_A r^3 / 3\bar{v}$, where $\mathcal{M}$ is the molar mass.) Assume $T = 298$ K.

The diffusion coefficient of hemoglobin in water is found in Table 21.3 as $D = 0.069 \times 10^{-9}$ m^2 s^{-1}. Since, according to Equation 21.23, $D = k_B T / f$,

$$f = \frac{k_B T}{D} = \frac{\left(1.381 \times 10^{-23} \text{ J K}^{-1}\right)(298 \text{ K})}{0.069 \times 10^{-9} \text{ m}^2 \text{ s}^{-1}} = 5.96 \times 10^{-11} \text{ N m}^{-1} \text{ s}$$

Finding the Stokes law frictional coefficient requires the radius of (assumed) spherical hemoglobin, which has a molar mass of 64450 g mol^{-1} (Table 22.2). The partial specific volume is assumed to be 0.74 mL g^{-1}. From the equation suggested in the problem,

$$r = \left(\frac{3\bar{v}M}{4\pi N_A}\right)^{1/3}$$

$$= \left[\frac{3\left(0.74 \text{ mL g}^{-1}\right)\left(64450 \text{ g mol}^{-1}\right)}{4\pi\left(6.022 \times 10^{23} \text{ mol}^{-1}\right)}\right]^{1/3}$$

$$= 2.66 \times 10^{-7} \text{ cm}$$

$$= 2.66 \times 10^{-9} \text{ m}$$

Using this radius in Stokes law with the viscosity of water taken from Table 21.1,

$$f_0 = 6\pi\eta r = 6\pi\left(0.00101 \text{ N s m}^{-2}\right)\left(2.66 \times 10^{-9} \text{ m}\right) = 5.06 \times 10^{-11} \text{ N m}^{-1}\text{ s}$$

The ratio of the two frictional coefficients is

$$\frac{f}{f_0} = \frac{5.96 \times 10^{-11} \text{ N m}^{-1}\text{ s}}{5.06 \times 10^{-11} \text{ N m}^{-1}\text{ s}} = 1.2$$

This is quite close to unity and implies that it is a fairly good assumption that hemoglobin is spherical.

For fibrinogen, the given diffusion coefficient is assumed to be in water, and Equation 21.23 gives,

$$f = \frac{k_B T}{D} = \frac{\left(1.381 \times 10^{-23} \text{ J K}^{-1}\right)(298 \text{ K})}{1.98 \times 10^{-7} \text{ cm}^2\text{ s}^{-1}\left(\frac{1 \text{ m}}{100 \text{ cm}}\right)^2} = 2.078 \times 10^{-10} \text{ N m}^{-1}\text{ s}$$

Assuming fibrinogen to be spherical, the radius is found as

$$r = \left(\frac{3\bar{v}M}{4\pi N_A}\right)^{1/3}$$

$$= \left[\frac{3\left(0.725 \text{ mL g}^{-1}\right)\left(339700 \text{ g mol}^{-1}\right)}{4\pi\left(6.022 \times 10^{23} \text{ mol}^{-1}\right)}\right]^{1/3}$$

$$= 4.605 \times 10^{-7} \text{ cm}$$

$$= 4.605 \times 10^{-9} \text{ m}$$

Using this radius in Stokes law with the viscosity of water taken from Table 21.1,

$$f_0 = 6\pi\eta r = 6\pi\left(0.00101 \text{ N s m}^{-2}\right)\left(4.605 \times 10^{-9} \text{ m}\right) = 8.767 \times 10^{-11} \text{ N m}^{-1}\text{ s}$$

The ratio of the two frictional coefficients is

$$\frac{f}{f_0} = \frac{2.078 \times 10^{-10} \text{ N m}^{-1}\text{ s}}{8.765 \times 10^{-11} \text{ N m}^{-1}\text{ s}} = 2.37$$

This is a significant deviation from unity, consistent with the rodlike shape of fibrinogen.

22.16   Referring to Figure 22.28, state whether the β structure in ribonuclease is parallel or antiparallel.

The β structure in ribonuclease is antiparallel.

22.17   How does hydrophobic interaction differ from both covalent and noncovalent bonds? What role does it play in protein structure and stability?

The hydrophobic interaction differs from the other types of inter- and intramolecular forces because it does not arise from the interaction between a pair of atoms or two groups of a molecule; instead, it involves a considerable number of water molecules. The hydrophobic interaction is largely an entropy-driven process resulting from the ordering of water molecules around a hydrocarbon or non-polar group in a molecule. The liberation of these water molecules surrounding the non-polar species and their return to bulk water increases the entropy and promotes the aggregation of the non-polar entities.

22.18   As Figure 22.17 shows, the average pitch of an α helix is 5.4 Å. Assuming this pitch to be the same for human hair and that hair grows at the rate of 0.6 inch month^{-1}, how many turns of the α helix are generated each second? (1 month = 30 days)

The number of turns of the α helix generated each second is

$$\left(\frac{0.6 \text{ inch month}^{-1}}{5.4 \times 10^{-8} \text{ cm turn}^{-1}} \right) \left(\frac{1 \text{ month}}{30 \text{ days}} \right) \left(\frac{1 \text{ day}}{24 \text{ hrs}} \right) \left(\frac{1 \text{ hr}}{3600 \text{ s}} \right) \left(\frac{2.54 \text{ cm}}{1 \text{ in}} \right) = 11 \text{ turns s}^{-1}$$

22.19   Hair contains keratins made of α helixes coiled to form a superhelix. The disulfide bonds linking the α helixes together are largely responsible for the shape of the hair. Based on this information, explain how "permanent waves" are formed.

In the "permanent wave" process, the disulfide bonds between α helixes are first reduced, and the hair is then shaped by curlers or other means. This is followed by oxidation to reform the disulfide bonds which will then hold the hair in its new style.

22.20   The α helical structure of poly-L-lysine is formed at pH 10, whereas that of the random coil is formed at pH 7. Account for the pH-dependent structural change.

The side chain of lysine contains an amino group that is protonated ($-NH_3^+$) at pH = 7. The electrostatic repulsion between these charged groups prevents the formation of an α helix. At pH = 10, the amino groups become deprotonated, allowing the α helix to form.

22.21   Proteases such as trypsin and carboxypeptidase catalyze the hydrolysis of peptide bonds of proteins (as in digestion). Explain how such an enzyme might bind to the substrate protein so that its main chain becomes fully extended in the vicinity of the targeted peptide bond. What kind of structure would it resemble?

One possibility would be that a segment of the main chain of the protease forms hydrogen bonds to the main chain of the substrate to make a structure resembling either a parallel or antiparallel β sheet.

22.22 Proteins denatured by compounds such as urea or SDS can be renatured when the denaturants are removed by dialysis. On the other hand, thermal denaturation of proteins is often irreversible. Explain. (*Hint*: Consider the *trans* configuration of the peptide bond.)

Heating can result in the *trans*→*cis* isomerization of the peptide bond. Upon cooling, many such isomerizations would be "frozen" in place, and these *cis* peptide bonds would have a major impact on the protein's structure. Neither urea nor SDS causes these geometric isomerizations.

22.23 Proteins generally have widely different structures, whereas nucleic acids have quite similar structures. Explain.

There are 20 different amino acids and only 4 different bases in either DNA or RNA. Furthermore, in DNA, the base pairing is very specific (A–T and C–G) whereas the amino acids in proteins can interact non-specifically. Consequently, there is a much wider variety of possible structures for proteins than for nucleic acids.

22.24 A compact disc (CD) stores about 4.0×10^9 bits of information. This information is stored as a binary code; that is, every bit is either 0 or 1. **(a)** How many bits would it take to specify each nucleotide pair in a DNA sequence? **(b)** How many CDs would it take to store the information in the human genome, which consists of 3×10^9 nucleotide pairs?

(a) There are four nucleotides in DNA so that two bits are required to specify each nucleotide pair. (Note that although there are only two pairings, A–T must be distinguished from T–A and C–G from G–C, so that four combinations, such as 00, 01, 10, and 11, are required.)

(b) It would require 2 bits for each pair, or $2 \times 3 \times 10^9 = 6 \times 10^9$ bits of information to store the human genome. This would require

$$\frac{6 \times 10^9 \text{ bits}}{4.0 \times 10^9 \text{ bits CD}^{-1}} = 1.5 \text{ CD}$$

Thus, 2 CD's will suffice with left over room for additional information such as error correction bits.

22.25 Referring to Figure 22.26, if after one generation, the DNA sample was thermally denatured and then the sample was annealed, how many bands would appear in the CsCl gradient?

When the $^{15}N/^{14}N$ is denatured and the resulting single stranded DNA randomly anneals, three different isotopic variants are made, $^{14}N/^{14}N$, $^{15}N/^{14}N$, and $^{15}N/^{15}N$, which appear as three separate bands in the gradient.

22.26 The following T_m data were obtained for double-stranded DNA in certain buffer solutions:

Sample	Percent (C + G)	$T_m/°\text{C}$
1	40.0	86.6
2	49.0	90.0
3	62.0	95.0
4	71.0	98.4

(a) Derive an equation relating the percent (C + G) to T_m. **(b)** Calculate the percent (C + G) content for a sample whose $T_m = 88.3°\text{C}$.

(a) A plot of % (C + G) vs T_m gives a straight line.

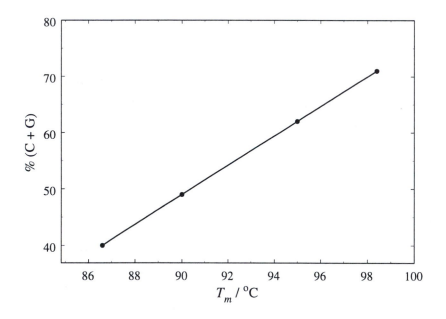

The equation for the line is $y = 2.623x - 187.13$ where y is the % of (C + G) and x is T_m in °C.

(b) For $x = 88.3$,

$$y = (2.623)(88.3) - 187.13 = 44.5$$

Thus, the % (C + G) is 44.5% for a sample with $T_m = 88.3$ °C.

22.27 The enthalpy change in the denaturation of a certain protein is 125 kJ mol^{-1}. If the entropy change is 397 J K^{-1} mol^{-1}, calculate the minimum temperature at which the protein would denature spontaneously.

The protein would denature spontaneously when

$$\Delta G = \Delta H - T\Delta S < 0$$

Thus, the minimum temperature is

$$T = \frac{\Delta H}{\Delta S} = \frac{125 \times 10^3 \, \text{J mol}^{-1}}{397 \, \text{J K}^{-1} \, \text{mol}^{-1}} = 315 \, \text{K}$$

22.28   Consider the formation of a dimeric protein

$$2P \rightarrow P_2$$

At 25°C, we have $\Delta_r H° = 17 \, \text{kJ mol}^{-1}$ and $\Delta_r S° = 65 \, \text{J K}^{-1} \, \text{mol}^{-1}$. Is the dimerization favored at this temperature? Comment on the effect of lowering the temperature. What general conclusion can you draw about the so-called "cold-labile" enzymes?

$$\Delta_r G° = \Delta_r H° - T\Delta_r S°$$
$$= 17 \, \text{kJ mol}^{-1} - (298 \, \text{K}) \left(65 \times 10^{-3} \, \text{kJ K}^{-1} \, \text{mol}^{-1}\right)$$
$$= -2 \, \text{kJ mol}^{-1}$$

Since $\Delta_r G° < 0$, the dimerization is favored at standard conditions and 25°C. As the temperature is lowered, $\Delta_r G°$ becomes less negative so that the dimerization is less favored. At low enough temperature ($T < 262 \, \text{K}$), the reaction becomes spontaneous in the reverse direction and denaturation occurs. For an enzyme to be cold-labile, it must have $\Delta_r H° > 0$ and $\Delta_r S° > 0$ for folding to the native state so that below a certain temperature, the enthalpy term dominates, and denaturation occurs spontaneously.

22.29   The cause and properties of sickle-cell hemoglobin (HbS) were discussed in Chapter 16. Under certain conditions, the aggregates of the HbS molecules formed at body temperature break apart as the temperature is lowered. Explain.

This is similar to the case of the cold-sensitive enzymes discussed with regard to the hydrophobic interaction in Section 22.4. The aggregation of HbS has a positive ΔH and a positive ΔS. At higher temperatures the $-T\Delta S$ dominates making ΔG negative, and the aggregation occurs spontaneously, but as the temperature is lowered, the entropy term becomes less important, and the ΔH term causes ΔG to be positive so that aggregation is thermodynamically unfavorable.

22.30   In this chapter, an expression was derived showing that the number of ways N sulfhydryl groups can form a linkage is $(N-1)(N-3)(N-5)...1$ if N is even. Derive an expression if N is odd.

If N is odd, then one sulfhydryl group will remain unlinked, and the number of ways the remaining $N-1$ groups can form a linkage is given by the expression for an even number of groups (if N is odd, $N - 1$ must be even), or $[(N - 1) - 1][(N - 1) - 3][(N - 1) - 5]\ldots 1 = (N - 2)(N - 4)(N - 6)\ldots 1$. There are N possibilities, however, for which sulfhydryl group remains unlinked, so the total number of ways N sulfhydryl groups can form a linkage with N odd is $N(N - 2)(N - 4)(N - 6)\ldots 1$.

22.31   What assumptions must be made in the study of protein folding using the hydrogen–deuterium exchange technique?

The technique depends critically on the assumption that the rate of H/D isotope exchange is much faster than the rate of protein folding.

22.32   A denatured protein contains 10 cysteine residues. Upon oxidation, what fraction of the molecules will randomly form the correct linkages if the native protein has **(a)** five disulfide bonds, and **(b)** three disulfide bonds.

(a) With ten cysteine residues, five disulfide bonds represents complete linkage, and there are $9 \times 7 \times 5 \times 3 \times 1 = 945$ ways to form the five linkages. Since only one of these represents the correct linkage in the native form, the fraction is $1/945$.

(b) With only three disulfide bonds, four –SH groups remain unlinked. The number of ways of picking 4 –SH groups out of the 10 is given by the binomial coefficient, since the order of picking does not matter.

$$\text{\# of ways of picking 4 groups out of 10} = \frac{10!}{(10 - 4)!4!} = \frac{10!}{6!4!} = 210$$

Of the remaining six –SH groups, the number of ways to form three disulfide bonds is

$$\text{\# of ways of forming 3 disulfide bonds} = 5 \times 3 \times 1 = 15$$

Therefore the total number of ways of leaving four groups unlinked and linking the remaining six is

$$210 \times 15 = 3150$$

Since only one choice is the correct choice for the native protein, the fraction with the correct linkages is $1/3150$. (Note that this is calculated assuming that all the randomly formed molecules formed only three disulfide bonds. More likely, there will be a mixture with various numbers of disulfide bonds forming, and the fraction with the correct linkages will be even smaller.)

CHAPTER 23

Statistical Thermodynamics

PROBLEMS AND SOLUTIONS

23.1 Suppose we have 8 distinguishable particles that have a total energy equal to 6 units (see Figure 23.3). The distributions for this system are labeled (n_0, n_1, n_2, n_3). Calculate the number of microstates (W) for each of the following distributions: **(a)** (6, 0, 0, 2), **(b)** (5, 1, 1, 1), bffont**(c)** (4, 2, 2, 0), **(d)** (2, 6, 0, 0).

The number of microstates is

$$W = \frac{n!}{n_0! \, n_1! \, n_2! \, n_3!}$$

(a) $W = \dfrac{8!}{6! \, 0! \, 0! \, 2!} = 28$ **(b)** $W = \dfrac{8!}{5! \, 1! \, 1! \, 1!} = 336$ **(c)** $W = \dfrac{8!}{4! \, 2! \, 2! \, 0!} = 420$

(d) $W = \dfrac{8!}{2! \, 6! \, 0! \, 0!} = 28$

23.2 Calculate the number of microstates (W) when 10 molecules are equally distributed among five energy levels. What is the value of W if one molecule is removed from one state and added to another?

When the molecules are equally distributed among five energy levels,

$$W = \frac{10!}{2! \, 2! \, 2! \, 2! \, 2!} = 113400$$

If one molecule is removed from one state and added to another,

$$W = \frac{10!}{1! \, 3! \, 2! \, 2! \, 2!} = 75600$$

23.3 A useful estimate of the enormity of the number of microstates associated with the most probable macrostate for a macroscopic system is the Poincaré recurrence time, which is the average time it takes for a system to return to any microstate that it has once occupied. The magnitude of Poincaré recurrence time can be appreciated by doing the following exercise. Consider a deck of cards dealt out in any specific order (both in numbers and in suits) as defining a microstate. **(a)** How many microstates are there? **(b)** If we could shuffle the cards and deal one per second, how long would we have to wait before a duplicate of that order is dealt again?

(a) There are 52 cards. Assuming the cards are shuffled well after each deal, there are $52! (= 8.07 \times 10^{67})$ possible microstates, all presumably equally likely.

(b) If we could shuffle the cards and deal one per second, the time it would take before a duplicate of that order is dealt again is 8.07×10^{67} s, or

$$\left(8.07 \times 10^{67} \text{ s}\right) \left(\frac{1 \text{ hr}}{3600 \text{ s}}\right) \left(\frac{1 \text{ day}}{24 \text{ hr}}\right) \left(\frac{1 \text{ yr}}{365 \text{ days}}\right) = 2.56 \times 10^{60} \text{ years}$$

The universe is only 10 billion years, so, on average, it will take about 2.56×10^{50} times the age of the universe to repeat a deal.

23.4 The population ratio between two energy levels separated by 1.5×10^{-22} J is 0.74. What is the temperature of the system?

$$\frac{n_2}{n_1} = e^{-\Delta\epsilon/k_B T}$$

$$T = -\frac{\Delta\epsilon}{k_B \ln \frac{n_2}{n_1}} = -\frac{1.5 \times 10^{-22} \text{ J}}{\left(1.381 \times 10^{-23} \text{ J K}^{-1}\right) \ln (0.74)} = 36 \text{ K}$$

23.5 What is the high temperature limit (that is, as $T \to \infty$) of Equation 23.25?

$$\lim_{T \to \infty} \frac{n_2}{n_1} = \lim_{T \to \infty} \frac{g_2}{g_1} e^{-\Delta\epsilon/k_B T} = \frac{g_2}{g_1} e^0 = \frac{g_2}{g_1}$$

At the limit of infinite temperature, the probability that any state is occupied becomes equal, regardless of their energy. Thus each state has the same population, and the relative population of any two energy levels is just given by the ratio of the number of states with the respective energies, which is the ratio of the degeneracies.

23.6 Given that the bond length is 1.128 Å, calculate the ratio of $J = 1$ to $J = 0$ populations for carbon monoxide at **(a)** 300 K and **(b)** 600 K. **(c)** What is the limiting value of this ratio as $T \to \infty$? (*Hint*: See Example 17.1 on p. 714.)

The energy of the J level is

$$E_J = \frac{J (J + 1) h^2}{8\pi^2 I}$$

$I = 1.46 \times 10^{-46}$ kg m^2 (Example 17.1). Thus,

$$E_0 = 0$$

$$E_1 = \frac{2\left(6.626 \times 10^{-34} \text{ J s}\right)^2}{8\pi^2 \left(1.46 \times 10^{-46} \text{ kg m}^2\right)} = 7.617 \times 10^{-23} \text{ J}$$

$$\Delta\epsilon = E_1 - E_0 = 7.617 \times 10^{-23} \text{ J}$$

The degeneracy of each level is $2J + 1$. Thus, the degeneracy of the $J = 0$ level is 1 while that of the $J = 1$ level is 3. The population ratio is

$$\frac{n_1}{n_0} = 3 \exp\left(-\frac{\Delta\epsilon}{k_B T}\right)$$

(a) At 300 K,

$$\frac{n_1}{n_0} = 3 \exp\left[-\frac{7.617 \times 10^{-23} \text{ J}}{\left(1.381 \times 10^{-23} \text{ J K}^{-1}\right)(300 \text{ K})}\right] = 2.95$$

(b) At 600 K,

$$\frac{n_1}{n_0} = 3 \exp\left[-\frac{7.617 \times 10^{-23} \text{ J}}{\left(1.381 \times 10^{-23} \text{ J K}^{-1}\right)(600 \text{ K})}\right] = 2.97$$

(c) As $T \to \infty$,

$$\lim_{T \to \infty} \frac{n_1}{n_0} = \lim_{T \to \infty} 3 \exp\left(-\frac{\Delta\epsilon}{k_B T}\right) = 3e^0 = 3$$

23.7 The fundamental vibrational wavenumber for N_2 is 2360 cm^{-1}. For 1 mole of the molecules, calculate the number of N_2 molecules in the $v = 0$ and $v = 1$ levels at (a) 298 K and (b) 1000 K.

The ratio between the number of molecules in the $v = i$ level and the total number of molecules is

$$\frac{n_i}{N} = \frac{e^{-\Delta\epsilon_i/k_B T}}{\sum e^{-\Delta\epsilon_i/k_B T}} = \frac{e^{-\Delta\epsilon_i/k_B T}}{q_{vib}}$$

Thus, the number of molecules in the $v = i$ level is

$$n_i = N\left(\frac{e^{-\Delta\epsilon_i/k_B T}}{q_{vib}}\right)$$

$$= N\left(\frac{e^{-\Delta\epsilon_i/k_B T}}{\frac{1}{1-e^{-hv/k_B T}}}\right)$$

$$= N \exp\left(-\frac{\Delta\epsilon_i}{k_B T}\right)\left[1 - \exp\left(-\frac{hv}{k_B T}\right)\right]$$

v is related to the vibrational wavenumber:

$$v = (2360 \text{ cm}^{-1}) (3.00 \times 10^{10} \text{ cm s}^{-1}) = 7.080 \times 10^{13} \text{ s}^{-1}$$

The energy of the v level relative to the $v = 0$ level is vhv. Thus,

$$E_0 = 0$$

$$E_1 = hv = (6.626 \times 10^{-34} \text{ J s}) (7.080 \times 10^{13} \text{ s}^{-1}) = 4.691 \times 10^{-20} \text{ J}$$

(a) At 298 K,

$$1 - \exp\left(-\frac{hv}{k_B T}\right) = 1 - \exp\left[-\frac{(6.626 \times 10^{-34} \text{ J s}) (7.080 \times 10^{13} \text{ s}^{-1})}{(1.381 \times 10^{-23} \text{ J K}^{-1}) (298 \text{ K})}\right] = 1.000$$

Thus,

$$n_0 = (6.022 \times 10^{23}) \left[\exp(0)\right] (1.000) = 6.02 \times 10^{23}$$

$$n_1 = (6.022 \times 10^{23}) \left[\exp\left(-\frac{(4.691 \times 10^{-20} \text{ J})}{(1.381 \times 10^{-23} \text{ J K}^{-1}) (298 \text{ K})}\right)\right] (1.000) = 6.75 \times 10^{18}$$

(b) At 1000 K,

$$1 - \exp\left(-\frac{hv}{k_B T}\right) = 1 - \exp\left[-\frac{(6.626 \times 10^{-34} \text{ J s}) (7.080 \times 10^{13} \text{ s}^{-1})}{(1.381 \times 10^{-23} \text{ J K}^{-1}) (1000 \text{ K})}\right] = 0.9665$$

Thus,

$$n_0 = (6.022 \times 10^{23}) \left[\exp(0)\right] (0.9665) = 5.82 \times 10^{23}$$

$$n_1 = (6.022 \times 10^{23}) \left[\exp\left(-\frac{(4.691 \times 10^{-20} \text{ J})}{(1.381 \times 10^{-23} \text{ J K}^{-1}) (1000 \text{ K})}\right)\right] (0.9665) = 1.95 \times 10^{22}$$

23.8   A system consists of three energy levels: a ground level ($\epsilon_0 = 0$, $g_0 = 4$); a first excited level ($\epsilon_1 = k_B T$, $g_1 = 2$); and a second excited level ($\epsilon_2 = 4k_B T$, $g_2 = 2$). Calculate the partition function of the system. What is the probability for the second energy level?

The partition function is

$$q = g_0 e^{-\Delta\epsilon_0/k_B T} + g_1 e^{-\Delta\epsilon_1/k_B T} + g_2 e^{-\Delta\epsilon_2/k_B T}$$

$$= 4e^{-0/k_B T} + 2e^{-k_B T/k_B T} + 2e^{-4k_B T/k_B T}$$

$$= 4 + 2e^{-1} + 2e^{-4}$$

$$= 4.772$$

The probability for the second energy level is

$$\frac{n_1}{N} = \frac{g_1 e^{-\Delta\epsilon_1/k_B T}}{q}$$

$$= \frac{2e^{-1}}{4.772}$$

$$= 0.154$$

23.9 Explain why q_{trans} increases with **(a)** m and **(b)** T.

(a) As m increases, the spacing between translational energy levels decreases, so that there are more levels at lower energies, and more levels available to the system. Thus, the partition function increases.

(b) As T increases, more levels with energies above the ground state energy will be populated, and the partition function will increase.

23.10 Starting with the relation $P = -(\partial A/\partial V)_T$ [See Equation 4 on p. 191], show that $P = k_B T \,(\partial \ln Q/\partial V)_T$. Using argon as an example of an ideal monatomic gas, derive the ideal-gas equation ($PV = nRT$).

From Equation 23.71, the Helmholtz energy is

$$A = U_0 - k_B T \ln Q$$

Pressure can be expressed as

$$P = -\left(\frac{\partial A}{\partial V}\right)_T = -\left[\frac{\partial\left(U_0 - k_B T \ln Q\right)}{\partial V}\right]_T = k_B T \left(\frac{\partial \ln Q}{\partial V}\right)_T$$

Argon has only translational motion. Therefore, Q is (Equation 23.60)

$$Q = \left[\frac{\left(2\pi m k_B T\right)^{1/2} eV}{Nh^3}\right]^N$$

and

$$\ln Q = N\left[\ln V + \ln \frac{\left(2\pi m k_B T\right)^{1/2} e}{Nh^3}\right]$$

The pressure for an ideal monatomic gas is

$$P = k_B T \left(\frac{\partial \ln Q}{\partial V}\right)_T = k_B T N \frac{1}{V}$$

Rearranging this equation,

$$PV = k_B T N = \left(\frac{N}{N_A}\right)(k_B N_A) T = nRT$$

which is the ideal gas equation.

23.11 Calculate the entropy of HCl at 298 K and 1 bar, given that the bond length is 1.275 Å and the masses of ^{1}H and ^{35}Cl are 1.008 amu and 34.97 amu, respectively. The frequency of vibration in wavenumbers is 2886 cm^{-1}.

The contributions to entropy are translational, rotational, and vibrational.

<u>Translational contribution</u>

$$S_{\text{trans}} = R \ln\left[\frac{(2\pi m k_B T)^{3/2}}{h^3}\frac{k_B T}{P}e^{5/2}\right]$$

$$= (8.314\ \text{J K}^{-1}\,\text{mol}^{-1})$$

$$\ln\left\{\frac{\left[2\pi\,(35.98\ \text{amu})\,(1.661\times 10^{-27}\ \text{kg amu}^{-1})\,(1.381\times 10^{-23}\ \text{J K}^{-1})\,(298\ \text{K})\right]^{3/2}}{(6.626\times 10^{-34}\ \text{J s})^3}\right.$$

$$\left.\frac{(1.381\times 10^{-23}\ \text{J K}^{-1})\,(298\ \text{K})}{(1\ \text{bar})\left(\frac{10^5\ \text{N m}^{-2}}{1\ \text{bar}}\right)}e^{5/2}\right\}$$

$$= 153.5\ \text{J K}^{-1}\,\text{mol}^{-1}$$

<u>Rotational contribution</u>

To calculate the rotational contribution to the entropy, the moment of inertia of HCl must be found, which in turn requires the reduced mass of the molecule.

$$I = \mu r^2$$

$$= \frac{m_H m_{Cl}}{m_H + m_{Cl}}$$

$$= \frac{(1.008\ \text{amu})\,(34.97\ \text{amu})}{1.008\ \text{amu} + 34.97\ \text{amu}}(1.661\times 10^{-27}\ \text{kg amu}^{-1})\,(1.275\times 10^{-10}\ \text{m})^2$$

$$= 2.6455\times 10^{-47}\ \text{kg m}^2$$

This is used in the expression for the rotational contribution to the entropy

$$S_{\text{rot}} = R \ln q_{\text{rot}} + R = R \ln\frac{8\pi^2 I k_B T}{\sigma h^2} + R$$

$$= (8.314\ \text{J K}^{-1}\,\text{mol}^{-1})\ln\frac{8\pi^2\,(2.6455\times 10^{-47}\ \text{kg m}^2)\,(1.381\times 10^{-23}\ \text{J K}^{-1})\,(298\ \text{K})}{(1)\,(6.626\times 10^{-34}\ \text{J s})^2}$$

$$+ 8.314\ \text{J K}^{-1}\,\text{mol}^{-1}$$

$$= 33.04\ \text{J K}^{-1}\,\text{mol}^{-1}$$

<u>Vibrational contribution</u>

First calculate $h\nu/k_B T$:

$$\frac{h\nu}{k_B T} = \frac{\left(6.626 \times 10^{-34}\,\text{J s}\right)\left(2886\,\text{cm}^{-1}\right)\left(3.00 \times 10^{10}\,\text{cm s}^{-1}\right)}{\left(1.381 \times 10^{-23}\,\text{J K}^{-1}\right)\left(298\,\text{K}\right)} = 13.94$$

The vibrational contribution to entropy is

$$S_{\text{vib}} = -R \ln\left(1 - e^{-h\nu/k_B T}\right) + R\frac{h\nu}{k_B T}\frac{1}{e^{h\nu/k_B T} - 1}$$

$$= -\left(8.314\,\text{J K}^{-1}\,\text{mol}^{-1}\right) \ln\left(1 - e^{-13.94}\right) + \left(8.314\,\text{J K}^{-1}\,\text{mol}^{-1}\right)(13.94)\left(\frac{1}{e^{13.94} - 1}\right)$$

$$= 1.097 \times 10^{-4}\,\text{J K}^{-1}\,\text{mol}^{-1}$$

As expected, vibrational motion makes a negligible contribution to entropy at 298 K.

The entropy of HCl is

$$S = \left(153.5 + 33.04 + 1.097 \times 10^{-4}\right)\,\text{J K}^{-1}\,\text{mol}^{-1} = 186.5\,\text{J K}^{-1}\,\text{mol}^{-1}$$

23.12   Calculate the temperature at which $q_{\text{vib}} = 5.0$ for carbon monoxide. The vibrational frequency is $\tilde{\nu} = 2135\,\text{cm}^{-1}$.

The vibrational partition function is

$$q_{\text{vib}} = \frac{1}{1 - e^{-h\nu/k_B T}} = 5.0$$

Thus,

$$1 - e^{-h\nu/k_B T} = 0.20$$

$$e^{-h\nu/k_B T} = 0.80$$

$$\frac{h\nu}{k_B T} = -\ln 0.80 = 0.223$$

$$T = \frac{h\nu}{0.223 k_B} = \frac{\left(6.626 \times 10^{-34}\,\text{J s}\right)\left(2135\,\text{cm}^{-1}\right)\left(3.00 \times 10^{10}\,\text{cm s}^{-1}\right)}{(0.223)\left(1.381 \times 10^{-23}\,\text{J K}^{-1}\right)} = 1.4 \times 10^4\,\text{K}$$

CO would not survive at this temperature!

23.13   Calculate the translational partition function of helium at 1 bar in a 1.00-m^3 container. The large value of q_{trans} means that this motion can be treated classically. When $q_{\text{trans}} \leq 10$, however, the motion must be treated quantum mechanically. Calculate the temperature at which this change occurs.

The translational partition function is

$$q_{trans} = \frac{(2\pi m k_B T)^{3/2} V}{h^3}$$

$$= \frac{\left[2\pi (4.003 \text{ amu}) \left(1.661 \times 10^{-27} \text{ kg amu}^{-1}\right) \left(1.381 \times 10^{-23} \text{ J K}^{-1}\right) (298 \text{ K})\right]^{3/2} (1.00 \text{ m}^3)}{\left(6.626 \times 10^{-34} \text{ J s}\right)^3}$$

$$= 7.75 \times 10^{30}$$

When $q_{trans} = 10$,

$$q_{trans} = \frac{(2\pi m k_B T)^{3/2} V}{h^3} = 10$$

$$T^{3/2} = \frac{10 h^3}{(2\pi m k_B)^{3/2} V}$$

$$= \frac{10 \left(6.626 \times 10^{-34} \text{ J s}\right)^3}{\left[2\pi (4.003 \text{ amu}) \left(1.661 \times 10^{-27} \text{ kg amu}^{-1}\right) \left(1.381 \times 10^{-23} \text{ J K}^{-1}\right)\right]^{3/2} (1.00 \text{ m}^3)}$$

$$= 6.638 \times 10^{-27} \text{ K}^{3/2}$$

$$T = 3.53 \times 10^{-18} \text{ K}$$

23.14 Calculate the value of q_{vib} for the water molecule at 298 K. (*Hint*: See Figure 17.14.)

Water has three vibrational modes. The fundamental wavenumbers are 3760 cm^{-1}, 3650 cm^{-1}, 1595 cm^{-1}. The values of $h\nu/k_B T$ corresponding to these wavenumbers are

$$3760 \text{ cm}^{-1}: \quad \frac{h\nu}{k_B T} = \frac{\left(6.626 \times 10^{-34} \text{ J s}\right) \left(3760 \text{ cm}^{-1}\right) \left(3.00 \times 10^{10} \text{ cm s}^{-1}\right)}{\left(1.381 \times 10^{-23} \text{ J K}^{-1}\right) (298 \text{ K})} = 18.16$$

$$3650 \text{ cm}^{-1}: \quad \frac{h\nu}{k_B T} = \frac{\left(6.626 \times 10^{-34} \text{ J s}\right) \left(3650 \text{ cm}^{-1}\right) \left(3.00 \times 10^{10} \text{ cm s}^{-1}\right)}{\left(1.381 \times 10^{-23} \text{ J K}^{-1}\right) (298 \text{ K})} = 17.63$$

$$1595 \text{ cm}^{-1}: \quad \frac{h\nu}{k_B T} = \frac{\left(6.626 \times 10^{-34} \text{ J s}\right) \left(1595 \text{ cm}^{-1}\right) \left(3.00 \times 10^{10} \text{ cm s}^{-1}\right)}{\left(1.381 \times 10^{-23} \text{ J K}^{-1}\right) (298 \text{ K})} = 7.704$$

The vibrational partition function is

$$q_{vib} = \left(\frac{1}{1 - e^{-18.16}}\right) \left(\frac{1}{1 - e^{-17.63}}\right) \left(\frac{1}{1 - e^{-7.704}}\right) = 1.000453$$

23.15 List the symmetry number (σ) for each of the following molecules: Cl_2, N_2O (NNO), H_2O, HDO, BF_3, CH_4, CH_3Cl. (*Hint*: For CH_4, note that each of the four C–H bonds represents a three-fold symmetry axis about which three successive 120° indistinguishable rotations are possible.)

Cl_2: 2; N_2O: 1; H_2O: 2; HDO: 1; BF_3: 3; CH_4: $4 \times 3 = 12$; CH_3Cl: 3.

23.16 Calculate the equilibrium constant for the following reaction at 1274 K:

$$I_2(g) \rightleftharpoons 2I(g)$$

The bond length of I_2 is 2.67 Å and the vibrational frequency is 213.7 cm^{-1}. The bond dissociation energy of I_2 is 149.0 kJ mol^{-1}. [*Hint*: To calculate the degeneracy of the iodine atom in its ground electronic state, note that there is an unpaired electron in the $5p$ orbital. The degeneracy is given by $(2J + 1)$, where J, the total angular momentum, is given by the sum of the orbital angular momentum and spin angular momentum.]

The equilibrium constant is

$$K = \frac{q_I^2}{q_{I_2}} N_A^{-\Delta n} e^{-\Delta U_0 / RT} \tag{23.16.1}$$

First calculate the partition functions for I and I_2.

I atom

The partition functions consists of two parts: translational and electronic.

$$q_{trans} = \frac{(2\pi m k_B T)^{3/2} V}{h^3} = \frac{(2\pi m k_B T)^{3/2} \frac{nRT}{P}}{h^3}$$

V can be calculated using the ideal gas law. (The standard pressure of 1 bar is used, since equilibrium constants derive from $\Delta G°$.)

$$V = \frac{nRT}{P} = \frac{(1 \text{ mol}) (8.314 \text{ J K}^{-1} \text{ mol}^{-1}) (1274 \text{ K})}{10^5 \text{ Pa}} = 0.10592 \text{ m}^3$$

Thus,

$$q_{trans} = \frac{\left[2\pi (126.90 \text{ amu}) (1.661 \times 10^{-27} \text{ kg amu}^{-1}) (1.381 \times 10^{-23} \text{ J K}^{-1}) (1274 \text{ K})\right]^{3/2} (0.10592 \text{ m}^3)}{(6.626 \times 10^{-34} \text{ J s})^3}$$

$$= 1.2950 \times 10^{33}$$

The unpaired electron is in the $5p$ orbital. Thus, $l = 1$, $s = 1/2$, and $J = l + s = 3/2$. The degeneracy is $2J + 1 = 4$ and $q_{elec} = 4$.

The partition function of I is

$$q_I = q_{trans} q_{elec} = 5.1800 \times 10^{33} \tag{23.16.2}$$

I$_2$ molecule

The partition functions consists of three parts: translational, rotational, and vibrational.

The translational partition function for I$_2$ is the same as that for I, except the mass of I$_2$ is twice that of I. Thus, for the molecule,

$$q_{trans} = 2^{3/2} \left(1.2950 \times 10^{33}\right) = 3.6628 \times 10^{33}$$

To calculate the rotational partition function, the moment of inertia must be determined.

$$\mu = \frac{m_I m_I}{m_I + m_I} = \frac{m_I}{2} = \frac{(126.90 \text{ amu}) \left(1.661 \times 10^{-27} \text{ amu kg}^{-1}\right)}{2} = 1.0539 \times 10^{-25} \text{ kg}$$

$$I = \mu r^2 = \left(1.0539 \times 10^{-25} \text{ kg}\right) \left(2.67 \times 10^{-10} \text{ m}\right)^2 = 7.513 \times 10^{-45} \text{ kg m}^2$$

The rotational partition function is

$$q_{rot} = \frac{8\pi^2 I k_B T}{\sigma h^2}$$

$$= \frac{8\pi^2 \left(7.513 \times 10^{-45} \text{ kg m}^2\right) \left(1.381 \times 10^{-23} \text{ J K}^{-1}\right) (1274 \text{ K})}{(2) \left(6.626 \times 10^{-34} \text{ J s}\right)^2}$$

$$= 1.189 \times 10^4$$

To calculate the vibrational partition function, the value of $h\nu/k_B T$ must be determined.

$$\frac{h\nu}{k_B T} = \frac{\left(6.626 \times 10^{-34} \text{ J s}\right) \left(213.7 \text{ cm}^{-1}\right) \left(3.00 \times 10^{10} \text{ cm s}^{-1}\right)}{\left(1.381 \times 10^{-23} \text{ J K}^{-1}\right) (1274 \text{ K})} = 0.24144$$

The vibrational partition function is

$$q_{vib} = \frac{1}{1 - e^{-h\nu/k_B T}} = \frac{1}{1 - e^{-0.24144}} = 4.6619$$

Therefore, the partition function for I$_2$ is

$$q_{I_2} = q_{trans} q_{rot} q_{vib} = \left(3.6628 \times 10^{33}\right) \left(1.189 \times 10^4\right) (4.6619) = 2.030 \times 10^{38} \tag{23.16.3}$$

Substituting Equations 23.16.2 and 23.16.3, and the values for the bond dissociation energy and Δn into Equation 23.16.1,

$$K = \frac{\left(5.1800 \times 10^{33}\right)^2}{2.030 \times 10^{38}} \left(6.022 \times 10^{23}\right)^{-1} e^{-149.0 \times 10^3 \text{ J mol}^{-1} / \left[\left(8.314 \text{ J K}^{-1} \text{ mol}^{-1}\right)(1274 \text{ K})\right]} = 0.171$$

which is very close to the experimental value of 0.167.

23.17   Calculate the approximate value of $\Delta_r S^\circ$ for

$$^{16}O_2(g) + {}^{18}O_2(g) \rightarrow 2\,{}^{16}O^{18}O(g)$$

Assume that differences in molar masses, moments of inertia, and vibrational frequencies are negligible.

The major contribution to $\Delta_r S^\circ$ is the symmetry factor. Assuming that any differences in molar masses, moments of inertia, and vibrational frequencies are negligible, only the symmetry factor in the rotational contribution to entropy need to be considered.

$$S_{\text{rot}} = R \ln q_{\text{rot}} + R = R \ln \frac{8\pi^2 I k_B T}{\sigma h^2} + R$$

Let $A = \frac{8\pi^2 I k_B T}{h^2}$. A is the same for the reactants and product in the reaction considered. Then the entropy expression becomes

$$S_{\text{rot}} = R \ln A - R \ln \sigma + R$$

The entropy of reaction is

$$\Delta_r S^\circ = 2 S_{\text{rot}}({}^{16}O^{18}O) - S_{\text{rot}}({}^{16}O_2) - S_{\text{rot}}({}^{18}O_2)$$

$$= 2 \left[R \ln A - R \ln \sigma({}^{16}O^{18}O) + R \right]$$

$$- \left[R \ln A - R \ln \sigma({}^{16}O_2) + R \right] - \left[R \ln A - R \ln \sigma({}^{18}O_2) + R \right]$$

$$= (-R) \left\{ 2 \left[\ln \sigma({}^{16}O^{18}O) \right] - \ln \sigma({}^{16}O_2) - \ln \sigma({}^{18}O_2) \right\}$$

$$= \left(-8.314 \, \text{J K}^{-1} \, \text{mol}^{-1} \right) [2 \ln 1 - \ln 2 - \ln 2]$$

$$= 11.53 \, \text{J K}^{-1} \, \text{mol}^{-1}$$

Values of Some Fundamental Constants

Constant	Value
Avogadro's constant (N_A)	6.0221367×10^{23} mol^{-1}
Bohr radius (a_o)	$5.29177249 \times 10^{-11}$ m
Boltzmann constant (k_B)	1.380658×10^{-23} J K^{-1}
Electron charge (e)	1.602177×10^{-19} C
Electron mass (m_e)	$9.1093897 \times 10^{-31}$ kg
Faraday constant (F)	96485.309 C mol^{-1}
Gas constant (R)	8.314510 J K^{-1} mol^{-1}
Neutron mass (m_N)	1.674928×10^{-27} kg
Permittivity of vacuum (ε_0)	8.854×10^{-12} C^2 N^{-1} m^{-2}
Planck constant (h)	6.626075×10^{-34} J s
Proton mass (m_P)	1.672623×10^{-27} kg
Rydberg constant (R_H)	109737.31534 cm^{-1}
Speed of light in vacuum (c)	299792458 m s^{-1}

Pressure of Water Vapor at Various Temperatures

Temperature/°C	Water Vapor Pressure/mmHg
0	4.58
5	6.54
10	9.21
15	12.79
20	17.54
25	23.76
30	31.82
35	42.18
40	55.32
45	71.88
50	92.51
55	118.04
60	149.38
65	187.54
70	233.7
75	289.1
80	355.1
85	433.6
90	525.76
95	633.90
100	760.00